CELL CYCLE CONTROL: NEW RESEARCH

CELL CYCLE CONTROL: NEW RESEARCH

NATHAN H. LEROY
AND
NOAH T. FOURNIER
EDITORS

Nova Biomedical Books
New York

For permission to use material from this book please contact us:
Telephone 631-231-7269; Fax 631-231-8175
Web Site: http://www.novapublishers.com

Library of Congress Cataloging-in-Publication Data

Cell cycle control : new research / Nathan H. Leroy and Noah T. Fournier (editor).
p. ; cm.
Includes bibliographical references and index.
ISBN 978-1-60456-796-0 (softcover)
1. Cellular control mechanisms. 2. Cell cycle. I. Leroy, Nathan H. II. Fournier, Noah T.
[DNLM: 1. Cell Cycle--physiology. 2. Cell Cycle Proteins--physiology. QU 375 C39256 2008]
QH604.C38 2008
571.8'4--dc22 2008023085

Published by Nova Science Publishers, Inc. ✦ New York

Contents

Preface

This book provides leading edge research on a cell cycle, which is an ordered and highly controlled set of events that leads to cell growth and proliferation. Cell cycle progression is driven by changes in the substrate specificity and subcellular localization of cyclin-dependent kinases (Cdks), which in turn are modulated by a collection of cyclins, Cdk-activating and Cdk-inhibiting kinases, and Cdk inhibitors (CDKIs). Regulation of the cell cycle is critical for the normal development of multicellular organisms and dysregulation of cell cycle could lead to cancer, a disease where normal cell growth and behavior are lost. Cell cycle regulation is tightly controlled by both synthesis and degradation of short-lived proteins, such as cyclins and CDKIs, and degradation of these proteins is mainly mediated by the ubiquitin-dependent proteasome pathway.

Short Communication - Microenvironmental stimuli cause multiple signal transduction pathways which determine cellular response. Mitogen-activated protein kinases (MAPKs) are pivotal signal cascades to transduce signals into dynamic gene expression waves in the nucleus. Gene expression patterns in given cells are totally dependent on the combined signals and integrity of activated transcription factors. The activating protein-1 (AP-1) is a classical transcription factor that can transduce MAPK signals into the gene expression pattern changes and this cellular system has a large impact on pathophysiological aspects of cells. Indeed, AP-1 is involved in a wide range of cellular events including cell proliferation, transformation, apoptosis, and stress response. Jun joins in with the AP-1 transcription complex as a homodimer or heterodimer, and has been identified as a mammalian cellular homologue of retroviral onco-proteins. Jun is known as an immediate early gene and its activity is largely regulated by MAPK signals. Expression of Jun can be induced by oncogenic stress. Taken together, these facts indicate that Jun is responsible for certain part of the AP-1 capability to modulate cellular response. It is also defined by the abundance of distinct members of AP-1 components within a given cell type and their surrounding stimuli. In this review, the authors focus on AP-1 function during cell fate determination. In particular, they review how Jun senses cellular stimuli and functions in a wide range of pivotal cellular processes.

Chapter I - This chapter discusses the role of the insulin-like growth factor (IGF) family in the proliferative regulation of vertebrate cells *in vitro* and *in vivo*. The initial part of this chapter peruses the fundamental knowledge in the regulation of cell cycle regulation, its gatekeeper checkpoints, and markers commonly utilized for assessing and quantifying its progression. Recent findings in genomics and post-genomics position characteristics of the architecture and assembly mechanism of chromatin at a critical and pivotal point in the commitment of the cell to continue to pursue its proliferation schedule through the cell cycle, or to engage in apoptotic programmed cell death. The complex family of insulin-like growth factors (e.g., IGF-1 & 2), their receptors (i.e., IGF-1R, IGF-2R), and binding proteins (i.e., IGFBP's) together form a complex interactomic system, whose role in regulating traversal of the cell cycle is increasingly characterized, particularly in terms of its synergy with other cell proliferative factors. The chapter concludes with a discussion of the translational implications of this fundamental *in vitro* knowledge in the context of patients with diabetes, with cancer, and aging-neuro-degenerative diseases.

Chapter II - Cell cycle control is a fundamental process for maintenance of the genome stability. An aberrant cell cycle eventually transforms the cell into a malignant phenotype. Among cell cycle controllers, p53, c-Myc, and E2F1 are pivotal transcription factors, as they determine the cell fate during cell damage. MicroRNAs (miRNAs) are emergent members of small and non protein-coding RNA molecules of about 22 nucleotides in length that repress translation and promote mRNA degradation by sequence-specific interaction with mRNA. More than four hundred miRNAs have been identified so far in humans (http://microrna.sanger.ac.uk). Deregulation of miRNAs expression has been shown in human pathogenesis especially for many cancers; however, the overall picture as to what mechanisms deregulate miRNAs during tumor formation remains unclear. To better understand the molecular etiology of cancer, we need to take a count of miRNAs. In this review, the authors focus on recent advances concerning the functional link between miRNAs and transcriptional factors relevant to the cell cycle control and carcinogenesis.

Chapter III - All living entities, from cells to the human body, experience periodic functional and morphologic abnormalities, intermingled with the state of health. If and how the various states are related, and what factors assume greater or lesser significance in this process, however, is subject to discussion. This chapter considers that all the various states, experienced by biologic entities during their life cycles, are reflections of their compliance with systems principles. This view is based on the concept of cycles and systems inter-relatedness.

Systems science, complexity, and chaos theories were employed to construct a conceptual model applicable to living entities. For each state, representative systems' characteristics were matched to examples of approximate biologic equivalents. This model conceptualizes *a zone of order* and a *zone of chaos* with an overlapping but fragile *health territory*, the heart of this schema. The *health territory* is proposed to be a dynamic amalgamation of the proximate *outer core* of the *zone of order* and the *inner edge* of the *zone of chaos*. In addition, two extremes states are postulated as well: the *outer edge* of the *zone of chaos* and the *zone of entropy* which predominates within the *inner core* of the *zone of order*.

Examples of plausible correlations were found between cancer and the *outer edge* of the *zone of chaos*. A similar association was found between degenerative/inflammatory diseases, including aging, and the *zone of entropy*. Based on this model, biologic entities which stray into the postulated extreme states of *chaos* or *entropy* may still reverse back to the *health territory* by restoring their *organized complexity*. The exponential cellular growth within the *zone of chaos,* characteristic of cancer, could be changed through a process of cyto-education which reverses *disorganized complexity* to an organized one by creating an environment which highly favors cellular differentiation instead of blind multiplication. The path back to the *health territory* from the second extreme, the system-engulfing *entropy*, would involve a process of increasing system's *complexity.*

Applying systems science principles to biologic entities, as attempted in this model, not only allows for the appreciation of disease states in a different light but also suggests what steps could be taken to return the entire system into the state of health.

Chapter IV - Signaling from Transforming Growth Factor-β (TGFβ) through its transmembrane receptor serine-threonine kinases plays an essential role in cell proliferation, differentiation, apoptosis, and the pathogenesis of many human diseases in various tissues including hematopoietic cells. One of the major functions of TGFβ in hematopoiesis is its inhibitory effect on cell proliferation. Downregulation of multiple cell cycle-regulatory molecules appears to be a dominant event in TGFβ-mediated growth inhibition of several human myeloid leukemia cell lines. At least, the inhibition of some of the molecules is regulated at mRNA level. Both G1-and G2-check point protein kinases participate in the regulation of TGFβ-mediated growth inhibition, whereas Cdk inhibitor p27-induced by TGFβ may have both a positive and negative role. TGFβ also rapidly inhibits phosphorylation of pRb at several serine and threonine residues and regulates the association of a repressor E2F transcription factor, E2F-4, with pRb family proteins (pRb, p107, and p130). Major p107-E2F complexes are found in proliferating cells whereas the under-phosphorylated pRb and p130 are associated with E2F-4 in G1 phase. Since TGFβ inhibits synthesis of p107 and has no effect on the expression of E2F-4, it suggests that E2F-4 is released from p107 and switched to pRb and p130 in response to TGFβ stimulation. Western blot and DNA binding experiments suggest that p130 and p107, but not pRb, play a critical role in regulating E2F response genes in TGFβ-mediated cell cycle control in human hematopoietic cells. Although the activation of SMAD-independent p44/42 MAPK signal by TGFβ has been recently reported in various cell types, which is linked to TGFβ-induced proliferation, apoptosis, and growth inhibition, TGFβ does not activate p44/42 MAP kinase pathways in human myeloid leukemia cells whose growth is inhibited by TGFβ, but turns off GM-CSF-induced p44/42 MAPK signal via inhibition of PI3-K-Akt pathway.

Chapter V - Cell cycle is dependent on correct cycling and activation of Cyclin Dependent Kinases (CDKs)/Cyclin pairs. Their protein levels and regulation mechanisms are altered in transformed cells and CDKs are considered attractive targets for cancer therapy. Data from knockout and siRNA experiments have shown a redundancy in CDKs functions. Recent clinical trials and preclinical studies have identified the possibility to use inhibitors specific for subclasses of cell-cycle CDKs (CDK1/2 or CDK4/6) as single agents for a few specific cancer types that show a more pronounced dependency on CDKs functions. Pan-CDK inhibitors are giving, instead, more promising results. They inhibit both cell-cycle CDKs and transcriptional CDKs by combining cytostatic and apoptotic effects in cancer cells. Apoptosis is induced through depletion of anti-apoptotic proteins, sensitive to inhibition of transcription because encoded by mRNAs with short half-lifes and rapid turnover. Recently, new crystal structures of transcriptional CDKs (CDK7 and CDK9) have provided molecular basis for targeting this subclass. Combinations of pan-CDK inhibitors with traditional chemotherapic agents have shown a synergistic cytotoxic effect in a variety of cancers and the restoration of susceptibility in cancers that developed resistance to traditional therapies. Flavopiridol, a pan-CDK inhibitor, is also being used in clinical trials in combination with Histone deacetylase (HDAC) inhibitors, which are also being developed as anti-neoplastic drugs. Evidence has been reported of cross-talk between CDKs and HDACs that indicates the existence of a complex cross-regulation mechanism. This cross-talk between CDKs and HDACs mediated by Retinoblastoma (pRb) and/or CDK inhibitors (CKI) is of particular pharmacological relevance. It engenders the expectation that the combined pharmacological inhibition of CDKs and HDACs might have a higher therapeutic profile than the single treatment.

Chapter VI - Down syndrome or Trisomy 21 is the most frequent genetic cause (1/700 new birth) of the mental retardation. It is characterized by the triplication of human chromosome 21 in all cells and by overdosage of genes on this chromosome in all cells. Thus, trisomy 21 could provide a genetic model for the role of aneuploidy in leukemogenesis, tumorigenesis, cancerigenesis and neurogenesis abnormalities, and to decipher the molecular and cellular mechanisms involved in these different abnormalities. Genetic overdosage determines transcriptional alteration of most of the chromosome 21 genes. Thus, trisomic cells suffer of dosage imbalance of a lot of proteins that can potentially impair all molecular pathways, including cell cycle control. Trisomy 21 is a unique example of the parallel amplification of multiple genes, which may act cooperatively in the same leukemogenic, tumorigenic, cancerigenic or neurogenic pathway. A better knowledge of the function of genes mapping to chromosome 21 may contribute to the understanding of the several mechanisms of diseases, acting in patients with DS, such as leukaemia, tumors, cancer, and neurogenesis alterations related to cell cycle control alterations. Studies of the three oncogenes AML1, ERG and ETS2, higly related to the genesis of leukemias, are of the most interest to contribute to the general understanding of the oncogenic mechanisms of chromosomal aneuploidies, the most common abnormalities in cancer. In addition, the direct involvement of the human chromosome 21 Single minded 2 gene (SIM2) in the inhibition of gene expression, growth inhibition, inhibition of tumor growth as well as induction of apoptosis that could involve a block of cell cycle, establish SIM2 gene as a molecular target for cancer therapeutics and have additional important implications for the future diagnosis and treatment of specific solid tumors as well as for further understanding the cancer risk in Down syndrome patients. In addition, some members of the phosphorylation pathways that control the cell cycle are encoded by specific chromosome 21 genes, particularly the minibrain gene Dyrk1A, suggesting some alteration in cell cycle control in Down syndrome individuals. The involvement of Dyrk1A gene in the regulation of neural proliferation on the basis of alterations in the proliferation centres of brain of Drosophila minibrain (MNB) mutants and the altered levels of cell cycle regulators in transgenic mice overexpressing Dyrk1A as well as the phosphorylation of cell cycle regulators by DYRK1A support the evolutionary conservation of DYRK1A role as regulator of proliferation in neural progenitor and its role in cell cycle regulation. All these genes localized in the human chromosome 21 play important roles in cell cycle and cell growth and, with the recent developments in genomics and proteomics, could promise significant advancements in Down syndrome knowledge and considerable progresses to decode the molecular and cellular mechanisms involved in the DS associated diseases related to cell cycle alterations in view of new development of drugs and treatments.

Chapter VII - Progression through each phase of the cell cycle (G1, S, G2 and M) is precisely controlled by the activity of distinct cyclin-dependent kinases (CDKs) and their associated regulatory subunits known as cyclins. The composition and regulation of CDKs vary according to cell type and environment. The transition from the first growth G1 phase to the DNA synthesis S phase is regulated by the phosphorylation of retinoblastoma protein (pRb). The cyclin D1/CDK4/6 complex phosphorylates pRb, while $p16^{INK4a}$, a CDK inhibitor that interacts with CDK4, inhibits this process. Further CDK inhibitors $p15^{INK4b}$, $p21^{CIP1/WAF1}$, and $p27^{KIP1}$ affect pRb. The tumor suppressor activity of TGFβ depends on its ability to regulate the expression of a number of proteins that play key roles in the control of cell cycle progression from the G1 phase to the S phase. TGFβ is able to induce growth arrest in a variety of cell types including epithelial, endothelial, hematopoietic, neural, and certain types of mesenchymal cells. TGFβ induces the transcription of genes for $p15^{INK4b}$ and $p21^{CIP1/WAF1}$. The $p15^{INK4b}$ specifically blocks the activation of CDK4 and CDK6, whereas $p21^{CIP1/WAF1}$ and $p27^{KIP1}$ inhibit the activity of CDK2. Induction of $p15^{INK4b}$ and $p21^{CIP1/WAF1}$ is important for TGFβ mediated growth arrest in epithelial cells, but not in hematopoietic cells where upregulation of $p57^{KIP2}$ is essential for TGFβ-induced growth arrest. In epithelial cells in addition to regulating the activity of CDK complexes, TGFβ also inhibits the transcription of the protooncogene c-myc that plays a role in controlling cell cycle entry into S phase in higher eukaryotes. Downregulation of c-myc by TGFβ is frequently impaired in cancer cells. The identification and characterization of TGFβ signaling pathway increased our understanding of the dual role of TGFβ as a tumor suppressor and prooncogenic factor. In late stages of epithelial tumorigenesis tumor cells become resistant to growth inhibition by TGFβ due to inactivation of TGFβ signaling pathway or aberrant regulation of the cell cycle. Solid tumors frequently harbor mutations in members of the TGFβ signaling pathway, including in Smad4 and TβRII. On the other hand, hematologic cancers generally become TGFβ resistant through decreased expression of TGFβ signaling by oncoproteins including Evi1, Tax and fusion proteins Runx1-Evi1, Runx1-ETO and PML-RARα. Tumor cells often show increased expression or production of TGFβ. TGFβ can regulate extracellular matrix composition and degradation that play complex roles in tumor invasion and metastasis.

Chapter VIII - The CCAAT/enhancer binding protein alpha (C/EBPα) is the founding member of a family of related leucine zipper transcription factors that play important roles in differentiation of liver, lung, adipocyte and myeloid cells. C/EBPα controls differentiation-dependent gene expression and inhibits proliferation in these types of cells. The capacity of C/EBPα to promote growth arrest has been studied *in vitro*, by analysis of knockout mice and by examination of leukemic cells. Several models of C/EBPα-induced growth arrest have been described. These include C/EBPα-mediated (1) stabilization of the cyclin-dependent kinase 2 (CDK2) inhibitor, $p21^{WAF1/CIP1}$, (2) regulation of growth-inhibiting Rb-E2F complexes, (3) interaction with free E2F, inhibition of E2F activity and down-regulation of the E2F target gene c-Myc, (4) interaction with Max, a member of the basic region –helix-loop-helix-leucine zipper proteins, that belongs to a network of transcription factors, which includes the Myc and Mad families of protein, (5) inhibition of CDK2 and CDK4 activity, and (6) interaction with the SWI/SNF chromatin remodelling complexes. C/EBPα may promote growth arrest by cell specific mechanisms. This means that different models of C/EBPα-mediated cell cycle arrest operate in different cell types. Different regions of C/EBPα are involved in growth inhibition. According to the third model, C/EBPα interacts with free transcription factor E2F through the non-DNA binding surface of its basic region and N-terminal region of the C/EBPα also plays role in growth inhibition. The basic DNA-binding region of the C/EBPα is also involved in the C/EBPα and Max interaction. In the CDK2/CDK4 inhibition model, C/EBPα interacts with and inhibits the activity of CDK2/CDK4 through proline- and histidine-rich region located in the central part of C/EBPα. According to the SWI/SNF recruitment model, C/EBPα interacts with SWI/SNF components through a centrally located 75-amino acid region overlapping with the CDK2/CDK4 binding region. Acute myeloid leukemias (AMLs) are clonal disorders that are characterized by a block in differentiation along one or more haematopoietic lineage. Molecular abnormalities are frequently detected in AML. The genetic alterations in AML often affect transcription factors that also have an important role in normal hematopoiesis. C/EBPα mutations have been observed in AML patients with the frequency approximately 5-14% and are often associated with the induction of proliferation.

Chapter IX - Renal fibrosis is a common pathway of progressive renal disease and parallels the reduction in organ function that necessitates dialysis or transplantation. In the last decades, accumulating experimental and clinical evidence convincingly suggests that the renin-angiotensin system (RAS) is a key player in the progression of chronic kidney disease. Angiotensin II (ANG II), the main peptide of the RAS, was classically viewed as a vasoactive factor participating in the local and systemic hemodynamic regulation. With the progression of researches, it has emerged as a multifunctional cytokine exhibiting many non-hemodynamic effects such as growth, profibrotic, matrix metabolism and even proinflammation. These new understandings of ANG II have important clinical implications. It predicted and explained why blockade of the RAS would significantly slow the progression of fibrotic disease. However, it further suggested that higher doses and/or a combination of ANG II blockade with another agent or agents might truly halt the progressive fibrosis.

This review will focus on the non-hemodynamic aspect of ANG II and its contribution in renal fibrogenesis.

Chapter X - Aging is a complex set of phenotypes characterized by reduced repair and/or regeneration of lost or damaged cells. Mammalian ageing is associated with reduced regenerative capacity in tissues that contain stem cells. It has been proposed that this is at least partially caused by the senescence of progenitors with age. Mitogen-dependent progression through the first gap phase (G_1) and initiation of DNA synthesis (S phase) during the mammalian cell division cycle are cooperatively regulated by several classes of cyclin-dependent kinases (CDKs) whose activities are in turn constrained by CDK inhibitors (CKIs). CKIs are important negative regulators of the cell cycle. The Ink4a/Arf locus encodes 2 tumor suppressor molecules, p16INK4a and Arf, which are principal mediators of cellular senescence. Expression of the INK4a/ARF locus, is not only a major suppressor of cancer, but also an effector of mammalian aging. Ink4a/Arf expression is a biomarker of aging. The cell cycle inhibitor p16 (INK4A) is expressed in vivo in many tissues with age. The exposure of certain chronic stresses can trigger p16 (INK4A) expression and a senescence-like phenotype. Recent data suggest that expression of p16 (INK4a) induces an age-dependent decrease in the proliferative capacity of certain tissue-specific stem cells and unipotent progenitors. Telomere shortening limits the proliferative lifespan of human cells by activation of DNA damage pathways, including upregulation of the cell cycle inhibitor p21. Cellular senescence induced by different stresses and telomere shortening appears to play an important role in the aging process.

Chapter XI - Maintenance of genomic integrity is critical for prevention of a wide variety of adverse cellular effects including apoptosis, cellular senescence, and malignant cell transformation. Under stress conditions and even during an unperturbed cell cycle, checkpoint proteins play the key role in genome maintenance by and mediating cellular response to DNA damage, and represent an essential part of the "cellular stress response proteome". Intact checkpoint signal transduction cascades check the presence of genome damage, trigger cell cycle arrest, and forward the information to the protein core of cell cycle machinery, replication apparatus, repair, and/or apoptotic protein cores. Genetic checkpoint defects lead to syndromes that demonstrate chromosomal instability, increased sensitivity to genotoxic stress, tissue degeneration, developmental retardation, premature aging, and cancer predisposition that is most extensively studied for the ATM-checkpoint mutated in *Ataxia telangiectasia*. Tissue specific epigenetic control over the function of cell cycle checkpoints can be, further, deregulated by aberrant DNA methylation status. The consequent checkpoint deregulation may result in tissue specific degenerative processes such as degeneration and calcification of heart aortic valves, diabetic cardiomyopathy, hyperhomocysteinemic cerebrovascular, peripheral vascular and coronary heart diseases, neurodegenerative disorders (Alzheimer and Parkinson diseases, amyotrophic lateral sclerosis, glaucoma), and premature aging frequently accompanied with cancer. This chapter focuses on the checkpoints shown to be crucial for unperturbed cell cycle regulation, deregulation of which might be considered as a potential molecular marker for early diagnosis of and therapy efficiency in neurodegenerative, cardiovascular and cancer diseases. An application of the most potent detection technologies such as "Clinical Proteomics" and "Disease-specific Transcriptomics" also considered here, allows a most specific selection of diagnostic markers.

Chapter XII - Dysregulation of cell cycle control is of major importance in cancer pathogenesis. Consequently, the identification of genetic and epigenetic changes in genes encoding cell cycle control proteins is instrumental to understand how dysregulation of the cell cycle machinery contributes to the proliferation of tumour cells. Coiled-coil domain containing 94 (CCDC94) was recently characterized as a novel human nuclear protein that plays a role in cell cycle control. Ectopic expression of CCDC94 protein *in vitro* results in cell cycle arrest whereas suppression of *CCDC94* mRNA induces a dramatic increase in cell proliferation. *CCDC94* was originally identified as a target gene of an acquired chromosome translocation in a patient with acute myeloid leukemia (AML) that resulted in its haplo-insufficiency in leukemic cells. *CCDC94* adds to a number of other genes involved in the cell cycle machinery whose expression is frequently altered in hematological malignancies. Chromosome rearrangements and deletions involving D-type cyclins, cyclin-dependent kinases (CDKs) or cyclin-dependent kinase inhibitors (CDKIs) are frequent in lymphoid tumours. In particular, chromosome translocations cause overexpression of D-type cyclins in mature B-cell neoplasms such as mantle cell lymphoma and multiple myeloma whereas different types of rearrangements lead to overexpression of CDK6 in mature T-cell malignancies. In addition, deletion or promoter hypermethylation of *CDKN2A* and/or *CDKN2B* genes, which encode 2 members of the INK4A family of CDKIs, $p16^{INK4A}$ and $p15^{INK4B}$ respectively, are recurrent events in acute lymphoblastic leukemias. In contrast, patients with myelodysplastic syndrome and AML frequently present with hypermethylation of *CDKN2B* alone. This review summarizes the current knowledge on the molecular changes involving D-type cyclins, CDKs and CDKIs in several hematological neoplasms and addresses the most relevant effects of these alterations in the control of cell cycle.

Chapter XIII - Cell cycle is an ordered set of processes by which one cell grows and divides into two daughter cells. Drosophila is an attractive organism that has been used as a model for the study of the mitotic cell cycle. The fruit fly has not only been extensively used for genetics studies but also its developmental biology is highly suitable for studies of the mitotic cycle. The mitotic division in the early embryos consists of rapid successions of Mitosis (M) and DNA synthesis (S) phases with no discernible G1 or G2 (Gap) phases as found at later stages of development. In the subsequent development of the organism, two different types of tissue arise, having either proliferating diploid cells or cells that do not divide but become highly polyploid. This review presents the various modes of cell cycling that occur during Drosophila development and discusses their regulation in light of the latest findings.

Chapter XIV - This chapter introduces the concept of cell cycle control and the many enzymes involved. The authors provide a brief history on the first research and continue with their most recent research on this subject.

Chapter XV - Community pharmacists are integral members of the health care team who are ideally placed to offer and provide health education to patients in a timely and regular manner. Community pharmacists are highly skilled and knowledgeable health care professionals and, as with other members of the health care team, are assumed to have the skills to transfer their knowledge through effective education to the patient. However, knowledge associated with health and disease management is ever changing, and the need for pharmacists to effectively uptake knowledge and then transfer it to patients is one of the keys to effective disease management in the community setting.

This article reports on an innovative approach to pharmacist education, utilizing the "Train the Trainer" educational model, targeting appropriate inhaler use by people with asthma in the community. The authors highlight the way in which an educational module based on theory can be effectively delivered to community pharmacists and feasibly implemented into practice. Following the implementation process, the innovative educational module resulted in improved clinical and humanistic health outcomes for people with asthma in the community.

Chapter XVI - In recent years, several dietary compounds or nutritional agents, i.e. isoflavone genistein, indole-3-carbinol (I3C), 3,3'-diindolylmethane (DIM), curcumin, (-)-epigallocatechin-3-gallate (EGCG), resveratrol, lycopene, etc, have been recognized as cancer chemopreventive agents because of their anti-carcinogenic activity. Moreover, these compounds also exert their anti-tumor activities through the regulation of different cell signaling pathways, resulting in the cell cycle arrest and subsequent cell growth inhibition. The cell cycle is known to be tightly regulated by different cyclins and cyclin dependent kinases (CDKs), which form key regulatory complex, in different phases of the cell cycle.

The activity of cyclins/CDKs complex is negatively regulated by several CDK inhibitors (CDKIs) including $p21^{WAF1}$, $p27^{KIP1}$, $p16^{INK4a}$, etc. In addition, the molecules which interact with the key regulatory complex and CDKIs also influence the progression of cell cycle. The data from the authors' laboratory and others have demonstrated that dietary chemopreventive agents inhibit the kinase activity of CDKs and regulate the expression of cyclins, CDKs, and CDKIs, leading to the cell cycle arrest at different phases of cell cycle. A number of molecules, which are critical for the cell cycle control, have been found to be altered by chemopreventive agents in the cell cycle synchronization assay. Moreover, the authors and other investigators have utilized gene expression profiling showing alteration in the expression of genes that are related to cell cycle progression by chemopreventive agents. In summary, emerging studies using genomics and proteomics is likely to advance our knowledge regarding the molecular mechanism of cell cycle regulation as a target for the prevention and/or treatment of human cancers by nutritional agents.

Chapter XVII - Chromatin is the intricate complex of proteins and DNA that is involved in the control of the cell cycle and the modulation of gene expression in higher organisms. Methods in biochemistry and molecular biology have made it possible to identify and isolate many of the chromatin-related molecules vital to these processes. A wealth of information on the structure and function of these molecules and larger complexes is increasingly being elucidated by biophysical techniques, such as: mass spectrometry for highly accurate protein identification; electron microscopy (EM), nuclear magnetic resonance (NMR), small-angle X-ray and neutron scattering (SAXS & SANS) and in particular X-ray crystallography at third generation synchrotron radiation sources for solving complete three-dimensional static structures of macromolecules to atomic resolution.

In this review a non-exhaustive overview of the current status and recent developments in these and other biophysical techniques in relation to the structure and function of chromatin-related molecules is given. Some of the possibilities for this crucial area of research with single-molecule methods and future fourth-generation light sources are also discussed.

In: Cell Cycle Control: New Research
Editors: N. H. Leroy, N. T. Fournier
ISBN: 978-1-60456-796-0

Short Communication

Regulation of Cell Fate Determination by Jun, a Component of Activator Protein 1

Kenichi Yoshida *
Department of Life Sciences, Meiji University,
1-1-1 Higashimita, Tama-ku, Kawasaki, Kanagawa 214-8571, Japan

Abstract

Microenvironmental stimuli cause multiple signal transduction pathways which determine cellular response. Mitogen-activated protein kinases (MAPKs) are pivotal signal cascades to transduce signals into dynamic gene expression waves in the nucleus. Gene expression patterns in given cells are totally dependent on the combined signals and integrity of activated transcription factors. The activating protein-1 (AP-1) is a classical transcription factor that can transduce MAPK signals into the gene expression pattern changes and this cellular system has a large impact on pathophysiological aspects of cells. Indeed, AP-1 is involved in a wide range of cellular events including cell proliferation, transformation, apoptosis, and stress response. Jun joins in with the AP-1 transcription complex as a homodimer or heterodimer, and has been identified as a mammalian cellular homologue of retroviral onco-proteins. Jun is known as an immediate early gene and its activity is largely regulated by MAPK signals. Expression of Jun can be induced by oncogenic stress. Taken together, these facts indicate that Jun is responsible for certain part of the AP-1 capability to modulate cellular response. It is also defined by the abundance of distinct members of AP-1 components within a given cell type and their surrounding stimuli. In this review, we focus on AP-1 function during cell fate determination. In particular, we review how Jun senses cellular stimuli and functions in a wide range of pivotal cellular processes.

* Correspondence: Kenichi Yoshida, D.V.M., Ph.D. Department of Life Sciences, Meiji University, 1-1-1 Higashimita, Tama-ku, Kawasaki, Kanagawa 214-8571, Japan. Tel. & Fax.: +81-44-934-7107, e-mail: yoshida@isc.meiji.ac.jp

Introduction

Regulation of cell fate determination is a central issue towards understanding cell physiology and pathology. Environmental stimuli surrounding cells can trigger cellular responses mainly through signaling cascades. To date, various signaling pathways have been identified and characterized. In most cases, signals eventually lead to the activation of transcription factors to promote or repress the target gene expression. Spatio-temporal gene expression changes regulated by transcriptional networks restrict the cell fate according to extracellular signals. The activator protein 1 (AP-1) has been shown to regulate numerous genes important for cell proliferation, apoptosis, and inflammation [1, 2]. In general, AP-1 is a symbolic term referring to dimeric transcription factors composed of Jun (alias c-Jun), Fos (alias c-Fos), or ATF subunits. Jun and Fos are proto-type components of AP-1, and they can form homodimers and heterodimers, but Fos cannot form homodimers. Notably, oncogenic Ras phosphorylates Jun, through activation of specific MAPKs (mitogen-activated protein kinases), which is considered as an essential mechanism for subsequent target gene activation and thereby tumorigenesis when the transcriptional activity of Jun is deregulated [3]. Jun has opposing effects on cells, namely promotion of cell proliferation in certain circumstances but on the other hand can induce apoptosis. These different functions of Jun can be partly explained by Jun's distinct upstream regulators and distinct dimerizing partners [4]. We review the recent understanding about how Jun senses upstream signals and regulates target genes properly. In particular, we summarize the AP-1 function during cell fate determination.

1. MAPK Signals and Jun

MAPK cascades can be characterized as serine threonine kinases that are sequentially phosphorylated by upstream kinases MAPKKK and MAPKK. Depending on stimuli responses, MAPKs can be divided into three pathways, namely extracellular signal-regulated kinases (ERKs 1 and 2, also known as p44/p42 MAPK), c-Jun-NH_2-terminal kinases (JNKs 1, 2, and 3, which are also known as stress-activated protein kinases SAPKs), and p38 kinases (alpha, beta, gamma, and delta) [5-7]. These signaling pathways lead to the promotion of the ability of Jun and Fos family members, which are components of the AP-1 transcription factor [8]. Post-translational modification is the most important step for Jun to gain transcriptional activity. For example, Jun activation has been widely recognized as a consequence of phosphorylation of the N-terminal activation domain of Jun at Ser63 and Ser73 by JNKs [9, 10]. Because of the JNKs' ability to sense oncogenic or DNA damaging signals, it is considered that JNKs are promising drug developmental targets in the area of cancer, inflammation, and diabetes [11]. By using Jun null fibroblasts, Jun is shown to be required for the G1 phase progression of cell cycle and this involves direct transcriptional control of the cyclin D1 gene; however, these processes are independent of Ser63 and Ser73 phosphorylation. On the other hand, Jun has been shown to require Ser63 and Ser73 phosphorylation to protect cells from UV-induced death [12]. This strongly suggests that the differential phosphorylation status enables specific Jun target gene expression. On the other hand, C-terminal Src kinase (CSK) phosphorylates Jun at Y26 and Y170 and this is known to promote Jun degradation [13]. In case of Fos, p38 MAPKs have been shown to play a pivotal role in mediating stress-induced Fos phosphorylation and consequent gene transcription activation [14]. The critical kinases responsible for the control of Fos activity remain elusive.

2. AP-1 Transcription Factor

Jun and Fos family members activated by specific MAPKs can form a variety of homodimers and heterodimers via their leucine zipper structure, which can bind to target DNA sequences [15]. These family members are expanding ones as they include JunB and JunD as the Jun family, and in the same manner FosB, Fra1, and Fra2 as the Fos family. In addition, ATF (ATFa, ATF2, and ATF3) and JDP (JDP1 and JDP2) subfamilies have also been recognized [16]. Recent studies highlight the importance of the relative abundance of each family member of Jun, Fos, or ATF in the composition of AP-1 dimers [17]. The abundance of AP-1 subunits is thought to be critically regulated by cellular stimuli in a given cell type and thus influences cell fates. Indeed, in an adipocyte differentiation model, compositional changes of Jun, JunB, and JunD in AP-1 resulted in changes of the DNA binding activities of AP-1 [18].

Initially, AP-1 was found as an activator of the transcription of human metallothionein (hMTIIA) [19]. hMTIIA is one of the phorbol diester (12-O-tetradecanoylphorbol-13-acetate, hereafter referred to as TPA)-inducible genes and TPA-induced AP-1 can bind to TPA-responsive elements (TREs) located in target gene promoters [20]. TREs are preferentially bound by Jun and Fos heterodimers and have been recognized as the AP-1 consensus sequence 5'-TGA(G/C)TCA-3' [15], whereas Jun and ATF dimers prefer the 5'-TGACGTCA-3' sequence known as the cyclic AMP response element (CRE). Besides TPA, AP-1 has been shown to be induced by a wide range of stimuli such as growth factors, cytokines, neurotransmitters, and UV irradiation; however, specific MAPKs cascades are largely responsible for the transcriptional ability of Jun as described above. Cellular stimuli can quickly induce Jun and Fos transcription independent of *de novo* protein synthesis, so they are referred to as immediate early genes [21], indicating AP-1 potentially sits on the summit of transcriptional cascades. Indeed, AP-1 transcription factor regulates a wide range of cellular processes, including cellular proliferation, differentiation, programmed cell death, survival, and inflammation [1, 2]. This could be partly explained by the notion that a combination of distinct AP-1 dimers can execute specific cellular programs in given cell types or cell cycle timing. For example, Jun, JunD, and Fra2 show a high level of expression in growing NIH 3T3 fibroblasts [22]. In Swiss 3T3 cells, each member of the Jun and Fos families exists in different levels depending on the cell cycle [23]. Phosphorylation levels of Jun and JunD are also cell cycle dependent [24].

3. Transcriptional Target of AP-1

AP-1 regulates cell proliferation largely by mediating transcriptional control of cyclins and cyclin-dependent kinase (Cdk) inhibitors such as cyclin D1, cyclin A, cyclin E, $p21^{Cip1}$, p16^{Ink4a} as well as p53 and p14^{ARF} [1, 2]. Jun null mouse fibroblasts show a proliferation defect accompanying inefficient G1-S phase progression. This can be explained by the fact that p53 expression is directly suppressed by Jun [25]. Particular AP-1 members, such as Jun and Fra1, can upregulate p14^{ARF} expression [26]. AP-1 suppresses transcriptional activity of the human but not mouse telomerase catalytic subunit (hTERT) [27]. Cooperative interaction between AP-1 and beta-catenin has been shown to synergistically activate serum-induced cyclin D1 transcription [28]. Unexpectedly, p73, a structural and functional homologue of the tumor suppressor p53, induces cyclinD1 expression in a Jun-dependent manner [29]. These Jun-regulated factors mainly function during the G1 to S phase of cell cycle transition, senescence, and immortalization. AP-1 also regulates the E2F and its target gene expression and this leads to promotion of breast cancer cell proliferation [30]. JNK can function in mitosis by promoting the expression of Aurora B following phosphorylation of Histone H3 [31], indicating that Jun can regulate mitosis as well as the G1/S phases.

4. Pathophysiological Function of AP-1

In undifferentiated and aggressive human sarcomas, Jun is highly amplified and over-expressed which resulted in inhibition of an early step of adipocyte differentiation [32], suggesting Jun involvement in tumorigenesis and the suppressive effect against differentiation. Jun has been shown as an essential factor in cellular migration via the induction of stem cell factor (SCF) [33], indicating that Jun participates not only in cellular proliferation but also in cell migration and angiogenesis, which are characteristic features of malignant cancerous cells. Angiopoietin-1 (Ang-1)-dependent migration and proliferation of endothelial cells are promoted through the release of interleukin-8 (IL-8) in an autocrine manner. In this context, the Ang-1-bound Tie-2 receptor triggers ERK1/2 and SAPK/JNK, leading to the phosphorylation of Ser63 and Ser73 of c-Jun. The IL-8 promoter has been shown to be regulated by AP-1 [34].

Jun plays a key role during hepatocyte proliferation, because liver-specific Jun knockout mice display impaired liver regeneration after partial hepatectomy [35]. Hepatocytes failed to regenerate accompanied by increased cell death and inefficient G1-S phase progression. Jun regulates liver regeneration partly through p53, p21^{Cip1}, and the stress kinase p38alpha [36]. As well as in hepatocytes, Jun plays a central role in the proliferation of keratinocytes [17]. AP-1 has also been implicated in the regulation of postnatal mammary gland growth and development [37]. Jun knockout mice died around mid-gestation concomitantly with impaired hepatogenesis, and the growth rate of null fibroblasts was repressed [38, 39]. On the other hand, JunB deficiency in postnatal mice resulted in a myeloproliferative disorder and revealed a stem cell-specific role for JunB in normal and leukemic hematopoiesis [40]. From the analysis of JunD null mice, JunD regulates proliferation during the process of chronic renal diseases [41]. These mice genetic approaches indicate a typical example of a distinct dimer of AP-1 even if Jun family members are involved in distinct roles in given cell types. During muscle differentiation, different properties of the AP-1 subunit composition, containing Jun, JunD, and Fra2, have been characterized [42]. Interestingly, JunB and JunD have similar target genes involved in cell cycle regulation similar to Jun. For example, JunB regulates cell proliferation partly through up-regulation of cyclin A and down-regulation of p16^{Ink4a} [43, 44] and JunD regulates cyclin D1 [45]. Beyond the dimer composition of AP-1, the interaction of NFAT transcription factor and AP-1 has been shown to influence T cell anergy [46]. Finally, Jun and JunB have an antagonistic function against each other in the fibroblast-mediated paracrine control of keratinocyte proliferation and differentiation [47]; however, a JunB knock-in strategy clearly showed that JunB can substitute for Jun loss, indicating that the spatio-temporal gene expression controlled by AP-1 may have significant effects on cell fate determination compared with the dimer components of AP-1 [48].

Conclusion

Accumulating bodies of evidence suggest that AP-1 has a dual function as it regulates normal cell proliferation and apoptosis but in other circumstances it promotes transformation and maintains cell survival. The capability of AP-1 to modulate cell fate determination is primary regulated by microenvironment stimuli, but selective MAPK cascades and the variation or abundance of distinct dimers of AP-1 are responsible for cellular response. Therefore, understanding the full picture of the dimer composition of AP-1 in a given cell is important to realize how AP-1 exerts its transcriptional activity to determine cell fate. Studies on AP-1 have generated huge amounts of knowledge to enable understanding of transcriptional regulation. Continuous studies on functional analyses of AP-1 have to open a window to elucidate the sophisticated systems that execute the program of cell fate determination.

References

[1] Shaulian, E., and Karin, M. (2001) AP-1 in cell proliferation and survival. *Oncogene. 20*, 2390-2400.

[2] Shaulian, E., and Karin, M. (2002) AP-1 as a regulator of cell life and death. *Nat. Cell Biol. 4*, E131-136.

[3] Weiss, C., and Bohmann, D. (2004) Deregulated repression of c-Jun provides a potential link to its role in tumorigenesis. *Cell Cycle. 3*, 111-113.

[4] Leppä, S., and Bohmann, D. (1999) Diverse functions of JNK signaling and c-Jun in stress response and apoptosis. *Oncogene. 18*, 6158-6162.

[5] Davis, R.J. (1994) MAPKs: new JNK expands the group. *Trends Biochem. Sci. 19*, 470-473.

[6] Marshall, C.J. (1995) Specificity of receptor tyrosine kinase signaling: transient versus sustained extracellular signal-regulated kinase activation. *Cell. 80*, 179-185.

[7] Cobb, M.H., and Goldsmith, E.J. (1995) How MAP kinases are regulated. *J. Biol. Chem. 270*, 14843-14846.

[8] Karin, M. (1995) The regulation of AP-1 activity by mitogen-activated protein kinases. *J. Biol. Chem. 270*, 16483-16486.

[9] Hibi, M., Lin, A., Smeal, T., Minden, A., and Karin, M. (1993) Identification of an oncoprotein- and UV-responsive protein kinase that binds and potentiates the c-Jun activation domain. *Genes Dev. 7*, 2135-2148.

[10] Dérijard, B., Hibi, M., Wu, I.H., Barrett, T., Su, B., Deng, T., Karin, M., and Davis, R.J. (1994) JNK1: a protein kinase stimulated by UV light and Ha-Ras that binds and phosphorylates the c-Jun activation domain. *Cell. 76*, 1025-1037.

[11] Karin, M., and Gallagher, E. (2005) From JNK to pay dirt: jun kinases, their biochemistry, physiology and clinical importance. *IUBMB Life. 57*, 283-295.

[12] Wisdom, R., Johnson, R.S., and Moore, C. (1999) c-Jun regulates cell cycle progression and apoptosis by distinct mechanisms. *EMBO J. 18*, 188-197.

[13] Zhu, F., Choi, B.Y., Ma, W.Y., Zhao, Z., Zhang, Y., Cho, Y.Y., Choi, H.S., Imamoto, A., Bode, A.M., and Dong, Z. (2006) COOH-terminal Src kinase-mediated c-Jun phosphorylation promotes c-Jun degradation and inhibits cell transformation. *Cancer. Res 66*, 5729-5736.

[14] Tanos, T., Marinissen, M.J., Leskow, F.C., Hochbaum, D., Martinetto, H., Gutkind, J.S., and Coso, O.A. (2005) Phosphorylation of c-Fos by members of the p38 MAPK family. Role in the AP-1 response to UV light. *J. Biol. Chem. 280*, 18842-18852.

[15] Angel, P., and Karin, M. (1991) The role of Jun, Fos and the AP-1 complex in cell-proliferation and transformation. *Biochim. Biophys. Acta. 1072*, 129-157.

[16] Jochum, W., Passegué, E., and Wagner, E.F. (2001) AP-1 in mouse development and tumorigenesis. *Oncogene. 20*, 2401-2412.

[17] Hess, J., Angel, P., and Schorpp-Kistner, M. (2004) AP-1 subunits: quarrel and harmony among siblings. *J. Cell Sci. 117*, 5965-5973.

[18] Wang, H., Xie, Z., and Scott, R.E. (1996) Differentiation modulates the balance of positive and negative Jun/AP-1 DNA binding activities to regulate cellular proliferative potential: different effects in nontransformed and transformed cells. *J. Cell Biol. 135*, 1151-1162.

[19] Lee, W., Haslinger, A., Karin, M., and Tjian, R. (1987) Activation of transcription by two factors that bind promoter and enhancer sequences of the human metallothionein gene and SV40. *Nature. 325*, 368-372.

[20] Angel, P., Imagawa, M., Chiu, R., Stein, B., Imbra, R.J., Rahmsdorf, H.J., Jonat, C., Herrlich, P., and Karin, M. (1987) Phorbol ester-inducible genes contain a common cis element recognized by a TPA-modulated trans-acting factor. *Cell. 49*, 729-739.

[21] Thomson, S., Mahadevan, L.C., and Clayton, A.L. (1999) MAP kinase-mediated signalling to nucleosomes and immediate-early gene induction. *Semin. Cell Dev. Biol. 10*, 205-214.

[22] Lallemand, D., Spyrou, G., Yaniv, M., and Pfarr, C.M. (1997) Variations in Jun and Fos protein expression and AP-1 activity in cycling, resting and stimulated fibroblasts. *Oncogene. 14*, 819-830.

[23] Kovary, K., and Bravo, R. (1992) Existence of different Fos/Jun complexes during the G0-to-G1 transition and during exponential growth in mouse fibroblasts: differential role of Fos proteins. *Mol. Cell Biol. 12*, 5015-5023.

[24] Bakiri, L., Lallemand, D., Bossy-Wetzel, E., and Yaniv, M. (2000) Cell cycle-dependent variations in c-Jun and JunB phosphorylation: a role in the control of cyclin D1 expression. *EMBO J. 19*, 2056-2068.

[25] Schreiber, M., Kolbus, A., Piu, F., Szabowski, A., Möhle-Steinlein, U., Tian, J., Karin, M., Angel, P., and Wagner, E.F. (1999) Control of cell cycle progression by c-Jun is p53 dependent. *Genes Dev. 13*, 607-619.

[26] Ameyar-Zazoua, M., Wisniewska, M.B., Bakiri, L., Wagner, E.F., Yaniv, M., and Weitzman, J.B. (2005) AP-1 dimers regulate transcription of the p14/p19ARF tumor suppressor gene. *Oncogene. 24*, 2298-2306.

[27] Takakura, M., Kyo, S., Inoue, M., Wright, W.E., and Shay, J.W. (2005) Function of AP-1 in transcription of the telomerase reverse transcriptase gene (TERT) in human and mouse cells. *Mol. Cell Biol. 25*, 8037-8043.

[28] Toualbi, K., Güller, M.C., Mauriz, J.L., Labalette, C., Buendia, M.A., Mauviel, A., and Bernuau, D. (2007) Physical and functional cooperation between AP-1 and beta-catenin for the regulation of TCF-dependent genes. *Oncogene. 26*, 3492-3502.

[29] Vikhanskaya, F., Toh, W.H., Dulloo, I., Wu, Q., Boominathan, L., Ng, H.H., Vousden, K.H., and Sabapathy, K. (2007) p73 supports cellular growth through c-Jun-dependent AP-1 transactivation. *Nat. Cell Biol. 9*, 698-705.

[30] Shen, Q., Uray, I.P., Li, Y., Krisko, T.I., Strecker, T.E., Kim, H.T., and Brown, P.H. (2008) The AP-1 transcription factor regulates breast cancer cell growth via cyclins and E2F factors. *Oncogene. 27*, 366-377.

[31] Oktay, K., Buyuk, E., Oktem, O., Oktay, M., and Giancotti, F.G. (2008) The c-Jun N-terminal kinase JNK functions upstream of Aurora B to promote entry into mitosis. *Cell Cycle. 7*, 533-541.

[32] Mariani, O., Brennetot, C., Coindre, J.M., Gruel, N., Ganem, C., Delattre, O., Stern, M.H., and Aurias, A. (2007) JUN oncogene amplification and overexpression block adipocytic differentiation in highly aggressive sarcomas. *Cancer Cell. 11*, 361-374.

[33] Katiyar, S., Jiao, X., Wagner, E., Lisanti, M.P., and Pestell, R.G. (2007) Somatic excision demonstrates that c-Jun induces cellular migration and invasion through induction of stem cell factor. *Mol. Cell Biol. 27*, 1356-1369.

[34] Abdel-Malak, N.A., Kristof, A.S., Magder, S.A., Hussain, S.N., Srikant, C.B., and Di Battista, J.A. (2008) Angiopoietin-1 promotes endothelial cell proliferation and migration through AP-1-dependent autocrine production of interleukin-8. *Blood. 111*, 4145-4154.

[35] Behrens, A., Sibilia, M., David, J.P., Möhle-Steinlein, U., Tronche, F., Schütz, G., and Wagner, E.F. (2002) Impaired postnatal hepatocyte proliferation and liver regeneration in mice lacking c-jun in the liver. *EMBO J. 21*, 1782-1790.

[36] Stepniak, E., Ricci, R., Eferl, R., Sumara, G., Sumara, I., Rath, M., Hui, L., and Wagner, E.F. (2006) c-Jun/AP-1 controls liver regeneration by repressing p53/p21 and p38 MAPK activity. *Genes Dev. 20*, 2306-2314.

[37] Shen, Q., Zhang, Y., Uray, I.P., Hill, J.L., Kim, H.T., Lu, C., Young, M.R., Gunther, E.J., Hilsenbeck, S.G., Chodosh, L.A., Colburn, N.H., and Brown, P.H. (2006) The AP-1 transcription factor regulates postnatal mammary gland development. *Dev. Biol. 295*, 589-603.

[38] Hilberg, F., Aguzzi, A., Howells, N., and Wagner, E.F. (1993) c-jun is essential for normal mouse development and hepatogenesis. *Nature. 365*, 179-181.

[39] Johnson, R.S., van Lingen, B., Papaioannou, V.E., and Spiegelman, B.M. (1993) A null mutation at the c-jun locus causes embryonic lethality and retarded cell growth in culture. *Genes Dev. 7*, 1309-1317.

[40] Passegué, E., Wagner, E.F., and Weissman, I.L. (2004) JunB deficiency leads to a myeloproliferative disorder arising from hematopoietic stem cells. *Cell. 119*, 431-443.

[41] Pillebout, E., Weitzman, J.B., Burtin, M., Martino, C., Federici, P., Yaniv, M., Friedlander, G., and Terzi, F. (2003) JunD protects against chronic kidney disease by regulating paracrine mitogens. *J. Clin. Invest. 112*, 843-852.

[42] Andreucci, J.J., Grant, D., Cox, D.M., Tomc, L.K., Prywes, R., Goldhamer, D.J., Rodrigues, N., Bédard, P.A., and McDermott, J.C. (2002) Composition and function of AP-1 transcription complexes during muscle cell differentiation. *J. Biol. Chem. 277*, 16426-16432.

[43] Passegué, E., and Wagner, E.F. (2000) JunB suppresses cell proliferation by transcriptional activation of p16(INK4a) expression. *EMBO J 19*, 2969-2979.

[44] Andrecht, S., Kolbus, A., Hartenstein, B., Angel, P., and Schorpp-Kistner, M. (2002) Cell cycle promoting activity of JunB through cyclin A activation. *J. Biol. Chem. 277*, 35961-35968.

[45] Toualbi-Abed, K., Daniel, F., Güller, M.C., Legrand, A., Mauriz, J.L., Mauviel, A., and Bernuau, D. (2008) Jun D cooperates with p65 to activate the proximal kappaB site of the cyclin D1 promoter: role of PI3K/PDK-1. *Carcinogenesis. 29*, 536-543.

[46] Macián, F., García-Cózar, F., Im, S.H., Horton, H.F., Byrne, M.C., and Rao, A. (2002) Transcriptional mechanisms underlying lymphocyte tolerance. *Cell. 109*, 719-731.

[47] Szabowski, A., Maas-Szabowski, N., Andrecht, S., Kolbus, A., Schorpp-Kistner, M., Fusenig, N.E., and Angel, P. (2000) c-Jun and JunB antagonistically control cytokine-regulated mesenchymal-epidermal interaction in skin. *Cell. 103*, 745-755.

[48] Passegué, E., Jochum, W., Behrens, A., Ricci, R., and Wagner, E.F. (2002) JunB can substitute for Jun in mouse development and cell proliferation. *Nat. Genet. 30*, 158-166.

In: Cell Cycle Control: New Research
Editors: N. H. Leroy, N. T. Fournier

ISBN: 978-1-60456-796-0

Chapter I

Insulin-Like Growth Factor Regulation of Cell Proliferation

***Francesco Chiappelli*[1,*]*, Alberto Angeli*[2]*, Frank Calzone*[3]*, Elaina Cajulis*[3]*, Javier Iribarren*[4]*, Jackie Kaslow*[1]*, Yue Liu*[1]*, Ercolano Manfrini*[5]*, David Moradi*[1]*, Negoita Neagos*[6] *and Paolo Prolo*[1]**

[1]UCLA School of Dentistry, Division of Oral Biology & Medicine (Oral Biology), USA;
[2]Internal Medicine, University of Turin, Orbassano, Italy;
[3]Amgen, Inc., Thousand Oaks, CA, USA;
[4]Spanish Speaking Mental Health Clinic, UCLA School of Medicine (Psychiatry);
[5]Neurology, University of Ancona, Italy;
[6]Glendale Memorial Hospital, Glendale CA, USA.

Abstract

This chapter discusses the role of the insulin-like growth factor (IGF) family in the proliferative regulation of vertebrate cells *in vitro* and *in vivo*. The initial part of this chapter peruses the fundamental knowledge in the regulation of cell cycle regulation, its gatekeeper checkpoints, and markers commonly utilized for assessing and quantifying its progression. Recent findings in genomics and post-genomics position characteristics of the architecture and assembly mechanism of chromatin at a critical and pivotal point in the commitment of the cell to continue to pursue its proliferation schedule through the cell cycle, or to engage in apoptotic programmed cell death. The complex family of insulin-like growth factors (e.g., IGF-1 & 2), their receptors (i.e., IGF-1R, IGF-2R), and binding proteins (i.e., IGFBP's) together form a complex interactomic system, whose role in regulating traversal of the cell cycle is increasingly characterized, particularly in terms of its synergy with other cell proliferative factors. The chapter concludes with a

---

*Correspondence concerning this article should be addressed to: Professor Francesco Chiappelli, Ph.D., Division of Oral Biology & Medicine, UCLA School of Dentistry, Center for the Health Sciences, 63-090, Los Angeles, CA 90095-1668. T: 310-794-6625; F: 310-794-7109; E: fchiappelli@dentistry.ucla.edu.

discussion of the translational implications of this fundamental *in vitro* knowledge in the context of patients with diabetes, with cancer, and aging-neuro-degenerative diseases.

Keywords: Insulin-like growth factor (IGF), IGF receptor (IGF-R), IGF binding proteins (IGFBP's), cyclins, cyclin-dependent kinases (Cdk), chromatin, apoptosis, traumatic brain injury, PTSD, Alzheimer's disease, cancer, metastasis, genome, proteome, metabolome, interactome, physiolome, post-genomics.

1. Cell Cycle Regulation: Stages, and Check Points

A. The Stages and Checkpoints of the Eukaryotic Cell Cycle

The cell cycle refers to the series of processes that take place in a eukaryotic cell, and lead to its replication. It is the product of the totality of the interactive, biochemical and molecular events that results in the full readiness of the cell to divide. In the current conceptualization of cell biology, the regulation of the cell cycle is one predominant domain of the cytome[1], the overall functionality of the cell as a complex system (Lodish et al, 2000).

The cell cycle is divided in two broad periods. During interphase, the cell grows in size and complexity, elaborating the complex architecture of its cell organelles, accumulating nutrients, transcribing its RNA, translating its proteins during Gap 1 (G1) and G2 (each about 8-10 h), and replicating DNA during the DNA synthesis (S, about 6-8 h) phase. The relatively short mitosis (M, 30 min-2 h), the period of the cell cycle during which the cell splits itself into two distinct "daughter cells", consists of two tightly coupled processes. The first involves the condensation and division of the cell's chromosomes; and the second, cytokinesis proper, which results in a quasi-equal distribution of the cytoplasmic materials to the emerging daughter cells (Glotzer, 2001; Lodish et al, 2000).

The cell-division cycle is a vital and process that mirrors a series of tightly regulated *sine qua non* steps. Activation of each step and of each phase is dependent on the proper progression and completion of the previous one in domino-type sequence, rather than a clock-like timed process. When these biochemical requirements are not met, progress through the cell cycle is temporarily hindered, or reversibly stopped. Cells do not divide, but enter a transitional state of quiescence, the G0 phase. G0 is generally considered a pause during traversal of G1. The longer the cells remain in G0, the less likely it is that the quiescent state will be reversible. Cells that are in an irreversible G0 state do not grow or proliferate, and

---

[1] The suffix "-om-" originates from the Greek stem "σωμ(ατ)-" (body), and has been extended in biology to pertain to wholeness or completed entity. Omes and omics convey implications of complex biological systems and networks in practical and theoretical science. They were first introduced in bioinformatics and molecular biology in the 1980's-90's (hence, genome, proteonome, interactome, cytome, physiome, metabolome). The trend has now become widely accepted. In the context of this discussion, cytomics refers to the complex processes involved in cell-cycle regulation, as the study of cell systems at a single cell level. It combines all the bioinformatic knowledge in an attempt to understand the molecular architecture and functionality of the cell system (cytome) by means of cellular, biochemical, molecular and microscopic techniques that allow the various components of a cell to be visualized as they interact in vivo and in situ.

most often die[2]. Certain cell types, upon reaching their final maturation stage, remain quiescent for long periods of time, possibly indefinitely, as is the case for neurons and lymphocytes. These cells are generally fully differentiated, and do not engage in necrotic or apoptotic cell death simply by the fact of being arrested in G0: rather, they are fully functional in that state (Lodish et al, 2000). This has important implications in clinical cases, such as traumatic brain injury (TBI) or neuro-degenerative diseases, including Alzheimer's disease (*vide infra*).

The various stages of interphase are not usually morphologically distinguishable, except in some cases (e.g., lymphocytes) for an evident enlargement of overall cell size due to increased cytoplasmic content and complexity. These differences can be reliably measured by fractal analysis[3]. Moreover, these changes are clearly evidenced by dual fluorescence flow cytometry to monitor overall cell size (e.g., propidium iodine, PI) and DNA content, of which mitochondrial DNA is a small component (e.g., bromo deoxy-uridine, BrdU) (Figure 1) (Chiappelli, 2007).

Each phase of the cell cycle has distinct checkpoints[4]: sets of specialized biochemical processes that prepare the cell for initiation of cell division. Two key classes of regulatory molecules, cyclins and cyclin-dependent kinases (CdKs), determine a cell's progress through the cell cycle. These proteins and protein pathway specialized in safeguarding cell cycle traversal are, for the most part, well conserved in the evolution of simple eukaryotes (e.g., yeast) to humans, and across the genealogic tree of vertebrates (Chiappelli & Liu, 1999). Many of the relevant genes were first identified in yeast (e.g., *Saccharomyces cerevisiae*), and genetic nomenclature in yeast dubs many of these genes cell division cycle gene products, followed by an identifying number (e.g., cdc25).

In brief, the eukaryotic cell cycle is regulated by a complex set of intertwined and interdependent gate-keeper proteins (i.e., cyclins), cyclin-dependent kinases, (CdK), and adaptor (CdK's inhibitors) proteins, which together play salient roles in the maintenance of the integrity of the cell cycle traversal process during interphase. The study of cell cycle regulation is principally the study of interphase[5] (Lodish et al, 2000; Karp, 2005).

---

2 Necrosis (Gk, νεκρός - dead) refers to the process of "accidental" and "disorderly" cell death, in contrast to the orderly process of apoptosis (Gk: απο - from, πτοσισ - falling), programmed cell death. Necrosis can arise from severe deregulation or damage to one essential system in the cell (e.g., cell-cycle regulation), which leads to secondary damage to other systems in a cascade of effects and the release of lysosomal enzymes that engage in progressively digesting cell components. In the context of tissues and organs, partially necrosed cells induce a characteristic inflammatory process.

3 Fractal analysis is a reliable procedure for quantifying the complexity of cellular or nuclear membranes. The fractal dimension of a cell, a reflection of its plasma membrane complexity, is a non-integer measure of how "complicated" the cells, or any "self-similar" object is. Fractal dimension is generally larger than the integer derived from a topological dimension based on Euclidean geometry (the Euclidean dimension) by a factor of 5 up to 20%, depending on the cell growth and maturation stage. The fractal dimension allows the quantitative comparison of the objects (here, cells) under study, whereas the Euclidean dimension describes the number of coordinates required to specify the object. Fractal analysis permits to quantify cell membrane complexity (fractal dimension) and topology (Euclidean dimension) (Chiappelli et al, 2005)

4 Several checkpoints are designed to ensure that damaged or incomplete DNA is not passed on to daughter cells. Two main checkpoints exist: the G1/S checkpoint and the G2/M checkpoint. p53 plays an important role in triggering the control mechanisms at both G1/S and G2/M checkpoints. G1/S transition is a rate-limiting step in the cell cycle and is also known as restriction point (*vide infra*).

5 Cyclins form the regulatory subunits and CdKs the catalytic subunits of an activated heterodimer. Cyclins have no catalytic activity and CdKs are inactive in the absence of a partner cyclin. When activated by a bound

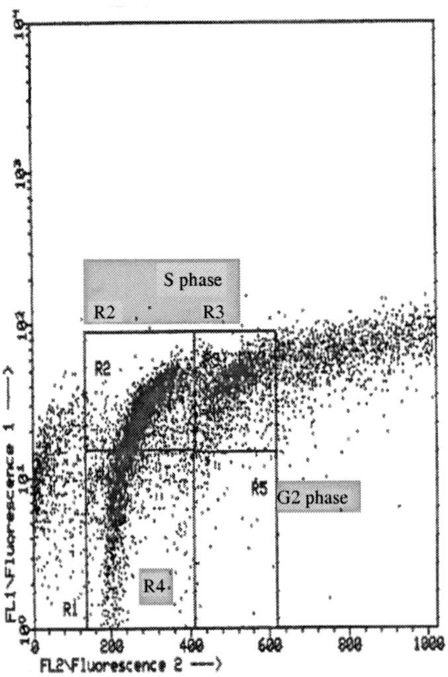

Figure 1. PI/BrDU Dual Flow Cytometry of the Cell Cycle[6].

B. Cell cycle traversal: G1 and G1/S transition

Interphase commences with G1. During this initial stage of cell growth, the biosynthetic activities of the cell, which had been considerably slowed down during M phase, resume and result in a marked surge in RNA (i.e., transcription) and protein synthesis (i.e., translation), and post-translational modifications of the proteins required for continued successful progression through the cell cycle. New organelles are generated, which require that the cell produce structural, regulatory and functional proteins (i.e., enzymes) (Lodish et al, 2000; Karp, 2005).

This initial phase of the cell cycle G1 consists of four distinct sub-phases:

- Competence (G1a)
- Entry (G1b)
- Progression (G1c)
- Assembly (G1d)

---

cyclin, CdKs become kinases, and by means of phosphorylation activate or inactivate target proteins to orchestrate coordinated transition to the next phase of the cell cycle. Different cyclin-CdK combinations determine the downstream proteins targeted. CdKs are constitutively expressed, but the expression of cyclins is typically cell cycle stage-specific. A variety of signals impact upon the compacted genome during interphase (i.e., eu- and heterochromatin; *vide infra*) to regulate and modulate specific cyclin expression.

6 Adapted from Chiappelli, 2007. The gated regions R1-R5 correspond respectively to G1/G0, S, G2 & M

Coordinated progression through these sub-phases is critical, lest the cell become arrested. Progression is regulated by limiting growth factors and nutrient supply in the extracellular environment, and intracellular regulatory factors. G1 cyclins and CdK's activate the transcription factors for a variety of genes responsible for nuclear DNA synthesis proteins and G1/S transition. G1 cyclin-CdK complexes also promote the degradation of molecules that function as S phase inhibitors, and target them for ubiquitination[7]. DNA damage is assessed before engaging the nuclear DNA replicative machinery, and if DNA damage is detected by specialized sensor mechanisms, the cell either stalls or enters a DNA repair mode[8]. If repairs are impossible, the cell enters apoptotic cell death (Lodish et al, 2000; Karp, 2005).

A restriction point[9] toward the end of G1 ensures that the cell is ready to engage in the DNA replicative phase. If, indeed, the DNA is intact, and the cytome and cellular physionome are operating normally, then the cell can commit to entering S. The restriction point is principally controlled by action of the CdK inhibitor p16 (CdKI- p16) protein. By inhibiting CdK4 and CdK6, it voids potential interaction with cyclin D1 and cell cycle progression through G1. Active, non-CKI-p16-inhibitted CdK4/6 form complexes with cyclin D1, and mediate phosphorylation of the tumor suppressor retinoblastoma (Rb), which then relieves the inhibition of the transcription factor E2F[10]. E2F permits expression of cyclin E later in G1, which interacts with CdK2 and initiation of irreversible G1/S phase transition (Knudsen & Wang, 1996; Lodish et al, 2000; Karp, 2005).

---

7 Once a protein has been ubiquitinated, it is targeted for proteolytic degradation by the proteasome pathway.

8 DNA damage activates signaling cascades driven by the ATM-ATR (cf., footnote 13) protein kinases, culminating in the recruitment of repair proteins to the damage site. ATM-dependent phosphorylation of mediator and chromatin-associated proteins at sites of double-strand breaks promotes ubiquitylation of local nucleosomes and elicit recruitment of repair complexes to the damage site (Bennet & Harper, 2008).

9 The restriction point, the G1 phase checkpoint at which time the cells commits to S, was first described by Arthur Pardee (1921 -), now Emeritus Professor at Dana-Farber Cancer Institute, in 1974. It is the point at which eukaryotic cells typically arrest the cell cycle if progression is deemed impossible, and the cell enters into G0.

10 Cyclin D is the first cyclin produced in the cell cycle, in response to extracellular signals (e.g., growth factors). Cyclin D binds to existing CdK4, forming the active cyclin D-CdK4 complex, which phosphorylates Rb. Hyperphosphorylated Rb (pRb) dissociates from the E2F/DP1/Rb complex, which blocked the E2F responsive genes from transcription. Release of pRb from the complex in effect frees and activates E2F, thus resulting in transcription of various genes that push the cell cycle forward. Cyclin E is one such gene whose expression is consequential to the phosphorylation of Rb-activation of E2F event. Cyclin E thus produced binds to CdK2, forming the cyclin E-CdK2 complex, which pushes the cell from G1 to S phase (G1/S transition). E2F codifies a family of transcription factors involved in cell cycle and DNA synthesis regulation in eukaryotes (e.g., cyclin E, cyclin A, for regulation of DNA polymerase, thymidine kinase , dihydrofolate reductase , chromosomal replication origin-binding protein [HsOrc1] and MCM 5. E2F1,2 and E2F3a are activators of cell cycle traversal, but E2F3b, E2F4-8 suppress the cell cycle. In the absence of the phosphorylated form of Rb (pRb), E2F-1 and its binding partner DP-1 mediate the trans-activation of E2F-1 target genes that facilitate G1/S transition and S phase progression. When bound to E2F-3a, pRb represses E2F-3a target genes by recruiting chromatin remodeling complexes and histone modifying activities such as histone deacetylase (HDAC) to the promoter (*vide infra*). E2F3b, E2F4, E2F5 are expressed in quiescent cells and are associated with E2F-binding elements on E2F-target promoters during G0, thus putatively acting as sink for E2F binding sites.

It is important to note that two families of genes, the Cip/Kip family[11] and the INK4a/ARF (Inhibitor of Kinase 4/Alternative Reading Frame) prevent the progression of the cell cycle. Because these genes are instrumental in prevention of tumor formation, they are known as tumor suppressors. The cip/kip family includes the genes p21, p27 and p57, and halt the cell cycle in G1 phase, by binding to, and inactivating, cyclin-CdK complexes. p21 is activated by p53, which, in turn, is triggered by DNA damage (e.g., consequential to radiation). p27 is activated by the growth inhibitor transforming growth factor(TGF)-β. By contrast, the INK4a/ARF family, which includes p16INK4a, binds to CdK4 and arrests the cell cycle in G1. The ARF p14/ARF inhibits Mdm2, an important negative regulator of p53, thus preventing tumor p53 degradation (Knudsen & Wang, 1996; Freedman et al, 1999; Lodish et al, 2000; Karp, 2005).

C. Cell Cycle Traversal: S and S/G2 Transition

The S phase is the period in the cell cycle, between G1 and G2, during which DNA replication occurs. Each chromosome, composed of a coiled DNA double helix molecule (i.e., chromatid), is duplicated. As the cell is to exit S, each chromosome has two identical coiled DNA double helix molecules (i.e., sister chromatids[12]). The centrosome[13] is also duplicated during S. These two events are unconnected, but require many of the same factors to progress. The duplicated genetic material in the cell will eventually be divided during mitosis.

Damage to DNA may takes place during this replicative phase. DNA repair is often required. Cellular senescence is a state that occurs in response to DNA damage or degradation that would make a cell's progeny nonviable; it is often a biochemical alternative to the self-destruction of such a damaged cell by apoptosis (Lodish et al, 2000).

A key enzyme in base excision repair pathway that reconstitutes deaminated cytosines and 5-methyl-cytosines is the DNA repair enzyme thymine DNA glycosylase. Incomplete DNA repair flags cell cycle post-replication checkpoints to halt cell cycle progression.

---

11 The Cip/Kip family is well conserved phylogenetically, suggesting that it is biologically important. Despite structural and biochemical similarities among the Cip/Kip members, the phenotypes of each Cip/Kip member are surprisingly different, indicating a variety of physiological functions (Nakayama & Nakayama, 1998).

12 Sister chromatid cohesion is mediated by cohesin, but the process of cohesin binding to chromatin is dynamic in G1 and requires continued activity of the cohesin loader Mis4/Ssl3. In fact, cohesin instability in G1 depends on wpl1, the fission yeast ortholog of mammalian Wapl. The association becomes stabilized during S-phase, as cohesin stability increases in a reaction that can be uncoupled from DNA replication (Bernard et al, 2008).

13 First described in 1888 by Theodor Boveri (1862-1915), the centrosome is a unique constituent of animal cells, as opposed to other eukaryotes (i.e., plant cells). It consists of the principal microtubule organizing center of the cell, and the main regulator of cell cycle progression. It comprises nine triplet microtubule blades in a pinwheel structure, and two orthogonally organized centrioles surrounded by the amorphous pericentriolar material, rich in the proteins responsible for microtubule nucleation and anchoring. It is fixed to the nuclear membrane during interphase. As the nulclear membrane comes apart during mitosis, the centrosome-nucleated microtubules organize the chromosomes into the mitotic spindle. The centrosome is duplicated only once per cell cycle so that each daughter cell inherits one centrosome, containing two centrioles (Rodrigues-Martins et al, 2007).

(Callegari & Kelly, 2007). When the check point kinase 1 (*chk1*) gene[14], a typical example of a post-replication checkpoint that is well conserved from yeast to humans, is disrupted, cells halt cell cycle traversal, exhibit gross chromosomal damage, and die. Proteins required for *chk1* phosphorylation include the single-stranded DNA binding protein replication protein A (RPA), topoisomerase II binding protein 1 (TopBP1), Claspin, the Rad17/replication factor C (Rad17/RFC) complex, and the Rad1/Hus1/Rad9 (RHR/9-1-1) complex[15] following DNA damage. The BRCA1 tumor suppressor also contributes to the modulation of activation of human chk1 gene, suggesting an important role for post-replication checkpoints in tumorogenesis. (Yarden et al, 2002; Guan et al, 2007; Rodriguez-Bravo et al, 2007).

In general, active S cyclin-CdK complexes phosphorylate proteins that make up the pre-replication complexes on DNA replication origins assembled during G1 phase. The phosphorylation activates each already-assembled pre-replication complex, and prevents new complexes from forming. This ensures that every portion of the cell's genome will be replicated once and only once. The reason for prevention of gaps in replication is so that daughter cells that are missing all or part of crucial genes die. However, because of consequences related to gene copy number, possession of extra copies of certain genes may prove deleterious to the daughter cells. If the cell successfully completes the S phase, it is most likely that it will continue on to complete successfully the cell cycle (Lodish et al, 2000; Karp, 2005).

D. Cell Cycle Traversal: G2 and G2/M Transition

The second growth phase of the cell cycle (G2) follows S. G2 is usually the shortest phase of interphase, during which the cell completes its period of growth and protein synthesis and post-translational modification to prepare for mitosis (M). It follows S only if successful DNA synthesis and chromosomal replication was completed. At the end of this gap/growth phase is a control checkpoint (G2 checkpoint) that determines if the cell can proceed to enter the M phase and divide. The G2 checkpoint helps maintain genomic stability by preventing cells from entering mitosis with DNA damaged since the last division, and by

---

14 Claspin is an example of an essential upstream regulator of checkpoint kinase 1, triggers a checkpoint arrest of the cell cycle in the presence of DNA templates (Lin et al, 2004; Chini & Chen, 2005; Rodríguez-Bravo et al, 2007).

15 The 9-1-1 complex is the principal surveillance machinery involved in the preservation of genome stability by acting both as a damage sensor, and in regulating the DNA glycosylase in the base excision repair pathway. The 9-1-1 checkpoint protein complex on chromatin (*vide infra*) is also apparently involved in its phosphorylation, which is required for the DNA-damage-induced cell cycle G2 arrest. Rad17 is a cell cycle checkpoint gene required for cell cycle arrest and DNA damage repair in response to DNA damage. It shares strong similarity with RFC, and can form a complex with it. It binds to chromatin prior to DNA damage and is phosphorylated by ATR. Ataxia telangiectasia mutated (ATM) and Rad3-related (ATR) is a critical regulator of DNA replication and damage checkpoint responses, and possesses a binding partner called ATRIP, which associates with DNA. ATR and ATRIP are mutually dependent on one another for stable binding to DNA. ATRIP must associate with ATR in order for ATR to carry out the phosphorylation of Chk1 effectively. This function of ATRIP does not involve its ability to mediate the stable binding of ATR to defined checkpoint-inducing DNA templates (Kim et al, 2005). Replication protein A (RPA), a protein complex that associates with single-stranded DNA (ssDNA), is required for the recruitment of ATR to sites of DNA damage and for ATR-mediated Chk1 activation in human cells (Zhou & Elledge, 2003).

providing a last opportunity for DNA repair. Transition out of G2 signifies termination of the interphase, and onset of mitosis[16] (M).

E. Cell Cycle Traversal: M and Cytokinesis

M is a relatively short but highly complex and well-orchestrated period of the cell cycle. It is the time during which cytokinesis occurs, which divides the nuclei, cytoplasm, organelles and cell membrane into two quasi-identical daughter cells containing roughly equal shares of these cellular components. In animal cells, contrary to yeast, fungi and plant cells, the nuclear envelope breaks down before the chromosomes condense, attach to microtubullar fibers and separate (Uhlmann et. al, 1999; Lodish et al, 2000; Karp, 2000).

Mitotic cyclin-CdK complexes, which are synthesized but inactivated during the S and G2 phases, promote the initiation of mitosis by stimulating downstream proteins involved in chromosome condensation and mitotic spindle assembly. For instance, Cyclin A and CdK2 form the cyclin A-CdK2 complex, which initiates the G2/M transition. Cyclin B-CdK1 complex activation initiates breakdown of the nuclear envelope and initiation of prophase.

The complex and regulated sequence of events during M is divided into subphases:

- Prophase: the condensin complexes[17] mediate condensation of chromatin (*vide infra*) into the highly ordered structure, called the chromosome, which can then be identified based on size, morphology, and characteristic giemsa staining[18] and G-banding. The genetic material having been duplicated during S, two identical sets of each chromosome, the sister chromatids, emerge, and are attached to each other at the DNA centromere in part via the cohesin protein. The centrosome, and its replicated copy, recruit γ-tubulin and increase their microtubule-nucleating activity. They are displaced, each pushed apart to opposite ends of the cell nucleus by the action of the microtubules. The nuclear envelope breaks down, and the microtubules

---

16 Because cytokinesis usually occurs in conjunction with mitosis, "mitosis" is often used interchangeably with "mitotic phase", M. However, mitosis proper (i.e., the process of chromosome duplication) and cytokinesis (i.e., the process of dividing the nuclear contents and cytoplasm) are distinct, yet conjoined events, which can occur separately, resulting in single cells with multiple nuclei.

17 Two condensin complexes, condensin I and condensin II, have been identified to date. Whereas they share the same pair of core structural maintenance of chromosomes (SMC), SMC2 and SMC4, that have ATPases activity, each complex also contains distinct non-SMC regulatory subunits. Condensin I, of which at least one subunit is a CdK target, is present in the cytoplasm during interphase, and gains access to the chromosomes after the nuclear envelope breaks down at the end of prophase. Condensin II is present within the cell nucleus during interphase and is involved in an early stage of chromosome condensation within the prophase nucleus. During prometaphase and metaphase, both condensin I and condensin II contribute to the assembly of condensed chromosomes, in which the sister chromatids are fully resolved. Both complexes remain associated with the chromosomes until the end of anaphase.

18 Giemsa stain (1904), named after Gustav Giemsa (1867-1948), is a notable improvement over the original Romanowsky stain (Eosin Y and Methylene Blue) by stabilization with glycerol for reproducible staining of cells for microscopy

reach the kinetochores[19] on the chromosomes. The nuclear envelope break down marks the end of prophase.

- Prometaphase: The chromosomes are captured by the microtubules at the opposite poles of the spindle, as the nuclear envelope continues to fragment, and to disappear. Forces exerted by protein motors (e.g., myosin, dynein and kinesin; energy in the form of ATP) associated with spindle microtubules move the chromosomes toward the center of the cell. Prometaphase ends when the kinetochore microtubules are secured[20] to their kinetochores. Metaphase can begin.
- Metaphase: The centromeres of the chromosomes come to rest on the metaphase plate, an imaginary line equidistant from the two centrosome poles. Most cohesin is removed, except for some at the centromere.
- Anaphase begins abruptly with the regulated activation of the anaphase promoting complex, which cleaves and inactivates the metaphase cyclin required for the function of metaphase CdK's. This complex is a ubiquitin[21] ligase that also acts to promote degradation of structural proteins associated with the chromosomal kinetochore (e.g., securin is ubiquitinated, then proteasome degraded, releasing separase, which hydrolyzes the remaining cohesin). The sister chromatids separate from each other. The individual kinetochores on each chromatid drive their movement to the spindle poles that will define the two new daughter cells. The anaphase-promoting complex also targets the mitotic cyclins for degradation, ensuring that telophase and cytokinesis can proceed.
- Telophase (Gk, τελος, end; φασις, stage) consists in the final process of reversal of the events precipitated in prophase and prometaphase (e.g., inactivation of cyclin B-CdK1). A new nuclear envelope is formed using fragments of the parent cell's nuclear membrane, and both sets of now newly encapsulated chromosomes, de-condense into chromatin (*vide infra*). Telophase cleans up the secondary after-effects of mitosis.

Cytokinesis proper begins shortly after the onset of sister chromatid separation in anaphase, and progresses usually at the same time the nuclear envelope is reforming during telophase. A contractile ring, comprised of non-muscle myosin II and actin filaments, assembles equatorially (i.e., in the middle of the cell) at the cell cortex (i.e., adjacent to the cell membrane). Myosin II uses the free energy released when ATP is hydrolyzed to move along these actin filaments, constricting the cell membrane to form a cleavage furrow. Continued hydrolysis causes this cleavage furrow to ingress and to move inwards to form the

---

19 The kinetochore, a complex of close to 50 well-conserved proteins, assembles as a punctate structure on the centromere to link the chromosome to microtubule polymers from the mitotic spindle. The inner kinetochore tightly attaches to the centromere DNA via a specialized histone H3 (*vide infra*) variant (CenH3), while its outer core interacts with the microtubules. Monocentric organisms, such as vertebrates, have a single centromeric region on each chromosome that ties to one kinetochore. The sister chromatids, held together each with its own kinetochore, face and link to opposing poles of the mitotic spindle.

20 An unattached kinetochore results in a non-aligned chromosome, and triggers the spindle checkpoint signal, which prevents further progression to anaphase by inhibiting the anaphase-promoting complex.

mid-body structure, which is cleaved along its rough midpoint during the process of abscission. This sequence of events is regulated by septin filaments that provide the scaffolding of the cleavage furrow[22]. Following cytokinetic division, non-kinetochore microtubules reorganize and disappear into a new cytoskeleton as the two daughter cells initiate their cycle in interphase (G1a) (Longtine et al, 1996; Glotzer, 2001).

2. Chromatin Regulation of Cell Cycle Traversal

A. Chromatin Architecture

The single most critical set of events in the process of cell cycle traversal pertains to the packaging, replication, and distribution of the genomic material to the daughter cells. This material is packaged into chromosomes, translocated and distributed during M. During interphase the chromosome material is found within the nuclear compartment in its uncondensed chromatin structure. A complete understanding of the regulation of the cell cycle can only be obtained by incorporating a perspective on the regulatory systems of chromatin architecture, assembly and repair.

Chromatin[23] is the complex packaging of DNA and protein that makes up chromosomes, and is relevant as well to the regulation of the processes of DNA replication and DNA repair. It has an elaborate architecture, which is determined to a large extent by the histone and the non-histone proteins[24]. Changes in chromatin architecture result mainly from methylation of DNA and proteins, and from phosphorylation and acetylation of proteins (*vide infra*) (Turner, 2002; Roloff & Nuber, 2005; Quina et al, 2006; Kornberg & Lorch, 2007).

In its most tightly compacted form, heterochromatin is silent, and transcriptionally inactive. The alternative form of chromatin, euchromatin, has a looser architecture characterized by a 30 nm fiber arranged into loops along a central protein scaffold, and is transcriptionally active. Different levels of chromatin compaction define distinct stages of the cell cycle, although both active and inactive genes are now known to be found within either

---

21 Ubiquitin is a highly conserved small regulatory protein that is ubiquitous in eukaryotes. A prominent role for ubiquitination, a post-translational modification by means of covalent attachment of one or more ubiquitin monomers, is proteasomal degradation (Purvis & Singleton, 2008).

22 In the budding yeast *Saccharomyces cerevisiae*, a leading model organism for studying cell cycle regulation, polarity relates to the organization of the actin cytoskeleton, which guides secretion towards the budding site or tip of the growing bud, resulting in polarized cell growth. CdK's and their cyclin partners play a critical role in cell polarity: G1 cyclins trigger polarized growth, M cyclins reverse this process. Inactivation of CdK's in telophase triggers redistribution of the actin cytoskeleton to the neck region for cytokinesis (Lew & Reed, 1993).

23 Walther Flemming (1843-1905), considered by most as the founder of the study of cytogenetics, introduced term "chromatin" in 1882. Making use of aniline dyes, he observed a structure that strongly absorbed basophilic stains, and named it chromatin.

24 Roger Kornberg (1947 -; presently, professor of structural biology at Stanford University School of Medicine) was awarded the Nobel Prize in Chemistry in 2006 for his studies of the molecular basis of eukaryotic transcription. He is the son of Arthur Kornberg (1918 – 2007), Nobel Prize in Physiology/ Medicine (1959) for discovering the mechanisms in the biological synthesis of deoxyribonucleic acid (DNA). In 1974, Roger Kornberg published the first report that the nucleosome consists of a tetramer of histones H3 and H4 and two dimers of histones H2A and H2B (Kornberg, 1974; Kornberg & Lorch, 1999, 2007).

hetero- or euchromatin regions (Turner, 2002; van Driel et al, 2003; Dillon, 2004; Sexton et al, 2007).

The structure of chromatin varies considerably as the cell progresses through the cell cycle. Chromatin architecture during interphase is optimized to allow easy access of transcription and DNA repair factors to the DNA while compacting the DNA into the nucleus. Genes that require regular access by RNA polymerase require the looser structure in euchromatin. The changes in structure are required to allow the DNA to be used and managed, whilst minimizing the risk of damage (Turner, 2002; Paweletz, 2001; Quina et al, 2006).

The fundamental structural unit of chromatin is the pattern of DNA wrapping around nucleosomes - the so-called "beads on a string" structure: electron microcopy reveals a 30 nm condensed chromatin fiber of repeating nucleosomes with a 10-nm-fiber intervening "linker" DNA form. The structure is though to range from a packing ratio of ~6, compared to "free" DNA (per nm length) in a helical solenoid, zigzag ribbon structure, and to having no regular structure with a packing ratio of ~40. Crystallographic studies, which have been obtained thus far with a resolution 2.0-2.8 Å., are optimized for four nucleosome aggregates, the tetranucleosome (Asturias et al, 2002, 2004; Turner, 2002; Kornberg & Lorch, 2007).

The evidence suggests that the nucleosome repeats, with some variations and exceptions, roughly every 200 base pairs (bp) throughout chromatin. The nucleosome core particle consists of 146 bp double stranded DNA (dsDNA) wrapped in 1.65 left-handed superhelical turns around four identical pairs of histone proteins[25]. This unit is the histone octamer (Turner, 2002; Asturias et al, 2002, 2004; Roloff & Nuber, 2005; Quina et al, 2006; Kornberg & Lorch, 2007).

Within the octamer, each of the four-histone dimers interacts with the dsDNA at three distinct sites. Centrally, the site is formed by a hydrogen bond between the α helix from each histone and a single phosphate group on the dsDNA. At positions 10 bp away on either side, a loop from both histones in the pair converges to hydrogen bond to another single phosphate groups on the dsDNA. Thirdly, a set of interactions, for a total of 14 involve the interaction of histone tails from each of the H3 histones at the entry and exit points of the dsDNA wrapping around the nucleosome (Figure 2). Together, these anchors provide critical structural, albeit relative[26] rigidity to the unit by helping clamp these regions onto the core particle (Asturias et al, 2002, 2004; Quina et al, 2006).

---

[25] The four histones (H2A, H2B, H3, and H4) each have three α helices separated by loops. In solution, they spontaneously dimerize, and H3 and H4 further assemble into tetramers. These interactions involve buried charged groups of the same α helix forming hydrogen bond to each other (Aragay et al, 1988)

[26] Curvature into the superhelix comes primarily when either the minor or the major groove faces the octamer. This tends to occur in clumps of roughly 5 bp. While the major groove bending around the octamer occurs smoothly, the minor groove bending must be facilitated by arginine side chains inserted into the groove. Adenosine and thymine are more favorably compressed into the inner minor grooves. This occurs more smoothly around the H3-H4 tetramer, but is forced into a kink about the H2A/H2B dimers. That dsDNA is most tightly constrained in regions where it interacts with the double loop structures of the histone dimers implies that there is more variability in how dsDNA interacts with the double α helix structures of the histone dimers in order to bind to different sequences (Heng et al, 2004).

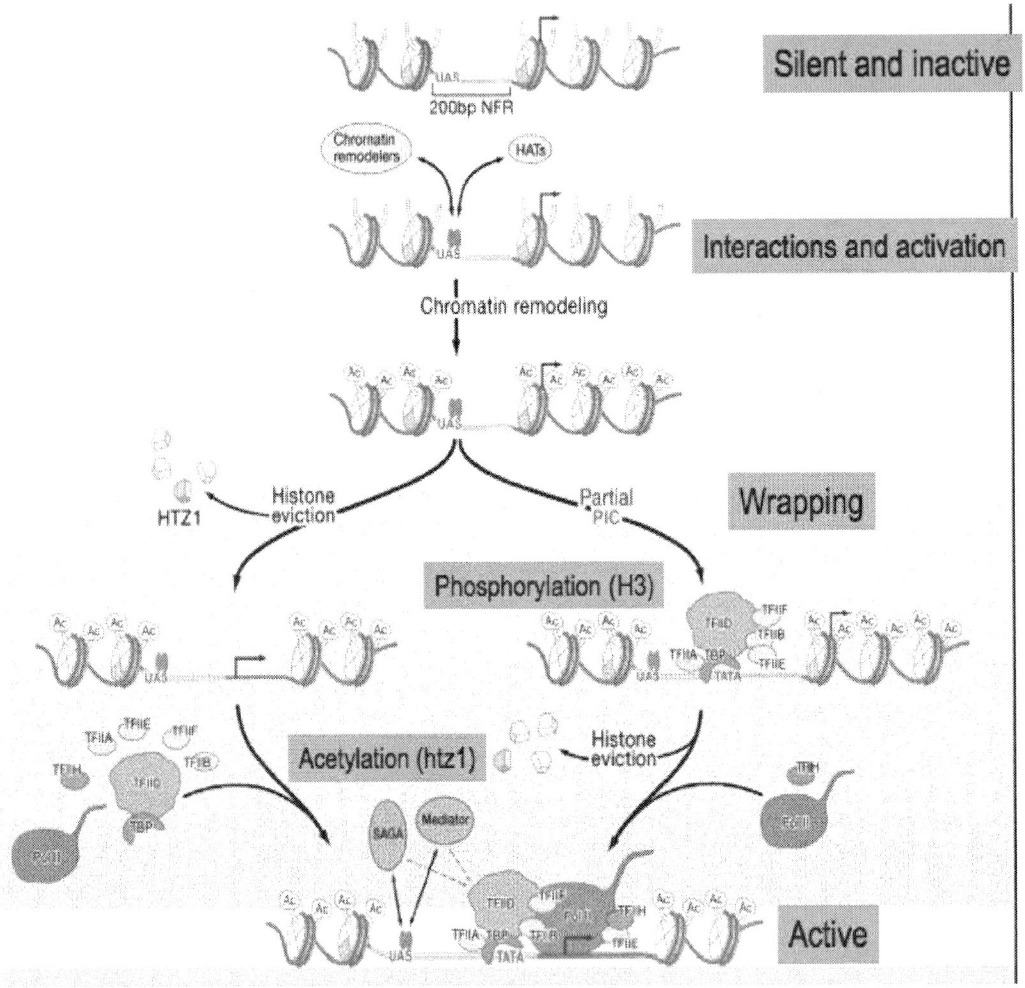

Figure 2. Model of chromatin assembly[27].

The remaining 50 bp of the repeating unit consists of "linker DNA", dsDNA which separates the core particles where histone H1 lays.

The crystal structure of chromatin reveals a two-start helix model, in which nucleosomes are assembled in a zigzag ribbon that twists or supercoils, as noted, and with histone protein H1 involved in stabilizing this 30 nm twist structure (Asturias et al, 2002, 2004; Olins & Olins, 2003; Kornberg & Lorch, 2007).

---

27 At the silent inactive promoter, nucleosomes flank a 200 bp NFR on both sides. Upon targeting to the upstream-activation sequence (UAS), activators recruit various coactivators. This recruitment further increases the binding of activators, particularly for those bound within nucleosomal regions. More importantly, histones are acetylated at promoter-proximal regions, and these nucleosomes become much more mobile. In one model (left), a combination of acetylation and chromatin remodeling directly results in the loss of Htz1-containing nucleosome. In the other model (right), which represents the remodeled state, partial PICs could be assembled at the core promoter without loss of Htz1. (mod. from Bing L. et al. Cell 128: 707-719; 2007)

B. Chromatin Assembly

The *crux* of cell cycle regulation lies imbedded in its genome. And, the *crux* of the regulation of the expression of the cell's genome throughout interphase, and of the processes that lead to its packaging into condensed chromosomes during mitosis is the high degree of regulation of chromatin architecture, assembly and repair during cell cycle traversal.

The process of assembly of the nucleosome core particle is the fundamental initial step in chromatin architecture during interphase. It first involves the attachment of the H3-H4 tetramer onto the dsDNA, and subsequently the association of two separate H2A-H2B dimers. This finely regulated pathway occurs in a cooperative manner such that the H2A-H2B dimers assemble onto the tetramer structure as one carefully timed event in the concerted interphase traversal process. The fine regulatory signals of chromatin assembly remain, however, to be fully elucidated (Kornberg & Lorch, 2007).

What is known is that nucleosomes package DNA about the four pairs of histones H2A, H2B, H3 and H4, and provide the lowest level of compaction, which is required to fit dsDNA into the cell nucleus. Assembled chromatin is promptly repaired in the event of unforeseen damage or alterations during cell cycle traversal. The discrete regulation of chromatin architecture, assembly and repair during G1, S and G2, and the delicate pattern of events that directs chromatin condensation in chromosomes in prophase, and decondensation back into the chromatidic scaffolding in anaphase underscore the critical role this cellular structure plays during the cell cycle. It is not surprising, therefore, that evidence increasingly shows the importance of chromatin structure in genomic, protonomic, metabolomic, interactomic[28], and cytomic properties and function in health, disease and senescence (Turner, 2002; Ausió & Abbott, 2004; Quina et al, 2006; Fernandez & Anderson, 2007; Kornberg & Lorch, 2007; Zhang & Adams, 2007).

The fundamental nucleosomic pattern provides regulation of transcription, for instance, by preventing RNA polymerase from unnecessarily accessing the promoter regions of genes, not needed at a given moment of the cell cycle. As the cell cycle progresses, and requirements of the cell change, chromatin-remodeling factors[29] can alter, remove or change

---

[28] Interactome is the whole set of molecular interactions in cells, usually displayed as a directed graph. When spoken in terms of proteomics, it refers to protein-protein interaction network (PPI), or protein network (PN).The principal interactome databases include: Database of interacting protein (DIP) (Xenarios et al. 2000); GRID database; MIPS database; PSIMAP database: The first protein structural interactome DB InterPare database: A structural protein interfaceome DB; Biomolecular Interaction Network Database (BIND) (Bader et al, 2003); Online Predicted Human Interaction Database (OPHID) (Brown and Jurisica, 2005); Human Protein Reference Database (HPRD) (Peri, S. et al, 2003); Himap(Human Interactome Map); HPID; MINT; PINdb; IntAct - The Molecular Interaction Database; APID - Agile Protein Interaction DataAnalyzer APID is interactive bioinformatic web-tool that integrates and unifies main known experimentally validated protein-protein interactions.

[29] Enzymes responsible for post-translational modification of histone and non-histone proteins on chromatin. Whereas compaction of chromatin into the cell nucleus is not yet fully understood in the complexity of its biochemical regulation, it is clear that several enzymes (e.g.,RSC, SWI/SNF, SWI/SNF related, matrix associated, actin dependent regulator of chromatin, subfamily α, member 5 [SMARCA5] which has both helicase and ATPase activities and regulates transcription by altering the chromatin structure around target genes) work in a concerted action to change the position of nucleosomes, thus exposing genetic information held within the nucleosome core particle, or concealing DNA stretches from RNA polymerase, when it is

the position of the nucleosome to allow access to RNA polymerase[30]. The processes involved in nucleosome structure plasticity/remodeling require biological energy in the form of ATP. They provide the fundamental epigenetic information[31] and memory of the cell and its progeny, and act as the major carriers of epigenetically inherited information (Turner, 2002; Roloff & Nuber, 2005; Quina et al, 2006; Whitehouse & Tsukiyama, 2006; Kornberg & Lorch, 2007; Zhang & Adams, 2007).

C. Chromatin Plasticity in the Cell Cycle

Non-histone proteins[32] and chromatin-remodeling factors (e.g., ISWI/II family) significantly contribute to the genomic regulatory function of chromatin. Histone and non-histone proteins are modified post-translationally, which determine their ability to modify and to modulate DNA packing compaction, and therefore gene expression. Whereas our knowledge of these processes is still incomplete, what is clear is that the key mechanism to these chromatin modifications (i.e., acetylation and deacetylation, phosphorylation, methylation, ubiquitination and ATP-dependent chromatin remodeling) requires the flexible and charged structure of histone tails that extend from the hydrophobic nucleosome core[33]. When thus modified, these tails regulate chromatin function throughout the phases of the cell

---

required by the cell. Chromatin remodeling is the *sine qua non* genomic and post-genomic frontiers of research for increased understanding of cell cycle regulation (Hara & Sancar, 2002; Kornberg & Lorch, 2007).

30 Case in point, the novel transcription regulatory factor DREF (DNA replication-related element-binding factor), first characterized in Drosophila and later found to have orthologues in other species including humans, binds specifically to the DRE sequence (5'-TATCGATA) in the promoters of many DNA replication/cell proliferation-related genes to activate their transcription, and the N-terminal region of DREF. Both Drosophila and human DREF's interact with key factors related to chromatin structure regulation. DREF is likely to activate the expression of proliferation-related genes through modification of the 3-D conformation of DNA (Matsukage et al, 2007).

31 Epigenetics refers to features such as chromatin and DNA modifications that are stable over rounds of cell division but do not involve changes in the underlying DNA sequence of the organism. Epigenetic changes regulate cellular differentiation and maturation, allowing cells to retain acquired characteristics, which can be inherited when cells divide. Specific epigenetic processes include, in addition to post-translational modification of histone and non-histone proteins, paramutation, bookmarking, imprinting, gene silencing, position effect, reprogramming, etc, and can lead to pathology of the cell population or the organ (e.g., cancer) (Ducasse & Brown, 2006; van Beekum & Kalkhoven, 2007; Kornberg & Lorch, 2007).

32 Non-histone chromosomal proteins are found associated with isolated chromatin, and fall into several functional categories: chromatin-bound enzymes (e.g., DNA transcription, replication and repair, nucleases, proteases, acetylation/deacetylation, methylation/demethylation, kinases/phosphatases), high mobility group (HMG) proteins [HMG's are distinguished from other nonhistone chromatin proteins by their ability to preferentially recognize the structure of the narrow minor groove of A.T-sequences. During the cell cycle, the HMG-I(Y) protein localizes in the A.T-rich G/Q bands and in the "scaffold-associated regions" (SAR's) of metaphase chromosomes, and function as structural transcription factors during G1 and G2 by introducing additional bends and supercoils in dsDNA], transcription factors, scaffold proteins (e.g., chromatin proteins such as insulators, domain boundary factors and cellular memory modules), transition proteins (testis specific proteins), protamines (present in mature sperm) (Reeves & Nissen, 1995; Rice & Allis, 2001; Ausió & Abbott, 2004; Kornberg & Lorch, 2007).

33 Chromatin immunoprecipitation experiments have revealed critical recruitment of 14-3-3 proteins (Winter et al, 2008). 14-3-3 proteins (the name derives from the characteristic electrophoretic migration pattern) constitute a family of conserved regulatory molecules in eukaryotes, which bind a variety of signaling proteins (e.g., kinases, phosphatases, transmembrane receptors) along a specialized amphipathic binding cleft (i.e., possessing both hydrophilic and hydrophobic properties).

cycle – and play a critical role as triggers of cell cycle traversal regulators. They modulate and regulate fundamental cytomic pathways of transcription, DNA-dependent chromatin assembly, DNA repair, mitosis and silencing by heterochromatin (Cheung et al, 2000; Wu & Grunstein, 2000; Turner, 2002; Fazzio & Tsukiyama, 2003; Roloff & Nuber, 2005; Quina et al, 2006; Kirmizis et al, 2007).

- Phosphorylation can occur at the nucleic acid, and modulates covalent bonds between the α helix and the DNA, and the coiling of the double strand. Phosphorylation of histone and nonhistone proteins is usually mediated by cyclin dependent kinases (e.g., CdK2[34]), and generally reduces the binding affinity of these proteins to dsDNA (Nissen et al, 1991). Phosphorylation of chromatin protein can also involve the mitogen-activated protein kinase (MAPK) family of serine/threonine kinases, which include the discrete and interrelated steps of activation by upstream kinases, sub-cellular and sub-nuclear relocation and translocation, recognition, binding and phosphorylation of their protein target substrates. MAPK activity in chromatin usually involves the scaffolding molecules, transcription factors, and chromatin-remodeling molecules, thereby altering nuclear regulatory molecules during cell cycle traversal (Turjanski et al, 2007). Another important role of histone phosphorylation is to modulate receptor function. Case in point, protein-kinase-C-related kinase(PrK)-1 phosphorylates threonine at position 11 of H3 (H3T11) upon ligand-dependent recruitment to androgen receptor target genes. When PrK-1 is blocked, androgen-induced H3T11 phosphorylation is abrogated, androgen-induced H3 demethylation is inhibited, serine-5-phosphorylated RNA polymerase II is no longer observed at androgen receptor target promoters, and the proliferation of androgen-dependent tumor cell lines is blunted. By contrast, PrK-1-mediated H3T11 phosphorylation favors demethylation by the Jumonji C (JmjC)-domain[35]-containing protein JMJD2C[36], and establishes a novel chromatin mark for gene activation (Metzger et al, 2007).
- The two enzymes Dnmt3a and Dnmt3b catalyze DNA methylation in embryonic development. Dnmt3a and Dnmt3b form the core components of a stable complex and directly interact with and stimulate each other independently of the enzymes' catalytic activity. In embryonic mice, the two enzymes work synergistically to methylate the promoters of the Oct4 and Nanog genes. DNA methylation by Dnmt3a and Dnmt3b is essential for the normal expression of Oct4 and Nanog in pluripotent

---

34 Cyclin E/Cdk2 acts at G1/S to promote the E2F transcriptional program (*vide supra*), which initiates DNA synthesis. To control cell cycle phase events, cyclin E/Cdk2 phosphorylates the nuclear protein ataxia-telangiectasia (NPAT) protein p220(NPAT) at discrete nuclear foci. Cells in G0 and G1 contain two p220 foci, S- and G2-phase cells usually contain four such foci, whereas cells in M (i.e., metaphase and telophase) have none. p220 foci contain cyclin E and are coincident with Cajal bodies [aka, nucleolar accessory bodies: spherical nuclear sub-organelles observed in cells actively engaged in cell cycle traversal, and putatively responsible for the transcription machinery of the nucleus; Cremer & Cremer, 2001]. Five cyclin E/Cdk2 phosphorylation sites have been characterized in p220, and p220 phosphorylation correlates with the appearance of cyclin E at the G1/S boundary and is sustained until prophase (Ma et al, 2000).

35 The Jumonji C domain is a catalytic motif that mediates histone lysine demethylation.

36 Jumonji domain containing 2C, also known as GASC1, which promotes embryonic stem cells differentiation by acting as a histone demethylase.

stem cells as well (Li et al, 2007). When methylated, for example, chromatin proteins, both histones and non-histones hold DNA together tighter and more strongly: increased compaction further ensures restricted access to RNA polymerase for gene transcription, and other enzymes (DNA repair enzymes). For the most part, studies have implicated methylation (e.g., lysine27 on H3) in gene silencing by hindering transcription (Sampath et al, 2007). As expected, methylation depends upon cell activation and cell-cycle traversal. In unstimulated cells, continuous turnover of H3K9 acetylation occurs on all K4-trimethylated histone H3 tails; distribution of both modifications coincides across promoter and 5' part of the coding region. In contrast, K36- and K79-methylated H3 tails, which are not dynamically acetylated, are restricted to the coding regions of these genes. Upon stimulation, transcription-dependent increases in H3K4 and H3K36 trimethylation occur across coding regions, peaking at 5' and 3' ends, respectively. Huntingtin-interacting protein HYPB/Setd2 is responsible for virtually all global and transcription-dependent H3K36 trimethylation, but not H3K36-mono- or dimethylation, in an in vitro model of activated murine fibroblasts (Edmunds et al, 2008).

- Acetylation, which loosens chromatin[37] thus favoring replication and transcription, is usually a reversible addition of an acetyl group onto lysine residues within the N-terminal protein domains. The dynamic equilibrium of acetylation is maintained by two enzyme activities, histone acetyltransferase[38] and histone deacetylase. Recent data on histone deacetylases and on anionic motifs in chromatin- or DNA-binding regulatory proteins (e.g. transcription factors, nuclear proto-oncogenes) support the model that attributes a key function to histone deacetylation for the binding of regulatory proteins to chromatin by a transient, specific local increase of the positive charge in the N-terminal domains of nucleosomal core histones. The rapid deacetylation of distinct lysines (e.g., H2A and H2B) facilitates the association of anionic protein domains of regulatory nonhistone proteins to nucleosomes. In this fashion, histone deacetylase balance provides a uniquely critical regulatory mechanism in chromatin architecture, and function for gene activation throughout the cell cycle. The acetyltransferase activity provides stability, in chromatin repair and assembly, and for nucleosomal balance during cell cycle transition and the transcription process during G1 primarily and cG2 (López-Rodas et al, 1993).

---

[37] L-lysine has a positive charge on the nitrogen at its end, and binds tightly the negatively charged phosphates of the DNA backbone. This prevents them from repelling each other. The acetylation event converts the positively charged amine group on the side chain into a neutral amide linkage, and removes the positive charge causing the DNA to repel itself. When this occurs, the structure of chromatin becomes looser, and complexes like SWI/SNF and other transcriptional factors bind to the DNA, thus opening it up and exposing it to enzymes like RNA polymerase for transcription to commence.

[38] Several histone acetyltransferases (HAT's) enzymes have now been identified. The activity of many HAT's is dependent on HAT complexes, which consist of one HAT catalytic subunit, adapter proteins, other molecules of unknown function and a large protein, the TRansformation/tRanscription domain-Associated Protein (TRRAP). TRRAP, common to many HAT complexes, recruits these complexes to chromatin during transcription, replication and DNA repair. TRRAP mediates diverse cellular processes in a finely regulated, concerted and context-dependent recruitment of HAT's in order to coordinate distinct chromatin-based processes related to cell cycle events and traversal (Murr et al, 2007).

Histone and non-histone proteins modifications provide another dimension of chromatin fluidity. They are events, finely regulated themselves by transmenbrane and cytoplasmic enzymatic processes, that control and mediate the stringency and consistency of epigenetic regulation throughout interphase and the cell cycle at large (Turner, 2002; Fazzio & Tsukiyama, 2003; Roloff & Nuber, 2005; Quina et al, 2006; Kirmizis et al, 2007; Zhang & Adams, 2007).

Histone and non-histone proteins modifications are also emerging as having a critical role in the spatial organization of chromatin in the cell nucleus. The nucleus is structurally and functionally compartmentalized by eu- and heterochromatin. Specific regions of the chromatin are thought to be bound to the nuclear membrane, while other regions are bound together by protein complexes. On the inner face of the nuclear envelope, active genes localize to nuclear-pore structures whereas silent predominantly heterochromatin localizes to non-pore sites. Nuclear-pore components seem to not only recruit the RNA-processing and RNA-export machinery, but contribute a level of regulation that might enhance gene expression in a heritable manner. Epigenetic regulation of gene expression involves repositioning of loci in the nucleus through changes in large-scale chromatin structure in a finely regulated sequence throughout the individual cytomic moments of interphase (Turner, 2002; van Driel et al, 2003; Roloff & Nuber, 2005; Quina et al, 2006; Akhtar & Gasser, 2007; Sexton et al, 2007).

3. The Insulin-Like Growth Factor Family in the Cell Cycle

A. The Insulin-Like Growth Factor (IGF) Family

The insulin-like growth factor (IGF) family consists of two polypeptides with high sequence similarity to insulin, and high affinity to the insulin receptor, IGF-1[39] and IGF-2[40]. These ligands modulate almost every eukaryotic cell type, from muscle, cartilage, bone, liver, kidney, nerves, skin, and lungs, by binding either of two cell membrane receptors, IGF-1 receptor (IGF-1R) and IGF-2R[41]. The latter was identified as being the cation independent mannose-6- phosphate receptor. The IGF family also includes a group of six, to date, IGF-binding proteins(IGFBP)1-6, which act as protein carriers and prolong the half-life of circulating IGF1 or 2. (Pavelic et al, 2007).

39 IGF-1 (MW 7649d.) is a 70 amino acids, single chain peptide, with 3 intra-molecular disulfide bridges, produced primarily by the liver post-natally in response to growth hormone. Because IGF-1 levels can be measured in the blood (range: 10-1000 ng/ml), and circadian variations are minimal, it is often used as a biomarker for growth hormone deficiency and excess. But, confounding variables, including a wide normal range, variations by age (the lowest levels occur in infancy and in old age) and by sex, can complicate the interpretations of the results. IGF-1 is structurally related to insulin, and binds the insulin receptor, albeit at much lower affinity than insulin (*vide infra*).

40 Aka, somatomedin A. IGF-2 (67 aa.) acts primarily as a growth-promoting hormone during gestation.

41 Crystallographic studies have revealed that Phe_{19} and Leu_{53} of IGF-2 lock into a hydrophobic pocket unique to domain 11 of mammalian IGF-2R (Brown et al, 2008).

In brief, therefore, the relevance of the IGF family in translational research is unquestionable. It provides the instrument for going from the clinical observation in the patients to fundamental cell biology (hence, regulation of cell cycle traversal), and returning with experimental research evidence of the cytomic (e.g., on the cell cycle) effects of IGF's to better engage and resolve clinical cases, (*vide infra*).

The inter-relationship between the IGF family and growth hormone regulation[42] (i.e., the growth hormone/IGF1 axis) is one important aspect of the physiolome, and, by regulating cell cycle traversal within given and organs, significantly modulate their physiologic functions. The IGF family plays an important role in the regulation of cellular processes, including metabolism and cell cycle traversal by promoting cell proliferation (i.e., mitogenic activity) and inhibiting of programmed cell death (i.e., apoptosis)[43].

The primary actions of the IGF family are mediated by the IGF receptors present on most cell types in almost every tissue and organ. The data in Table 1 lists a variety of cell lines and IGF receptor levels.

Both IGF-1R and IGF-2R belong to a tyrosine kinase receptors receptor-family that also include the insulin receptor. Whereas there is a 60% homology between IGF-1R and the insulin receptor, IGF-1 activates the insulin receptor at only 0.1x the potency of insulin. Following binding of the IGF ligands, the signal is transduced via IGF-1R that possesses a transmembrane tail[44]. IGF-2R does not, and is simplistically thought to act as a binding "sink" for clearance of excess circulating IGF's. However, the role of IGF-2R is emerging as putatively being more intricate. Recent evidence indicates the affinity of IGF-2R for four different classes of ligands (i.e., mannose 6-phosphate-containing lysosomal enzymes, non-glycosylated IGF-2, retinoic acid, and urokinase-type plasminogen activator receptor). IGF-2R is now understood to interact with signal structures in the cytosolic and membrane-bound domains. Thus, it contributes significantly to an array of cytomic events, including, it is fair to presume (although hard evidence in this respect is still scant), the regulation of cell cycle traversal (Braulke, 1999; Brown et al, 2007; Pavelic et al, 2007).

IGF-1R consists of two α subunits and two β subunits linked by disulfide bonds, and which traverse the plasma membrane. Binding of IGF-1 or IGF-2 to IGF-1R engages the insulin receptor substrate(IRS)-1/4 and the Shc (Src homology collagen) proteins (p46/p52/p66), which serve as signaling nodes for distinct intracellular pathways, and contribute to the maintenance of glucose homeostasis. Phosphorylated IRS's recruit and activate class 1a phosphoinositide 3-kinase (PI 3-kinase) via the two SH2 domains of the adaptor protein p85, leading to synthesis of membrane-associated phosphorylated inositols. These molecules, in turn, recruit and activate phosphoinositide-dependent kinases (PdKs),

---

42 Growth hormone (i.e., somatotropin, 191 aa., MW: 22,000d) is synthesized in response to (among others) growth hormone releasing hormone from the arcuate nucleus of the hypothalamus, ghrelin and arginine. Growth hormone is produced, stored, and secreted by the somatotroph cells in the lateral wings of the anterior pituitary gland. Negative regulators of growth hormones include hypothalamic somatostatin from the periventricular nucleus, circulating growth hormone and IGF-1 in a negative feedback loop, and output of the hypothalamic-pituitary-adrenal axis (i.e., glucocorticoids).

43 Case in point the effects of IGF-1 on the modulation of growth and development of condylar cartilage *in vivo*, for example (Yokota et al, 2008).

44 Both IGF-1 and IGF-2 bind to IGF-1R, but IGF-2 alone binds to IGF-2R, which lowers the levels of IGF-2 available for binding with IGF-1R.

which then phosphorylate and activate other protein kinases, including protein kinase Cζ (PK Cζ), phosphatidyl inositol-3 kinase (PI3K)-serine/threonine protein kinase B(PKB)/AKT[45] - target of rapamycin (ToR) system and serine/threonine kinase (RAF)-mitogen activated protein kinase (MAPK) systems[46].

Table 1. IGF-1R & IR Measurements[47] from Selected Mammalian Cell Lines

Line	Origin/Description	IGF-1R levels		(mean)	(SD)	Insulin R
Colo 205	Human (Hu) colon carcinoma	14,792	29,433	22112.5	10352.75	7,129
U266-B1	Hu IgE λ secreting myeloma	7,272	5,863	6567.5	996.31	
RPMI 8226	Hu Plasmacytoma myeloma B-lymphocyte	7,176	12,120	9648	3495.946	
HL-60	Promyelocytic leukemia	1,919	3,832	2875.5	1352.70	
HL-60 Clone 15	Hu Promyelocytic leukemia	8,057	5,465	6761	1832.82	
K-562	Hu chronic myelogenous leukemia	4,048	4,490	4269	312.54	
Jurkat	Hu T-cell leukemia	2,659	3,229	2944	403.06	
Daudi	Hu Burkitt Lymphoma	-	6,196	6196	-	
Ramos	Hu Burkitt Lymphoma B cell	-	5,994	5994	-	
Namawla	Hu Burkitt lymphoma, B cell EBV+	-	6,499	6499	-	
Calu 6	Hu lung carcinoma	40,045	-	40045	-	
NCI 1299	Hu lung large carcinoma	42,790	-	42790	-	
MDA- MB 231	Hu breast adenocarcinoma	22,403	-	22403	-	
MDA- MB 468	Hu breast adenocarcinoma	37,868	-	37868	-	
MIA PaCa2	Hu pancreatic carcinoma	12,113	-	12113	-	
ZR-75	Hu breast carcinoma	19,408	-	19408	-	
MDA MB 435 S	Hu breast ductal carcinoma	17,189	-	17189	-	
T47D	Hu breast, ductal carcinoma, pleural effusion	90,408	-	90408	-	
BT474	Hu breast ductal carcinoma	63,712	-	63712	-	
RH-1	Hu malignant embryonal rhabdomycosarcoma; bone marrow; metastasis	-	8,113	8113	-	14,973
A-673	Hu muscle rhabdomyosarcoma	-	8,990	8990	-	2,143
SK-ES-1	Hu anaplastic osteosarcoma or Ewing's sarcoma	-	17,949	17949	-	12,842
SK-NEP-1	Human kidney, Wilms' tumor, pleural effusion	-	27,166	27166	-	11,056
MCF 7	Hu breast ER+ luminal	85,538	90,607	88,073	3584.32	7,844
HT 29	Hu colorectal adenocarcinoma	-	17,291	17291	-	4,279

45 Akt (protein kinase B) is a serine/threonine protein kinase, a downstream effector of phosphoinositide 3-kinase (PI3K), that functions as a signaling intermediate linked to multiple signaling programs involved in survival, inflammation, and growth. Akt is closely associated with key membrane-bound receptors and represents a convergent integration point for multiple stimuli. The Akt/mammalian target of rapamycin (mTOR)/ribosomal protein S6 kinase (p70S6K) pathway (*vide infra*) is a central regulator of protein synthesis and of cell cycle traversal, cell differentiation (which, as noted above, directs and maintains the cell in G0), and survival (and inhibitors of the PI3K/Akt/mTOR pathway [e.g., LY294002, rapamycin] abrogate survival and cell cycle traversal at G1).

46 It must be noted that Akt can inhibit the Raf protein kinase by phosphorylation on Ser259, leading to interaction of Raf with 14-3-3 proteins and inhibition of the Ras/ERK signal transduction pathway.

47 IGF-1R levels were established by means of the Quantum Simply Cellular® Microbeads (bangslab.com) for the quantification of the IGF-1R as the cellular antigen expression, using specific monoclonal antibody, followed by flow cytometry analysis. The data are expressed as Antibody Binding Capacity (ABC) units, and transformed in number of antigen units (i.e., receptor epitope) per cell, and analyzed by Kolmorov-Smirnov analysis.

Table 1. (Continued)

Line	Origin/Description	IGF-1R levels		(mean)	(SD)	Insulin R
SJSA-1	Osteosarcoma Cell line	-	14,682			348
BXPC3	Pancreatic adenocarcinoma	-	31,987	31,987	-	514
Mia Paca-2	Pancreatic carcinoma					

Protein kinase B (PKB), which has sequence homology to both protein kinase A and C, is identical to the kinase Rac, and is the cellular homologue of the transforming Akt[48]. Activation of the Akt-dependent pathways results in variety of responses, including cell cycle traversal, cell proliferation, differentiation, migration, and protection from apoptosis. IGF-1 is one of the most potent natural activators of the AKT signaling pathway, and utilizes principally this mechanism in its role as stimulator/regulator of cell growth, replication and proliferation, and a potent inhibitor of programmed cell death (Doepfner et al, 2007; Kulkarni, 2007; Zeng & Yee, 2007).

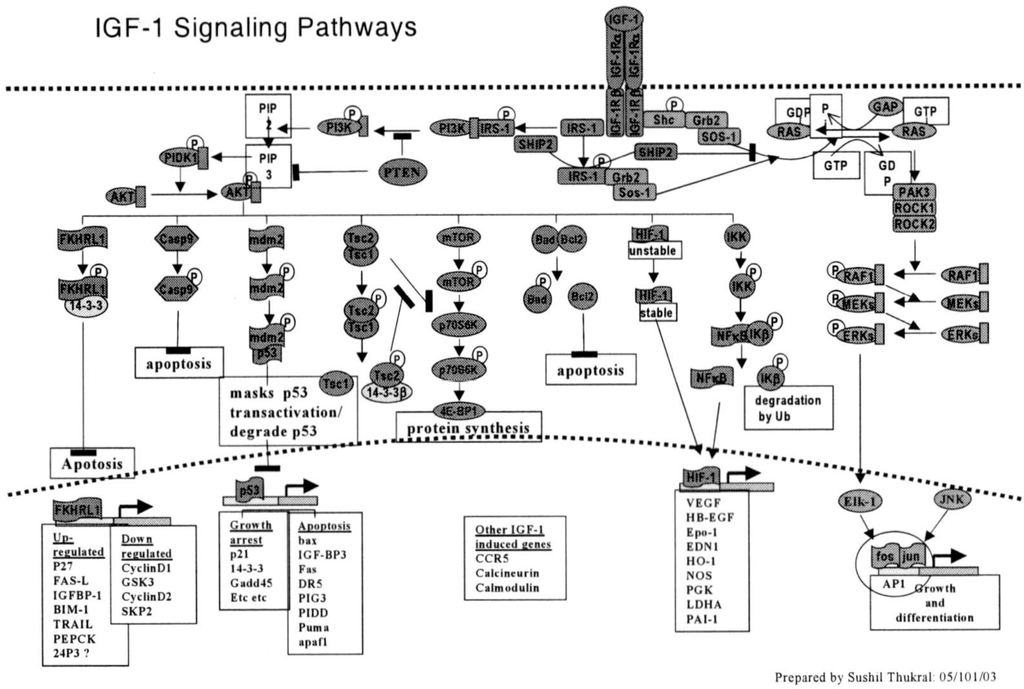

Figure 3. IGF-1R signaling pathways[49].

[48] The "Ak" in Akt derives from a temporary classification of the original mouse strain (AKR) used in early studies of spontaneous thymic lymphomas, and the "t" stands for 'transforming'. The first isolated transforming retrovirus was in fact from the Ak strain (Akt-8), and the oncogene encoded in this virus was eventually designated *v-Akt* (Bellacosa et al, 1991).

[49] The three signaling processes initiated by stimulation of IGF-1R are illustrated. The fundamental signaling cascades are: (1) the Akt Pathway activated by IRS-1/PI3Kinase, (2) the MAPK pathway that is strongly dependent on SHC and involves RAS, and (3) the 14.3.3 pathway that promotes mitochondrial translocation of Raf-1 kinase. Apoptosis is prevented by phosphorylation of BAD and sequestration by 14-3-3 adapters. The

Exposure of cells (e.g., astrocytes) to IGF-1 leads to the reduction in Phosphatase and Tensin (PTEN)[50] mRNA and protein levels. PTEN is a lipid phosphatase that counteracts the PI3K/Akt pathway. It inhibits the entry into the cell cycle by blocking G1/S transition by down-regulating cyclin D and, putatively, cyclin E. Inhibition of PTEN involves activation of the PI3K/protein kinase C (PKC) pathway, and decrease, in a proteasome-dependent step, the levels of the transcription factor Egr-1[51]. This leads to decreased binding of Egr-1 to the PTEN promoter. The sharp proliferative rate of certain cancer cells can be traced to alteration in PTEN regulation (e.g., endometrial carcinoma cells often lack endogenous wild-type PTEN, PTEN activity decreases expression of cyclins D3 and hypo-phosphorylation of Rb in glioma cell lines) (Fernandez et al, 2007).

An alternative IGF signaling pathway also boosts cell cycle traversal, and is coupled to the recruitment to both IRS-1 and Shc of the guaninenucleotide-exchange factor Sos, via the SH2 domain of the Grb2 adaptor[52]. This results in G-protein Ras activation, which leads to the activation of the protein serine kinase Raf, and of the extracellular-signal-regulated kinase (ERK) cascade (Shelton et al, 2004; Bertrand et al, 2006).

Thirdly, IGF-1R activation is coupled to the stimulation of MAP kinases that include Jun kinase(JNK)-1 and -2 and p38 MAP kinase. JNK phosphorylates the amino terminus of c-Jun[53], which favors transcription activation during G1 and G2 (O'Connor, 2003; Leahy et al, 2004).

IGF-1 induces tyrosine phosphorylation of the cytoplasmic protein Crk[54], which shares some sequence homology with Grb2 and Nck. Crk associates with the guanine nucleotide exchange factor, SOS, already bound to GRB2 in the MAPK/ERK pathway through the SH3 domain. Thusly, it modulates the IGF-1R-mediated regulation of the Ras-MAP kinase signaling pathway. Crk contributes by enhancing IGF receptor-dependent mitogenesis (Karras et al, 2001; Klammt et al, 2002).

Akt, MapK and 14.3.3 pathways can stimulate mitogenesis through MAPK (ERK1/2) while also inhibiting apoptosis through BAD inactivation.

50 The tumor suppressor, PTEN, is a ubiquitous phosphatase. Specifically, in fact, it acts in most systems as a phosphodiesterase and an inhibitor of the phospho-AKT pathway by removing the 3' phosphate group of phosphatidylinositol (3,4,5)-trisphosphate (PtdIns [3,4,5]P3). PTEN generally favors signaling cells to stop dividing and to undergo programmed cell death (apoptosis, *vide supra*) (Leslie & Downes, 2004).

51 The tumor suppressor early growth response-1, Egr-1, is a key regulator of PTEN levels

52 Shc competes with IRS-1 for a limited cellular pool of Grb2, and the extent of Shc/Grb2 binding correlates with the amount of insulin-activated ERK and c-fos transcription. Shc isoforms have opposite effects on the ERK signaling cascade. The Shc proteins originate by alternative use of three distinct translation starting points on a longer transcript (p46Shc, p52Shc, and p66Shc) and two translation starting points on a shorter transcript (p46Shc, p52Shc). The mRNA transcripts are generated by alternative splicing from a single gene. The three Shc isoforms are tyrosine-phosphorylated upon growth factor stimulation, But p46/p52Shc are coupled to growth and survival signals, and p66Shc, which also undergoes serine phosphorylation, mediates pro-apoptotic responses.

53 C-jun, the putative transforming gene of avian sarcoma virus 17, forms the AP-1 early response transcription factor in combination with c-Fos. AP-1 interacts directly with specific target DNA sequences to regulate gene expression.

54 The Crk gene belongs to a cytomic adapter protein family that typically binds to proteins that are phosphorylated on tyrosine residues. Crk is endowed with src-homology(SH)2 and SH3 domains, and plays a regulatory role in signaling pathways, and cell growth and cell cycle traversal, by recruiting cytoplasmic proteins in the vicinity of tyrosine kinases. The N-terminal SH2 domain functions as a positive regulator, and the C-terminal SH3 domain is a negative regulator of cell transformation.

The activity of IGF1 and IGF-2 is modulated by association with the IGF binding proteins[55](IGFBP) 1-6 (*vide supra*). By binding to IGF-1 & 2, IGFBP's limit both the growth-promoting effects, and the availability of these ligands for the receptors, while simultaneously increasing the half-life and bio-availability of circulating IGF's. IGFBP proteases cleave intact IGFBP's into smaller molecular weight peptides with altered association properties to IGF's. Whereas IGFBP-3 is the principal binding protein for IGF-1, IGFBP-4 and IGFBP-5 are most abundant in the osteoimmune system, and play a relevant role in bone pathology and metabolism[56] (*vide infra*) (Mohan et al, 1995; Sitar et Al, 2006; Mukherjee & Rotwein, 2007; Pavelic et al, 2007).

B. Cytomic Modulation by the IGF Family

The IGF family is intimately intertwined in the regulation of a vast number of cytomic events. Both IGF-1 and –2 are potent modulators of estrogen (E2)-dependent breast-cancer cells synergistically or additively at the G1/S and the G2/M phases of the cell cycle. IGF's acts directly on key factors in cell cycle regulation (e.g., expression of cyclin D1 and cyclin E, phosphorylation of Rb, cf., footnote 9), and these effects are potentiated by E2. Whereas IGF-I increases the protein level of p21 and the transcriptional activity of the p21 promoter, it reduces the protein level of p27 without affecting p27 promoter activity. E2 does not affect p21 expression, but it does decrease the protein level of p27 and the p27 promoter activity[57]. Co-treatment with IGF-1 and E2 synergistically stimulates G1/S transition, coincidentally with synergistic activation of the cyclin E/CdK2 complex lacking p21 (Lai et al, 2001).

In the context of brain cell and glial development, IGF-1 and fibroblast growth factor(FGF)-2 combine in a synergestic fashion to induce cyclin D1 in oligodendrocyte progenitors. This pathway is dependent upon multiple concerted signals that involve the p44/p42 MAPK signaling pathway arising from FGF-2 induction and the IGF-1-induced the

---

55 The insulin-like growth factor binding proteins (IGFBP's) are cysteine-rich proteins, with conserved N- and C-domains that bind insulin-like growth factors (IGF's) with high affinity. In addition, another group of cysteine-rich proteins exist that share important structural similarities with the IGFBP's, but have demonstrably lower affinity for IGF's, the IGFBP-related proteins (IGFBP-rP's). IGFBP-rP's and IGFBP's constitute the IGFBP superfamily (Hwa et al, 1999).

56 Bone metabolism refers to the complex composite of cellular, biochemical and molecular events that signifies the metabolic balance (i.e., homeostasis) or drive to regain metabolic balance (i.e., allostasis) between bone formation by osteoblasts, and bone resorption by osteoclasts. Key to these events is the ability and drive of the osteoblast and osteoclast populations to growth, and to proliferate. In as much as cell cycle traversal of these two bone cell population is regulated by cytokines and other growth factors produced by resident and transient immune cells (e.g., pro-inflammatory cytokines by myeloid cells; TH1, TH17 cytokines by regulatory T [Tregs], naïve and memory T cells), the emerging interactomic domain is referred to as "osteoimmunology". One domain of this field with particular relevance to the present discussion, from the viewpoint of translational implications of the modulation of cell cycle traversal by the IGF family pertains to bone erosion and healing in patients with diabetes or the 'metabolic syndrome' (insulin resistance syndrome, syndrome X) (*vide infra*). In this context, we must note that the term 'metabolic syndrome' dates back to the late 1950s and came into common usage in the late 1970s to describe various associations of risk factors with diabetes, that had been noted for at least 5 decades. It now defines a constellation of abnormalities associated with increased risk for the development of type 2 diabetes and atherosclerotic vascular disease (e.g. heart disease and stroke).

57 We must note that increased expression of p21 leads to increased p21-associated CdK2, and decreased p27 expression leads to decreased p27-associated CdK2.

PI3-kinase/Akt pathway to lead to phosphorylation of GSK-3β[58], enhanced cyclin D1 expression, cyclin E/CdK2 complex activation, and G1/S transition. IGF-2 also plays an important role is allowing nuclear localization of cyclin D1, which is also dependent upon PI3K, and contributes to up-regulating DNA synthesis in S. (Frederick & Wood, 2004; Frederick et al, 2007). In neural progenitor cells, stimulation of the ERK pathway by FGF-2 is key for the induction of the expression of cyclin D1. IGF-I, in the presence of FGF-2, significantly up-regulates the expression of cyclin D1 (Kalluri et al, 2007).

The research evidence supports the dogma that IGF-1 and –2 and their receptors are implicated in the regulation of various facets of cytomic regulation and protein turnover, and exert potent mitogenic and differentiating effects on most cell types. IGF-1R as a target for intervention is validated by its primary pathway affiliation – its linkage to Akt activation through the IRS adapter proteins. The adapter proteins that link IGF-1R activation to the cytoplasmic signal transduction machinery that regulate these cellular processes are listed in Table 2 (Adams et al, 2000). IGF-1 mitogenic actions have been well documented for a wide variety of developing and mature cell types, including stem and progenitor cells of the neural and erythropoetic lineages, fibroblasts, pituitary cells, and uterine and mammary epithelial cells.

Table 2.

Protein	Function	Site
IRS-1-4	adaptor	P-Tyr ^{950}
Shc	adaptor	P-Tyr ^{950}
CrkII	adaptor	P-Tyr ^{943}/P-Tyr 950
SH2-B	adaptor	P-Tyr^{950}/P-Tyr^{1316}
p85	regulatory subunit of PI 3-kinase	P-Tyr^{1316} and other residues
	ATP binding	Lys 1003
GAP	GTPase-activating protein	P-Tyr ^{950}
Syp/SHP-2	Tyrosine phosphatase	P-Tyr^{1316}
CSK	Cytoplasmic tyrosine kinase	P-Tyr^{943}/P-Tyr^{1316}
	Autophosphorylation	P-Tyr^{1131,1135,1136}
Grb10	adaptor	aa 1229-1245
14.3.3	Adaptor scaffolding proteins	P-Ser ^{1272}/P-Ser^{1283}

The mechanism by which IGF-1 and IGF-2 induce cell proliferation and cell cycle traversal involves regulation of the cell replicative machinery at several levels. For example,

58 Which is then involved in proteasome-mediated cyclin D1 degradation.

before entering the cycle, cells must accumulate sufficient translational machinery, mainly ribosomes, to ensure the rapid processing of transcripts through the cycle. Ribosome biosynthesis is a key process in cell proliferation. This is accomplished at least in part by the phosphorylation of the ribosomal S6 protein by the P70S6 kinase[59], which is activated by growth factors including IGF-1 downstream of Akt (*vide supra*). Following IGFR-1 stimulation, Akt activates the mTOR cascade leading to p70S6K activation. This cascade pathway has been well characterized in muscle cells, for example, where it is clear that in response to IGF-1, Akt activates downstream effectors, mTOR and p70S6K to stimulate RNA transcription and protein synthesis thereby increasing the cytoplasmic compartment in muscle fibers (Guttridge, 2004).

In brief, IGF-1, by activating, P70S6 kinase facilitates G0/G1 transition and progression through G1 phase. G1/S transition requires the activity of CdK4/6 and CdK2: CdK4 and CdK6 regulate early events in G1, and CdK2 acts to bring G1 to a close and to actualize transition to S (*vide supra*). The CdK's serine/threonine kinase activity mediates active heterodimeric complexes following binding to their specific cyclin ligands, which act as regulatory subunits (*vide supra*). IGF-1 induces the expression of the genes for the D family of type cyclins (cyclins D1-3), which associate with CdK4/6. Thusly activated, CdK4/6 hyper-phosphorylates Rb, which is then released of the transcription factor E2F (*vide supra*). Free E2F then contributes to the expression of late G1 cyclins (e.g., cyclin E), which engage G1/S transition. The IGF family, in this manner, intimately contributes to the modulation of G1 traversal and G1/S transition by playing a critical role in the up-regulation of cyclin D and cyclin E expression (Baserga et al, 1993; Dupont et al, 2003).

It can also be the case that IGF-1 contributes to the down-regulation of cell cycle traversal by means of induction of counter signaling pathways, or CdK inhibitors (*vide supra*). Case in point, in certain breast cancer cell lines, IGF-1-induced PI3K/Akt pathway can both induce elevation of cyclin D1 protein and mRNA, as noted above, and lead to a detrimental accumulation of cyclin D1 in the nucleus during G1, hindering G1/S transition (Hamelers et al, 2002). IGF-1 and –2, the three members of the Cip/Kip family (*vide supra*), and increased levels and nuclear redistribution of p27Kip1, a critical mediator of PTEN, mediate G1 arrest. IGF-1 can also block G1/S transition by increasing the expression of p21Cip1, such as for example in the A549 human lung adenocarcinoma line (Kodama et al, 2002).

---

[59] p70S6 Kinase (p70S6K) is a serine/threonine kinase, whose target substrate is the S6 ribosomal protein. Phosphorylation of S6 induces protein synthesis at the ribosome. P70S6K is a component of the mammalian target of rapamycin (mToR), a serine/threonine protein kinase itself that regulates cell growth and cell proliferation, as well as other cytomic functions including migration and survival, protein synthesis, and transcription (Wullschleger et al, 2006).

4. Implications for Translational Research

The regulation of the cell cycle is critical to the maintenance of processes, pathways and events within the normative physiolomic balance, which in its complex interactions maintains the healthy state of the organism. There are several clinical conditions characterized by alterations in cell cycle regulation and associated with, or consequential to deregulations in the IGF family. Translational inquiry yields novel understanding of fundamental findings pertaining to cell cycle traversal regulated by the IGF family. Three principal domains are briefly outlined below, and serve as conclusive remarks for this discussion.

A. Diabetes[60] & Osteoimmunology

Providing an alternative to fixed and removable bridges, partials, and dentures, dental implants look, feel, and function just like natural teeth and have therefore become very popular throughout the field of dentistry (Moradi et al, 2006). Despite the effectiveness of replacing missing teeth, the predictability of dental implants relies on the successful osseointegration[61] of the implant with its surrounding bone. Osseointegration depends upon bone cell proliferation[62] about the implant material, and is defined by the direct structural and

---

60 Affecting over 140 million individuals, diabetes mellitus is one of the world's greatest chronic health problems with a prevalence of approximately 3%. Diabetes mellitus is a multi-factorial metabolic disorder affecting people on a global scale. According to the World Health Organization (WHO), "*there is an emerging global epidemic [...] that can be traced back to rapid increases in overweight, obesity and physical inactivity*". Diabetes is characterized by an increase in blood glucose levels. Insulin dependent type I diabetes is different from insulin independent type II diabetes in that the former occurs when there is a deficiency of insulin in the blood while in the latter, glucose fails to enter the cells. Type I diabetes is an auto-immune disease characterized by the degeneration of pancreatic beta cells which produce insulin. Patients with this disease must be supported with exogenous insulin (hence insulin dependent) to ensure survival. Type II diabetes, on the contrary, is a multi-factorial disease that results from environmental effects on genetically predisposed individuals and is normally associated with obesity and age. This disease is characterized by defective insulin secretion and can be treated with exercise and careful diet maintenance (hence insulin independent). In the United States, type II diabetes was the sixth leading cause of death in 2002. As of 2005 the total population prevalence was 7%, representing approximately 21 million people with the disease. American Indians and African Americans. are 2.2 and 1.8 times at a greater risk respectively for the disease, than non-Latino whites, with 13% (African American) and 15% (American Indian) of their populations affected (CDC, 2005; ADA,2008). Both types of diabetes results in high blood glucose due to missing or resistant insulin (Valero et al, 2007). The consequent increase in blood glucose concentrations is linked to several complications associated with diabetes, including poor wound healing, cardiovascular disease, adverse pregnancy outcomes, periodontal disease, and increased frequency of oral infections (Farzad et al, 2002).

61 The time period required for osseointegration takes about three to six months and is formerly known as the healing period. During this time, the implant must remain load-free until osseointegration fully occurs (Moradi et al, 2006).

62 Bone is a metabolically active organ that undergoes continuous remodeling throughout development and aging. Remodelling of the bone matrix is induced by bone cells the osteoblasts and osteoclasts, and depends on local factors that include cytokines, growth factors and other mediators of cell-to-cell and matrix-to-cell communication. Matrix-producing osteoblasts and bone-resorbing osteoclasts maintain bone homeostasis. Osteoclasts are multinucleated, giant cells of hematopoietic origin formed by the fusion of mononuclear pre-osteoclasts derived from myeloid cells. Fusion-mediated giant cell formation is critical for osteoclast maturation; without it, bone resorption is inefficient. The d2 isoform of vacuolar (H^+) ATPase (v-ATPase) V0 domain (Atp6v0d2) is one the principal regulators of osteoclast fusion and bone formation. The bone

functional connection between living bone with the surface of the dental implant (Fiorellini & Nevins, 2000). One important factor that must be considered for successful osseointegration is the overall health of the individual patient, and diseases such as osteoporosis and diabetes significantly affect osseointegration and therefore the success of the dental implant.

The synthesis of hepatic IGF-1 is stimulated by growth hormone (GH) release in the pituitary in response to hypothalamic GHRH (GH releasing hormone). The serum concentration of IGF-1 increases over 100 fold between ages 5-15 in humans[63]. Insulin and GH are mutual antagonists whereas IGF-1 is an insulin sensitizer (Mauras & Haymond, 2005). The GH/IGF axis acts in concert with insulin in the pathophysiology of type II diabetes. As noted above, IGF-1 has a relatively high sequence homology to both pro-insulin and insulin molecules (49%) conferring biologically relevant binding affinity to the insulin receptor (IR). Insulin binds with high affinity to IR, but with much lower affinity to IGF-1R and IGF-2R. IGF's and insulin play different biological roles: insulin regulates a variety of metabolic processes, whereas IGF's promotes cellular growth (Clemmons et al, 2005).

In brief, in addition to several systemic complications, diabetes also induces changes in bone formation and IGF's, as we as by altering bone metabolism[64] (Valero et al, 2007).

---

remodeling cycle involves a complex series of sequential steps that are highly regulated. The "activation" phase of remodeling is dependent on the effects of local and systemic factors on mesenchymal cells of the osteoblast lineage. These cells interact with hematopoietic precursors to form osteoclasts in the "resorption" phase. Subsequently, there is a "reversal" phase during which mononuclear cells are present on the bone surface. They may complete the resorption process and produce the signals that initiate formation. Finally, successive waves of mesenchymal cells undergo regulated cell cycle traversal, and differentiate into functional osteoblasts, which lay down matrix in the "formation" phase (Raisz, 1999; Hadjidakis & Androulakis, 2006). The regulation of bone remodeling is both systemic and local. The major systemic regulators include parathyroid hormone (PTH), calcitriol, and other hormones such as growth hormone, glucocorticoids, thyroid hormones, and sex hormones, prostaglandins, tumor growth factor (TNF)-β, bone morphogenetic proteins (BMP), Through the RANK/receptor activator of NF-□B ligand (RANKL)/osteoprotegerin (OPG) system the processes of bone resorption and formation are tightly coupled to IGF regulation, allowing a wave of bone formation to follow each cycle of bone resorption, thus maintaining skeletal integrity (Hadjidakis & Androulakis, 2006). The intimate interactions in the process of prliferative regulation between osteoblasts and osteoclasts on the one hand, and immune cells present in the bone matrix on the other hand defines the osteoblastic niche. This intercellular relationship arises at the very onset of development and cell cycle traversal of bone and hematopoietic stem cells with multi-lineage potentials (Calvi, 2006). Stem cell fate is influenced by specialized microenvironments of the niche that are finely modulated by a myriad of factors, including Parathyroid hormone (PTH), through activation of the PTH/PTHrP receptor (PTH1R) and bone morphogenetic protein (BMP) through BMP receptor type IA (BMPRIA) (Calvi, 2006; Arai & Suda, 2007).

63 Primary IGF-1 deficiency arising form partial gene deletions, and secondary IGF-1 deficiency resulting from defects in GH production or signaling are not lethal. The affected individuals exhibit growth retardation at birth, grow slowly and can face certain CNS abnormalities (Woods & Savage, 1999).

64 Bone is a metabolically active organ that undergoes continuous remodeling throughout development and aging. Remodelling of the bone matrix is induced by bone cells the osteoblasts and osteoclasts, and depends on local factors that include cytokines, growth factors and other mediators of cell-to-cell and matrix-to-cell communication. Matrix-producing osteoblasts and bone-resorbing osteoclasts maintain bone homeostasis. Osteoclasts are multinucleated, giant cells of hematopoietic origin formed by the fusion of mononuclear pre-osteoclasts derived from myeloid cells. Fusion-mediated giant cell formation is critical for osteoclast maturation; without it, bone resorption is inefficient. The d2 isoform of vacuolar (H+) ATPase (v-ATPase) V0 domain (Atp6v0d2) is one the principal regulators of osteoclast fusion and bone formation. The bone remodeling cycle involves a complex series of sequential steps that are highly regulated. The "activation" phase of remodeling is dependent on the effects of local and systemic factors on mesenchymal cells of the osteoblast lineage. These cells interact with hematopoietic precursors to form osteoclasts in the "resorption"

Insulin and IGF's play an instrumental role in modulating bone cell prliferation and growth (Canalis, 1980). Insulin and IGF-1 affect bone by stimulating bone matrix synthesis, stimulates the formation of the osteoblastic matrix, osteoblasts proliferation and differentiation, and controls and regulates IGF-1 production by hepatocytes (Fiorellini & Nevins, 2000).

During the healing period, the osseointegration of the implant surface with its surrounding bone requires the formation of osteoblasts and bone matrix. In the diabetic patient[65], however, there are deficiencies in bone remodeling processes and mineralization due to the lack of insulin (Valero et al, 2007). Diabetics treated with exogenous insulin have growth in bone matrix and osteoid formation similar to that of control patients (Locatto et al, 1993).

B. Cancer & Cancer Metastasis[66]

It is now well recognized that IGF-1R signaling promotes, and is required for oncogenic transformation and cell-cycle progression following growth factor stimulation. Expression of IGF-1R in tumors is rarely, if ever, lost, and the level of receptor generally ranges from 10^3 to 10^5 molecules per cell. An intimate physiolomic role of IGF is evident in tumorogenesis, and in breast cancer for instance, estrogen and IGF seem to cooperate in the maintenance or progression of well-differentiated tumor. The schematic model presented in Figure 4 is derived from an experimental model derived from studies with embryonic fibroblasts isolated from IGF-1R (-/-) knockout mice (Baserga 2000). One view of the invariant presence of IGF-1R in the context of cancer biology is that its activity is vital to tumor maintenance and progression.

phase. Subsequently, there is a "reversal" phase during which mononuclear cells are present on the bone surface. They may complete the resorption process and produce the signals that initiate formation. Finally, successive waves of mesenchymal cells undergo regulated cell cycle traversal, and differentiate into functional osteoblasts, which lay down matrix in the "formation" phase (Raisz, 1999; Hadjidakis & Androulakis, 2006).

65 In type I diabetes or insulin deficiency, GH levels increase, and IGF-1 levels decrease. In obesity, enhanced insulin signaling leads to increased IGF-1 generation, reduced IGFBP-1, and a rise in free IGF-1, which leads to reduced pituitary GH secretion. In type II diabetes, GH & IGF-1 change in tandem, and low circulating IGF-1 and GH levels are found in response to chronic hyper-insulinaemia (Segev et al, 2007). In type I diabetes, there is reduction in bone mineral density, in bone formation, and an increase in bone resorption (Levin et al, 1976). In type II diabetes, both bone formation and bone resorption diminish, and the fundamental process of osseointegration is hampered (Fiorellini & Nevins, 2000).

66 Metastasis (Gk: μετά=next + στάσις=placement), the process by which cancer cells can "break away", "leak", or "spill" from a primary tumor, enter lymphatic and blood circulation, and establish new foci of cell growth and proliferation in ectopic sites, typically normal tissues elsewhere in the body. It is a complex process that requires a series of steps: initially, malignant cells must break away from the primary tumor, attach to and degrade proteins that make up the surrounding extracellular matrix (ECM), thus escape usually to "sentinel" lymph nodes, and eventually where the lymphatic and blood circulation transports them . Metastasis is combatted by a variety of mechanisms by means of a class of endogenous metastasis suppressor genes and their proteins products (e.g., nm23; NME1, Stegg et al, 1988) (Yoshida et al, 2000).

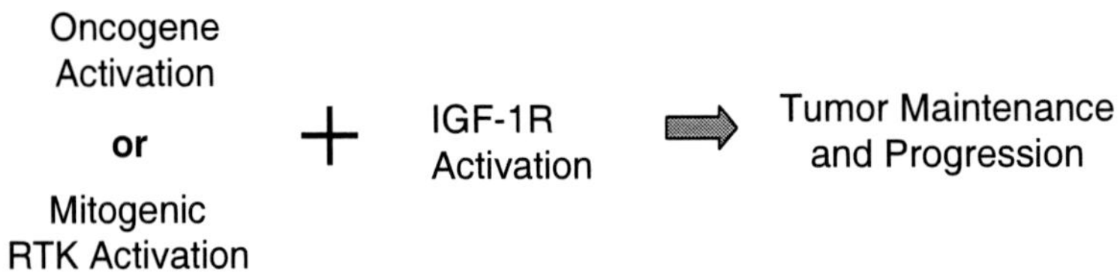

Figure 4. IGF-1R signaling mediates oncogenic[67] transformation.

The utility of IGF-1R as a potential therapeutic target is supported by abundant animal tumor data obtained with receptor-directed inhibitors. Anti-IGF-1R monoclonal antibodies that inhibit receptor activation and/or promote its internalization and degradation are effective in reducing tumor initiation and growth in a variety of murine tumor models (Furlanetto et al, 1993; Kalebic et al, 1994; Zia et al, 1996; Hailey et al, 2002). The use of IGF-1R antisense RNA to eliminate receptor expression has also proven successful in generating IGF-1R targeted reagents without physical cross-reactivity with the insulin receptor (Resnicoff et al, 2004). Restriction of IGF-1 bioavailability by means of truncated receptors containing part or all of the IGF-1R extracellular domain, or IGF binding proteins (IGFBP's) generated significant reductions in tumor growth, and metastasis across a variety of experimental models (Reiss et al, 1998; Samani et al, 2004). Indeed, tumor growth inhibition is also obtained by interfering with the GH/IGF axis (e.g., GH antagonists; Dagnaes-Hansen et al, 2004).

C. Neurobiopathology & Aging

The IGF family is expressed in the CNS in a temporal–spatial manner during development. Furthermore, the expression of IGF-1 occurs in coordination with rapid neural growth. A large body of experimental evidence demonstrates that IGF-1 plays an important role in CNS development by promoting neural cell proliferation, survival, and differentiation, where it likely acts primarily in a paracrine or autocrine fashion (Ye & D'Ercole, 2006). IGF-1 and IGF-2 can increase the number of neural cells, including neurons, astrocytes, and oligodendrocytes, in vivo and in vitro in part by increasing the proliferation of precursors (D'Ercole et al, 1996; Hodge et al, 2004). Case in point, IGF-1 induces an increase in NSC/progenitor proliferation in Tg mice[68] via decreased duration of the cell cycle and an increased percentage of daughter cells re-entering the cell cycle. The process results from a

---

[67] Oncogenes require an additional step, such as mutations in another gene, or environmental factors such as viral infection, to induce transformation and cancer. Note: Receptor tyrosine kinases (RTK) are key regulators of normal cellular processes, and play a critical role in the development and progression of many types of cancer.

[68] Popken et al. (2004) have generated a line of transgenic (Tg) mice (IGF-Inestin Tg mice) that express a nestin-driven IGF-I transgene in the CNS beginning at least as early as embryonic day (E) 13. The data show a 15% increase in BrdU-labeled cells in the cerebral cortical ventricular zone (VZ) and subventricular zone (PVZ) of E14 IGF-Inestin Tg mice. Given that proliferating cells in VZ and PVZ at this developmental stage are predominantly NSCs and neuron progenitors, these data confirm that IGF-1 stimulates proliferation of NSC and neuron progenitor cells.

shortened G1 phase by up to 2 hours, but no significant change in the length of other cell cycle phases (Hodge et al, 2004). In brief, IGF-1 promotes cell cycle progression by modulating the length of G1 phase or the G0/G1 phase transition (Loef et al, 1982; Stull et al, 2002).

During development and aging, IGF-1 plays an important role CNS vascular remodeling by regulating cell cycle traversal of the endothelium. From the translational perspective, this represents an important issue in traumatic brain injury (TBI), and in neurodegenerative disease, such as senile dementia of the Alzheimer's type (sDAT). The aging brain requires IGF-1 for reactive vessel remodeling in the traumatized or degenerative brain[69]. This pro-angiogenic[70] activity of IGF-1 is likely to contribute to its neuroprotective actions. Better vascularization implies proper nutrient and oxygen supply to neurons. Low IGF-1 input to the brain is altering the angiogenic response, and also has severe clinical consequences (Bondanelli et al., 2004). It is also clear that, not only local IGF-1 in the CNS, but also systemic IGF-1 is a trophic factor for the adult brain (Torres-Alenan, 2000).

Post-traumatic stress disorder (PTSD) provides an interesting clinical example of the degree to which the biological inter-relationships are intertwined in the cultural and ethnic domains of the patients (Pole et al, 2005; Loo, 2007)[71]. Hispanic Americans may be exposed to a larger number of traumatic events and of greater severity (Brewin et al, 2000). In addition, they may, as a group, have culture-sensitive approaches[72] in resolving conflicts and in seeking health[73], that are not propitious for ameliorating traumatic stress events (Perilla et al, 2002; Pole et al, 2005). These and other psychological and cultural reasons may explain why Hispanic Americans have higher rates of PTSD (Pole et al, 2005), when compared to non-Hispanic Caucasians and Black Americans and other ethnic groups, across different walks of life[74].

This clinical observation is of great significance because Hispanics are the fastest growing minority in the U.S. (US Bureau of the Census, 2003). Furthermore, and in the context of this discussion, it is important to note that 10.2% of Hispanic Americans suffer from diabetes (NIH statistics, 2000). Hispanic women, as well as other minority women (i.e., African American, Native American [*vide supra*], and Asian/Pacific Islander) are affected by diabetes 2 to 4 times higher the rate of White American women; and 25% of Hispanic women between 65 to 74 years of age suffer from type 2 diabetes (Beckless & Thomson-Reid, 2001).

---

69 Clinically, the observation of poor or delayed reaction times, decreased exercise, decreased ability to concentrate, and inappropriate behavior seriously suggests traumatic brain injury and low IGF-1 (Ghigo et al., 2007).

70 Studies show that IGF-1 modulates vessel formation during brain development, and that IGF-1 is involved in diabetic retinal neovascularization and possibly in age-related changes in brain vasculature. (Lopez-Lopez et al., 2004).

71 Cf. footnote 57

72 These include, but are not limited to more wishful thinking (Pole et al, 2005), lower/different social support (Escobar et al, 1983; Pole et al, 2005), preater perception of racism and psycho-social discrimination (Ruef et al, 2001; Pole et al, 2005).

73 Prolo Iribarren & Chiappelli, In preparation

74 Hispanic police officers (Pole et al, 2001, 2005), Hispanic victims of ecological disasters (Perilla et al, 2002) or terrorist attacks (Galea et al, 2002), and Hispanic Vietnam veterans (Kulka et al., 1990 show greater prevalence of PTSD than other groups. In fact, in a study that examined the issue of readjustment after military service, Hispanics displayed higher rates of PTSD than African Americans or White Americans (Kulka et al., 1990).

Diabetes (i.e., type 2) and aging are intertwined, and during aging, there is a progressive loss of sensitivity to IGF-1 and also to insulin, as the levels of circulating IGF-1 decline (Breese et al., 1991). Both the choroid plexus and the brain vasculature manifest structural disturbances (*vide supra*: role of IGF-1 in angiogenesis of the CNS) that affect their barrier properties and their general functioning (Strazielle & Ghersi-Egea, 2000). Therefore, in normal aging, there is a progressive loss of IGF-1 levels, sensitivity and signaling, and a concomittant blood–brain barrier dysfunction. These events, when coupled, increase the propensity for inflammatory processes within the CNS, such as those associated with deposition of β-Amyloid protein in sDAT, and lower brain amyloid clearance (Weller et al., 2002). Indeed, even normal older individuals show moderate brain amyloidosis as compared to younger adults. It follows that a promising candidate mechanism leading to amyloid deposition in sDAT is associated with, perhaps dependent from abnormal function of the IGF family[75] (Gasparini et al, 2001). Altogether, this supports the notion that a pathogenic event in Alzheimer's disease is the development of IGF-1 resistance at the blood–brain barrier. In conclusion, IGF-1 is produced throughout life, as measured by its concentration in plasma and other body fluids, but the rate of its production decreases significantly with age: low serum IGF-1 levels relate to the cognitive deficits seen in aging and putatively in sDAT (Tei et al, 2007). Taken together these lines of evidence converge to suggest that the observed deregulation of IGF-1 in sDAT may characterize type-3 form of diabetes distinct from the metabolic syndrome[76] (de la Monte et al, 2005; Chiappelli et al, 2007; Prolo et al, 2007).

Acknowledgements

The authors thank the students, trainees, co-workers and junior faculty who have contributed to various aspects of the material presented in this chapter, as well as extended collaborations with Drs. Grasso, Moy, and Taylor over the years. The authors also express gratitude for the individuals who have served as subject in the control of patient groups for the studies we describe. Lastly, the authors acknowledge support for these studies by the UCLA School of Dentistry Faculty Pilot Grant Program; the Alzheimer's Association; the Fondazione Cassa di Risparmio di Cuneo; the Fondazione Cassa di Risparmio di Saluzzo; the Wilshire Rotary Foundation and the National Institutes of Health.

---

[75] Disrupted IGF-1 input to the brain appears to be involved in the pathogenesis of amyloidosis (Carro et al, 2002). IGF-1 appears to stimulate brain amyloid elimination by means of stimulation of neuronal amyloid release and subsequent clearance from the brain parenchyma. IGF-1 mimics insulin in its ability to stimulate amyloid protein release from neurons (Gasparini et al., 2001), and IGF-1 increases neuronal excitability and release of the amyloid (Kamenetz et al., 2003). At the same time, IGF-1 favors clearance of the amyloid protein by enhancing brain levels of transthyretin, albumin (Carro et al., 2002), apolipoprotein J, and other as yet uncharacterized amyloid carriers, which implement an ''amyloid sink'' whereby amyloid is cleared out of the brain into the circulation. In fact, Alzheimer's patients have lower CSF transthyretin levels than normal aged subjects (Serot et al., 1997). That CSF transthyretin levels are controlled by IGF-1 acting at the choroid plexus (Carro et al., 2002) suggests that IGF-1 signaling is more depressed in the choroid plexus of Alzheimer's patients than in normal subjects.

[76] Most patients with the metabolic syndrome are midle-aged or older, over-weight or obese, sedentary with excess caloric intake, and have a subclinical degree of insulin resistance.

References

Adams TE, Epa VC, Garrett TP, Ward CW. Structure and function of the type 1 insulin-like growth factor receptor. *Cell Mol Life Sci.* 2000, 57:1050-93.

Akhtar A, Gasser SM. The nuclear envelope and transcriptional control. *Nat Rev Genet.* 2007 8:507-17.

American Diabetes Association: http://diabetes.org/diabetes-statistics.jsp.

Aragay AM, Diaz P, Daban JR. Association of nucleosome core particle DNA with different histone oligomers. Transfer of histones between DNA-(H2A,H2B) and DNA-(H3,H4) complexes. *J Mol Biol.* 1988 204:141-54.

Arai F, Suda T.Regulation of hematopoietic stem cells in the osteoblastic niche. *Adv Exp Med Biol.* 2007;602:61-7.

Asturias FJ, Chung WH, Kornberg RD, Lorch Y. Structural analysis of the RSC chromatin-remodeling complex. *Proc Natl Acad Sci U S A.* 2002 99:13477-80.

Asturias FJ, Ezeokonkwo C, Kornberg RD, Lorch Y.Electron microscopic analysis of the RSC chromatin remodeling complex. Methods Enzymol. 2004, 376:48-62.

Ausió A, Abbott WD. What happens to nucleosomes during transcription? *New Comp Biochem*, 2004, 39:467-91.

Baserga R. The contradictions of the insulin-like growth factor 1 receptor. *Oncogene.* 2000, 19:5574-81

Bellacosa, A., J. R. Testa, S. P. Staal, and P. N. Tsichlis. A retroviral oncogene, akt, encoding a serine-threonine kinase containing an SH2-like region. *Science* 1991, 254:274-7.

Bernard P, Schmidt CK, Vaur S, Dheur S, Drogat J, Genier S, EkwallK, Uhlmann F, Javerzat J-P. Cell-cycle regulation of cohesin stability along fission yeast chromosomes. The EMBO J, 2008, 27:111–21.

Baserga R, Porcu P, Rubini M, Sell C. Cell cycle control by the IGF-1 receptor and its ligands. *Adv Exp Med Biol.* 1993;343:105-12.

Beckles GLA, Thompson-Reid PE. *Diabetes and women's health across the life stages: a public health perspective.* Atlanta, GA: U.S. Department of Health and Human Services, Centers for Disease Control and Prevention, National Center for Chronic Disease Prevention and Health Promotion, Division of Diabetes Translation; 2001.

Bertrand FE, Steelman LS, Chappell WH, Abrams SL, Shelton JG, White ER, Ludwig DL, McCubrey JA. Synergy between an IGF-1R antibody and Raf/MEK/ERK and PI3K/Akt/mTOR pathway inhibitors in suppressing IGF-1R-mediated growth in hematopoietic cells. *Leukemia.* 2006, 20:1254-60.

Bennett EJ & Harper JW. DNA damage: ubiquitin marks the spot. Nature Struct Mol Biol, 2008, 15:20-2.

Bondanelli M, De Marinis L, Ambrosio MR, Monesi, M, Valle D, Zatelli MC, Fusco A, Bianchi A, Farneti M, Degli Uberti E. Occurrence of Pituitary Dysfunction following Traumatic Brain Injury. *J Neurotrauma*, 2004, 21:685-96.

Braulke T. Type-2 IGF receptor: a multi-ligand binding protein. *Horm Metab Res.* 1999, 31:242-6.

Breese CR, Ingram RL, Sonntag WE. Influence of age and long-term dietary restriction on plasma insulin-like growth factor-1 (IGF-1), IGF-1 gene expression, and IGF-1 binding proteins. *J Gerontol.* 1991, 46:B180-7.

Brewin CR, Andrews B, Valentine JD. Meta-Analysis of risk factors for post-traumatic stress disorder in trauma-exposed adults. *J Consulting Clinical Psychol*, 2000, 68:748-66.

Brown J, Delaine C, Zaccheo OJ, Siebold C, Gilbert RJ, van Boxel G, Denley A, Wallace JC, Hassan AB, Forbes BE, Jones EY. Structure and functional analysis of the IGF-2/IGF-2R interaction. *EMBO J.* 2007, epub.; *The EMBO J,* 2008, 27:265–76

Callegari A, Kelly TJ. Shedding light on the DNA damage checkpoint. *Cell Cycle* 2007, 6:660-6.

Calvi LM.Osteoblastic activation in the hematopoietic stem cell niche. *Ann N Y Acad Sci.* 2006 1068:477-88.

Canalis E. Effect of insulin-like growth factor I on DNA and protein synthesis in cultured rat calvaria. *J Clin Invest*, 1980, 66:709-19.

Carro E, Trejo JL, Gomez-Isla T, LeRoith D, Torres-Aleman I. Serum insulin-like growth factor I regulates brain amyloid-beta levels. *Nat Med.* 2002, 8:1390-7.

Centers for Disease Control (CDC): http://apps.nccd.cdc.gov/ddtstrs/template/ndfs_2005.pdf

Clemmons DR. Involvement of insulin-like growth factor-I in the control of glucose homeostasis. *Cur Op Pharmacol*, 2006, 6:620-25.

Cheung P, C. David Allis CD, and Paolo Sassone-Corsi P. Signaling to Chromatin through Histone Modifications. *Cell*, 2000, 103:263-71

Chiappelli F, Immunity. In *Encyclopeadia of Stress II*, G Fink, Ed. In Chief, Academic Press, Oxford, 2007 (ISBN: 978-0-12-088503-9), Vol II, 485-93.

Chiappelli F, Liu NQ. Non-Mammalian Models of Neuroendocrine-Immune Modulation: Relevance for Research in Oral Biology and Medicine. *Int J Oral Biol* , 1999, 24:47-61

Chiappelli F, Alwan J, Prolo P, Christensen R, Fiala M, Cajulis OS, Bernard G. Neuro-immunity in Stress-Related Oral Ulcerations: A Fractal Analysis. *Front Bioscie* 2005 10:3034-41

Chiappelli F, Prolo P, Kristine D. Cajulis, Angeli A, Dovio A, Perotti P, Pautasso M, Sartori ML, Saba, L, Mussino S, Fraccalini T, Fanto' F, Manfrini E, Mocellini C, Rosso MG, Grasso E. Neuroendocrine-immunity in Patients with Alzheimer's Disease: Toward Translational Epigenetics. *Bioinformation* 2007, 2:1-4

Chini CC, Chen J. Claspin, a regulator of Chk1 in DNA replication stress pathway. *DNA Repair* (Amst.) 2005, 3:1033-7.

Cremer T, Cremer C. Chromosome territories, nuclear architecture and gene regulation in mammalian cells. *Nat Rev Genet* 2001, 2:292-301

Dagnaes-Hansen F, Duan H, Rasmussen LM, Friend KE, Flyvbjerg A. Growth hormone receptor antagonist administration inhibits growth of human colorectal carcinoma in nude mice. *Anticancer Res* 2004, 24:3735-42.

de la Monte SM, Wands JR. Review of insulin and insulin-like growth factor expression, signaling, and malfunction in the central nervous system: relevance to Alzheimer's disease. *J Alzheimers Dis*, 2005, 7:45-61.

D'Ercole AJ, Ye P, Calikoglu AS, Gutierrez-Ospina G. The role of the insulin-like growth factors in the central nervous system. *Mol Neurobiol* 1996,13:227–55.

D'Ercole AJ, Ye P, O'Kusky JR. Mutant mouse models of insulin-like growth factor actions in the central nervous system. Neuropeptides, 2002, 36:209–20.

Dillon N. Heterochromatin structure and function. *Biol Cell.* 2004 96:631-7.

Doepfner KT, Spertini O, Arcaro A. Autocrine insulin-like growth factor-I signaling promotes growth and survival of human acute myeloid leukemia cells via the phosphoinositide 3-kinase/Akt pathway. *Leukemia.* 2007 21:1921-30.

Ducasse M & Brown MA. Epigenetic aberrations and cancer. *Mol Cancer.* 2006; 5: 60. Dupont J, Pierre A, Froment P, Moreau C. The insulin-like growth factor axis in cell cycle progression. *Horm Metab Res.* 2003 35:740-50.

Edmunds J, Mahadevan L, Clayton AL. Dynamic histone H3 methylation during gene induction: HYPB/Setd2 mediates all H3K36 trimethylation. *The EMBO J,* 2008, 27: 406–20.

Escobar JI, Randolph ET, Puente G, Spiwak F, Asamen JK, Hill M, Hough RL. Posttraumatic stress disorder in Hispanic Vietnam veterans clinical phenomenology and sociocultural characteristics. *J Nervous Mental Dis,* 1983, 171:585-96.

Farzad P, Andersson, Nyberg J. Dental implant treatment in diabetic patients. *Implant Denstistry,* 2002, 11:262-7.

Fazzio TG, Tsukiyama T. Chromatin remodeling in vivo: evidence for a nucleosome sliding mechanism. *Mol Cell.* 2003 12:1333-40.

Fiorellini JP, Nevins ML. Dental implant considerations in the diabetic patient. *Periodontol,* 2000, 23:73-7.

Fernandez A, Anderson J. Nucleosome positioning determinants. *J Mol Biol.* 2007 371:649-68.

Fernández S, García-García M, Torres-Alemán I. Modulation by insulin-like growth factor I of the phosphatase PTEN in astrocytes. *Biochim Biophys Acta.* 2007 epub, Nov 13;

Freedman DA, Wu L, Levine AJ. Functions of the MDM2 oncoprotein. *Cell Mol Life Sci* 1999, 55:96-107.

Frederick TJ, Wood TL. IGF-I and FGF-2 coordinately enhance cyclin D1 and cyclin E-cdk2 association and activity to promote G1 progression in oligodendrocyte progenitor cells. *Mol Cell Neurosci.* 2004, 25:480-92.

Frederick TJ, Min J, Altieri SC, Mitchell NE, Wood TL. Synergistic induction of cyclin D1 in oligodendrocyte progenitor cells by IGF-I and FGF-2 requires differential stimulation of multiple signaling pathways. *Glia.* 2007, 55:1011-22.

Furlanetto RW, Harwell SE,Baggs RB. Effects of insulin-like growth factor receptor inhibition on human melanomas in culture and in athymic mice. *Cancer Res* 1993, 53:2522-6.

Galea S, Ahern J, Resnick H, Kilpatrick D, Bucuvalas M, Gold J, Vlahov D. Psychological sequelae of hte September 11 terrorist attacks in New York City. *NEJM,* 2002, 13:982-7.

Gasparini L, Gouras GK, Wang R, Gross RS, Beal MF, Greengard P, Xu H. Stimulation of beta-amyloid precursor protein trafficking by insulin reduces intraneuronal beta-amyloid and requires mitogen-activated protein kinase signaling. *J Neurosci.* 2001, 15;21:2561-70.

Ghigo E, Masel B, Aimaretti G, Léon-Carrión J, Casanueva FF, Dominguez-Morales MR, Elovic E, Perrone K, Stalla G, Thompson C, Urban R. Consensus guidelines on screening for hypopituitarism following traumatic brain injury. *Brain Injury*, 2007, 19:711–24.

Glotzer M. Animal cell cytokinesis. *Ann Rev Cell Biol* 2001, 17:35:390.

Guan X, Madabushi A, Chang DY, Fitzgerald ME, Shi G, Drohat AC, Lu AL. The human checkpoint sensor Rad9-Rad1-Hus1 interacts with and stimulates DNA repair enzyme TDG glycosylase. *Nucleic Acids Res*. 2007;35:6207-18.

Guttridge DC. Signaling pathways weigh in on decisions to make or break skeletal muscle. *Curr Opin Clin Nutr Metab Care*. 2004, 7:443-50.

Hailey J, Maxwell E, Koukouras K, Bishop WR, Pachter JA, Wang Y. Neutralizing anti-insulin-like growth factor receptor 1 antibodies inhibit receptor function and induce receptor degradation in tumor cells. *Mol Cancer Ther* 2002, 1:1349-53.

Hamelers IH, van Schaik RF, Sipkema J, Sussenbach JS, Steenbergh PH. Insulin-like growth factor I triggers nuclear accumulation of cyclin D1 in MCF-7S breast cancer cells. *J Biol Chem*. 2002, 277:47645-52.

Hara R & and Aziz Sancar A. The SWI/SNF Chromatin-Remodeling Factor Stimulates Repair by Human Excision Nuclease in the Mononucleosome Core Particle. *Mol Cell Biol*. 2002, 22: 6779–6787.

Heng HH, Goetze S, Ye CJ, Liu G, Stevens JB, Bremer SW, Wykes SM, Bode J, Krawetz SA. Chromatin loops are selectively anchored using scaffold/matrix-attachment regions. *J Cell Sci*. 2004 117:999-1008.

Hodge RD, D'Ercole AJ, O'Kusky JR. Insulin-like growth factor-I accelerates the cell cycle by decreasing G1 phase length and increases cell cycle reentry in the embryonic cerebral cortex. *J Neurosci* 2004, 24: 10201–10.

Hwa V, Oh Y, Rosenfeld RG. Insulin-like growth factor binding proteins: a proposed superfamily. *Acta Paediatr Suppl*. 1999 88:37-45.

Kalebic T, Tsokos M, Helman LJ. In vivo treatment with antibody against IGF-1 receptor suppresses growth of human rhabdomyosarcoma and down-regulates p34cdc2. *Cancer Res* 1994, 54:5531-4.

Kalluri HS, Vemuganti R, Dempsey RJ. Mechanism of insulin-like growth factor I-mediated proliferation of adult neural progenitor cells: role of Akt. *Eur J Neurosci*. 2007, 25:1041-8.

Kamenetz F, Tomita T, Hsieh H, Seabrook G, Borchelt D, Iwatsubo T, Sisodia S, Malinow R. APP processing and synaptic function. *Neuron*. 2003, 37:925-37.

Karas M, Koval AP, Zick Y, LeRoith D. The insulin-like growth factor I receptor-induced interaction of insulin receptor substrate-4 and Crk-II. *Endocrinol*. 2001, 142:1835-40.

Karp G. *Cell and Molecular Biology: Concepts and Experiments, 2005*, 4th Edition. Hoboken, NJ: John Wiley and Sons. Kim SM, Kumagai A, Lee J, Dunphy WG. Phosphorylation of Chk1 by ATM- and Rad3-related (ATR) in Xenopus egg extracts requires binding of ATRIP to ATR but not the stable DNA-binding or coiled-coil domains of ATRIP. *J Biol Chem*. 2005, 280:38355-64.

Kirmizis A, Santos-Rosa H, Penkett CJ, Singer MA, Vermeulen M, Mann M, Bähler J, Green RD, Kouzarides T. Arginine methylation at histone H3R2 controls deposition of H3K4 trimethylation. *Nature*. 2007 449:928-32.

Klammt J, Barnikol-Oettler A, Kiess W. Mutational analysis of the interaction between insulin receptor and IGF-I receptor with c-Crk and Crk-L in a yeast two-hybrid system. *Biochem Biophys Res Commun.* 2004, 325:183-90.

Knudsen ES, Wang JY. Differential regulation of retinoblastoma protein function by specific Cdk phosphorylation sites. *J. Biol. Chem.* 1996. 271: 8313-20.

Kodama Y, Baxter RC, Martin JL. Insulin-like growth factor-I inhibits cell growth in the a549 non-small lung cancer cell line. *Am J Respir Cell Mol Biol.* 2002, 27:336-44.

Kulka RA, Schlesenger WE, Fairbank JA, Hough RL, Jordan BK, Marmar CR, Weiss DS. *Trauma and the Vietnam war generation: Report of findings from the National Vietnam Veterans Readjustment Study.* New York: Brunner/Mazel, 1990.

Kulkarni RN. Receptors for insulin and insulin-like growth factor-1 and insulin receptor substrate-1 mediate pathways that regulate islet function. *Biochem Soc Trans.* 2002, 30:317-22.

Lai A, Sarcevic B, Prall OW, Sutherland RL. Insulin/insulin-like growth factor-I and estrogen cooperate to stimulate cyclin E-Cdk2 activation and cell Cycle progression in MCF-7 breast cancer cells through differential regulation of cyclin E and p21(WAF1/Cip1). *J Biol Chem.* 2001 276:25823-33.

Leahy M, Lyons A, Krause D, O'Connor R. Impaired Shc, Ras, and MAPK activation but normal Akt activation in FL5.12 cells expressing an insulin-like growth factor I receptor mutated at tyrosines 1250 and 1251. *J Biol Chem.* 2004, 279:18306-13.

Leof EB, Wharton W, Van Wyk JJ, Pledger WJ. Epidermal growth factor and somatomedin-C regulate G1 progression in competent balb-C 3T3 cells. *Exp Cell Res* 1982,141:107–15.

Leslie NR, Downes CP. PTEN function: how normal cells control it and tumour cells lose it. *Biochem J* 2004, 382:1–11.

Levin ME, Boisseau VC, Avioli LV. Effects of diabetes in bone mass in juvenile and adult onset-diabetes. *New Engl J Med*, 1976, 294:241-5.

Lew DJ & Reed SI. Morphogenesis in the yeast cell cycle: regulation by Cdc28 and cyclins. *J Cell Biol* 1993, 120:1305–20

Li JY, Pu MT, Hirasawa R, Li BZ, Huang YN, Zeng R, Jing NH, Chen T, Li E, Sasaki H, Xu GL. Synergistic function of DNA methyltransferases Dnmt3a and Dnmt3b in the methylation of Oct4 and Nanog. *Mol Cell Biol.* 2007, 27:8748-59.

Locatto ME, Abranzon H, Caferra D, Fernandez MC, Alloatti R, Puche RC. Growth and development of bone mass in untreated alloxan diabetic rats. Effects of collagen glycosylation and parathyroid activity on bone turnover. *Bone Miner.* 1993, 23:129-44

Longtine MS, DeMarini DJ, Valencik ML, Al-Awar OS, Fares H, De Virgilio C, Pringle,JR. The septins: roles in cytokinesis and other processes *Curr. Opin. Cell Biol.* 1996, 8:106-19

Lopez-Lopez C, LeRoith D, Torres-Aleman J. Insulin-like growth factor I is required for vessel remodeling in the adult brain. *PNAS*, 2004, 101:9833-8.

López-Rodas G, Brosch G, Georgieva EI, Sendra R, Franco L, Loidl P. Histone deacetylase : A key enzyme for the binding of regulatory proteins to chromatin. *FEBS Letters*, 1993, 317:175-80.

Ma T, Van Tine BA, Wei Y, Garrett MD, Nelson D, Adams PD, Wang J, Qin J, Chow LT, Harper JW. Cell cycle-regulated phosphorylation of p220(NPAT) by cyclin E/Cdk2 in Cajal bodies promotes histone gene transcription. *Genes Dev*. 2000, 14:2298-313.

Matsukage A, Hirose F, Yoo MA, Yamaguchi M. The DRE/DREF transcriptional regulatory system: a master key for cell proliferation. *Biochim Biophys Acta*. 2007 Dec 4; epub.

Mauras N, Haymond M. Are the metabolic effects of GH and IGF-I separable? Growth Hormone & IGF Research 2005, 15:19–27.

Metzger E, Yin N, Wissmann M, Kunowska N, , Fischer K, Friedrichs N, 2, Patnaik D, Higgins JMG, Potier N, Scheidtmann K-H, Buettner R, Schüle R. Phosphorylation of histone H3 at threonine 11 establishes a novel chromatin mark for transcriptional regulation. *Nature Cell Biol*, 2007, 10: 53-60.

Mohan S, Nakao Y, Honda Y, Landale E, Leser U, Dony C, Lang K, Baylink DJ. Studies on the mechanisms by which insulin-like growth factor (IGF) binding protein-4 (IGFBP-4) and IGFBP-5 modulate IGF actions in bone cells. *J Biol Chem*. 1995, 270:20424-31.

Moradi DR, Moy PK, Chiappelli F. Evidence-based research in alternative protocols to dental implantology: a closer look at publication bias. *J Calif Dent Assoc*. 2006, 34:877-86

Mukherjee A, Rotwein P. Insulin-like growth factor binding protein-5 in osteogenesis: facilitator or inhibitor? Growth Horm IGF Res. 2007, 17:179-85.

Murr R, Vaissière T, Sawan C, Shukla V, Herceg Z. Orchestration of chromatin-based processes: mind the TRRAP. *Oncogene*. 2007, 26:5358-72.

Nakayama K-I, Nakayama K. Cip/Kip cyclin-dependent kinase inhibitors: brakes of the cell cycle engine during development. *BioEssays* 1998, 20:1020-29.

Nissen MS, Langan TA, Reeves R. Phosphorylation by cdc2 kinase modulates DNA binding activity of high mobility group I nonhistone chromatin protein. *J Biol Chem*. 1991 266:19945-52.

O'Connor R. Regulation of IGF-I receptor signaling in tumor cells. *Horm Metab Res*. 2003, 35:771-7.

Olins DE, Olins AL. Chromatin history: our view from the bridge *Nat Rev Mol Cell Biol*. 2003 4:809-14.

Pardee AB. A restriction point for control of normal animal cell proliferation, *Proc Natl Acad Sci USA* 1974, 71:1286-90.

Perilla JL, Norris FH, Lavizzo EA. Ethnicity, culture and disaster response: Identifying and explaining ethnic differences in PTSD six months after Hurricane Andrew. *J Social Clinical Psychol*, 2002, 21:20-45,

Pavelić J, Matijević T, Knezević J. Biological & physiological aspects of action of insulin-like growth factor peptide family. *Indian J Med Res*. 2007, 125:511-22.

Paweletz N. W Flemming, Pioneer of Mitosis Research. *Nature Rev Mol Cell Biol* 2001, 2:72-5.

Pole N, Best SR, Weiss DS, Metzler T, Libermann AM, Fagan J. Marmar CR. Effects of gender and ethnicity on duty-related posttraumatic stress symptoms among urban police officers. *J Nervous Mental Dis*, 2001, 189:442-8.

Pole N, Best SR, Metzler T. Marmar. Why are Hispanics at greater risk for PTSD?. *Cultural Diversity Ethnic Minority Psychol*, 2005, 11:144-61.

Popken GJ, Hodge RD, Ye P, Zhang J, Ng W, O'Kusky JR, D'Ercole AJ. In vivo effects of insulin-like growth factor-I (IGF-I) on prenatal and early postnatal development of the central nervous system. *Eur J Neurosci* 2004,19:2056-68

Prolo P, Chiappelli F, Angeli A, Dovio A, Sartori ML, Fantó F, Neagos N, Manfrini E. Putative NeuroImmune Mechanisms in Alzheimer's Disease: Modulation by Cholinergic Anti-Inflammatory Reflex (CAIR*). Int J Integrative Biol*, 2007, 1:88-95

Purvis A. & Singleton MR. Insights into kinetochore–DNA interactions from the structure of Cep3. *EMBO*, 2008, 9:56-62.

Quina AS, Buschbeck M, Di Croce L. Chromatin structure and epigenetics. *Biochem Pharmacol.* 2006 72:1563-9.

Reeves R, Nissen MS. Cell cycle regulation and functions of HMG-I(Y). *Prog Cell Cycle Res.* 1995, 1:339-49.

Reiss K, D'Ambrosio C, Tu X, Tu C, Baserga R. Inhibition of tumor growth by a dominant negative mutant of the insulin-like growth factor I receptor with a bystander effect. *Clin Cancer Res* 1998, 4:2647-55.

Resnicoff M, Coppola D, Sell C, Rubin R, Ferrone S, Baserga R. Growth inhibition of human melanoma cells in nude mice by antisense strategies to the type 1 insulin-like growth factor receptor. *Cancer Res* 1994, 54:4848-50.

Rice JC, Allis CD. Histone methylation versus histone acetylation: new insights into epigenetic regulation. *Curr Opin Cell Biol.* 2001 13:263-73.

Rodrigues-Martins A, Riparbelli, Callaini MG, Glover DM, Bettencourt-Dias, M; Revisiting the Role of the Mother Centriole in Centriole Biogenesis; *Science* 2007, 316:1046-50.

Rodríguez-Bravo V, Guaita-Esteruelas S, Florensa R, et al. Chk1- and claspin-dependent but ATR/ATM- and Rad17-independent DNA replication checkpoint response in HeLa cells.. *Cancer Res.* 2007. 66:8672-9

Roloff TC, Nuber UA. Chromatin, epigenetics and stem cells. *Eur J Cell Biol.* 2005 84:123-35.

Ruef AM, Litz BT, Schlenger WE. Hispanic ethnicity and risk for combat-related posttraumatic stress disorder. *Cultural Diversity Ethnic Minority Psychol*, 2000, 6:235-51.

Sampath SC, Marazzi I, Yap KL, Sampath SC, Krutchinsky AN, Mecklenbräuker I, Viale A, Rudensky E, Zhou MM, Chait BT, Tarakhovsky A. Methylation of a histone mimic within the histone methyltransferase G9a regulates protein complex assembly. *Mol Cell.* 2007 27:596-608.

Samani AA, Chevet E, Fallavollita L, Galipeau J, Brodt P. Loss of tumorigenicity and metastatic potential in carcinoma cells expressing the extracellular domain of the type 1 insulin-like growth factor receptor. *Cancer Res* 2004, 64:3380-5.

Segev Y, Eshet R, Yakir O, Haim N, Phillip M, Landau D. Systemic and renal growth hormone–IGF1 axis involvement in a mouse model of type 2 diabetes. *Diabetologia*, 2007, 50:1327–34.

Serot JM, Christmann D, Dubost T, Couturier M. Cerebrospinal fluid transthyretin: aging and late onset Alzheimer's disease. *J Neurol Neurosurg Psychiat.* 1997, 63:506-8.

Sexton T, Schober H, Fraser P, Gasser SM. Gene regulation through nuclear organization. *Nat Struct Mol Biol.* 14:1049-1055.

Sitar T, Popowicz GM, Siwanowicz I, Huber R, Holak TA. Structural basis for the inhibition of insulin-like growth factors by insulin-like growth factor-binding proteins. *Proc Natl Acad Sci U S A.* 2006 103:13028-33

Steeg PS, Bevilacqua G, Kopper L, Thorgeirsson UP, Talmadge JE, Liotta LA, et al. Evidence for a novel gene associated with low tumor metastatic potential. *J Natl Cancer Inst* 1988;80:200–4.

Stull M, Richert M, Loladze A, Wood T. Requirement for IGF-I in epidermal growth factor cell cycle progression of mammary epithelial cells. *Endocrinol* 2002,143:1872-9.

Strazielle N, Ghersi-Egea JF. Choroid plexus in the central nervous system: biology and physiopathology. *J Neuropathol Exp Neurol.* 2000, 59:561-74.

Tei E, Yamamoto H, Watanabe T, Miyazaki A, Nakadate T, Kato N, Mimura M. Use of serum insulin-like growth factor-I levels to predict psychiatric non-response to donepezil in patients with Alzheimer's disease. *Growth Horm IGF Res.* 2007 Aug 20; Epub

Torres-Alaman I. Serum growth factors and neuroprotective surveillance: focus on IGF-1. *Mol Neurobiol.* 2000, 21:153-60.

Turjanski AG, Vaqué JP, Gutkind JS. MAP kinases and the control of nuclear events. *Oncogene.* 2007, 26:3240-53.

Turner B .*Chromatin and Gene Regulation: Mechanisms in Epigenetics.* Blackwell Pub, 2002.

Uhlmann F. Lottspeich F, Nasmyth K. Sister-chromatid Separation at anaphase-onset is promoted by cleavage of Scc1. *Nature* 1999, 400:37-42.

U.S. Bureau of the Census. (2003). Census Bureau releases population stimates by age, sex, race and Hispanic origin. http://www.census.gov/Press-Release/www/2003/cb03-16.html

US National diabetes statistics fact sheet: general information and national estimates on diabetes in the U S, 2000. NIH National Institute of Diabetes and Digestive and Kidney Diseases. http://www.niddk.nih.gov/health/diabetes/pubs/dmstats/dmstats.htm#10.

Valero AM, Garcia JCF, Ballester AH, Rueda CL. Effects of diabetes on the osseointegration of dental implants. *Med Oral Patol Oral Cir Bucal* 12: E38-43, 2007.

Van Beekum O, Kalkhoven E. berrant forms of histone acetyltransferases in human disease. *Subcell Biochem.* 2007, 41:233-62.

van Driel R, Fransz PF, Verschure PJ. The eukaryotic genome: a system regulated at different hierarchical levels. *J Cell Sci.* 2003 116:4067-75.

Weller RO, Massey A, Kuo YM, Roher AE. Cerebral amyloid angiopathy: accumulation of A beta in interstitial fluid drainage pathways in Alzheimer's disease. *Ann N Y Acad Sci.* 2000, 903:110-7.

Winter S, Simboeck E, Fischle W, Zupkovitz G, Dohnal I, Mechtler K, Ammerer, G, and Seiser C. 14-3-3 Proteins recognize a histone code at histone H3 and are required for transcriptional activation. *The EMBO J*, 2008, 27, 88–99

Whitehouse I, Tsukiyama T. Antagonistic forces that position nucleosomes in vivo. *Nat Struct Mol Biol.* 2006 13:633-40.

Woods KA, Savage MO 1999 IGF-Deficiency. In: RG Rosenfeld, CT Roberts, eds. *IGF in Health and Disease*, Totowa, USA The Humana Press Inc; 1999, pp. 651-74

Wua J Grunstein M. 25 years after the nucleosome model: chromatin modifications. *Trends Biochem Sciences*, 2000, 25:619-23

Wullschleger S, Loewith R, Hall M. TOR signaling in growth and metabolism. *Cell* 2006, 124: 471-84.

Yarden RI, Pardo-Reoyo S, Sgagias M, Cowan KH, Brody LC. BRCA1 regulates the G2/M checkpoint by activating Chk1 kinase upon DNA damage. Nat Genet 2002, 30:285-9.

Ye P, Joseph D'Ercole AJ. Insulin-Like Growth factor actions during development of neural stem cells and progenitors in the central nervous system. *J Neuroscie Res* 2006, 83:1–6

Yokota T, Shimokawa H, Shibata S, Itoh K, Baba Y, Ohya K, Ohyama K & Suzuki S, Insulin-like growth factor-1 regulates apoptosis in condylar cartilage. *J Dental Res* 2008, 87:159-63.

Yoshida BA, Sokoloff M, Welch DR, Rinker-Schaeffer CW. 2000. Metastasis-suppressor genes: a review and perspective on an emerging field. *J Nat Cancer Inst* 2000, 92:1717-30.

Zeng X, Yee D. Insulin-like growth factors and breast cancer therapy. *Adv Exp Med Biol.* 2007, 608:101-12.

Zhang R, Adams PD. Heterochromatin and its relationship to cell senescence and cancer therapy. *Cell Cycle*. 2007 6:784-9.

Zia F, Jacobs S, Kull F, Jr., Cuttitta F, Mulshine JL, Moody TW. Monoclonal antibody alpha IR-3 inhibits non-small cell lung cancer growth in vitro and in vivo. *J Cell Biochem* 1996, 24S:269-75.

Zou L, Elledge SJ. Sensing DNA damage through ATRIP recognition of RPA-ssDNA complexes. *Science*. 2003 300:1542-8

In: Cell Cycle Control: New Research
Editors: N. H. Leroy, N. T. Fournier
ISBN: 978-1-60456-796-0

Chapter II

The Beginning of the MicroRNA Era for Cell Cycle Control

*Kenichi Yoshida**
Department of Life Sciences, Meiji University School of Agriculture, 1-1-1 Higashimita, Tama-ku, Kawasaki, Kanagawa 214-8571, Japan.

Abstract

Cell cycle control is a fundamental process for maintenance of the genome stability. An aberrant cell cycle eventually transforms the cell into a malignant phenotype. Among cell cycle controllers, p53, c-Myc, and E2F1 are pivotal transcription factors, as they determine the cell fate during cell damage. MicroRNAs (miRNAs) are emergent members of small and non protein-coding RNA molecules of about 22 nucleotides in length that repress translation and promote mRNA degradation by sequence-specific interaction with mRNA. More than four hundred miRNAs have been identified so far in humans (http://microrna.sanger.ac.uk). Deregulation of miRNAs expression has been shown in human pathogenesis especially for many cancers; however, the overall picture as to what mechanisms deregulate miRNAs during tumor formation remains unclear. To better understand the molecular etiology of cancer, we need to take a count of miRNAs. In this review, we focus on recent advances concerning the functional link between miRNAs and transcriptional factors relevant to the cell cycle control and carcinogenesis.

Introduction

When we refer to miRNA deregulation in human cancers, we have to consider two axes of regulation. One is what kind of transcription factor regulates the expression of microRNA

* Correspondence concerning this article should be addressed to: Kenichi Yoshida, D.V.M., Ph.D. Associate Professor; Department of Life Sciences, Meiji University School of Agriculture, 1-1-1 Higashimita, Tama-ku, Kawasaki, Kanagawa 214-8571, Japan. Tel. & Fax.: +81-44-934-7107, e-mail: yoshida@isc.meiji.ac.jp.

(miRNA). In this context, the transcription factor could be an oncogene or antioncogene. Little is known about the transcription factors that regulate the expression of miRNA. It has been predicted that c-Myb, NF-Y, Sp-1, MTF-1, and AP-2alpha are master-regulators of miRNA expression [1]. In human cancers, miRNA levels tend to decrease, indicating that miRNA may have an intrinsic function in tumor suppression [2-5]. It has been revealed that almost half of the miRNAs' putative promoters contain potential p53-binding sites [6]. The other is what kind of target genes or mRNAs are post-transcriptionally regulated by miRNA. In this context, miRNA itself behaves like an oncogene or antioncogene. Those two important axes can be the basis for fabricating feedback/feed forward loops, which eventually contribute to the architecture of the combined transcriptional-post transcriptional regulatory network. We review the recent progress on miRNA during tumorigenesis, and further focus on the relationship between miRNA and important cell cycle regulators such as p53, c-Myc, and E2F transcription factors.

1. miRNA as an Oncogene or Antioncogene

Many miRNAs have been implicated in the regulation of cell growth, differentiation, and apoptosis [7-11]. They have also been thought to be involved in the regulation of the developmental and immune systems [12-19]. They are found in many plants, animals, and DNA viruses, suggesting their evolutional conservation [20-24]. Mechanism of miRNAs-mediated repression of gene expression has been elucidated [25-29]. Of note, mRNA is targeted for post-transcriptional repression by miRNA through pairing of the 3' untranslated region (UTR) sequences matches to bases at the 5' end of miRNA [30-32]. The study of miRNA has broadened our view that cancer-associated miRNAs can act as both tumor suppressors or oncogenes and are involved in the regulation of tumorigenesis [33-52]. In this concept, miRNAs can act like tumor suppressors or oncogenes by regulating transcription factors whose function is also categorized into oncogenes or antioncogenes. Those feedback/feed forward loops are set up for the architecture of the combined transcriptional-post transcriptional regulatory network [53-56].

2. miRNA Expression Profiling

It has been suggested that miRNA expression and its regulation can be applied to diagnostic and prognostic molecular markers as well as for eventual therapeutic intervention, by analyzing the miRNA expression in human tumors using an miRNA expression profiling method [57-73]. For example, preferential expression changes of miRNAs in cancerous cells were demonstrated for B cell chronic lymphocytic leukemia [74], glioblastoma [75], leukemia cell lines [76], breast and prostate cancer [77], cholangiocyte [78], lung cancer [79], hepatocellular carcinoma [80], breast tumor [81, 82], soft tissue sarcoma [83], prostate cancer [84,85], epithelial ovarian cancer [86], leiomyoma [87], colon cancer [88,89], colon cancer treated with 5-fluorouracil [90], cervical carcinoma [91], thyroid carcinoma [92-94], diffuse large B cell lymphoma [95], NCI-60 cancer cell panel [96,97], head and neck cancer cell lines [98], pancreatic cancer [99-101], acute myeloid leukemia [102], chronic lymphocytic leukemia [103], acute promyelocytic leukemia exposed to all-trans-retinoic acid [104], uterine leiomyoma [105], pancreatic ductal adenocarcinoma [106], pituitary adenoma [107], and germ cell tumors [108]. miRNA-regulated miRNA expression changes therefore need to be considered [109].

3. miRNA and Cancer Susceptibility

Although the expression change for miRNA is an important issue in understanding the properties of human cancers, there are some points to which attention is required. For example, single-nucleotide polymorphisms located at the miRNA-binding sites of the 3' UTR of mRNA are likely to affect the expression of the miRNA targets and are potentially associated with cancers [110]. In addition, chromosomal abnormalities, amplifications, and modifications of miRNA loci have been evoking attentions for tumorigenesis [111,112]. In humans and the mouse model, the chromosomal location of miRNA has been identified near the cancer susceptibility loci that influence the development of solid tumors [113-116], suggesting that mechanisms, apart from miRNA-mediated mRNA expression, should be considered for tumorigenesis. Sequence analyses of miRNA genes derived from chronic lymphocytic leukemia patients revealed considerable germ-line or somatic mutations [117]. In contrast, sequence variations in miRNA genes were not associated with hepatocarcinogenesis, whereas mutations in protein-coding genes played a key role during carcinogenesis [118]. Epigenetic alterations in cancer frequently harbor DNA hypermethylation of tumor suppressor genes and disruption of the histone modification patterns. In the case of miRNA, epigenetic aberrations could potentially be the case for cancerous cell proliferation [119-122]. Aberrant hypermethylation was shown as an epigenetic inactivation for miRNAs including miR-9-1 in human breast cancer [123].

4. Case Study for miRNAs

4.1. miR-15 and miR-16

miR-15 and miR-16 were first reported for their deregulation in cancerous tissues [124]. Half of the examined B cell chronic lymphocytic leukemia specimens lacked the segments including both miRNAs, both genes were seen to be down-regulated in the majority of B cell chronic lymphocytic leukemia, and this was also the case in pituitary adenomas [124-126]. miR-15 and miR-16 have been shown to induce apoptosis by targeting Bcl-2 in a leukemic cell line [125]. The suppressive role of miR-16 against Bcl-2 was also reported by another group [127]. Target genes regulated by miR-16 have been shown to be mainly involved in the regulation of cell cycle progression [128].

4.2. miR-17-92 Cluster

The miR-17-92 cluster presents the most intriguing evidence regarding the involvement of miRNA directly into the transcriptional network regulated by c-Myc and E2Fs. The miR-17-92 cluster together with c-Myc expression accelerated tumor development in a mouse B-cell lymphoma model, and is characterized by the absence of frequent apoptosis associated with c-Myc-induced lymphomas [129]. The miR-17-92 polycistronic cluster, which contains seven mature miRNAs (miR-17-5p, miR-17-3p, miR-18a, miR-19a, miR-20a, miR-19b, and miR-92-1) and resides in intron 3 of the C13orf25 gene at 13q31.3, was identified as the transcriptional target of c-Myc, and mir17-5p, and miR-20a in turn down-regulated the E2F1 protein [130]. The miR-17-92 cluster has other oncogenic properties such as promotion of vascularization [131].

Other than E2F1, miR-20a can suppress the translation of the E2F2 and E2F3 mRNAs [132]. It has also been reported that E2F1, E2F2, and E2F3 directly bound to the promoter of the mir-17-92 cluster, and resulted in the activation of mir-17-92 cluster transcription [132]. These results suggest that activator E2Fs and at least miR-20a within the mir-17-92 cluster form a feedback loop to maintain the physiological protein level of E2Fs to regulate proper cell proliferation. Among activator E2Fs (E2F1, E2F2, and E2F3), E2F1 uniquely functions in both cell proliferation and apoptosis. As mentioned above, the miR-17-92 cluster seems to suppress apoptosis; the main target of the miR-17-92 cluster could be elucidated as the apoptotic function of E2F1. Another group reported that E2F3 primarily binds to the promoter of the miR-17-92 cluster [133]. This report favors the idea that miR-17-92 promotes cell proliferation by down-regulating the apoptotic activity of E2F1 and shifting the E2F transcriptional balance toward the E2F3 transcriptional activity solely to promote cell proliferation. Rbl2, which belongs to the Rb family, was shown to be a target of miR-17-5p [134]. This is somewhat paradoxical, because Rb family proteins bind to E2Fs and repress the activity of E2Fs. The activity of activator E2Fs is tightly controlled by the Rb family of proteins, therefore, it should be checked whether Rb proteins are regulated by the mir-17-92 cluster or not. Rb was confirmed experimentally as targets of miR-106a [59]. Both miR-106a and miR-363 were highly over-expressed in T lymphoma [135,136]. As well as the miR-17-92 cluster, miR-106a was regulated by n-Myc and c-Myc. n-Myc is a member of the Myc family involved in neoplasia [137].

The miR-17-92 cluster is transcriptionally regulated by oncogenic c-Myc and E2F1 strongly suggesting that miR-17-92 potentially acts as a brake on the synergistic effects of c-Myc and E2F1, which is derived from the fact that the miR-17-92 cluster can repress the translation of E2F1. In this regard, the miR-17-92 cluster could be expected to be a tumor suppressor; however, it behaves like a promoter of proliferation and an inhibitor of apoptosis. Physiologically, inhibition of E2F1-induced apoptosis by miR-17-92 is regarded as an essential step for spermatocytes [138]. As shown in B-cell lymphoma, overexpression or amplification of the miR-17-92 cluster has also been involved in the development of lung cancers [139,140]. In addition, the genomic amplification has been reported of both c-Myc and miR-17-92 in lymphoma [141,142]. Moreover, the oncogenic role of miR-17-92 was also supported by the fact that inhibition of both miR-17-5p and miR-20a resulted in induction of apoptosis selectively in lung cancer cells over-expressing miR-17-92 [143]. The miRNA-17-

5p was also reported as a translational inhibitor of mRNA of AIB1 (amplified in breast cancer 1) [144].

4.3. miR-21

miR-21 was found to be expressed at high levels in glioblastomas and hepatocellular cancer. Down-regulation of miR-21 in cultured glioblastoma cells is said to trigger activation of caspases and increased apoptotic cell death [145]. Inhibition of miR-21 in hepatocellular cancer cells increased expression of the phosphatase and tensin homolog (PTEN) tumor suppressor [146]. miR-21 was highly over-expressed in breast tumors, and modulated tumorigenesis partly through regulation of Bcl-2 [147]. miR-21 induction by interleukin-6 (IL-6) was strictly Stat3-dependent [148]. Stat3 is implicated in the pathogenesis of many malignancies. miR-21 has also been shown to target tumor suppressor tropomyosin 1 (TPM1) [149]. In conclusion, taking all the above data into consideration, miR-21 can be considered as an oncogene. An miR-21 antagonist could therefore be useful for therapeutic intervention [150].

4.4. miR-34

The miR-34 cluster has been reported as a direct transcriptional target of p53 [151-155]. Comparing miRNA expression between wild-type and p53-deficient cells identified miR-34a~c as p53 target miRNAs, and the ectopic expression of miR-34 induced cell cycle arrest [153]. Ectopic miR-34a induced apoptosis and cell cycle arrest in the G1 phase of the cell cycle [155]. Loss of miR-34 can impair p53-mediated apoptosis [153]. Additionally, significant down-regulation of miR-34b and miR-34c was identified when p53 was impaired in primary mouse ovarian surface epithelium cells [152]. Down-regulation of miRNA-34b and miRNA-34c was also detected in p53 null human ovarian carcinoma cells [152]. miR-34a was absent in pancreatic cancer cells [151]. The miR-34 family regulates the Bcl-2 protein level, suggesting that inactivation of the miR-34 family might contribute to certain cancers [156]. Importantly, miR-34a is generally expressed at lower levels in neuroblastoma and targets the E2F3 protein [157].

4.5. let-7

let-7 was first demonstrated as a deregulated miRNA in human cancer [158]. Reduction of let-7 expression was manifested in lung cancer and overexpression of let-7 in A549 lung adenocarcinoma cell line inhibited lung cancer cell growth. It has been shown that Ras is a target of let-7 miRNA [159]. let-7 miRNA also down-regulates c-Myc [127]. In Burkitt's lymphoma, decreased Myc mRNA resulted in the induction of let-7a, let-7b, and miR-98, whereas mir-17-5p decreased. Overexpression of let-7a decreased the expression of Myc and its target genes and anti-proliferation in lymphoma cells [160]. let-7 miRNA expression was reduced in acinar adenocarcinoma and mucinous bronchioloalveolar carcinoma but was not implicated in the prognosis [161].

Disruption of the let-7 miRNA-targeted repression of HMG2A (high mobility group A2) by chromosomal translocations, which has been associated with human tumors, promoted oncogenic transformation [162]. HMGA2 is known as an oncogenic factor in a variety of tumors. Expression of let-7 and let-7 target HMGA2 has been considered as a better marker for less differentiated ovarian cancer [163]. Overexpression of let-7 reduced HMGA2 and cell proliferation in a lung cancer cell line [164]. This suggests that let-7 miRNA could function as a tumor suppressor in lung cells [165]. Expression of HMGA2 in head and neck squamous cell carcinoma cells is also regulated partly by another miRNA, miRNA-98 [166].

Hypomethylation of let-7a-3 was shown to correlate with a human lung cancerous cell phenotype [167]. let-7 was also demonstrated as a critical miRNA for PPARalpha agonist-induced liver proliferation and tumorigenesis [168]. IL-6 is known to contribute to tumor growth as described above. Over-expression of IL-6 enhanced tumor growth and inhibited apoptosis, which accompanied let-7a up-regulation [169]. This is a discrepancy in the concept that let-7 could potentially be an antioncogenic miRNA. On the other hand, overexpression of IL-6 resulted in reduction of miR-370 expression and supported the cell xenografts of human malignant cholangiocytes [170].

4.6. Other miRNAs

In this section, we introduce other miRNAs that have been strongly implicated in tumorigenesis. miR-10b, whose expression is induced by the transcription factor Twist, has been shown to be highly expressed in metastatic breast cancer cells and positively regulates cell migration and invasion [171]. The miR-29 clusters target DNA methyltransferase (DNMT)-3A and -3B in non-small-cell lung cancer by restoring the normal patterns of DNA methylation, inducing re-expression of methylation-silenced tumor suppressor genes, and inhibiting tumorigenesis [172]. miR-29 and miR-181 target Tcl1 expression in B-cell chronic lymphocytic leukemia [173]. miR-29b increased Mcl-1, an anti-apoptotic Bcl-2 family member, and reduced TRAIL-mediated apoptosis [174]. During the miRNA microarray screening, cyclin G1 was found as a target of miR-122a. miR-122a is considered as liver-specific and its down-regulation was detected in almost all hepatocellular carcinomas examined [175,176].

miR-125a and miR-125b induced mouse P19 embryonal carcinoma differentiation into neurons by targeting lin-28 mRNA [177]. ErbB2 and EreB3 oncogenes were found as targets of miR-125a and miR-125b [178]. miR-125a and miR-125b together with miR-9 were found to be down-modulated in primary neuroblastoma tumors, and repressed the truncated isoform of the neurotrophin receptor tropomyosin-related kinase C [179]. miR-127 seems to act as a tumor suppressor, potentially targeting Bcl-6 [180]. miRNA-143 and -145 are implicated in the progression of B-cell malignancies [181], and can potentially be identified in colon cancer [182]. miR-150 has been shown to regulate c-Myb expression, a transcription factor controlling lymphocyte development [183].

High levels of expression of primary miRNA BIC (pri-miR-155) have been reported in B cell lymphomas, especially in Hodgkin's lymphoma, and lack of expression in most non-Hodgkin's lymphoma including Burkitt's lymphoma [184-186]. miR155 transgenic mice exhibit B cell malignancy probably accompanying secondary genetic changes for full transformation [187]. Deregulation of miR-200c could be involved in epithelial-to-mesenchymal transition of the cells, indicating that miR-200c could play a significant role in the initiation of an invasive phenotype, through regulating its target E-cadherin [188]. miRNA-206 and estrogen receptor alpha have been linked as they formed a mutually inhibitory feedback loop [189].

Induction of miR-221 by n-Myc has been shown in neuroblastomas [190]. miR-221 and miR-222, encoded in tandem on chromosome X, were found to be highly expressed in aggressive prostate carcinomas when compared with slowly growing carcinomas. Overexpression of miR-221 and miR-222 suppressed the p27Kip1 protein, a cell cycle inhibitor and tumor suppressor, thus regarded as oncogenes [191]. Moreover, up-regulation of miR-221 and miR-222 has been detected in human thyroid papillary carcinomas and glioblastoma, and miR-221 and miR-222 again decreased p27Kip1, indicating that deregulated expression of miR-221 and 222 promoted cancerous growth by inhibiting the expression of p27Kip1 [192-194].

Finally, MUC1 and galectin-3 are implicated in the pathogenesis of human cancers, and both proteins function as part of a miR-322-dependent regulatory loop [195]. The miR-371~373 cluster, potential oncogenes in the presence of wild-type p53, is involved in the regulation of stem cell differentiation, and is involved in the development of human testicular germ cell tumors [196,197].

Conclusion

p53, c-Myc, and E2F1 oncogenic signaling share a common regulatory network, consisting of miRNA, mRNA, and protein levels, which governs cell fate [198-200]. Because c-Myc and E2F1 are known to activate each other [199,201] and both protein induce apoptosis as well as cell proliferation [202], the findings that the c-Myc (also E2F1)-regulated miR-17-92 cluster suppresses E2F1 (also E2F2 and E2F3) activity suggest a more complex picture of regulation but are worthy of note when thinking about the molecular mechanism of cancer development (Table). It has been shown that coordinated transcriptional and miRNA-mediated regulation could enhance the robustness of gene regulation in mammalian genomes [203]. In addition to the miR-17-92 cluster, accumulating bodies of evidence suggest that miR-34a could potentially be a tumor suppressor miRNA directly regulated by p53. Indeed, forced introduction of miR-34a into two human colon cancer cell lines caused the suppression of cell proliferation, induced senescence-like phenotypes, and down-regulated the E2F pathway (especially E2F3), but up-regulated the p53 pathway [204]. Because miR-34a targets the E2F3 protein, it is very intriguing that miR-34a can be integrated into the p53 and E2Fs pathways, as is the case for the c-Myc and E2Fs pathways integrated by the miR-17-92 cluster (Table). More attention must be paid to these processes, because transcriptional circuits regulated by transcription factors and miRNAs must be more complex than we can imagine. Indeed, another group has reported that miRNA-34a could be an oncogene instead of a tumor suppressor [205].

Table. Deregulated oncogenic and antioncogenic miRNAs participate in the gene transcriptional network mainly regulated by p53, c-Myc, and E2Fs transcription factors

	Expression in cancer	Upstream regulator	Downstream target	Apoptosis
Oncogenic miRNA				
miR-17-92 cluster	Up-regulated	c-Myc, E2F1 (E2F2, E2F3)	E2F1, E2F2, E2F3	Suppression
miR-21	Up-regulated		Bcl-2	Suppression
miR-106a	Up-regulated	c-Myc, n-Myc		
miR-221 & miR-222	Up-regulated	n-Myc	p27Kip1	
Antioncogenic miRNA				
miR-15 & miR-16	Down-regulated		Bcl-2	Induction
miR-34	Down-regulated	p53	Bcl-2, E2F3	Induction
let-7	Down-regulated		c-Myc, Ras, HMG2A	

In conclusion, p53, c-Myc, and E2F1 potentially are key players in oncogenic and antioncogenic signaling that are capable of cross-talk with miRNAs. We are just beginning to understand the importance of miRNAs in the context of cell cycle control and cancer regulation. We are currently investigating miRNAs regulated by c-Jun and oncogenic Ras. Using a miRNA microarray, we have confirmed that common miRNAs are similarly regulated both by c-Jun and oncogenic Ras. These observations will add a clue to dissect and recapture the complex network of the genome.

References

[1] Lee, J., Li, Z., Brower-Sinning, R., and John, B. (2007) Regulatory circuit of human microRNA biogenesis. *PLoS Comput Biol 3*, e67.

[2] Michael, M.Z., O' Connor, S.M., van Holst Pellekaan, N.G., Young, G.P., and James, R.J. (2003) Reduced accumulation of specific microRNAs in colorectal neoplasia. *Mol Cancer Res 1*, 882-891.

[3] Thomson, J.M., Newman, M., Parker, J.S., Morin-Kensicki, E.M., Wright, T., and Hammond, S.M. (2006) Extensive post-transcriptional regulation of microRNAs and its implications for cancer. *Genes Dev 20*, 2202-2207.

[4] Hammond, S.M. (2007) MicroRNAs as tumor suppressors. *Nat Genet 39*, 582-583.

[5] Kumar, M.S., Lu, J., Mercer, K.L., Golub, T.R., and Jacks, T. (2007) Impaired microRNA processing enhances cellular transformation and tumorigenesis. *Nat Genet 39*, 673-677.

[6] Xi, Y., Shalgi, R., Fodstad, O., Pilpel, Y., and Ju, J. (2006) Differentially regulated micro-RNAs and actively translated messenger RNA transcripts by tumor suppressor p53 in colon cancer. *Clin Cancer Res 12*, 2014-2024.

[7] Ambros, V. (2004) The functions of animal microRNAs. *Nature 431*, 350-355.

[8] Miska, E.A. (2005) How microRNAs control cell division, differentiation and death. *Curr Opin Genet Dev 15*, 563-568.

[9] O'Driscoll, L. (2006) The emerging world of microRNAs. *Anticancer Res 26*, 4271-4278.
[10] Bushati, N., and Cohen, S.M. (2007) microRNA Functions. *Annu Rev Cell Dev Biol 23*, 175-205.
[11] Chang, T.C., and Mendell, J.T. (2007) microRNAs in Vertebrate Physiology and Human Disease. *Annu Rev Genomics Hum Genet 8*, 215-239.
[12] Carrington, J.C., and Ambros, V. (2003) Role of microRNAs in plant and animal development. *Science 301*, 336-338.
[13] Harfe, B.D. (2005) MicroRNAs in vertebrate development. *Curr Opin Genet Dev 15*, 410-415.
[14] Alvarez-Garcia, I., and Miska, E.A. (2005) MicroRNA functions in animal development and human disease. *Development 132*, 4653-4662.
[15] Plasterk, R.H. (2006) Micro RNAs in animal development. *Cell 124*, 877-881.
[16] Cao, X., Pfaff, S.L., and Gage, F.H. (2007) A functional study of miR-124 in the developing neural tube. *Genes Dev 21*, 531-536.
[17] Hobert, O. (2007) miRNAs Play a Tune. *Cell 131*, 22-24.
[18] Neilson, J.R., Zheng, G.X., Burge, C.B., and Sharp, P.A. (2007) Dynamic regulation of miRNA expression in ordered stages of cellular development. *Genes Dev 21*, 578-589.
[19] Zhao, Y., and Srivastava, D. (2007) A developmental view of microRNA function. *Trends Biochem Sci 32*, 189-197.
[20] Doran, J., and Strauss, W.M. (2007) Bio-informatic trends for the determination of miRNA-target interactions in mammals. *DNA Cell Biol 26*, 353-360.
[21] Lee, C.T., Risom, T., and Strauss, W.M. (2007) Evolutionary conservation of microRNA regulatory circuits: an examination of microRNA gene complexity and conserved microRNA-target interactions through metazoan phylogeny. *DNA Cell Biol 26*, 209-218.
[22] Niwa, R., and Slack, F.J. (2007) The evolution of animal microRNA function. *Curr Opin Genet Dev 17*, 145-150.
[23] Siomi, H., and Siomi, M.C. (2007) Expanding RNA physiology: microRNAs in a unicellular organism. *Genes Dev 21*, 1153-1156.
[24] Zhang, B., Pan, X., and Cobb, G.P, (2007) Anderson TA. microRNAs as oncogenes and tumor suppressors. *Dev Biol 302*, 1-12.
[25] Cullen, B.R. (2004) Transcription and processing of human microRNA precursors. *Mol Cell 16*, 861-865.
[26] Chen, P.Y., and Meister, G. (2005) microRNA-guided posttranscriptional gene regulation. *Biol Chem 386*, 1205-1218.
[27] Engels, B.M., and Hutvagner, G. (2006) Principles and effects of microRNA-mediated post-transcriptional gene regulation. *Oncogene 25*, 6163-6169.
[28] Zeng, Y. (2006) Principles of micro-RNA production and maturation. *Oncogene 25*, 6156-6162.
[29] Pillai, R.S., Bhattacharyya, S.N., and Filipowicz, W. (2007) Repression of protein synthesis by miRNAs: how many mechanisms? *Trends Cell Biol 17*, 118-126.
[30] Chan, C.S., Elemento, O., and Tavazoie, S. (2005) Revealing posttranscriptional regulatory elements through network-level conservation. *PLoS Comput Bi*ol *1*, e69.

[31] Chaudhuri, K., and Chatterjee, R. (2007) MicroRNA detection and target prediction: integration of computational and experimental approaches. *DNA Cell Biol 26*, 321-337.

[32] Scherr, M., and Eder, M. (2007) Gene silencing by small regulatory RNAs in mammalian cells. *Cell Cycle 6*, 444-449.

[33] McManus, M.T. (2003) MicroRNAs and cancer. *Semin Cancer Biol 13*, 253-258.

[34] Chen, C.Z. (2005) MicroRNAs as oncogenes and tumor suppressors. *N Engl J Med 353*, 1768-1771.

[35] Mendell, J.T. (2005) MicroRNAs: critical regulators of development, cellular physiology and malignancy. *Cell Cycle 4*, 1179-1184.

[36] Garzon, R., Fabbri, M., Cimmino, A., Calin, G.A., and Croce, C.M. (2006) MicroRNA expression and function in cancer. *Trends Mol Med 12*, 580-587.

[37] Hammond, S.M. (2006) MicroRNAs as oncogenes. *Curr Opin Genet Dev 16*, 4-9.

[38] Hwang, H.W., and Mendell, J.T. (2006) MicroRNAs in cell proliferation, cell death, and tumorigenesis. *Br J Cancer 94*, 776-780.

[39] Kent, O.A., and Mendell, J.T. (2006) A small piece in the cancer puzzle: microRNAs as tumor suppressors and oncogenes. *Oncogene 25*, 6188-6196.

[40] Stahlhut Espinosa, C.E., and Slack, F.J. (2006) The Role of MicroRNAs in Cancer. *Yale J Biol Med 79*, 131-140.

[41] Zhang, L., and Coukos, G. (2006) MicroRNAs: a new insight into cancer genome. *Cell Cycle 5*, 2216-2219.

[42] Bandres, E., Agirre, X., Ramirez, N., Zarate, R., and Garcia-Foncillas, J. (2007) MicroRNAs as cancer players: potential clinical and biological effects. *DNA Cell Biol 26*, 273-282.

[43] Cho, W.C. (2007) OncomiRs: the discovery and progress of microRNAs in cancers. *Mol Cancer 6*, 60.

[44] Fabbri, M., Ivan, M., Cimmino, A., Negrini, M., and Calin, G.A. (2007) Regulatory mechanisms of microRNAs involvement in cancer. *Expert Opin Biol Ther 7*, 1009-1019.

[45] Giannakakis, A., Coukos, G., Hatzigeorgiou, A., Sandaltzopoulos, R., and Zhang, L. (2007) miRNA genetic alterations in human cancers. *Expert Opin Biol Ther 7*, 1375-1386.

[46] Lawrie, C.H. (2007) MicroRNA expression in lymphoma. *Expert Opin Biol Ther 7*, 1363-1374.

[47] Mathupala, S.P., Mittal, S., Guthikonda, M., and Sloan, A.E. (2007) MicroRNA and brain tumors: a cause and a cure? *DNA Cell Biol 26*, 301-310.

[48] Negrini, M., Ferracin, M., Sabbioni, S., and Croce, C.M. (2007) MicroRNAs in human cancer: from research to therapy. *J Cell Sci 120*, 1833-1840.

[49] Osada, H., and Takahashi, T. (2007) MicroRNAs in biological processes and carcinogenesis. *Carcinogenesis 28*, 2-12.

[50] Wiemer, E.A. (2007) The role of microRNAs in cancer: no small matter. *Eur J Cancer 43*, 1529-1544.

[51] Wu, W., Sun, M., Zou, G.M., and Chen, J. (2007) MicroRNA and cancer: Current status and prospective. *Int J Cancer 120*, 953-960.

[52] Zhang, B., Wang, Q., and Pan, X. (2007) MicroRNAs and their regulatory roles in animals and plants. *J Cell Physiol 210*, 279-289.

[53] Ambros, V. (2001) microRNAs: tiny regulators with great potential. *Cell 107*, 823-826.

[54] Esquela-Kerscher, A., and Slack, F.J. (2006) Oncomirs - microRNAs with a role in cancer. *Nat Rev Cancer 6*, 259-269.

[55] Shalgi, R., Lieber, D., Oren, M., and Pilpel, Y. (2007) Global and Local Architecture of the Mammalian microRNA-Transcription Factor Regulatory Network. *PLoS Comput Biol 3*, e131.

[56] Chen, K., and Rajewsky, N. (2007) The evolution of gene regulation by transcription factors and microRNAs. *Nat Rev Genet 8*, 93-103.

[57] Lu, J., Getz, G., Miska, E.A., Alvarez-Saavedra, E., Lamb, J., Peck, D., Sweet-Cordero, A., Ebert, B.L., Mak, R.H., Ferrando, A.A., Downing, J.R., Jacks, T., Horvitz, H.R., and Golub, T.R. (2005) MicroRNA expression profiles classify human cancers. *Nature 435*, 834-838.

[58] Calin, G.A., and Croce, C.M. (2006) MicroRNA signatures in human cancers. *Nat Rev Cancer 6*, 857-866.

[59] Volinia, S., Calin, G.A., Liu, C.G., Ambs, S., Cimmino, A., Petrocca, F., Visone, R., Iorio, M., Roldo, C., Ferracin, M., Prueitt, R.L., Yanaihara, N., Lanza, G., Scarpa, A., Vecchione, A., Negrini, M., Harris, C.C., and Croce, C.M. (2006) A microRNA expression signature of human solid tumors defines cancer gene targets. *Proc Natl Acad Sci U S A 103*, 2257-2261.

[60] Blenkiron, C., and Miska, E.A. (2007) miRNAs in cancer: approaches, aetiology, diagnostics and therapy. *Hum Mol Genet 16*, R106-113.

[61] Esau, C.C., and Monia, B.P. (2007) Therapeutic potential for microRNAs. *Adv Drug Deliv Rev 59*, 101-114.

[62] Goga, A., and Benz, C. (2007) Anti-oncomir suppression of tumor phenotypes. *Mol Interv 7*, 199-202.

[63] Hernando, E. (2007) microRNAs and cancer: role in tumorigenesis, patient classification and therapy. *Clin Transl Oncol 9*, 155-160.

[64] Jay, C., Nemunaitis, J., Chen, P., Fulgham, P., and Tong, A.W. (2007) miRNA profiling for diagnosis and prognosis of human cancer. *DNA Cell Biol 26*, 293-300.

[65] Jeyaseelan, K., Herath, W.B., and Armugam, A. (2007) MicroRNAs as therapeutic targets in human diseases. *Expert Opin Ther Targets 11*, 1119-1129.

[66] Lao, K., Xu, N.L., Sun, Y.A., Livak, K.J., and Straus, N.A. (2007) Real time PCR profiling of 330 human micro-RNAs. *Biotechnol* J *2*, 33-35.

[67] Nervi, C., Fazi, F., Rosa, A., Fatica, A., and Bozzoni, I. (2007) Emerging role for microRNAs in acute promyelocytic leukemia. *Curr Top Microbiol Immunol 313*, 73-84.

[68] Tricoli, J.V., and Jacobson, J.W. (2007) MicroRNA: Potential for Cancer Detection, Diagnosis, and Prognosis. *Cancer Res 67*, 4553-4555.

[69] Waldman, S.A., and Terzic, A. (2007) Translating MicroRNA discovery into clinical biomarkers in cancer. *JAMA 297*, 1923-1925.

[70] Wijnhoven, B.P., Michael, M.Z., and Watson, D.I. (2007) MicroRNAs and cancer. *Br J Surg 94*, 23-30.

[71] Wurdinger, T., and Costa, F.F. (2007) Molecular therapy in the microRNA era. *Pharmacogenomics J 7*, 297-304.

[72] Yu, S.L., Chen, H.Y., Yang, P.C., and Chen, J.J. (2007) Unique MicroRNA signature and clinical outcome of cancers. *DNA Cell Biol 26*, 283-292.

[73] Zhang, W., Dahlberg, J.E., and Tam, W. (2007) MicroRNAs in tumorigenesis: a primer. *Am J Patho*l *171*, 728-738.

[74] Calin, G.A., Liu, C.G., Sevignani, C., Ferracin, M., Felli, N., Dumitru, C.D., Shimizu, M., Cimmino, A., Zupo, S., Dono, M., Dell'Aquila, M.L., Alder, H., Rassenti, L., Kipps, T.J., Bullrich, F., Negrini, M., and Croce, C.M. (2004) MicroRNA profiling reveals distinct signatures in B cell chronic lymphocytic leukemias. *Proc Natl Acad Sci U S A 101*, 11755-11760.

[75] Ciafrè, S.A., Galardi, S., Mangiola, A., Ferracin, M., Liu, C.G., Sabatino, G., Negrini, M., Maira, G., Croce, C.M., and Farace, M.G. (2005) Extensive modulation of a set of microRNAs in primary glioblastoma. *Biochem Biophys Res Commun 334*, 1351-1358.

[76] Yu, J., Wang, F., Yang, G.H., Wang, F.L., Ma, Y.N., Du, Z.W., and Zhang, J.W. (2006) Human microRNA clusters: genomic organization and expression profile in leukemia cell lines. *Biochem Biophys Res Commun 349*, 59-68.

[77] Mattie, M.D., Benz, C.C., Bowers, J., Sensinger, K., Wong, L., Scott, G.K., Fedele, V., Ginzinger, D., Getts, R., and Haqq, C. (2006) Optimized high-throughput microRNA expression profiling provides novel biomarker assessment of clinical prostate and breast cancer biopsies. *Mol Cancer 5*, 24.

[78] Meng, F., Henson, R., Lang, M., Wehbe, H., Maheshwari, S., Mendell, J.T., Jiang, J., Schmittgen, T.D., and Patel, T. (2006) Involvement of human micro-RNA in growth and response to chemotherapy in human cholangiocarcinoma cell lines. *Gastroenterology 130*, 2113-2129.

[79] Yanaihara, N., Caplen, N., Bowman, E., Seike, M., Kumamoto, K., Yi, M., Stephens, R.M., Okamoto, A., Yokota, J., Tanaka, T., Calin, G.A., Liu, C.G., Croce, C.M., and Harris, C.C. (2006) Unique microRNA molecular profiles in lung cancer diagnosis and prognosis. *Cancer Cell 9*, 189-198.

[80] Murakami, Y., Yasuda, T., Saigo, K., Urashima, T., Toyoda, H., Okanoue, T., and Shimotohno, K. (2006) Comprehensive analysis of microRNA expression patterns in hepatocellular carcinoma and non-tumorous tissues. *Oncogene 25*, 2537-2545.

[81] Iorio, M.V., Ferracin, M., Liu, C.G., Veronese, A., Spizzo, R., Sabbioni, S., Magri, E., Pedriali, M., Fabbri, M., Campiglio, M., Ménard, S., Palazzo, J.P., Rosenberg, A., Musiani, P., Volinia, S., Nenci, I., Calin, G.A., Querzoli, P., Negrini, M., and Croce, C.M. (2005) MicroRNA gene expression deregulation in human breast cancer. *Cancer Res 65*, 7065-7070.

[82] Blenkiron, C., Goldstein, L.D., Thorne, N.P., Spiteri, I., Chin, S.F., Dunning, M.J., Barbosa-Morais, N.L., Teschendorff, A.E., Green, A.R., Ellis, I.O., Tavare, S., Caldas, C., and Miska, E.A. (2007) MicroRNA expression profiling of human breast cancer identifies new markers of tumour subtype. *Genome Biol 8*, R214.

[83] Subramanian, S., Lui, W.O., Lee, C.H., Espinosa, I., Nielsen, T.O., Heinrich, M.C., Corless, C.L., Fire, A.Z., and van de Rijn, M. MicroRNA expression signature of human sarcomas. *Oncogene 27*, 2015-2016.

[84] Ozen, M., Creighton, C.J., Ozdemir, M., and Ittmann, M. (2007) Widespread deregulation of microRNA expression in human prostate cancer. *Oncogene 27*, 1788-1793.

[85] Porkka, K.P., Pfeiffer, M.J., Waltering, K.K., Vessella, R.L., Tammela, T.L., and Visakorpi, T. (2007) MicroRNA expression profiling in prostate cancer. *Cancer Res 67*, 6130-6135.

[86] Iorio, M.V., Visone, R., Di Leva, G., Donati, V., Petrocca, F., Casalini, P., Taccioli, C., Volinia, S., Liu, C.G., Alder, H., Calin, G.A., Ménard, S., and Croce, C.M. MicroRNA signatures in human ovarian cancer. *Cancer Res 67*, 8699-8707.

[87] Marsh, E.E., Lin, Z., Yin, P., Milad, M., Chakravarti, D., and Bulun, S.E. (2007) Differential expression of microRNA species in human uterine leiomyoma versus normal myometrium. *Fertil Steril 89*, 1771-1776.

[88] Bandrés, E., Cubedo, E., Agirre, X., Malumbres, R., Zárate, R., Ramirez, N., Abajo, A., Navarro, A., Moreno, I., Monzó, M., and García-Foncillas, J. (2006) Identification by Real-time PCR of 13 mature microRNAs differentially expressed in colorectal cancer and non-tumoral tissues. *Mol Cancer 5*, 29.

[89] Lanza, G., Ferracin, M., Gafa, R., Veronese, A., Spizzo, R., Pichiorri, F., Liu, C.G., Calin, G.A., Croce, C.M., and Negrini, M. (2007) mRNA/microRNA gene expression profile in microsatellite unstable colorectal cancer. *Mol Cancer 6*, 54.

[90] Rossi, L., Bonmassar, E., and Faraoni, I. (2007) Modification of miR gene expression pattern in human colon cancer cells following exposure to 5-fluorouracil in vitro. *Pharmacol Res 56*, 248-253.

[91] Lui, W.O., Pourmand, N., Patterson, B.K., and Fire, A. (2007) Patterns of known and novel small RNAs in human cervical cancer. *Cancer Res 67*, 6031-6043.

[92] Cahill, S., Smyth, P., Finn, S.P., Denning, K., Flavin, R., O'Regan, E.M., Li, J., Potratz, A., Guenther, S.M., Henfrey, R., O'Leary, J.J., and Sheils, O. (2006) Effect of ret/PTC 1 rearrangement on transcription and post-transcriptional regulation in a papillary thyroid carcinoma model. *Mol Cancer 5*, 70.

[93] Pallante, P., Visone, R., Ferracin, M., Ferraro, A., Berlingieri, M.T., Troncone, G., Chiappetta, G., Liu, C.G., Santoro, M., Negrini, M., Croce, C.M., and Fusco, A. MicroRNA deregulation in human thyroid papillary carcinomas. *Endocr Relat Cancer 13*, 497-508.

[94] Visone, R., Russo, L., Pallante, P., De Martino, I., Ferraro, A., Leone, V., Borbone, E., Petrocca, F., Alder, H., Croce, C.M., and Fusco, A. (2007) MicroRNAs (miR)-221 and miR-222, both overexpressed in human thyroid papillary carcinomas, regulate p27Kip1 protein levels and cell cycle. *Endocr Relat Cancer 14*, 791-798.

[95] Lawrie, C.H., Soneji, S., Marafioti, T., Cooper, C.D., Palazzo, S., Paterson, J.C., Cattan, H., Enver, T., Mager, R., Boultwood, J., Wainscoat, J.S., and Hatton, C.S. MicroRNA expression distinguishes between germinal center B cell-like and activated B cell-like subtypes of diffuse large B cell lymphoma. *Int J Cancer 121*, 1156-1161.

[96] Blower, P.E., Verducci, J.S., Lin, S., Zhou, J., Chung, J.H., Dai, Z., Liu, C.G., Reinhold, W., Lorenzi, P.L., Kaldjian, E.P., Croce, C.M., Weinstein, J.N., and Sadee, W. MicroRNA expression profiles for the NCI-60 cancer cell panel. *Mol Cancer Ther 6*, 1483-1491.

[97] Gaur, A., Jewell, D.A., Liang, Y., Ridzon, D., Moore, J.H., Chen, C., Ambros, V.R., and Israel, M.A. (2007) Characterization of microRNA expression levels and their biological correlates in human cancer cell lines. *Cancer Res 67*, 2456-2468.

[98] Tran, N., McLean, T., Zhang, X., Zhao, C.J., Thomson, J.M., O'Brien, C., and Rose, B. MicroRNA expression profiles in head and neck cancer cell lines. *Biochem Biophys Res Commun 358*, 12-17.

[99] Roldo, C., Missiaglia, E., Hagan, J.P., Falconi, M., Capelli, P., Bersani, S., Calin, G.A., Volinia, S., Liu, C.G., Scarpa, A., and Croce, C.M. (2006) MicroRNA expression abnormalities in pancreatic endocrine and acinar tumors are associated with distinctive pathologic features and clinical behavior. *J Clin Oncol 24*, 4677-4684.

[100] Bloomston, M., Frankel, W.L., Petrocca, F., Volinia, S., Alder, H., Hagan, J.P., Liu, C.G., Bhatt, D., Taccioli, C., and Croce, C.M. (2007) MicroRNA expression patterns to differentiate pancreatic adenocarcinoma from normal pancreas and chronic pancreatitis. *JAMA 297*, 1901-1908.

[101] Lee, E.J., Gusev, Y., Jiang, J., Nuovo, G.J., Lerner, M.R., Frankel, W.L., Morgan, D.L., Postier, R.G., Brackett, D.J., and Schmittgen, T.D. (2007) Expression profiling identifies microRNA signature in pancreatic cancer. *Int J Cancer 120*, 1046-1054.

[102] Debernardi, S., Skoulakis, S., Molloy, G., Chaplin, T., Dixon-McIver, A., and Young, B.D. (2007) MicroRNA miR-181a correlates with morphological sub-class of acute myeloid leukaemia and the expression of its target genes in global genome-wide analysis. *Leukemia 21*, 912-916.

[103] Fulci, V., Chiaretti, S., Goldoni, M., Azzalin, G., Carucci, N., Tavolaro, S., Castellano, L., Magrelli, A., Citarella, F., Messina, M., Maggio, R., Peragine, N., Santangelo, S., Mauro, F.R., Landgraf, P., Tuschl, T., Weir, D.B., Chien, M., Russo, J.J., Ju, J., Sheridan, R., Sander, C., Zavolan, M., Guarini, A., Foà, R., and Macino, G. (2007) Quantitative technologies establish a novel microRNA profile of chronic lymphocytic leukemia. *Blood 109*, 4944-4951.

[104] Garzon, R., Pichiorri, F., Palumbo, T., Visentini, M., Aqeilan, R., Cimmino, A., Wang, H., Sun, H., Volinia, S., Alder, H., Calin, G.A., Liu, C.G., Andreeff, M., and Croce, C.M. (2007) MicroRNA gene expression during retinoic acid-induced differentiation of human acute promyelocytic leukemia. *Oncogene 26*, 4148-4157.

[105] Wang, T., Zhang, X., Obijuru, L., Laser, J., Aris, V., Lee, P., Mittal, K., Soteropoulos, P., and Wei, J.J. (2007) A micro-RNA signature associated with race, tumor size, and target gene activity in human uterine leiomyomas. *Genes Chromosomes Cancer 46*, 336-347.

[106] Szafranska, A.E., Davison, T.S., John, J., Cannon, T., Sipos, B., Maghnouj, A., Labourier, E., and Hahn, S.A. (2007) MicroRNA expression alterations are linked to tumorigenesis and non-neoplastic processes in pancreatic ductal adenocarcinoma. *Oncogene 26*, 4442-4452.

[107] Bottoni, A., Zatelli, M.C., Ferracin, M., Tagliati, F., Piccin, D., Vignali, C., Calin, G.A., Negrini, M., Croce, C.M., and Degli Uberti, E.C. (2007) Identification of differentially expressed microRNAs by microarray: a possible role for microRNA genes in pituitary adenomas. *J Cell Physiol 210*, 370-377.

[108] Voorhoeve, P.M., le Sage, C., Schrier, M., Gillis, A.J., Stoop, H., Nagel, R., Liu, Y.P., van Duijse, J., Drost, J., Griekspoor, A., Zlotorynski, E., Yabuta, N., De Vita, G., Nojima, H., Looijenga, L.H., and Agami, R. (2007) A genetic screen implicates miRNA-372 and miRNA-373 as oncogenes in testicular germ cell tumors. *Adv Exp Med Biol 604*, 17-46.

[109] Tuccoli, A., Poliseno, L., and Rainaldi, G. (2006) miRNAs regulate miRNAs: coordinated transcriptional and post-transcriptional regulation. *Cell Cycle 5*, 2473-2476.

[110] Yu, Z., Li, Z., Jolicoeur, N., Zhang, L., Fortin, Y., Wang, E., Wu, M., and Shen, S.H. (2007) Aberrant allele frequencies of the SNPs located in microRNA target sites are potentially associated with human cancers. *Nucleic Acids Res 35*, 4535-4541.

[111] Zhang, L., Huang, J., Yang, N., Greshock, J., Megraw, M.S., Giannakakis, A., Liang, S., Naylor, T.L., Barchetti, A., Ward, M.R., Yao, G., Medina, A., O'brien-Jenkins, A., Katsaros, D., Hatzigeorgiou, A., Gimotty, P.A., Weber, B.L., and Coukos, G. (2006) microRNAs exhibit high frequency genomic alterations in human cancer. *Proc Natl Acad Sci U S A 103*, 9136-9141.

[112] Calin, G.A., and Croce, C.M. (2007) Chromosomal rearrangements and microRNAs: a new cancer link with clinical implications. *J Clin Invest 117*, 2059-2066.

[113] Calin, G.A., Sevignani, C., Dumitru, C.D., Hyslop, T., Noch, E., Yendamuri, S., Shimizu, M., Rattan, S., Bullrich, F., Negrini, M., and Croce, C.M. (2004) Human microRNA genes are frequently located at fragile sites and genomic regions involved in cancers. *Proc Natl Acad Sci U S A 101*, 2999-3004.

[114] Lamy, P., Andersen, C.L., Dyrskjøt, L., Tørring, N., Ørntoft, T., and Wiuf, C. (2006) Are microRNAs located in genomic regions associated with cancer? *Br J Cancer 95*, 1415-1418.

[115] Huppi, K., Volfovsky, N., Mackiewicz, M., Runfola, T., Jones, T.L., Martin, S.E., Stephens, R., and Caplen, N.J. (2007) MicroRNAs and genomic instability. *Semin Cancer Biol 17*, 65-73.

[116] Sevignani, C., Calin, G.A., Nnadi, S.C., Shimizu, M., Davuluri, R.V., Hyslop, T., Demant, P., Croce, C.M., and Siracusa, L.D. (2007) MicroRNA genes are frequently located near mouse cancer susceptibility loci. *Proc Natl Acad Sci U S A 104*, 8017-8022.

[117] Calin, G.A., Ferracin, M., Cimmino, A., Di Leva, G., Shimizu, M., Wojcik, S.E., Iorio, M.V., Visone, R., Sever, N.I., Fabbri, M., Iuliano, R., Palumbo, T., Pichiorri, F., Roldo, C., Garzon, R., Sevignani, C., Rassenti, L., Alder, H., Volinia, S., Liu, C.G., Kipps, T.J., Negrini, M., and Croce, C.M. (2005) A MicroRNA signature associated with prognosis and progression in chronic lymphocytic leukemia. *N Engl J Med 353*, 1793-1801.

[118] Yang, J., Zhou, F., Xu, T., Deng, H., Ge, Y.Y., Zhang, C., Li, J., and Zhuang, S.M. (2007) Analysis of sequence variations in 59 microRNAs in hepatocellular carcinomas. *Mutat Res 638*, 205-209.

[119] Saito, Y., Liang, G., Egger, G., Friedman, J.M., Chuang, J.C., Coetzee, G.A., and Jones, P.A. (2006) Specific activation of microRNA-127 with downregulation of the

proto-oncogene BCL6 by chromatin-modifying drugs in human cancer cells. *Cancer Cell 9*, 435-443.

[120] Lujambio, A., and Esteller, M. (2007) CpG island hypermethylation of tumor suppressor microRNAs in human cancer. *Cell Cycle 6*, 1455-1459.

[121] Lujambio, A., Ropero, S., Ballestar, E., Fraga, M.F., Cerrato, C., Setién, F., Casado, S., Suarez-Gauthier, A., Sanchez-Cespedes, M., Git, A., Spiteri, I., Das, P.P., Caldas, C., Miska, E., and Esteller, M. Genetic unmasking of an epigenetically silenced microRNA in human cancer cells. *Cancer Res 67*, 1424-1429.

[122] Weber, B., Stresemann, C., Brueckner, B., and Lyko, F. (2007) Methylation of human microRNA genes in normal and neoplastic cells. *Cell Cycle 6*, 1001-1005.

[123] Lehmann, U., Hasemeier, B., Christgen, M., Müller, M., Römermann, D., Länger, F., and Kreipe, H. (2007) Epigenetic inactivation of microRNA gene hsa-mir-9-1 in human breast cancer. *J Pathol 214*, 17-24.

[124] Calin, G.A., Dumitru, C.D., Shimizu, M., Bichi, R., Zupo, S., Noch, E., Aldler, H., Rattan, S., Keating, M., Rai, K., Rassenti, L., Kipps, T., Negrini, M., Bullrich, F., and Croce, C.M. (2002) Frequent deletions and down-regulation of micro- RNA genes miR15 and miR16 at 13q14 in chronic lymphocytic leukemia. *Proc Natl Acad Sci U S A 99*, 15524-15529.

[125] Cimmino, A., Calin, G.A., Fabbri, M., Iorio, M.V., Ferracin, M., Shimizu, M., Wojcik, S.E., Aqeilan, R.I., Zupo, S., Dono, M., Rassenti, L., Alder, H., Volinia, S., Liu, C.G., Kipps, T.J., Negrini, M., and Croce, C.M. (2005) miR-15 and miR-16 induce apoptosis by targeting BCL2. *Proc Natl Acad Sci U S A 102*, 13944-13949.

[126] Bottoni, A., Piccin, D., Tagliati, F., Luchin, A., Zatelli, M.C., and degli Uberti, E.C. (2005) miR-15a and miR-16-1 down-regulation in pituitary adenomas. *J Cell Physiol 204*, 280-285.

[127] Koscianska, E., Baev, V., Skreka, K., Oikonomaki, K., Rusinov, V., Tabler, M., and Kalantidis, K. (2007) Prediction and preliminary validation of oncogene regulation by miRNAs. *BMC Mol Biol 8*, 79.

[128] Linsley, P.S., Schelter, J., Burchard, J., Kibukawa, M., Martin, M.M., Bartz, S.R., Johnson, J.M., Cummins, J.M., Raymond, C.K., Dai, H., Chau, N., Cleary, M., Jackson, A.L., Carleton, M., and Lim, L. (2007) Transcripts targeted by the microRNA-16 family cooperatively regulate cell cycle progression. *Mol Cell Biol 27*, 2240-2252.

[129] He, L., Thomson, J.M., Hemann, M.T., Hernando-Monge, E., Mu, D., Goodson, S., Powers, S., Cordon-Cardo, C., Lowe, S.W., Hannon, G.J., and Hammond, S.M. (2005) A microRNA polycistron as a potential human oncogene. *Nature 435*, 828-833.

[130] O'Donnell, K.A., Wentzel, E.A., Zeller, K.I., Dang, C.V., and Mendell, J.T. (2005) c-Myc-regulated microRNAs modulate E2F1 expression. *Nature 435*, 839-843.

[131] Dews, M., Homayouni, A., Yu, D., Murphy, D., Sevignani, C., Wentzel, E., Furth, E.E., Lee, W.M., Enders, G.H., Mendell, J.T., and Thomas-Tikhonenko, A. (2006) Augmentation of tumor angiogenesis by a Myc-activated microRNA cluster. *Nat Genet 38*, 1060-1065.

[132] Sylvestre, Y., De Guire, V., Querido, E., Mukhopadhyay, U.K., Bourdeau, V., Major, F., Ferbeyre, G., and Chartrand, P. (2007) An E2F/miR-20a autoregulatory feedback loop. *J Biol Chem 282*, 2135-2143.

[133] Woods, K., Thomson, J.M., and Hammond, S.M. (2007) Direct regulation of an oncogenic micro-RNA cluster by E2F transcription factors. *J Biol Chem 282*, 2130-2134.

[134] Lu, Y., Thomson, J.M., Wong, H.Y., Hammond, S.M., and Hogan, B.L. (2007) Transgenic over-expression of the microRNA miR-17-92 cluster promotes proliferation and inhibits differentiation of lung epithelial progenitor cells. *Dev Biol 310*, 442-453.

[135] Lum, A.M., Wang, B.B., Li, L., Channa, N., Bartha, G., and Wabl, M. (2007) Retroviral activation of the mir-106a microRNA cistron in T lymphoma. *Retrovirology 4*, 5.

[136] Landais, S., Landry, S., Legault, P., and Rassart, E. (2007) Oncogenic potential of the miR-106-363 cluster and its implication in human T-cell leukemia. *Cancer Res 67*, 5699-5707.

[137] Aubry, S., and Charron, J. (2000) N-Myc shares cellular functions wiht c-Myc. *DNA Cell Biol 19*, 353-364.

[138] Novotny, G.W., Sonne, S.B., Nielsen, J.E., Jonstrup, S.P., Hansen, M.A., Skakkebaek, N.E., Rajpert-De Meyts, E., Kjems, J., and Leffers, H. (2007) Translational repression of E2F1 mRNA in carcinoma in situ and normal testis correlates with expression of the miR-17-92 cluster. *Cell Death Differ 14*, 879-882.

[139] Hayashita, Y., Osada, H., Tatematsu, Y., Yamada, H., Yanagisawa, K., Tomida, S., Yatabe, Y., Kawahara, K., Sekido, Y., and Takahashi, T. (2005) A polycistronic microRNA cluster, miR-17-92, is overexpressed in human lung cancers and enhances cell proliferation. *Cancer Res 65*, 9628-9632.

[140] Venturini, L., Battmer, K., Castoldi, M., Schultheis, B., Hochhaus, A., Muckenthaler, M.U., Ganser, A., Eder, M., and Scherr, M. (2007) Expression of the miR-17-92 polycistron in chronic myeloid leukemia (CML) CD34+ cells. *Blood 109*, 4399-4405.

[141] Rinaldi, A., Poretti, G., Kwee, I., Zucca, E., Catapano, C.V., Tibiletti, M.G., and Bertoni, F. (2007) Concomitant MYC and microRNA cluster miR-17-92 (C13orf25) amplification in human mantle cell lymphoma. *Leuk Lymphoma 48*, 410-412.

[142] Tagawa, H., Karube, K., Tsuzuki, S., Ohshima, K., and Seto, M. (2007) Synergistic action of the microRNA-17 polycistron and Myc in aggressive cancer development. *Cancer Sci 98*, 1482-1490.

[143] Matsubara, H., Takeuchi, T., Nishikawa, E., Yanagisawa, K., Hayashita, Y., Ebi, H., Yamada, H., Suzuki, M., Nagino, M., Nimura, Y., Osada, H., and Takahashi, T. (2007) Apoptosis induction by antisense oligonucleotides against miR-17-5p and miR-20a in lung cancers overexpressing miR-17-92. *Oncogene 26*, 6099-6105.

[144] Hossain, A., Kuo, M.T., and Saunders, G.F. (2006) Mir-17-5p regulates breast cancer cell proliferation by inhibiting translation of AIB1 mRNA. *Mol Cell Biol 26*, 8191-8201.

[145] Chan, J.A., Krichevsky, A.M., and Kosik, K.S. (2005) MicroRNA-21 is an antiapoptotic factor in human glioblastoma cells. *Cancer Res 65*, 6029-6033.

[146] Meng, F., Henson, R., Wehbe-Janek, H., Ghoshal, K., Jacob, S.T., and Patel, T. (2007) MicroRNA-21 regulates expression of the PTEN tumor suppressor gene in human hepatocellular cancer. *Gastroenterology 133*, 647-658.

[147] Si, M.L., Zhu, S., Wu, H., Lu, Z., Wu, F., and Mo, Y.Y. (2007) miR-21-mediated tumor growth. *Oncogene 26*, 2799-2803.
[148] Löffler, D., Brocke-Heidrich, K., Pfeifer, G., Stocsits, C., Hackermüller, J., Kretzschmar, A.K., Burger, R., Gramatzki, M., Blumert, C., Bauer, K., Cvijic, H., Ullmann, A.K., Stadler, P.F., and Horn, F. (2007) Interleukin-6 dependent survival of multiple myeloma cells involves the Stat3-mediated induction of microRNA-21 through a highly conserved enhancer. *Blood 110*, 1330-1333.
[149] Zhu, S., Si, M.L., Wu, H., and Mo, Y.Y. (2007) MicroRNA-21 targets the tumor suppressor gene tropomyosin 1 (TPM1). *J Biol Chem 282*, 14328-14336.
[150] Corsten, M.F., Miranda, R., Kasmieh, R., Krichevsky, A.M., Weissleder, R., and Shah, K. (2007) MicroRNA-21 knockdown disrupts glioma growth in vivo and displays synergistic cytotoxicity with neural precursor cell delivered S-TRAIL in human gliomas. *Cancer Res 67*, 8994-9000.
[151] Chang, T.C., Wentzel, E.A., Kent, O.A., Ramachandran, K., Mullendore, M., Lee, K.H., Feldmann, G., Yamakuchi, M., Ferlito, M., Lowenstein, C.J., Arking, D.E., Beer, M.A., Maitra, A., and Mendell, J.T. (2007) Transactivation of miR-34a by p53 broadly influences gene expression and promotes apoptosis. *Mol Cell 26*, 745-752.
[152] Corney, D.C., Flesken-Nikitin, A., Godwin, A.K., Wang, W., and Nikitin, A.Y. (2007) MicroRNA-34b and MicroRNA-34c are targets of p53 and cooperate in control of cell proliferation and adhesion-independent growth. *Cancer Res 67*, 8433-8438.
[153] He, L., He, X., Lim, L.P., de Stanchina, E., Xuan, Z., Liang, Y., Xue, W., Zender, L., Magnus, J., Ridzon, D., Jackson, A.L., Linsley, P.S., Chen, C., Lowe, S.W., Cleary, M.A., and Hannon, G.J. (2007) A microRNA component of the p53 tumour suppressor network. *Nature 447*, 1130-1134.
[154] Raver-Shapira, N., Marciano, E., Meiri, E., Spector, Y., Rosenfeld, N., Moskovits, N., Bentwich, Z., and Oren, M. (2007) Transcriptional activation of miR-34a contributes to p53-mediated apoptosis. *Mol Cell 26*, 731-743.
[155] Tarasov, V., Jung, P., Verdoodt, B., Lodygin, D., Epanchintsev, A., Menssen, A., Meister, G., and Hermeking, H. (2007) Differential regulation of microRNAs by p53 revealed by massively parallel sequencing: miR-34a is a p53 target that induces apoptosis and G1-arrest. *Cell Cycle 6*, 1586-1593.
[156] Bommer, G.T., Gerin, I., Feng, Y., Kaczorowski, A.J., Kuick, R., Love, R.E., Zhai, Y., Giordano, T.J., Qin, Z.S., Moore, B.B., MacDougald, O.A., Cho, K.R., and Fearon, E.R. (2007) p53-mediated activation of miRNA34 candidate tumor-suppressor genes. *Curr Biol 17*, 1298-1307.
[157] Welch, C., Chen, Y., and Stallings, R.L. (2007) MicroRNA-34a functions as a potential tumor suppressor by inducing apoptosis in neuroblastoma cells. *Oncogene 26*, 5017-5022.
[158] Takamizawa, J., Konishi, H., Yanagisawa, K., Tomida, S., Osada, H., Endoh, H., Harano, T., Yatabe, Y., Nagino, M., Nimura, Y., Mitsudomi, T., and Takahashi, T. (2004) Reduced expression of the let-7 microRNAs in human lung cancers in association with shortened postoperative survival. *Cancer Res 64*, 3753-3756.
[159] Johnson, C.D., Esquela-Kerscher, A., Stefani, G., Byrom, M., Kelnar, K., Ovcharenko, D., Wilson, M., Wang, X., Shelton, J., Shingara, J., Chin, L., Brown, D., and Slack, F.J.

(2007) The let-7 microRNA represses cell proliferation pathways in human cells. *Cancer Res 67*, 7713-7722.

[160] Sampson, V.B., Rong, N.H., Han, J., Yang, Q., Aris, V., Soteropoulos, P., Petrelli, N.J., Dunn, S.P., and Krueger, L.J. (2007) MicroRNA Let-7a Down-regulates MYC and Reverts MYC-Induced Growth in Burkitt Lymphoma Cells. *Cancer Res 67*, 9762-9770.

[161] Inamura, K., Togashi, Y., Nomura, K., Ninomiya, H., Hiramatsu, M., Satoh, Y., Okumura, S., Nakagawa, K., and Ishikawa, Y. (2007) let-7 microRNA expression is reduced in bronchioloalveolar carcinoma, a non-invasive carcinoma, and is not correlated with prognosis. *Lung Cancer 58*, 392-396.

[162] Mayr, C., Hemann, M.T., and Bartel, D.P. (2007) Disrupting the pairing between let-7 and Hmga2 enhances oncogenic transformation. *Science 315*, 1576-1579.

[163] Shell, S., Park, S.M., Radjabi, A.R., Schickel, R., Kistner, E.O., Jewell, D.A., Feig, C., Lengyel, E., and Peter, M.E. (2007) Let-7 expression defines two differentiation stages of cancer. *Proc Natl Acad Sci U S A 104*, 11400-11405.

[164] Lee, Y.S., and Dutta, A. (2007) The tumor suppressor microRNA let-7 represses the HMGA2 oncogene. *Genes Dev 21*, 1025-1030.

[165] Johnson, S.M., Grosshans, H., Shingara, J., Byrom, M., Jarvis, R., Cheng, A., Labourier, E., Reinert, K.L., Brown, D., and Slack, F.J. (2005) RAS is regulated by the let-7 microRNA family. *Cell 120*, 635-647.

[166] Hebert, C., Norris, K., Scheper, M.A., Nikitakis, N., and Sauk, J.J. (2007) High mobility group A2 is a target for miRNA-98 in head and neck squamous cell carcinoma. *Mol Cancer 6*, 5.

[167] Brueckner, B., Stresemann, C., Kuner, R., Mund, C., Musch, T., Meister, M., Sültmann, H., and Lyko, F. (2007) The human let-7a-3 locus contains an epigenetically regulated microRNA gene with oncogenic function. *Cancer Res 67*, 1419-1423.

[168] Shah, Y.M., Morimura, K., Yang, Q., Tanabe, T., Takagi, M., and Gonzalez, F.J. (2007) Peroxisome proliferator-activated receptor alpha regulates a microRNA-mediated signaling cascade responsible for hepatocellular proliferation. *Mol Cell Biol 27*, 4238-4247.

[169] Meng, F., Henson, R., Wehbe-Janek, H., Smith, H., Ueno, Y., and Patel, T. (2007) The MicroRNA let-7a modulates interleukin-6-dependent STAT-3 survival signaling in malignant human cholangiocytes. *J Biol Chem 283*, 8256-8264.

[170] Meng, F., Wehbe-Janek, H., Henson, R., Smith, H., and Patel, T. (2007) Epigenetic regulation of microRNA-370 by interleukin-6 in malignant human cholangiocytes. *Oncogene 27*, 378-386.

[171] Ma, L., Teruya-Feldstein, J., and Weinberg, R.A. (2007) Tumour invasion and metastasis initiated by microRNA-10b in breast cancer. *Nature 449*, 682-688.

[172] Fabbri, M., Garzon, R., Cimmino, A., Liu, Z., Zanesi, N., Callegari, E., Liu, S., Alder, H., Costinean, S., Fernandez-Cymering, C., Volinia, S., Guler, G., Morrison, C.D., Chan, K.K., Marcucci, G., Calin, G.A., Huebner, K., and Croce, C.M. (2007) MicroRNA-29 family reverts aberrant methylation in lung cancer by targeting DNA methyltransferases 3A and 3B. *Proc Natl Acad Sci U S A 104*, 15805-15810.

[173] Pekarsky, Y., Santanam, U., Cimmino, A., Palamarchuk, A., Efanov, A., Maximov, V., Volinia, S., Alder, H., Liu, C.G., Rassenti, L., Calin, G.A., Hagan, J.P., Kipps, T., and

Croce, C.M. (2006) Tcl1 expression in chronic lymphocytic leukemia is regulated by miR-29 and miR-181. *Cancer Res 66*, 11590-11593.

[174] Mott, J.L., Kobayashi, S., Bronk, S.F., and Gores, G.J. (2007) mir-29 regulates Mcl-1 protein expression and apoptosis. *Oncogene 26*, 6133-6140.

[175] Kutay, H., Bai, S., Datta, J., Motiwala, T., Pogribny, I., Frankel, W., Jacob, S.T., and Ghoshal, K. (2006) Downregulation of miR-122 in the rodent and human hepatocellular carcinomas. *J Cell Biochem 99*, 671-678.

[176] Gramantieri, L., Ferracin, M., Fornari, F., Veronese, A., Sabbioni, S., Liu, C.G., Calin, G.A., Giovannini, C., Ferrazzi, E., Grazi, G.L., Croce, C.M., Bolondi, L., and Negrini, M. (2007) Cyclin G1 is a target of miR-122a, a microRNA frequently down-regulated in human hepatocellular carcinoma. *Cancer Res 67*, 6092-6099.

[177] Wu, L., and Belasco, J.G. (2005) Micro-RNA regulation of the mammalian lin-28 gene during neuronal differentiation of embryonal carcinoma cells. *Mol Cell Biol 25*, 9198-9208.

[178] Scott, G.K., Goga, A., Bhaumik, D., Berger, C.E., Sullivan, C.S., and Benz, C.C. (2007) Coordinate suppression of ERBB2 and ERBB3 by enforced expression of micro-RNA miR-125a or miR-125b. *J Biol Chem 282*, 1479-1486.

[179] Laneve, P., Di Marcotullio, L., Gioia, U., Fiori, M.E., Ferretti, E., Gulino, A., Bozzoni, I., and Caffarelli, E. (2007) The interplay between microRNAs and the neurotrophin receptor tropomyosin-related kinase C controls proliferation of human neuroblastoma cells. *Proc Natl Acad Sci U S A 104*, 7957-7962.

[180] Saito, Y., and Jones, P.A. (2006) Epigenetic activation of tumor suppressor microRNAs in human cancer cells. *Cell Cycle 5*, 2220-2222.

[181] Akao, Y., Nakagawa, Y., Kitade, Y., Kinoshita, T., and Naoe, T. (2007) Downregulation of microRNAs-143 and -145 in B-cell malignancies. *Cancer Sci 98,* 1914-1920.

[182] Akao, Y., Nakagawa, Y., and Naoe, T. (2007) MicroRNA-143 and -145 in colon cancer. *DNA Cell Biol 26*, 311-320.

[183] Xiao, C., Calado, D.P., Galler, G., Thai, T.H., Patterson, H.C., Wang, J., Rajewsky, N., Bender, T.P., and Rajewsky, K. (2007) MiR-150 Controls B Cell Differentiation by Targeting the Transcription Factor c-Myb. *Cell 131*, 146-159.

[184] Eis, P.S., Tam, W., Sun, L., Chadburn, A., Li, Z., Gomez, M.F., Lund, E., and Dahlberg, J.E. (2005) Accumulation of miR-155 and BIC RNA in human B cell lymphomas. *Proc Natl Acad Sci U S A 102*, 3627-3632.

[185] Kluiver, J., Poppema, S., de Jong, D., Blokzijl, T., Harms, G., Jacobs, S., Kroesen, B.J., and van den Berg, A. (2005) BIC and miR-155 are highly expressed in Hodgkin, primary mediastinal and diffuse large B cell lymphomas. *J Pathol 207*, 243-249.

[186] Kluiver, J., Haralambieva, E., de Jong, D., Blokzijl, T., Jacobs, S., Kroesen, B.J., Poppema, S., and van den Berg, A. (2006) Lack of BIC and microRNA miR-155 expression in primary cases of Burkitt lymphoma. *Genes Chromosomes Cancer 45*, 147-153.

[187] Costinean, S., Zanesi, N., Pekarsky, Y., Tili, E., Volinia, S., Heerema, N., and Croce, C.M. (2006) Pre-B cell proliferation and lymphoblastic leukemia/high-grade lymphoma in E(mu)-miR155 transgenic mice. *Proc Natl Acad Sci U S A 103*, 7024-7029.

[188] Hurteau, G.J., Carlson, J.A., Spivack, S.D., and Brock, G.J. (2007) Overexpression of the microRNA hsa-miR-200c leads to reduced expression of transcription factor 8 and increased expression of E-cadherin. *Cancer Res 67*, 7972-7976.

[189] Adams, B.D., Furneaux, H., and White, B.A. (2007) The micro-ribonucleic acid (miRNA) miR-206 targets the human estrogen receptor-alpha (ERalpha) and represses ERalpha messenger RNA and protein expression in breast cancer cell lines. *Mol Endocrinol 21*, 1132-1147.

[190] Schulte, J.H., Horn, S., Otto, T., Samans, B., Heukamp, L.C., Eilers, U.C., Krause, M., Astrahantseff, K., Klein-Hitpass, L., Buettner, R., Schramm, A., Christiansen, H., Eilers, M., Eggert, A., and Berwanger, B. (2007) MYCN regulates oncogenic MicroRNAs in neuroblastoma. *Int J Cancer 122*, 699-704.

[191] Galardi, S., Mercatelli, N., Giorda, E., Massalini, S., Frajese, G.V., Ciafrè, S.A., and Farace, M.G. (2007) miR-221 and miR-222 expression affects the proliferation potential of human prostate carcinoma cell lines by targeting p27Kip1. *J Biol Chem 282*, 23716-23724.

[192] Gillies, J.K., and Lorimer, I.A. (2007) Regulation of p27Kip1 by miRNA 221/222 in glioblastoma. *Cell Cycle 6*, 2005-2009.

[193] le Sage, C., Nagel, R., Egan, D.A., Schrier, M., Mesman, E., Mangiola, A., Anile, C., Maira, G., Mercatelli, N., Ciafrè, S.A., Farace, M.G., and Agami, R. (2007) Regulation of the p27(Kip1) tumor suppressor by miR-221 and miR-222 promotes cancer cell proliferation. *EMBO J 26*, 3699-3708.

[194] Visone, R., Pallante, P., Vecchione, A., Cirombella, R., Ferracin, M., Ferraro, A., Volinia, S., Coluzzi, S., Leone, V., Borbone, E., Liu, C.G., Petrocca, F., Troncone, G., Calin, G.A., Scarpa, A., Colato, C., Tallini, G., Santoro, M., Croce, C.M., and Fusco, A. (2007) Specific microRNAs are downregulated in human thyroid anaplastic carcinomas. *Oncogene 26*, 7590-7595.

[195] Ramasamy, S., Duraisamy, S., Barbashov, S., Kawano, T., Kharbanda, S., and Kufe, D. (2007) The MUC1 and galectin-3 oncoproteins function in a MicroRNA-dependent regulatory loop. *Mol Cell 27*, 992-1004.

[196] Voorhoeve, P.M., le Sage, C., Schrier, M., Gillis, A.J., Stoop, H., Nagel, R., Liu, Y.P., van Duijse, J., Drost, J., Griekspoor, A., Zlotorynski, E., Yabuta, N., De Vita, G., Nojima, H., Looijenga, L.H., and Agami, R. (2006) A genetic screen implicates miRNA-372 and miRNA-373 as oncogenes in testicular germ cell tumors. *Cell 124*, 1169-1181.

[197] Gillis, A., Stoop, H., Hersmus, R., Oosterhuis, J., Sun, Y., Chen, C., Guenther, S., Sherlock, J., Veltman, I., Baeten, J., van der Spek, P., de Alarcon, P., and Looijenga, L. (2007) High-throughput microRNAome analysis in human germ cell tumours. *J Pathol 213*, 319-328.

[198] Zindy, F., Eischen, C.M., Randle, D.H., Kamijo, T., Cleveland, J.L., Sherr, C.J., and Roussel, M.F. (1998) Myc signaling via the ARF tumor suppressor regulates p53-dependent apoptosis and immortalization. *Genes Dev 12*, 2424-2433.

[199] Matsumura, I., Tanaka, H., and Kanakura, Y. (2003) E2F1 and c-Myc in cell growth and death. *Cell Cycle 2*, 333-338.

[200] Coller, H.A., Forman, J.J., and Legesse-Miller, A. (2007) "Myc'ed messages": myc induces transcription of E2F1 while inhibiting its translation via a microRNA polycistron. *PLoS Genet 3*, e146.
[201] Leone, G., DeGregori, J., Sears, R., Jakoi, L., and Nevins, J.R. (1997) Myc and Ras collaborate in inducing accumulation of active cyclin E/Cdk2 and E2F. *Nature 387*, 422-426.
[202] Evan, G.I., Wyllie, A.H., Gilbert, C.S., Littlewood, T.D., Land, H., Brooks, M., Waters, C.M., Penn, L.Z., and Hancock, D.C. (1992) Induction of apoptosis in fibroblasts by c-myc protein. *Cell 69*, 119-128.
[203] Tsang, J., Zhu, J., and van Oudenaarden, A. (207) MicroRNA-mediated feedback and feedforward loops are recurrent network motifs in mammals. *Mol Cell 26*, 753-767.
[204] Tazawa, H., Tsuchiya, N., Izumiya, M., and Nakagama, H. (2007) Tumor-suppressive miR-34a induces senescence-like growth arrest through modulation of the E2F pathway in human colon cancer cells. *Proc Natl Acad Sci U S A 104*, 15472-15477.
[205] Dutta, K.K., Zhong, Y., Liu, Y.T., Yamada, T., Akatsuka, S., Hu, Q., Yoshihara, M., Ohara, H., Takehashi, M., Shinohara, T., Masutani, H., Onuki, J., and Toyokuni, S. (2007) Association of microRNA-34a overexpression with proliferation is cell type-dependent. *Cancer Sci 98*, 1845-1852.

In: Cell Cycle Control: New Research
Editors: N. H. Leroy, N. T. Fournier

ISBN: 978-1-60456-796-0

Chapter III

Cycles, Systems Science, and Biologic Entities: A Translational View

Ivo P. Janecka*
Health Research International, Helena Island. SC, USA.

Abstract

All living entities, from cells to the human body, experience periodic functional and morphologic abnormalities, intermingled with the state of health. If and how the various states are related, and what factors assume greater or lesser significance in this process, however, is subject to discussion. This chapter considers that all the various states, experienced by biologic entities during their life cycles, are reflections of their compliance with systems principles. This view is based on the concept of cycles and systems inter-relatedness.

Systems science, complexity, and chaos theories were employed to construct a conceptual model applicable to living entities. For each state, representative systems' characteristics were matched to examples of approximate biologic equivalents. This model conceptualizes *a zone of order* and a *zone of chaos* with an overlapping but fragile *health territory*, the heart of this schema. The *health territory* is proposed to be a dynamic amalgamation of the proximate *outer core* of the *zone of order* and the *inner edge* of the *zone of chaos*. In addition, two extremes states are postulated as well: the *outer edge* of the *zone of chaos* and the *zone of entropy* which predominates within the *inner core* of the *zone of order*.

Examples of plausible correlations were found between cancer and the *outer edge* of the *zone of chaos*. A similar association was found between degenerative/inflammatory diseases, including aging, and the *zone of entropy*. Based on this model, biologic entities which stray into the postulated extreme states of *chaos* or *entropy* may still reverse back to the *health territory* by restoring their *organized complexity*. The exponential cellular growth within the *zone of chaos,* characteristic of cancer, could be changed through a process of cyto-education which reverses *disorganized complexity* to an organized one by

* Correspondence concerning this article should be addressed to: Ivo P. Janecka, MD, MBA, PhD, 333 Westbrook Rd, St. Helena Island, SC 29920. janecka@post.harvard.edu; Telephone: 843-838-3602.

creating an environment which highly favors cellular differentiation instead of blind multiplication. The path back to the *health territory* from the second extreme, the system-engulfing *entropy*, would involve a process of increasing system's *complexity*.
Applying systems science principles to biologic entities, as attempted in this model, not only allows for the appreciation of disease states in a different light but also suggests what steps could be taken to return the entire system into the state of health.

Keywords: cycles, systems, complexity, chaos, entropy, cancer, cells.

"In all things there is a law of cycles."
Publius Cornelius Tacitus (56-117AD)

Introduction

It is through a process of convergent evolution of knowledge that we attempt to comprehend our world. Systems entities, from a single cell to the human body, from earthly ecology to celestial configurations, are involved in multiple interrelated rhythmic changes known as cycles. Responding with synchrony to such changes is a pre-requisite for the optimization of any individual system while simultaneously this harmony maximizes the potential influence of each cycle on related larger and smaller systems. Converse action opens the door to a cycle asynchrony and a system failure. One of the most familial generators of cycles is the electromagnetic radiation from the Sun which induces the oscillations of day and night. The related and multifaceted daily sleep and wake cycle as well as the longer multi-month seasonal changes are consequences of polar rotation of the Earth and its annual movement around the Sun. Together, these rhythmic changes influence major cycles of biologic entities.

Living entities experience and express various iterations of pendulum-like swings of their cycles in their physiology and morphology. These changes are controlled by resetting mechanisms which trigger alterations in the projected cycle trajectory. The resetting influence can cause a cycle to repeat itself, be elevated to a different but still rhythmic phase, analogous to Kohler's 1920 musical transposition (Drack 2007), or be moved into the abnormal zones of *chaos* or *entropy*. The resetting function is either inherent within the entity itself or is part of some larger external cycle. The resetting mechanism represents an important factor in cycle control. It is intimately related to the system's self-adaptation to its environment which in turn depends on the semipermeability of the system's boundary. Loss of this dynamic functionality threatens the entire system. The size of cycle movement also suggests the perspective from which it is best to consider it as disproportionate views may not make the pattern visible in its proper context. Outside of its physiologic range, a cycle morphs into dysfunction with increasing instability which leads to a greater evolutionary uncertainty.

Cycles do not exist in isolation as they are all part of a continuum, from small cycles to larger ones. Each cycle absorbs some characteristics of smaller cycles and provides components to larger ones as well. Cycles express relatedness, a key characteristic of systems, implying the interconnectedness and dependability of all components. This perception may be broadly seen as *Gestalt*, a perspective stressing that individual parts cannot stand alone or that a system cannot be understood through the simple amalgamation of its parts, but only in the context of the whole. In a well-functioning system, components are in relationships that optimize the entire system and not solely maximizing themselves.

The many examples cited in this text come from various fields of life sciences, covering a span from basic biology to clinical medicine. They are selected because they appear to support some important aspects of systems science and thus begin to form a translational pattern of cycle relatedness which is congruent with systems principles. As stated by Bertalanffy and Weiss, "Empirical data have to be ordered by laws." (Drack 2007) The emerging montage is of course incomplete, offering an opportunity for others to contribute vital pieces to this amazing convergent framework of systems and life sciences. Each scientific field will be enriched by reflecting the other.

Systems science has a cross-disciplinary reach encountering differing principles and uncommon nomenclature making each field isolated in its own technical terminology. In order to provide some degree of shared understanding, from a single cell processing to a human being, or from a society to the larger ecology, an attempt is made to find applicable metaphors. In this instance, metaphors substitute for the still to be developed unified language, critically important for unhindered systems' communication. The inter-relatedness of systems does generate fitting metaphors which could enhance the comprehension of the system-within-a system concept. For example, human body can be seen as a system; similarly a society could be seen as a biologic body with related functionality, needs, and aberrations. The allegorical process used in this text attempts to reflect not just the isomorphism of unique principles common to systems and biologic entities but also the unique way of human comprehension which is strongly based on created images.

A multi-level model is described below which attempts to assist in understanding of complex systems expressing cyclic dynamic states of biologic entities with transitions in and out of various boundaries. A single-level model would help to elucidate the function and structure of a component, a subsystem, but only the integration of all known levels provides a full system's view. Human body is a large system with a balanced hierarchy of a vast number of subsystems engaged in self-organization and self-adaptation.

It may seem incongruous to attempt to look at small and large cycles through the same prism of systems science at a time of incomplete basic science information in many areas of biology and often only existing generalities of systems principles. The potential benefit to both, even at this point of time, however, is quite substantial as the conceptual understanding of cycles and systems can greatly benefit all life sciences, by providing potentially a large blueprint awaiting detailed exploration. Similarly, systems science can greatly benefit from embracing biology which epitomizes the greatest systems in existence.

This model visualizes biologic entities, from their *initial state* to the *end state,* transitioning in a pendulum-like movement through the conceptualized *zone of order*, *chaos*, and *entropy*. A potential also exists in this model for aberrant paths leading to cancerous growth, autoimmune activity or even congenital deformities on one side, with aging, degenerative, and inflammatory changes on the other. The heart of this model lies in the *health territory*, which is straddling the *outer core* of the *zone of order* and the *inner edge of chaos*. The *health territory* is an active place of *self-organization* and *self-adaptation* within a nonlinear dynamical system following the principles of *organized complexity*; the stability of oscillations depends on the efficacy of resetting mechanisms and the specific functionality and structural integrity of system boundary. Outside of the limits of a well-functioning system's *health territory*, a biologic entity can enter either further into the *zone of chaos* or a *zone of entropy,* each dramatically affecting the system.

Winfree (1987) points out that "...living things on this planet are built of units, in each of which a copy of the organism's genetic code dominates chemical activities...These cells are units of function, not just structure...." (p. 128) Furthermore, he states that "Most living organisms are built of cells because one nuclear copy of the genetic code can efficiently dominate only a limited volume of the reacting gel around it." (p. 132) For example, in suspensions of yeast cells, "when the cells get more than about twenty diameters apart, the amplitude of their collective rhythm falls abruptly....". (p. 128) This fact may be the primary determinant of the size of any functioning biologic unit as it needs to be in a mathematical relationship to its information source, the genetic material.

A *singularity of oscillation* is an imaginary fixed point of the pendulum-like movements characterizing cycles. The renewal of embryonic stem cells may serve as an example of such repetitious cycle of multiplication. A *singularity of differentiation,* on the other hand, is conceptualized as a different point of oscillations which is characterized by a pre-set number of cell cycle repetitions and by some specificity of functioning. Progenitor cells may be such an example.

The described model begins with an *initial state*. It is a pre-*system* phase because a true *system* does not exist yet; there are only potentialities, among the multiple components, for the development new relationships, structures, and the creation of a system's *emergence*. This state is characterized by randomness and in system's terminology-*equipotentiality.* (Bertalanffy 1969) This state can advance to *self-organization*, a process of going "...toward higher differentiation." (Bertalanffy 1969) In human evolution, this phase could be analogous to the movement of eggs and sperms prior to fertilization.

The systems' *evolutionary phase* follows the completion of the *initial state* but prior to a stage of a fully-functioning *open system.* It can be epitomized by embryogenesis with the onset of *complexity* and *self-organization* materializing the development of one larger system's *emergence*, a new living entity. Complexity begins with the onset of relationships; and, with progressive layering of relationships, hierarchy emerges. Two basic forms of relationships exist in a system: static/structural, often defined by fractal geometry, and dynamic/functional relationships, characterized by feedback loops studied by the field of cybernetics. The primacy of relationships over components within a system is maintained even when a component is replaced; the relationships remain. Within biologic systems, function determines structure, providing that the system retains its open characteristics and its self-organization. Drack et al (2007) quote an observation by Weiss (1922) that "hierarchical order is necessary to allow actions...[of] a whole."

During a time-limited gestation, the emerging system retains the characteristics of a *closed system*; once enough *complexity* and *self-organization* develops, the c*losed system* must become an *open system* and this takes place during birth. In an *open system*, there is a gradient of energy and information between the intake and the output, creating system's functionality while subjecting it to epigenetic influences. Input of an open system should be: appropriate (matching system's needs and its capacity), adequate (matching volume/scope), timely, and sustainable. Embryogenesis, systematized as the evolution of multiple *subsystems* into a large *system*, is characterized by an exponential growth, the presence of significant angiogenesis, and temporary protection from immunosurveilance of a larger open system (the mother), among other features. This process is pre-programmed to end when the *closed system* needs to change into an *open system*. If a successful transition doesn't take place, the *closed system* of embryogenesis and ontogenesis spirals into maximum *entropy* and ceases to exist (e.g. a stillbirth). Such a process reflects the fact that the closed embryonic system became a failing system accumulating more and more of system's nonfunctional components. This fact leaves little for the development of system's actual functional adaptation necessary to allow the system the life-saving leap from a closed to an open system.

A fully functioning *open system* expresses a number of defining characteristics. Among the predominant features are: *relationships*, unified *communication*, *self-organization* and *self-adaptation*, and the potential to create a new *emergence*. System's stability depends on the quality, quantity, and diversity of patterned *relationships*. In an *open system*, the balance is not just a reflection of its internal *relationships* (*self-organization*) but also its external *relationships* to larger *systems* (*self-adaptation*). System's resiliency is related to its diverse redundancy.

The process of *self-organization* of biologic entities can be epitomized by a neural network which is constantly being changed through the ongoing construction and destruction of synaptic connections in the brain. (Satinover 2001) This neural complexity provides the framework which allows data to be organized and morphed into information and knowledge. Satinover makes an analogy between a neuron and an individual:

> Each individual neuron, like each individual in society, is itself likely to be a network composed of many individual processing elements...share a common computational and self-organizing capacity....[This] parallel computational capacity is woven into the very fabric of matter itself...and that intelligence at every level ...[is] a necessary concomitant....(p. 86) [Furthermore, there is a resemblance between how the brain and how a society is structured]..."The brain itself is composed of networked subunits. Human society is likewise divided into various groupings that define different scales." (p. 211)

Many system's attributes and *relationships* can be identified within a healthy human body which is an excellent example of a robust *open system*. By contrast, unhealthy human body exhibits inconsistent compliance with system's principles. Cancer/autoimmune diseases, for example, when seen from this system's perspective, initially exhibit features of a failing specific cellular *subsystem* but retain the potential to overwhelm the entire *system* through, what could be described as, a system's irrational exuberance. Aging, degenerative, and inflammatory diseases follow a mirror path in this schema. It is not a path of a positive exponential growth seen in cancer but a path of *entropy* with a negative exponential growth. We could also call this phase the system's irrational complacency and, at a certain point, *entropy* becomes an irreversible process. This state of *entropy* is accompanied by the loss of the favorable gradient of energy and information/mental acuity (in higher biologic entities), both essential to the functionality of an *open system*. Laboratory experiments demonstrate that reduction in caloric intake, for example, not only prolongs lifespan, reverses the age-related decline in melatonin production but concomitantly improves mental acuity followed by the reduction of the incidence of degenerative diseases. (Reiter and Robinson 1995 p. 151, 153) Lowering caloric intake of an open system, from *ad lib* levels, optimizes the energy gradient.

Chaos is characterized by exponential iterations with a potential for runaway growth acceleration. Waldrop (1992) describes *chaos* as a

> nonlinear phenomenon [where]...a tiny event over *here* can have an enormous effect over *there*...the flap of a butterfly's wings in Texas could change the course of a hurricane in Haiti a week later...everything is connected and often with incredible sensitivity. Tiny perturbations won't always remain tiny. Under the right circumstance [and critical timing], the slightest uncertainty can grow until the system's future becomes utterly unpredictable...chaotic...[assuming a] pattern of ever-increasing [disorganized] complexity.

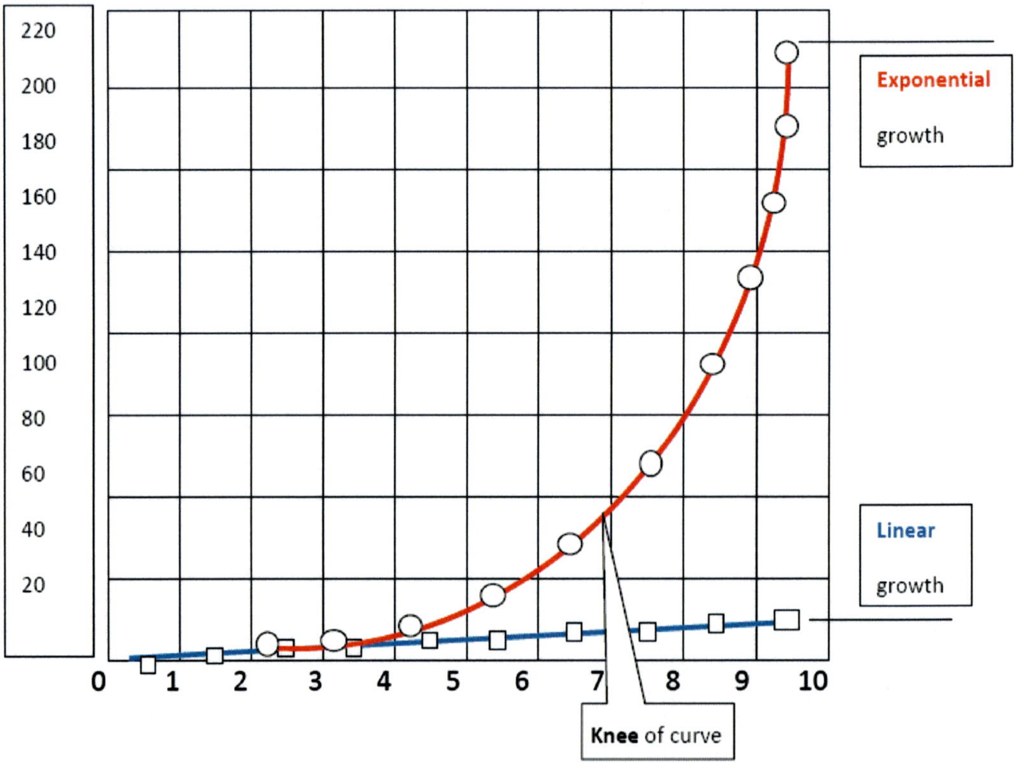

Figure 1. A comparison of a positive exponential curve with a linear one expresses the relationship between normal and malignant growth (Modified from Kurtzweil, 2005).

An exponential curve graphically illustrates the difference between the inner and the outer edge of chaos. The *inner edge of chaos* approximates the shape of the first segment of an exponential growth curve and runs in near proximity of a linear course (Figure 1) which represents a predictable form of a rhythmic growth. At an inflection point, the knee of the exponential curve, there is a significant divergence between longitudinal changes and the exponential ones; the *outer edge of chaos* is schematized to begin here and the exponential curve assumes a parabolic shape. The important ingredients of a functioning *system*, *relationships*, unified *communications*, and *organized complexity,* are breaking down along this parabolic path and, on a cellular level, the transcription of DNA to RNA is altered within the basic units of the *subsystems*, the stem cells. Kornberg (2006) stated that "Disturbances in the transcription process are involved in many human illnesses such as cancer, heart disease and various kinds of inflammation...[and that] the capacity of stem cells to develop into different types of specific cells with well-defined functions in different organs, is also linked to how the transcription is regulated." The structural and functional dilemmas of being in the *zone of chaos* have to be eventually resolved by living entities, either with the return to physiologic oscillation, a steady state, or by allowing the undifferentiated growth to continue into self-destructive status (cancer, autoimmune disease, etc.). (Fodde 2006) The importance of unified communication within a system is highlighted by Arevalo (2007) who reports that limited English proficiency (system's communication) greatly increases the risk of adverse events among hospitalized patients (system's components) when compared to English-speaking patients.

The final phase of a pendulum-like oscillating *system* of this model is the system's *devolutionary phase,* expressing *entropy.* It manifests features of degeneration, inflammation, and senescence. This phase also begins outside of the *health territory* and precedes the *end state.* It can be seen as an opposite image of the system's *evolutionary phase* because the *system* loses its *organized complexity*, *self-organization* and *self-adaptation*; it looks less and less like an *open system*. Randomness, volatility, and *disorganized complexity* with mutations can be observed; functionality is decreasing due to increasing *entropy* and the whole system is closing down. Laszo (1996) stated that "entropy can only increase in time in any isolated system [and that] such a system runs down [due to the fact that] the sum of the energies used up is always negative-more energy is used up than is generated." There exists a shortcut, possibly assisted by the process of inflammation, from *entropy*, negative exponential growth to a positive one leading to cancer, bridging the known cancer and aging association. Downregulation of sirtuins genes plays an important role in this process. (Ashraf et al 2006) Similarly, as observed by Richards et al (2007), declining levels of vitamin D which affect the telomere length of leukocytes, is a predictor of aging-related diseases. This process is heightened during inflammation. On the other hand, boosting the levels of vitamin D was found to be associated with longer leukocyte telomere length. There are additional examples illustrating how cellular processes of the *zone of entropy* can morph into the *zone of chaos*. For example, Peebles et at (2007) report that in lung cancer, inflammation can block apoptosis and concurrently stimulate the multiplication and invasiveness of mutated cells. The suppression of the immune system and increased angiogenesis further promotes neoplasia. Another illustration may come from psychiatry: Depression and dementia can be seen as examples of diseases reflecting system *entropy*. These clinical diagnoses reflect progressive neuronal loss likely due to inflammation. (Leonard 2007)

The *end state* follows system's *devolutionary phase* and is characterized by a negative exponential growth. Once an *end state* is reached, a *system* ceases to exist as an *open system* and begins to resemble a *closed system* with minimum of energy and information exchange and a non-permeable, rigid boundary. Even randomness ends due to the overwhelming *entropy*; the system ends. "A closed system *must*...eventually attain a time-independent state of equilibrium, defined by maximum entropy and minimum free energy...." (Bertalanffy 1969) An *end state* can also be observed at the termination of a positive exponential growth, characterizing the *zone of chaos*. At that point, the system boundary totally loses its semipermeability allowing entry and egress of uncontrolled streams of energy and information; what previously was defined as a system no longer possesses such characteristics. The system disintegrates into the external environment.

Human aging and degenerative/inflammatory diseases resemble the *devolutionary phase*. From system's perspective, aging, classically thought to parallel increasing frailty, involves a phase transition of life cycle into *entropy* which often does manifest degenerative/inflammatory diseases. Eskandari et al (2007) report on correlation, in premenopausal women, of low bone mineral density and depression which exemplify the process of *entropy* involving system's structure and function. Depression is known to be associated with physiologic alterations such as a hyperadrenergic state, hypercortisolemia, and an increase in inflammatory cytokines. The authors did find increased levels of pro-inflammatory cytokines (interleukins: IL-1β, IL-2, IL-6, and tumor necrosis factor α), and decreased levels of anti-inflammatory cytokines.

The process of dying of a biologic entity reflects changes from an *open* into a *closed system*. An active enhancement of longevity correspondingly suppresses cancer development. The sirtuins genes have been identified experimentally as proteins associated with increasing life span. Goymer (2006) stated that, the beneficial "mutations that extend lifespan in *Caenorhabditis elegans* also inhibit tumour growth."

System *boundaries* have anatomic and functional characteristics further defined by their internal and external specificity. The external boundary separates the entire *system* from its environment, exemplified by cellular membranes or skin. The internal boundaries refer to the individual subsystem perimeters, for example, from intracellular mitochondria to lining of internal organs. Each system *boundary* has a selective porosity also known as semipermeability. Under healthy conditions, such a *boundary* allows for specific forms of energy and information to enter and/or exit the system; a blood-brain barrier is an example of an extremely sophisticated and critically important subsystem boundary. In an *open system*, the competency of *self-organization* and *self-adaptation* affects the stability and integrity of its *boundary*. Once the defined functionality of the boundary is lost, the system becomes destabilized. In a cellular system, for example, free radicals increase fragility of membrane boundary while melatonin, a potent antioxidant and hormone, maintains system boundary. (Reiter and Robinson 1995) Subsystem pathological processes, such as malignancy, exhibit tendencies to invade surrounding structures by transgressing the system's failing boundaries. When observed in a clinical situation, such a development often heralds the onset of a metastatic process, most often seen in advanced and terminal cancer cases. By comparison, a process with characteristics of a temporary and physiologic delay of system resetting mechanism, while also allowing for some degree of local invasion, is observed in the early stages of gestation and during embryonic morphogenesis. Here, the distance of invasion and its duration are limited allowing for the timed embryogenesis of system's emergence. By comparison, the invasion by malignant cells, however, is not limited by distance or time.

All living systems have their specific *boundaries*. Retaining their integrity is critical to each system's health. For example, animal systems may contain various symbiotic bacteria/viruses which, however, may be pathogenic to humans. When the natural boundaries between humans and animals become defective, pathogens like Ebola, West Nile, AIDs, etc. can threaten human system. (Brilliant 2007)

A *boundary* is a reflection of the sophistication of the enclosed system, e.g. skin vs. blood-brain barrier, and its damage threatens the entire system. It is the boundary's semipermeability which maintains the system in its *health territory*. The destruction of a system boundary can come from the outside or from within the system. Becoming permeable/porous to energy/information, facilitates the development of *disorganized complexity* and *chaos* within a deteriorating *system*. For example, a failing blood-brain barrier promotes seizures, a true electrical chaos and disorganized complexity of the central nervous system. (Holmes 2007) Similarly, there is a well-known correlation between UV skin damage and cancer.

As previously mentioned, one of the most important cycles, that all living entities encounter, is the cycle of electromagnetic radiation from the Sun. It represents a fundamental example of interconnectedness of systems as this cycle has a profound effect on human body. The varied sleep/wake circadian cycle changes can be observed in individual cells and the entire human body which also experiences morphologic as well as cognitive changes. In humans, the genetically programmed internal clock is centered in the *suprachiasmatic nuclei*, which are programmed to transmit ongoing signals, via *norepinephrine,* to the pineal gland to produce melatonin. The genetic programming, set for a 25 hour cycle, is reset by light during the 24 hour Earth rotation. (Reiter and Robinson 1995 p.110-111)

The influence of this grand light/darkness cycle reaches deep into practically all biologic subsystems. Brown (2006) reviewed the literature regarding the connection between vitamin D, generally induced by sun exposure, and multiple sclerosis (MS). He found that a large prospective cohort study reported 40% reduction in the risk of developing MS with vitamin D supplementation, where inconsistent sun exposure exists. Similarly, in mice, vitamin D supplementation prevented the development and progression of autoimmune encephalitis, an animal model of MS. Low level of vitamin D has been linked to certain cancers. (van der Rhee et al 2006) On the other hand, higher levels of vitamin D favorably affect overall mortality. These findings indicate a broad impact of sunlight cycles on human health. (Autier and Gandini 2007) Within the described model, less than optimal synchronicity with the light cycle creates a system environment that is outside of the *health territory*.

The production of melatonin is intimately related to the light cycle and has a profound effect on numerous cycles of biologic entities. For example, "Animals 'interpret' the amount of time that melatonin is produced as the true sign of a seasonal cycle. If melatonin is produced for fourteen hours at a stretch…that signifies winter…If it is produced for only ten hours, it must be summer…melatonin [also] triggers seasonal changes in humans as well." (Reiter and Robinson 1995 p. 123) The significance of balance of the sleep/wake cycle is well known clinically. According to Reiter and Robinson (1995), it is the "melatonin's ability to lower body temperature that is one of the keys to its ability to enhance sleep…[as] temperature [should be] rising during the day and falling at night. "(p. 103) Cook and O'Brien (2003) explored some of the consequence of diminished/foreshortened sleep cycle.

They found that subjects who slept only four to six hours a night for fourteen consecutive nights, showed a significant deficit in their cognitive performance which is critical for the information processing within an open system.

The common age-related decrease in melatonin production alters the regularity and the length of the sleep/wake cycle with concomitant decrease in functionality of some internal subsystems such as immunity. Rodriquez et al (2007), for example, observed in a laboratory model an age-related rise in inflammatory cytokines [interleukin (IL)-1beta, IL-2, interferon (IFN)-gamma, tumor necrosis factor (TNF)-alpha, and granulocyte-macrophage colony-stimulating factor]. Chronic administration of melatonin, however, reversed this process. In another study, Gupta et al. (2004) observed in patients with epilepsy, which is known to manifest cellular oxidative stress, that melatonin significantly potentiated the antiepileptogenic effect of drug valproate by increasing the activity of antioxidant enzymes of glutathione peroxidase and reductase. Melatonin demonstrated to be a powerful neuroprotector due to its known antioxidant properties.

Petraschek et al describe a life-extending effect of antidepressant mianserin in their research nematode model, *Caenorhabditis elegans*. In humans, this drug blocks neural signaling by serotonin through altered synthesis, synaptic re-uptake. It also interferes with several serotonin receptors. The blocked pathway of neurotransmission is apparently the same one implicated to affect longevity with caloric-restriction. There is a close and reciprocal relationship between serotonin and melatonin in the circadian biologic rhythm. Reiter and Robinson (1995) state that "Serotonin keeps you alert and energized during the daytime; melatonin helps you rest and recuperate at night." (p. 164) Melatonin is made from serotonin in the pineal gland (p. 229) and serotonin is made from the amino acid tryptophan. (p. 194) Lambert et al (2002) report that the production of serotonin in the brain was directly related to the duration and intensity of bright sunlight and also that the metabolic turnover of serotonin in the brain was the lowest in winter, connecting serotonin with melatonin in the ongoing oscillating pattern of the circadian rhythm.

Complexity governs the myriad of interactions within a *system*. "Complexity…[is] a science of emergence" allowing system's ultimate mission, the creation of *emergence*. (Waldrop 1992) *Complexity* gradually appears within the system's *evolutionary phase*, reaches its greatest functionality, the *orderly complexity,* within the *health territory*, and then steadily declines during system's *devolutionary phase* expressing *disorderly complexity*. As Kurtzweil (2005) said, "complexity is a continuum" and depends on its functionality, which is a ration of the engaged *complexity* components vs. those that are not engaged in productive energy-generating and information-sharing interactions. The character of c*omplexity*, which can be either organized or disorganized, expresses the true functionality of a *system* and reflects the quality of *relationships* which produces *self-organization* and *self-adaptation. Complexity* begins to appear at the *initial state*, when randomness gives way to *self-organization*. It is the "…spontaneous emergence of order…[allowed by] a constant flow of energy and matter through the system…[with the] emergence of new structures and new forms of behavior…[which are] the hallmark of self-organization, [and] occurs only when the system is far from equilibrium." (Capra 1996) *Self-organization* reflects the evolving interactions of innumerable relationships of the system's components. As the end-product of *self-organization* is system's *emergence*, the opposite state reflects instability, increase of errors/mutations, and a potential for total randomness. Kurtzweil (2005) also stated that, "a key requirement for a self-organizing system is nonlinearity;" he pointed out that "...a lot of nature is *not* linear..." [and that] "The self-organizing systems...[are] governed by nonlinear dynamics...." (Waldrop 1992) In contrast stands the traditional linear system "in which the whole is precisely *equal* to the sum of its parts...[and its] plot is a straight line...." (Waldrop 1992) *Self-organization* expresses purposefulness in achieving ongoing steady state of internal vs. external energy/information gradient with the potential for the creation of a system's *emergence*.

Complex adaptive systems (CAS) express a high degree of resilience and robustness to environmental challenges through their *self-adaptation* and internal *self-organization*. "They actively try to turn whatever happens to their advantage...a kind of dynamism that makes them qualitatively different from static objects such as computer chips or snowflakes, which are merely complicated." (Waldrop 1992) One important characteristic of *CAS* is that "...the control of a complex adaptive system tends to be highly dispersed," (Waldrop 1992) implying a highly effective self-organization/horizontal hierarchy, with only limited vertical hierarchy. Lenski *et al* (2006) offer a discussion of robustness and evolvability: "One of the most important features of biology is the ability of organisms to persist in the face of changing conditions...To achieve this consistency, organisms must have a balance between robustness and evolvability, that is, between resisting and allowing change in their own internal states."

A modified image of the planet Saturn with its rings (Figure 2) provides a visual expression of the presented model with a superimposed pendulum, oscillating from a point of *singularity* and subject to resetting. The planet's rings represent the *inner* and *the outer edges of chaos*. The planet itself is imagined as the *zone of order* of this model with an *inner* and an *outer core*. The *inner core* represents a linear, rigid space in the model where the oscillating pendulum encounters friction which persistently tries to bring the pendulum to a halt. To continue its oscillation, pendulum needs ongoing inflow/outflow of energy and information involved in generating *complexity*. The basic level of *complexity* can be expressed here even in mechanistic terms, analogous to Newtonian physics, and allows for a linear prediction of future states with a great degree of uniformity but it lacks adequate adaptability.

The detrimental rigidity of the *inner core* of the *zone of order* can be observed clinically when spontaneous fluctuations in the heart electrical signal are severely limited. Both Ludwig (1847) and Goldberger *et al* (1990) demonstrated the need for a degree of *chaos* in a healthy heart. An electrocardiogram without a degree of *chaos* actually reflects an underlying cardiac abnormality, especially the lack of adaptability. These observations support the view that a perfect order is a non-physiological and a non-healthy state. A biologic entity in this range is in danger of *system* failure from any sudden change as its prevailing rigid predictability has suppressed its functional adaptation. Ben-Dov et al (2007) documented that the absence of dipping heart rate to sleep levels is directly related to all-cause mortality, signifying the importance of internal organ cycles to be in synchrony with the dominant circadian rhythm. A degree of variability is essential to maintain a *complex adaptive system*.

The *outer core* of the *zone of order* engages in some functional adaptation; the *system* is responsive to changes within a certain range. When the *outer core* combines with the *inner edge of chaos,* it forms the *health territory* where system's components are routinely replaced but the basic *relationships* and methods of *communication* persist. In this healthy state, living entities are undergoing cellular renewals; immunosurveilance is functioning well, and agniogenesis has its limits (Figures 3a, b).

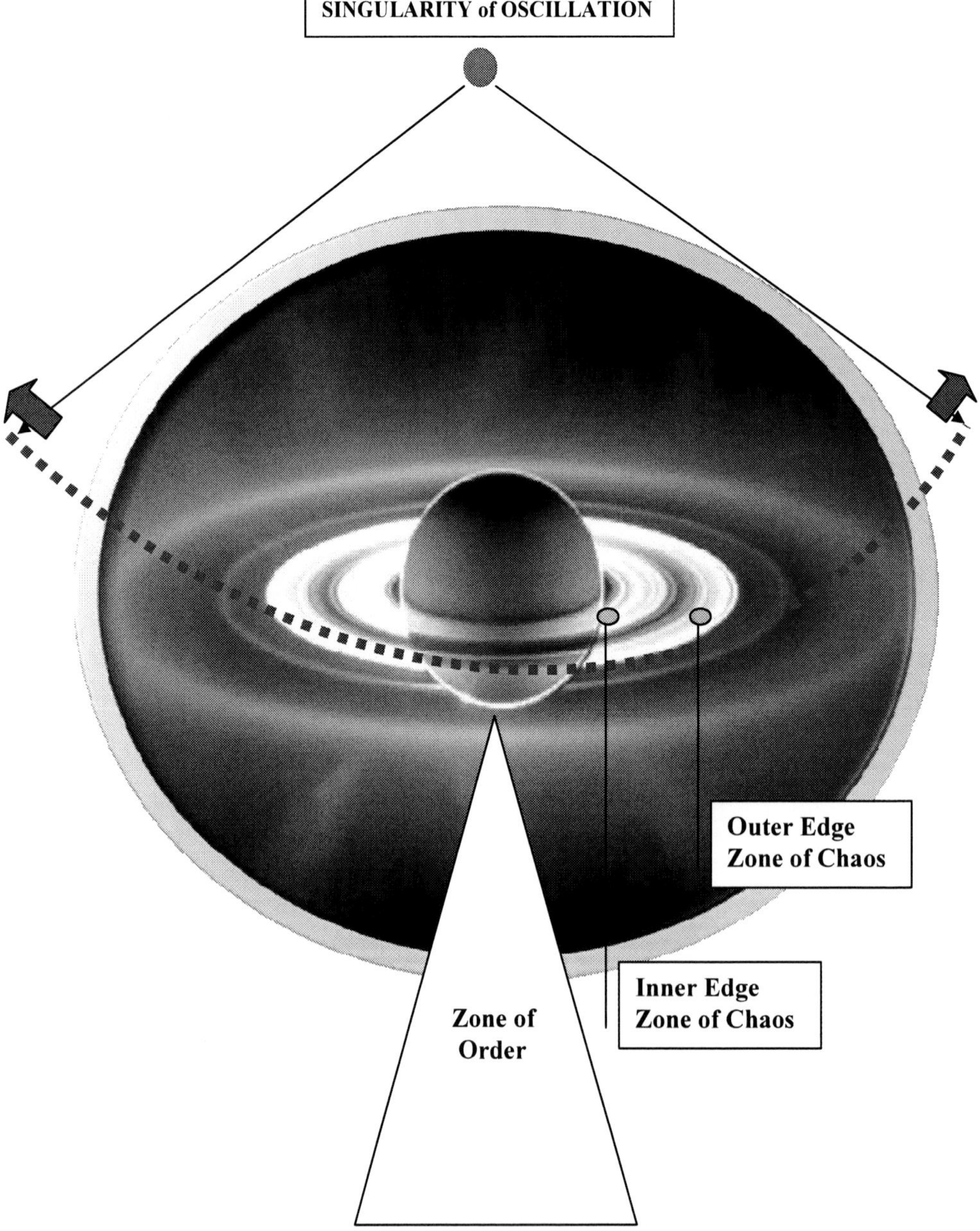

Figure 2. The model: pendulum-like transitions of biologic entities through *zones of order* and *chaos*. Adopted from www.nature.com/.../061009/full/061009-11.html

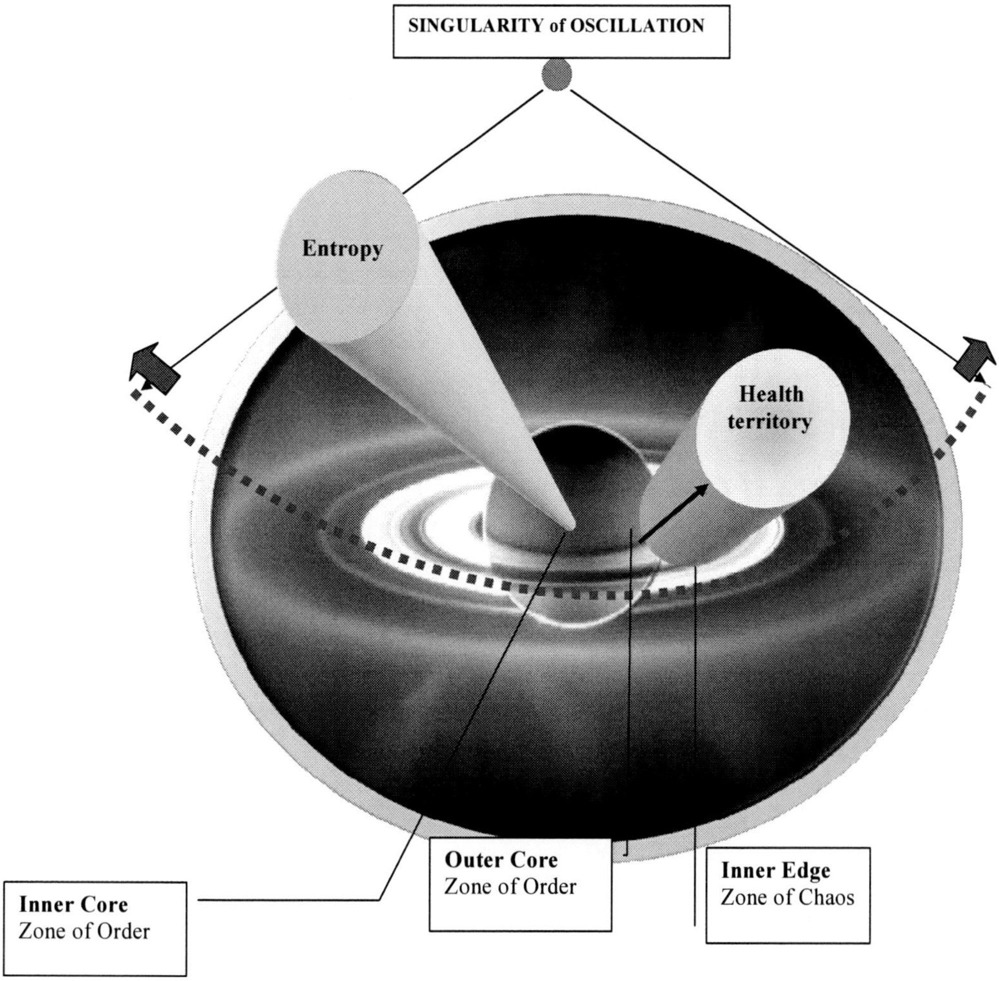

Figure 3a. The heart of this model lies in the *health territory*, which is straddling the *outer core* of the *zone of order* and the *inner edge of chaos*; it evolves from an initial state. The *inner core* of the *zone of order* favors *entropy* and an end state. Adopted from www.nature.com/.../061009/full/061009-11.html

The entire *zone of chaos* is expressed in fractals which is the architecture of multidimensional geometric patterns ubiquitous in nature. They express a feature of *self-affinity* and *self-similarity* at various scales (e.g. branching of trees or a pattern of angiogenesis). "Fractal geometry…[is a] mathematical language to describe the fine-scale structure of chaotic attractors…[Authored by] the French mathematician Benoit Mandelbrot…to describe and analyze the complexity of the irregular shapes in the natural world…." (Capra 1996) The branching of a fractal pattern is related to the sudden appearance of a new *chaotic attractor* causing the bifurcation of an established pattern of growth.

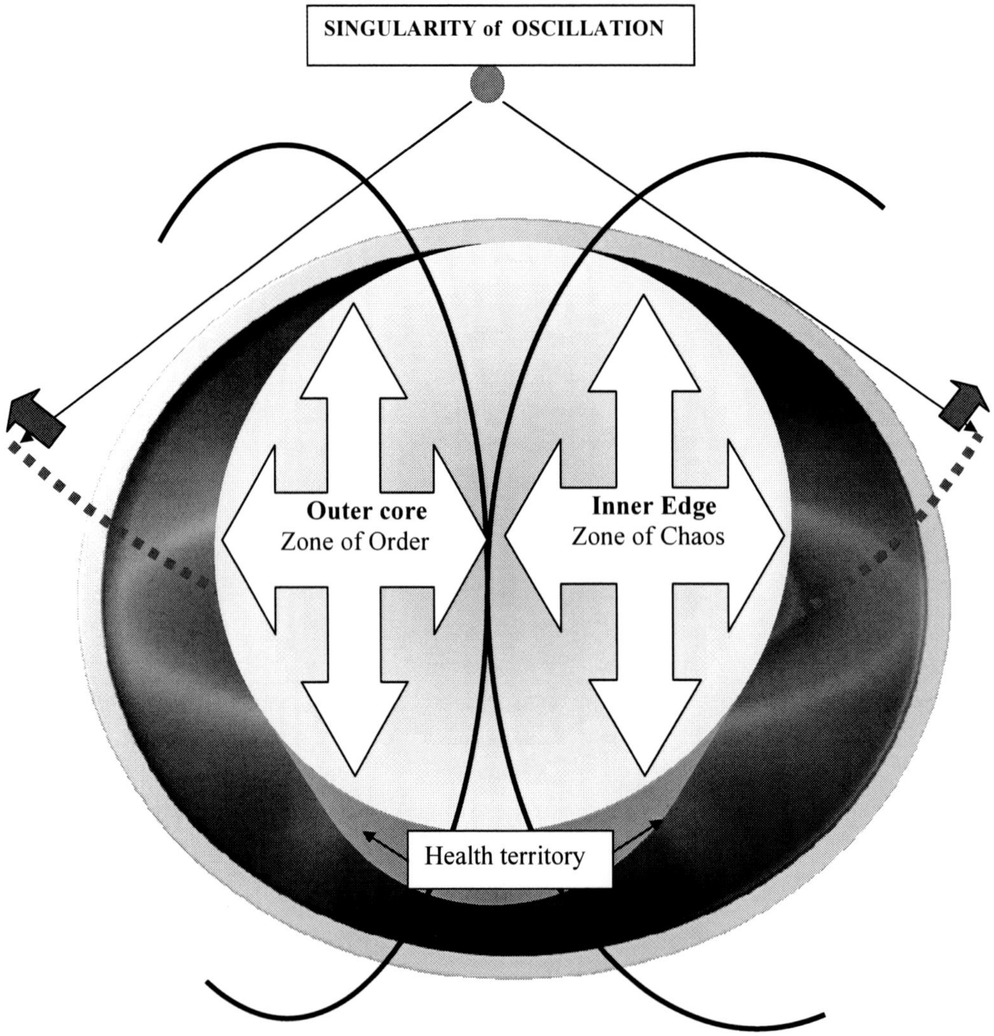

Figure 3b. Details of *health territory.* Adopted from www.nature.com/.../061009/full/061009-11.html

> All of these complex systems have somehow acquired the ability to bring order and chaos into a special kind of balance. This balance point...called *the edge of chaos*...where the components of a system never quite lock into place, and yet never quite dissolve into turbulence...where life has enough stability to sustain itself and enough creativity to deserve that name of life...where new ideas and innovative genotypes are forever nibbling away at the edge of the status quo.... (Capra 1996)

"The edge of chaos is the constantly shifting battle zone between stagnation and anarchy...place where a complex system can be spontaneous, adaptive, and alive." (Waldrop 1992) The similarity of a fractal pattern and cancer growth can be seen in a comparison of Mandelbrot's drawing and a histologic picture of Merkel cell carcinoma (Figure 4).

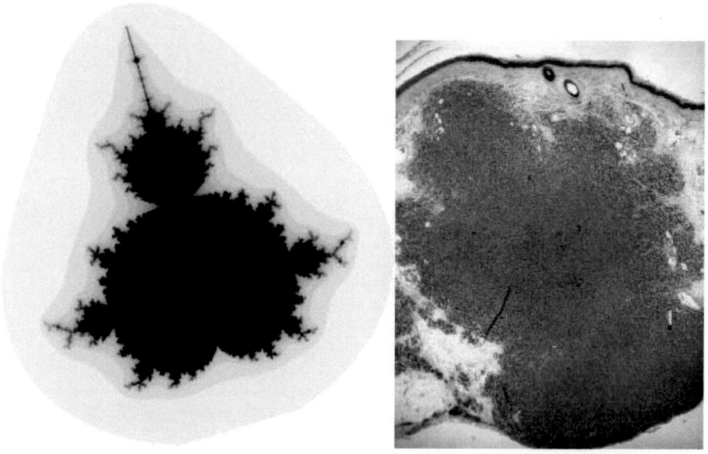

Source: (local.wasp.uwa.edu.au/.../fractals/fracintro/ and http://path.upmc.edu/cases/case398.html respectively)

Figure 4. Similarity between Mandelbrot fractal (geometric) pattern (left) and cancer (Merkel cell carcinoma; right).

The *outer edge of chaos* represents a *far from equilibrium* state where immunosurveilance is malfunctioning and neoagniogenesis has no limits. From here, cells may undergo exponential malignant growth. In this region of the model, cells basically stop communicating with the entire *system*, breaking off system-wide *relationships*, and grow at the expense of the larger *system*; some feedback loops persist but only in order for the tumor cells to harness the energy of the *system*. The transcription of DNA to RNA is altered (Figure 5).

Several options of progression exist for living entities functioning within the he*alth territory*. For example: cells can continue to divide as needed by the *system* (e.g., adult stem cells); cells can be programmed to undergo differentiation with a limited number of divisions (e.g., progenitor cells); cells can also undergo senescence with apoptosis. In this process, the p53 tumor suppressor gene plays a critical role. It is also possible that cells undergo an exponential malignant growth if their original programming, for a fixed number of cell divisions, fails. Cancer develops within the *outer edge* of the *zone of chaos* where cells exhibit the exponential growth phase; when plotted, the growth pattern approximates the parabolic phase of a positive exponential curve beyond its knee. It is here that *self-organizing* and *self-adapting relationships* of the entire *system* are visibly weakened making cancer a product of a failing *system*; the neoangiogenesis is activated and immunosurvailance fades. Malignant cells may also appear within the *inner edge of chaos* but their growth follows only the first segment of the exponential curve which approximates the longitudinal line. The integrity of the *system* is still intact and neoangiogenesis has not been activated yet; immunosurvailance is still active and clinical cancer has not developed.

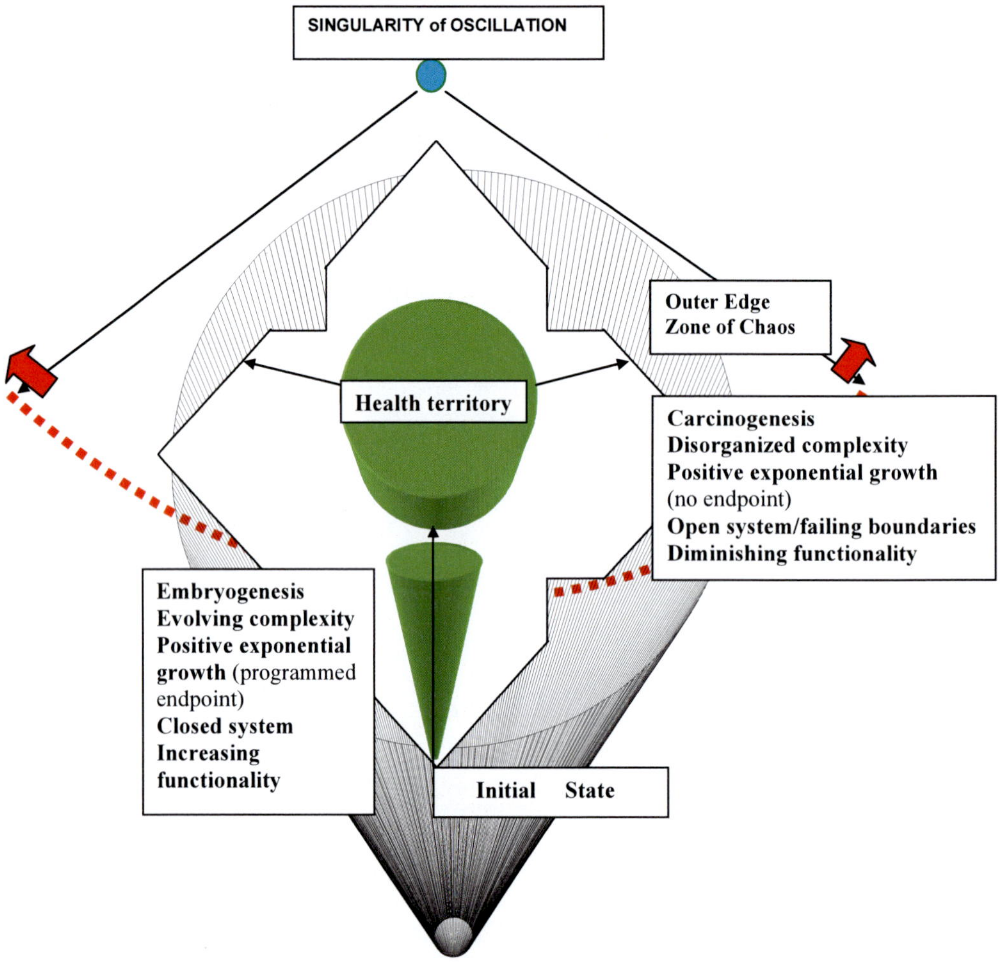

Figure 5. Evolution of a system: from the *initial state*, through embryogenesis to the *health territory*, and potentially to the *outer edge* of the *zone of chaos* exhibiting carcinogenesis.

On a cellular level, this model exemplifies the processes of adult stem cells. Tan *et al* Tan et al 2006) summarized that,

> Normal adult stem cells are tissue-specific…[with] the ability to self-renew and to differentiate…[they] can undergo an unlimited number of cell divisions, and with each division, they produce at least one daughter cell that maintains this indefinite capacity for cell division…they can also produce progenitor cells that have finite division capacity, ultimately differentiating into the mature cell types.

The distinction between the adult stem cell oscillations and the progenitor cell oscillations is evident in the cellular energy focus. Among the adult stem cells, the primary focus is on reproduction/cell division without much concern for other cellular functionality. On the other hand, progenitor cells, with their target-specific function, have only a limited potential for reproduction.

Mitchell (2007) favorably comments on research done by Eiges et al (Eiges et al 2007) who established a human embryonic stem cell line from a pre-implantation Fragile X-affected embryo (the most common genetic form of mental retardation). In spite of a full mutation, epigenetic silencing by DNA methylation and histone modification occurred upon differentiation. This study potentially expands the applicability of this systems' model beyond cancer to genetic diseases as well. It appears from Mitchell's (2007) description that even strong genetic determinism, in some cases, could be modified through epigenetic influences. Epigenetics is understood as a process of heritable traits carried out not through modification of DNA nucleotide structure but due to the susceptibility to environmental influences of various bonds formed during the creation of a new double helix.

The main feature of this model is its ability to express various cycles of living entities as an interrelated and a multifunctional *system*. It connects concepts of *systems, complexity,* and *chaos* with features of stability and instability, linearity and nonlinearity, bifurcation, attractors, etc., each constituting an important feature of one of the main theories. This model identifies an optimal zone of function, the *healthy territory,* where the principles of *deterministic chaos* of nonlinear dynamics operate. For vigorous functioning of a system, a healthy balance between order and variability needs to exist as a result of successful resolutions of recurring transitions from linearity to positive or negative exponentiality. Outside of this *health territory*, there exists a significant risk to a living system.

The principles of this model may be applicable to an entire range of living entities, from stem cells to the human body. By extrapolation, similarities may also be seen in the fields of physics, business, and psychology with model polarity encompassing inflation/deflation, anxiety/depression, peace/war, etc. as expressions of exponentiality/entropy. Even the science of organizational management embraced a biologic system to formulate the latest organizational theory, called an organismic theory. (Hatch 1997) Systems science also shares several features with quantum physics: the wave-particle duality, represented by component-relationship duality, and contextuality, which are inseparable from any entity existing in quantum physics or in systems. The features of indeterminism, absolute chance, and lack of trajectories (Satinover 2001, p. 130-131) may be found within systems when they reach the *outer edge of chaos*. The primary importance of *relationships* vs. *components* has been emphasized in physics. The *systems* science concept is close to Smolin's Loop Quantum Gravity theory (2003) which elevates relationships to the level of fundamental building units of the Universe. It postulates that "everything is built up from a network of relationships…[and that] when the network is tied in a braid it forms something like a particle." (Green 2006)

Biologic entities, ranging from cells to the human body, can express the key phases of this system model as a state of health (the *health territory*), cancer/autoimmune diseases/congenital deformities (the *zone of chaos*), and degenerative/inflammatory diseases/aging (the *zone of entropy*). The scope of this model can be clarified by connecting it to an analogy which offers familiar terminology and generally known experiences. The *zone of chaos*, for example, is of critical importance to a *system* as it represents a state of uncertainty and heightened potentialities. An analogy can be made with the human experience of "*relationships* in transition;" it is simply a confusing time for all parties involved. The ordinary rhythm of life and the integrity of a system boundary have been lost and replaced by unexpected reversals; one trend/cycle may be ending and another might be trying to emerge. This analogy focuses on *relationships* which are the foundation of systems but they also contain inherent fragility with the potential to bring the entire system into the *zone of chaos*. Not all is lost in this *zone,* however, as for example in the *inner edge of chaos,* just outside of the postulated *zone of order*, the probability still exists for a positive resolution of *relationships* in transition. With further progression into the *outer zone of chaos,* the chances for a total *system* failure, *relationships* and components/partners, increase exponentially. The *complexity* can be seen as patterns expressing the *relationships*, and *chaos* as the zone of *relationship* transition. As among human beings, there are ongoing *relationships* among cells which may be conceptualized to exist within similar framework of interactions: normal, abnormal but not cancerous, abnormal but cancerous, and/or abnormal but extinguishing. Of interest are observations by Wilkie (2007) who reports that identical mutation of the same gene can lead either to congenital malformation or to cancer depending upon the cellular and temporal context. This finding is in line with *chaos* theory of criticality of timing and context.

It is the General System Theory that offers a theoretical path to integration of advances in various disciplines through its concept of *isomorphism*. It implies that some principles discovered in one discipline can be transferred to another which is still searching for a similar understanding. Bertalanffy (1969) stated that "with the transfer of principles from one field to another…it will no longer be necessary to duplicate or triplicate the discovery of the same principles in different fields isolated from each other."

One of the key characteristics of a *system* is the focus on the importance of patterned *relationships*, each with a range of potentialities that eventually coalesce into a system's *emergence*. System's stability depends on diversity, the quality and the quantity of *relationship* interactions, reflecting the degree of relativity among various components. In an *open system*, its balance is not just a reflection of the quality of its internal *relationships* (*self-organization*) but also of its larger external *relationships* (*self-adaptation*) to a bigger *system*. Redundancy, reflecting the absolute number as well as the variability/diversity of components, is needed to maintain resiliency of a *system*. With diminishing redundancy, resiliency falters. In general, this is true in the early and the latter phases of biologic entity's oscillations expressed in this model. Redundancy is greatest in the *health territory* and diminishes toward the periphery of this model.

Any growth of a system should be proportionate to its simultaneous functional network expansion which should be supported by an appropriate relationship between vertical and horizontal *hierarchy*. That means that there is a limit on the size of a system and its functionality. The network of system's *relationships* resembles neural network and for that reason, clues as to the size-to-function relationship, could be derived from neuroscience. Leingartner et al (2007) report that mammalian neocortex area size can dictate performance at modality-specific behaviors. They suggest that areas have an optimal size which is influenced by parameters of its neural system in order to achieve maximum behavioral performance. According to the researchers, the area size that gives optimal performance is the one that is best tuned to the context of the neural system within which that area functions. The cortex, in this instance the larger system, needs to appropriately fit the functional profile of the subsystem handling the energy/information. If the proportions between the system and the subsystem are not of a favorable ration, the energy/information processing will not be effective and the overall system will lack in its performance. This observation does reflect the concept of a "system-within-a system" relationship and the dependency, in terms of its optimal size and function, of a subsystem on its larger system. As mentioned above, a similar relationship was observed by Winfree (1987) who stated that "Most living organisms are built of cells because one nuclear copy of the genetic code can efficiently dominate only a limited volume of the reacting gel around it." (p. 132)

In an optimally functioning *system*, there is a reciprocal relationship between horizontal and vertical *hierarchy*. *Horizontal hierarchy* reflects the degree of system's *self-organization*; *vertical hierarchy* resembles top-down dominant chain of command. As the horizontal hierarchy increases, the need for vertical/top-down hierarchy diminishes. These hierarchical relationships should be seeking a mutual balance. Favoring one over the other destabilizes the entire system, with a negative effect either on the internal *self-organization* or the external *self-adaptation*. This framework of hierarchy can be seen as one of the conduits open to the epigenetic influences on a given system. The reciprocal relationship of system's hierarchies may be schematically seen in Maslow's pyramid of human needs (Maslow 1968). When there is a predominant need for "safety," a fundamental need, there is also likely to be a strong demand for vertical hierarchy. On the other hand, once the higher stage of "self-actualization" is reached, there should only be minimal need for top-down hierarchy due to the fact that self-organization has reached lofty complexity and differentiation.

The renewal of a cell cycle through the process of duplication is directly tied to the length of its telomere. Under normal circumstances, there is an end point to cellular divisions. Each replication shortens the telomere and without the telomere-preserving enzyme telomerase, the cell cycle eventually ends. The shortening of a telomere, which caps each chromosome, parallels senescence. The *relationship* between a telomere and a telomerase, which maintains its stability by adding nucleotides to it, may also serve as an example of system's *resetting* process. The general blueprint for telomerase functioning is to remain silent in most tissues in order to allow cells to go through their programmed number of divisions. A major exception to this rule occurs in embryogenesis, when the enzyme telomerase is active, and is easily identified in cells. During this process, telomerase maintains the length of a telomere which then allows ongoing cell proliferation but still with a defined end point. A similar process seems to be repeated in carcinogenesis but this time without the intrinsic end point. The ongoing malignant exponential cellular growth continues unchecked with the capacity to overwhelm the entire system. As expected, telomerase, according to Shay (2005), the "ribonucleoprotein (RNP) reverse transcriptase, is not detected in most normal human tissues but is almost universally expressed in human cancers." The commonality between embryogenesis and carcinogenesis is not only in the rate of cellular divisions but also in their cells of origin, the stem cells. Another similarity but with a distinction, between embryogenesis and carcinogenesis, is their common propensity for tissue invasion. The distinction lies in the fact that during embryogenesis the cellular proliferation is directional, tissue specific, and has an end point. In carcinogenesis, this process is without any seeming purpose except to allow its own growth at the expense of the system. (Ferreti *et al.* 2006)

Richard et al (2007) studied the changing length of telomeres in leukocytes as a proxy for the state of inflammation, a process that fits in the *zone of entropy* of this model. The authors state that leukocyte telomere length (LTL) is a predictor of age-related diseases and increased inflammation, and that vitamin D is a strong inhibitor of proinflammatory response. Among women that the authors studied, they found that age was negatively correlated with LTL while serum vitamin D levels were positively associated with LTL. The difference, between the highest and the lowest tertile of vitamin D, was equivalent to 5 years of telomeric aging. These observations provide an interesting explanation about how one system can be impacted by a larger system, in this instance, the cycle of changing electromagnetic radiation. Sunlight generates vitamin D which in turn is an inflammation antagonist and telomere protector.

The clinical relevance of the cited examples indicates that there are already some biologic markers which allow objective determination of a cycle phase of a given system. It is plausible that all the various phases of abnormal cell life-cycles and their fractal travels through zones described in this schematic model are expressed in additional phase-specific cell surface characteristics or markers. This feature should help to identify, and potentially offer ways to correct, an aberrant cellular path.

A disease, including cancer, can be conceptualized as a "*system*-in-failing." A localized cancer, for example, represents a *subsystem* failure and a metastatic cancer can be seen as a total large *system* failure heading for its termination. Cancer, known to express an exponential growth of cells, is placed graphically in this model in the *outer edge* of the *zone of chaos*. At this point, the entire system's survival is being threatened. From this perspective, cancer cure would require not just successful cancer therapy of the gross disease, in order to limit the total cancer burden of the *system*, but also simultaneously understand and correct the "why" and "where" the entire biologic *system* went awry. Otherwise, the likelihood of developing another clinically apparent cancer is high.

The difference between the current single-target cancer therapy, focusing primarily on destroying the multiplying cells of a failing sub-system, and a comprehensive all-system approach, can be highlighted by Ashby's theorem of requisite variety that states that "...for effective control the variety available to the controller should be the same as the variety available to the system to be controlled." (Warfield 2004) The implications of this theorem are significant due to the fact that a complex *system* "...to be controlled has [almost infinite] *n* variables [and therefore] the controller must be able to control all *n* variables in order to avoid the significant risk of leaving some subset of those variables uncontrolled." (Warfield 2004)

The enormity of controlling all cancer variables, as looked at through Ashby's theorem, can be seen in considering the number of potential genetic and proteomic variations (30,000 genes and thousands of potential proteins from each gene). This astronomic number does not even include the essential relationships among genes, proteins, and so on.

How does this model fit with other observed biologic phenomena, for example, the observed doubling of secondary cancers in patients following stem cell transplantation for leukemia or myelodysplastic syndrome? (Tanne 2006) This model would indicate that the introduction of stem cells, with their dual growth potential, into a larger *system* could, on one hand, shift the balance of system's oscillations toward the *outer edge of chaos* and increase the potential for cancer growth, and, on the other hand, it could choose the path of stability through differentiation.

Systems in *chaos* are amenable to a stabilizing change through a process of differentiation. Similarly, *systems* descending into *entropy* could be rescued by increasing their *organized* complexity. The described model offers potential corrective pathways out of the system's undesirable states. First, there is a need, however, to recognize where a *system* is within the model. Is it in *chaos, zone of order,* or *entropy*? Determining a given system's location within this model, allows for the selection of the most appropriate framework for looking at the observed phenomena: linear thinking is best suited for the *zone of order* but nonlinear thinking is most appropriate for the *zone of chaos* and *entropy*.

To stop the undifferentiated exponential growth leading to cancer and the *disorderly complexity* of the *zone of chaos*, the *system* has to be reset to induce differentiation and be brought back to *orderly complexity*. In order to restrain the limitless multiplication of cancer stem cells, which are devoting all of their energy endowment to reproduction, there needs to be a change from their *singularity of oscillation* to a new *singularity of differentiation*. This would divert the finite cellular energy from pure reproduction to a better balance between function and repair/multiplication. This step could be accomplished through a process of increasing cellular function leaving only limited energy for reproduction. Such a concept would imply that there is a need for reorientation of treatment principles from cellular killing (cyto-toxic therapies) to cellular retraining (cyto-education). Cyto-toxic therapies, capable of eliminating 100% of cancer stem cells, have so far proven elusive. Physical activity and melatonin have been shown to facilitate the change-over process to differentiation. (Sainz et al 2003, Slattery et al 2004, Blask et al 2005) Laufs *et al* (2004) stated that "Physical activity increases the production and circulating numbers of endothelial progenitor cells...[and has] antiapoptotic effect." Also, Takahashi and Kubotu (2005) confirmed that "Exercise plays a pivotal role in myoblast differentiation." From the system's model point of view, it seems that physical activity, through its rhythmic, coordinated movement, cardiovascular/respiratory/metabolic adjustments, thermoregulation, etc., increases organized complexity.

Additional experimental evidence also suggests that a process of changing the undifferentiated growth of cancer stem cells to differentiated normal cells, may be a valid concept. Piccirily and Vescovi (2006) and Vescovi *et al* (2006) demonstrated that bone morphogenetic proteins are capable of arresting the development of brain tumors. They showed that cancer stem cells can be redirected to develop into normal supportive cells. Autie and Gandine (2007) revealed that vitamin D may not only reduce cellular proliferation but also activate cellular differentiation. In addition, Adams et al (2007) reported that "alternating electrical fields can help control cell identity, cell number, position and movement, which is relevant to everything from embryonic development to regeneration to cancer and almost any biomedical phenomenon." All three processes, embryogenesis, regeneration, and cancer share some of the basic properties and characteristics. For example, they all exhibit rapid cellular growth following the initiating stimulus. The first two processes previously listed, embryogenesis and regeneration, however, take place within a defined time period, e.g. nine months in human embryogenesis and likewise normal tissue regeneration usually ends through cellular communication called contact inhibition. Not all cells, however, show the same propensity for function and regeneration.

Using the described model, it is also possible to conceptualize that an alternating current, with its fixed rate of oscillations, changes disorganized complexity of cancer cells into organized complexity. Wodarz and Nathke (2007) report that the loss of cell polarity can be observed in tumors especially those which show the propensity to invade and metastasize. They studied the tumor suppressor genes which control apical-basal cell polarity in *Drosophila melanogaster*. Colen (2007) reports on a very imaginative reprogramming of stem cells described by Yamanaka and Takahashi (2006) who inserted just 4 genes into mouse skin cells and achieved some cellular pluripotency. Another pathway open to controlling the exponential growth of cancer cells may lie, as Serrano (2007) comments on, in inducing cellular senescence, accomplished in mice through the activation of a single gene, *Tp53*, which encodes the protein p53. Subsequent commentary by Katsnelson (2007), on the additional work by Yamanaka et al (2006), revealed that it is possible to lower the incidence of tumor formation with adult-to-embryonic stem cell reprogramming. They accomplished that by eliminating one of the four original genes (c-Myc) used during the initial experiment. What was also observed was the fact that not only tumorigenesis was suppressed but also the stem cell pluripotency was markedly reduced. These observations may indicate that c-Myc is an important gene involved in transforming cellular systems into an exponential growth phase, a characteristic not only of embryonic stem cells but also of a neoplastic growth.

The question of what influences dividing cells to choose the path of differentiation or undifferentiation was addressed by Wang et al (2007). They showed, in their experiment with *Drosophila* neuroblasts, that the asymmetric cell division is mediated by the protein Numb which is unequally partitioned as well. The daughter cell that receives Numb will likely differentiate, while the other daughter cell continues as an undifferentiated progenitor retaining its capacity self-renewal.

To stop the link between senescence *entropy* and its diseases, the *system* has to improve its *orderly complexity* and *self-organization*. Some evidence indicates that upregulating genes associated with longevity, the sirtuins, not only prolongs the life cycle but simultaneously suppresses oncogenes leading to lower cancer incidence. Experimental evidence already demonstrated that upregulation of the longevity genes downregulated tumor genes, opening a potential avenue for modifying the historic correlation between aging and cancer incidence. (Goymer 2006) Other studies showed that longevity genes can be upregulated with caloric restriction, resveratrol, and physical activity. (Ingram et al 2006, Keilbronn et al 2006) It has been demonstrated experimentally that aging-associated diminution of progenitor cell divisions in the brain, can be reversed by "encourage[ing] stem cells in the brain to divide…using behavioral modification techniques, such as physical exercise." (Hattiangady and Shetty 2006)

If we accept that cancer is a reflection of a failing *system*, preventive steps should involve rebalancing the entire *system* through the following: lowering the *disorderly complexity* and lessening *entropy,* while simultaneously optimizing *self-organization* and *orderly complexity*. To interrupt the strong correlation between aging and cancer, a system's view would encourage several steps: 1-focus on the ongoing creation of *organized complexity*; 2- the system must remain functioning as an *open system*. It is the *organized complexity* that allows *self-organization* process to take place; *disorganized complexity* does not. A healthy life needs a great variety with some degree of unpredictability in order to keep the *complex adaptive system* adapting. Along the same venue, from stem cells to whole organs, biologic entities constantly undergo a process of renewal. But, it is not enough just to 'renew,' a balancing 'differentiation' must also take place as well. A division without the diversity of differentiation increases the risk of systems' failure. Furthermore, all healthy biologic systems need to express synchronization of relevant internal as well as external cycles in order to remain within the dynamic *health territory* of this model.

The compartmentalized accomplishments in the numerous fields of life sciences, in spite of the separateness of their research concepts and methodologies, ultimately, have the same goals: reaching better understanding of living entities and facilitating the health of their cycles. To accomplish the unifying goals, the maintenance of health and the disease management of living entities, these goals need to be contextualized by integrating them into the broader, systems science-based framework.

Conclusion

Biologic cycles express profound interrelationships when examined in the relevant context. Multiple cycles that are germane to cells and the human body have been studied in order to construct an integrated translational view. Utilizing principles of systems science, complexity, and chaos theories, a dynamic model was developed. It identifies the domain of health, the *health territory*, and disease states named the *zone of chaos*, and the *zone of entropy*. The *zone of health* expresses organized complexity, self-organization with adaptation, a common language of communication, and the presence of well-functioning resetting mechanisms. From this perspective, the abnormal biological states of cancer/autoimmunity or degenerative/inflammatory diseases, including aging, can be considered expressions of systems-in-failing.

The system's science translational view, expressed in this model, also points to a potentially useful therapeutic paradigm. When biologic entities are outside of the *health territory*, in the *zones of chaos* or *entropy*, the focus of intervention should be on enhancing *organized complexity* and *self-organization*. These corrective pathways would allow a failing biologic system to return to the steady state of the *health territory* thus favoring morphologic and physiologic health.

Correlative biologic examples that are cited in this chapter demonstrate the potential usefulness of this model. Cancer control efforts, for example, could benefit from focusing on processes that enhance stem cell differentiation instead of seeking non-specific cellular death with cytotoxic therapies.

The compliance of biologic entities with systems' principles highly favors congruence of their internal and external cycles as a pre-requisite to reach optimal functionality conceptualized in this model as the *health territory*. Integration of all the inter-related systems optimizes ecologic and evolutionary developments of biologic entities. This is reflected in unhindered exchange of energy and information among the systems, in spite of the Darwin's "hostile forces of nature" and the Mendel's genetic determinism. Cognition, however, leaves biologic systems also vulnerable to the epigenetic "hostile forces of nurture."

Acknowledgment

Adopted in part from Janecka IP: Cancer control through principles of systems science, complexity, and chaos theory: A model. *Int J Med Sci 2007*; 4:164-173. http://www.medsci.org/v04p0164.htm

References

Arevalo, J.D. (2007). Language barriers increase risk for adverse events and effects. Retrieved 1-3-08 http://www.amnhealthcare.com/News.aspx?ID=17279.

Adams, D.S., Masi, A. and Levin, M. (2007). H^+ pump-dependent changes in membrane voltage are an early mechanism necessary and sufficient to induce *Xenopus* tail regeneration. *Development, 134*, 1323-1335.

Ashraf, N., Zino, S., MacIntyre, A., Kingsmore, D., Payne, A.P., George, W.D., and Shiels, P.G. (2006). Altered sirtuin expression is associated with node-positive breast cancer. *Br J Cancer, 95*: 1056-1061.

Autier, P. & Gandini, S. (2007). Vitamin D supplementation and total mortality. *Arch Intern Med, 167*:1730-1737.

Ben-Dov, I,, Kark, J.D., Ben-Shay, D., Mekler, J., Den-Arie, L., Bursztyn, M. (2007). Blunted Heart Rate Dip During Sleep and All-Cause Mortality. *Arch Intern Med, 167*:2116-2121.

Bertalanffy L. (1969). *General system theory*. New York, USA: George Braziller.

Blask, D., Dauchy, R., and Sauer, L. (2005). Putting cancer to sleep at night: The neuroendocrine/circadian melatonin signal. *Endocrine, 27*(2):179-188.

Brilliant, L. (2007). The Spreading Epidemic. Retrieved: 5/7/07 http://www.forbes.com/free_forbes/2007/0507/173.html.

Brown, S.J. (2006). The role of vitamin D in multiple sclerosis. *Ann Pharmacother, 40*(6):1158-1161.

Capra F. (1996). *The Web of Life*. New York, USA: Anchor Books.

Colen, B.D. (2007). Major progress toward cell reprogramming. Harvard Univ Gazette Online. http://www.news.harvard.edu/gazette/2007/06.07/99-stemcell.html.

Cook, L. & O'Brien, E. (2003). Sustained reduced sleep can have serious consequences. NIH News/ Sleep Disorders. National Institute of Nursing Research. Retrieved March 12, 2003 from http://www.nih.gov/news/pr/mar2003/ninr-12.htm.

Cromie, W.H. (2007). Sleep found to repair and reorganize the brain. Why mice sleep longer than humans. Harvard Univ Gazette Online, Retrieved March 15, 2007 http://www.news.harvard.edu/gazette/2007/03.15/01-sleepsize.html.

De Vogli, R., Chandola, T., Marmot, M.G. (2007). Negative aspects of close relationships and heart disease. *Arch Intern Med, 167*(18):1951-157.

Drack, M., Apfalter, W., Pouvreau, D. (2007). On the making of a system theory of life: Paul A Weiss and Ludwig von Bertalanffy's conceptual connection. *The Quarterly Review of Biology, 82*(4):349-373.

Editorial. (2007). Patching together a world view. Nature (2007), 761. Retrieved Dec 6, 2007. http://www.nature.com/nature/journal/v450/n7171/full/450761a.html.

Eiges, R., Urbach, A., Malcov, M., Frumkin, T., Schwartz. T., Amit, A., Yaron. Y., Eden, A., Yanjka, O., Benvenisty, N., Ben-Yosef, D. (2007). Developmental study of Fragile X Syndrone using human embryonic stem cells derived from preimplantation genetically diagnosed embryos. *Cell Stem Cell, 1*,568-577.

Eskandari, F., Martinez, P., Torvik, S., Phillips, T.M,, Sternberg, E.M., Mistry, S., Ronsaville, D., Wesley, R., Toomey, C., Sebring, N., Reynold, J.C., Blackman, M.R., Calis, K.A., Gold, P.W., Cizza, G. (2007). Low bone mass in premenopausal women with depression. *Arch Intern Med, 167*(21):2329-2336.

Fodde, R. (2006). Stem Cells and Metastatic Cancer: Fatal Attraction? *PLoS Med, 3*(12): e482.

Goldberger, A.L., Rigney, D.R. and West, B.J. (1990). Chaos and Fractals in Human Physiology. *Scientific American, 262*(2): 42-9.

Goymer, P. (2006). Ageing: Longevity mutations inhibit tumours. *Nature Reviews Genetics, 7*: 742-743.

Greene, B. (2006). The Universe on a String. Accessed, December 23, 2006. http://www.nytimes.com/2006/10/20/opinion/20greenehed.html?th&emc=th.

Hongyan, W, Yingshi, O., Somers, W.S., William Chia, W. & Lu, B. (2007). Polo inhibits progenitor self-renewal and regulates Numb asymmetry by phosphorylating Pon. *Nature, 449*, 96-100.

Hatch M. (1997). *Organization theory*. New York, USA: Oxford University Press.

Hattiangady, B., & Shetty, A. (2006). Aging does not alter the number or phenotype of putative stem/progenitor cells in the neurogenic region of the hippocampus. *Neurobiology of Aging, 6.*

Holmes, G.L. (2007). Blood-Brain Barrier Leakage in Temporal Lobe Epilepsy. *Brain, 130*:521-534.

Ingram, D.K., Zhu, M., Mamczarz, J., Zou, S., Lane, M.A., Roth, G.S., deCabo, R. (2006). Calorie restriction mimetics: an emerging research field. *Aging Cell, 5*(2): 97-108.

Keilbronn, L.K., de Jonge, L., Frisard, M.I., DeLany, J.P., Larson-Meyer, D.E., Rood, J., Nguyen, T., Martin, C.K., Volaufova, J., Most, M.M., Greenway, F.L., Smith, S.R., Deutsch, W.A., Williamson, D.A., Rayussin, E., Pennington, CALERIE Team. (2006). Effect of 6-month calorie restriction on biomarkers of longevity, metabolic adaptation, and oxidative stress in overweight individuals: A randomized controlled trial. *JAMA, 295*(13): 1539-48.

Katsnelson, A. (2007). Pluripotency without tumors? The Scientist. Retrieved Nov 30, 2007 http://www.the-scientist.com/blog/display/53946/.

Kendall, P. (2007). Financial manias and the trade of a lifetime. Stocks, Futures, and Options Magazine, October 2007. http://www.elliottwave.com/club/protected/pdf/Financial_ Manias.pdf.

Kornberg, R. (2006). Nobel Prize in Chemistry 2006. Accessed December 23, 2006. http://nobelprize.org/nobel_prizes/chemistry/laureates/2006/press.html.

Kurzweil R (2005). *The singularity is near*. New York, USA: Penguin Group.

Lambert, G.W., Reid C., Kaye D.M., Jennings G.L. & Esler M.D. (2002). Effect of sunlight and season on serotonin turnover in the brain. *Lancet, 7*, 360(9348), 1840-1842.

Laszo E. (1996). The systems view of the world. Cresskill, NJ: Hampton Press.

Laufs, U., Werner, N., Link, A., Endres, M., Wassmann, S., Jurgens, K., Miche, E., Bohm, M., Nickening, G. (2004). Physical training increase endothelial progenitor cells, inhibits neointima formation, and enhances angiogenesis. *Circulation, 109*(2): 220-6.

Leaf, C. (2004). Why we're losing the war on cancer. *Fortune,* 76-92.

Leingartner, A., Thuret, S., Kroll, T.K., Chou, S.J., Leasure, J.L., Gage, F.H., and O'Leary, D.M. (2007). Cortical area size dictates performance at modality-specific behaviors. Proc Natl Acad Sci, Retrieved Fed 27, 2007 at http://www.pnas.org/cgi/content/abstract/0611723104v1?etoc.

Leonard, B.E. (2007). Inflammation, depression and dementia: are they connected? *Neurochem Res, 32*(10):1749-56.

Ludwig, C. (1847). Beiträge zur Kenntnis des Einflusses der Respirationsbewegungen auf den Blutlauf im Aortensystem. *Arch Anat Physiol Leipzig, 13*: 242-302.

Ludwig, D. (2007). Childhood obesity-the shape of things to come. *NEJM, 357*(23): 2325-2327.

Malsow A. H. (1968). Toward a Psychology of Being. Princeton, New Jersey: D.Van Nostrand Co.

Mitchell, S. (2007). Stem cells model disease. The Scientist. Retrieved Nov 14, 2007 http://www.the-scientist.com/news/home/53859/.

Novak, B., Tyson, J.J., Gyorffy, B., and Csikasz-Nagy, A. (2007). Irreversible cell-cycle transitions are due to systems-level feedback (Commentary). Nature Cell Biology 9(7):721-853 http://ealerts.nature.com/cgi-bin24/DM/y/eere0SozZh0HjD0BWWm0EC.

Peebles, K.A., Lee, J.M., Mao, J.T., Hazra, S., Reckamp, K.L., Krysan, K., Dohadwala, M., Heinrich, E.L., Walser, T.C., Cui, X., Baratelli, F.E., Garon, E., Sharma, S., Dubinett, S.M. (2007). Inflammation and lung carcinogenesis: applying findings in prevention and treatment. *Expert Rev Anticancer Ther, 7*(10):1405-21.

Petraschek, M., Ye, X., Buck, L.B. (2007). An antidepressant that extends lifespan in adult Caenorhabditis elegans. *Nature, 450*(7169):553-556.

Piccirillo, S.G.M. & Vescovi, A. L. (2006). Bone morphogenetic proteins inhibit the tumorigenic potential of human brain tumour-initiating cells. *Nature, 444*: 761–765.

Reiter, R.J. & Robinson, J. (1995). *Melatonin*. New York: Bantam Books.

Richards, J.B., Valdes, A.M., Gardner, J.P., Paximadas, D., Kimure, M., Nessa, A., Lu, X., Surdulescu, G.L., Swaminathan, R., Spector, T.D., Aviv, A. (2007). Higher serum vitamin D concentrations are associated with longer leukocyte telomere length in women *Am J Clin Nutr, 86*(5): 1420-1425.

Ritchie, K., Carriere, I., deMendonca, A., Portet, F., Dartigues, J.F., Rouaud, O., Bargerger-Gateau, P., Ancelin, M.I. (2007). The neuroprotective effects of caffeine: a prospective population study (the Three City Study). *Neurology, 69*(6):536-545.

Rodriguez, M.I., Escames, G., Loez, L.C., Lopez, A., Garcia, J.A., Ortiz, F., Acuna-Castroviejo, D. (2007). Chronic melatonin treatment reduces the age-dependent inflammatory process in senescence-accelerated mice. *Journal of Pineal Research, 42*(3):272-279.

Sainz, R., Mayo, J., Tan, D., Lopez-Burillo, S., Natarajan, M., and Reiter, R. J. (2003). Antioxidant activity of melatonin in Chinese hamster ovarian cells: Changes in cellular proliferation and differentiation. *Biochemical and Biophysical Research Communications, 302*(3):625-634.

Satinover, J. (2001). *The Quantum Brain. The Search for freedom and the next generation of man*. New York: John Wiley & Sons, Inc.

Schmidt, B., Roberts, R.S., Davis, P., Doyle, L.W., Barrington, K.J., Ohlsson. A., Solimano, A., Tin, W. (2007). Long-term effects of caffeine therapy for apnea of prematurity. *NEJM, 357*(19):1093-1902.

Shi, Y., Feng, Y., Kang, J., Liu, C., Li, Z., Li, D., Cau, W., Qiu, J., Guo, Z., Bi, E., Zang, L., Lu, C., Zhang, J.Z. and Pei, G. (2007). Critical regulation of $CD4^+$ T cell survival and autoimmunity by β-arrestin 1. *Nature Immunology, 8*:817-824.

Slattery, M., Murtaugh, M., Caan, B., Ma, K., Wolff, R., and Samowitz, W. (2004). Associations between BMI, energy intake, energy expenditure, *VDR* genotype and colon and rectal cancers (United States). *Cancer Causes and Control, 15*: 863-872.

Smolin, L. (2006). Loop quantum gravity. Accessed December 23, 2006. http://www.edge.org/3rd_culture/smolin03/smolin03_index.html.

Takahashi, M., Kubota, S. (2005). Exercise-related novel gene is involved in myoblast differentiation. *Biomed Res, 26*(2):79-85.

Tan, B.T., Park, C.Y., Ailles, L., Weissman, I.L. (2006). The cancer stem cell hypothesis: A work in progress. *Laboratory Investigation, 86*: 1203–1207.

Tanne, J.H. (2006). Second cancers more common in people who receive donor stem cell transplants. *Br Med J, 333*: 1140.

van der Rhee, H.J., de Vries, E., Coebergh, J.W. (2006). Does sunlight prevent cancer? A systematic review. *Eur J Cancer, 42*:2222-22226.

Vescovi, A. L., Galli, R. and Reynolds, B. A. (2006). Brain tumour stem cells. *Nature Rev. Cancer, 6*: 425–436.

Warfield, J. (2004). Linguistic adjustments: Precursors to understanding complexity. *Systems Research and Behavioral Science, 21*:128.

Waldrop, M. (1992). *Complexity*. New York: Simon & Schuster.

Weiss, P.A. (1922). *Richtungbestimmende Einflusse außerer Faktoren. Die Ruhestellung der Vanessiden*. PhD diss., University of Vienna.

Wilkie, A.O. (2007). Cancer drugs to treat birth defects. *Nature Genetics, 39*(9):1057-1059.

Wilson, R.S., Schneider, J.A., Arnold, S.E., Bienias, J.L., Bennett, D.A. (2007). Conscientiousness and the incidence of Alzheimer disease and mild cognitive impairment. *Arch Gen Psychiatry, 64*(10):1204-12.

Wodarz, A. & Nathke, I. (2007). Cell polarity in development and cancer. *Nature Cell Biology, 9*:1016-1024.

Woodfield, J.C., Beshay, N.M., Pettigrew, R.A., Plank, L.D., and vanRii, A.M. (2007). American Society of Anesthesiologists classification of physical status as a predictor of wound infection. *ANZ J of Surgery 77*(9):738-741.

Yamanaka, S. & Takahash, K. (2006). Induction of pluripotent stem cells from mouse fibroblast cultures. *Tanpakushitsu Kakusan Koso, 51*(15):2346-2351.

In: Cell Cycle Control: New Research
Editors: N. H. Leroy, N. T. Fournier

ISBN: 978-1-60456-796-0

Chapter IV

TGFβ: a Unique and Powerful Negative Regulator of the Cell Cycle in Human Myeloid Leukemia Cells

***Xiao Tang Hu*[1,*] *and Kenneth S. Zuckerman*[2]**

[1]School of Natural and Health Science, Barry University, Miami Shores, FL;
[2]Departments of Oncology, Internal Medicine, and Molecular Medicine, University of South Florida, and H. Lee Moffitt Cancer Center and Research Institute, Tampa, FL.

Abstract

Signaling from Transforming Growth Factor-β (TGFβ) through its transmembrane receptor serine-threonine kinases plays an essential role in cell proliferation, differentiation, apoptosis, and the pathogenesis of many human diseases in various tissues including hematopoietic cells. One of the major functions of TGFβ in hematopoiesis is its inhibitory effect on cell proliferation. Downregulation of multiple cell cycle-regulatory molecules appears to be a dominant event in TGFβ-mediated growth inhibition of several human myeloid leukemia cell lines. At least, the inhibition of some of the molecules is regulated at mRNA level. Both G1-and G2-check point protein kinases participate in the regulation of TGFβ-mediated growth inhibition, whereas Cdk inhibitor p27-induced by TGFβ may have both a positive and negative role. TGFβ also rapidly inhibits phosphorylation of pRb at several serine and threonine residues and regulates the association of a repressor E2F transcription factor, E2F-4, with pRb family proteins (pRb, p107, and p130). Major p107-E2F complexes are found in proliferating cells whereas the under-phosphorylated pRb and p130 are associated with E2F-4 in G1 phase. Since TGFβ inhibits synthesis of p107 and has no effect on the expression of E2F-4, it suggests that E2F-4 is released from p107 and switched to pRb and p130 in response to TGFβ stimulation. Western blot and DNA binding experiments suggest that p130 and p107, but not pRb, play a critical role in regulating E2F response genes in TGFβ-

* Correspondence concerning this article should be addressed to: Xiaotang Hu, Ph.D. School of Natural and Health Sciences, Barry University, 11300 Northeast Second Avenue, Miami Shore, FL 33161. Phone: (305)899-3295; Fax: (305)899-3225; E-mail: xthu@mail.barry.edu.

mediated cell cycle control in human hematopoietic cells. Although the activation of SMAD-independent p44/42 MAPK signal by TGFβ has been recently reported in various cell types, which is linked to TGFβ-induced proliferation, apoptosis, and growth inhibition, TGFβ does not activate p44/42 MAP kinase pathways in human myeloid leukemia cells whose growth is inhibited by TGFβ, but turns off GM-CSF-induced p44/42 MAPK signal via inhibition of PI3-K-Akt pathway.

Introduction

Transforming growth factor-β (TGFβ) plays a key role in cell growth, differentiation, tissue homeostasis, and the pathogenesis of many human diseases in various tissues including blood cells (Ewen ME. 1993, Kehrl JH. 1986, Knabbe C 1987, Agrotis A. 2005, Hu X 1999). The inhibitory effect has been observed in epithelial cells, fibroblasts, and hematopoietic lineages. As an inhibitor of cell proliferation TGFβ is particularly attractive due to its profound inhibitory effect at a very low concentration, even in the presence of growth factors. Mutations in TGFβ or the components of TGFβ signaling pathway are linked to a number of human diseases including cancers. In many cases malignant transformation occurs when normal progenitor cells lose control of signaling pathway that regulate responses to TGFβ (Hinshelwood RA 2007, Guasch G 2007). The current model for intracellular inhibitory effects produced by TGFβ is a linear signaling pathway, which is initiated when a ligand binds to type II TGFβ receptor followed by phosphorylation of type I receptor. The phosphorylated type I receptor then signals to Smad proteins. Subsequently, the latter traslocate into the nucleus where they interact with transcription factors and regulate DNA transcription (Attisano L & Wrana JL 2002, Massagué J, 2000). Due to its diverse activities and importance in the understanding human diseases, over past 20 years, TGFβ signaling has been a major area of investigation and considerable insights into its functions have been learned. We will review these major achievements in next several sections.

Hematopoiesis is the formation and development of blood cells. In human, hematopoiesis occurs in the yolk sac and the aorto-gonadal-mesonephros (AGM) region, then liver, followed by the bone marrow. In normal adults, the bone marrow becomes the almost exclusive hematopoietic organ. All blood cells develop from pluripotent hematopoietic stem cells. The pluripotent cells differentiate into multipotent and then unipotent lineage committed progenitor cells that further differentiate to mature blood cells (Figure 1). Hematopoiesis is a tightly regulated process that is controlled by complex interactions between growth stimulatory and inhibitory cytokines. Many interleukins and cytokines, such as stem cell factor (SCF), granulocyte-macrophage colony-stimulating Factor (GM-CSF), flt-3 ligand, thrombopoietin, erythropoietin, G-CSF, and interleukine-3 (IL-3), are required for hematopoietic stem and progenitor cell proliferation and differentiation, whereas only a few cytokines, such as tumor necrosis factor (TNFα) and TGFβ, are known to be physiologic negative regulators of the proliferation of hematopoietic cells.

In this review we first outline the principal aspects of the cell cycle control and TGFβ signal pathway based on the studies from epithelial and fibroblast cells and then highlight some recent progress in understanding the effect of TGFβ on several important cell cycle regulatory molecules and Smad-dependent and -independent pathways, with a focus on human hematopoietic cells.

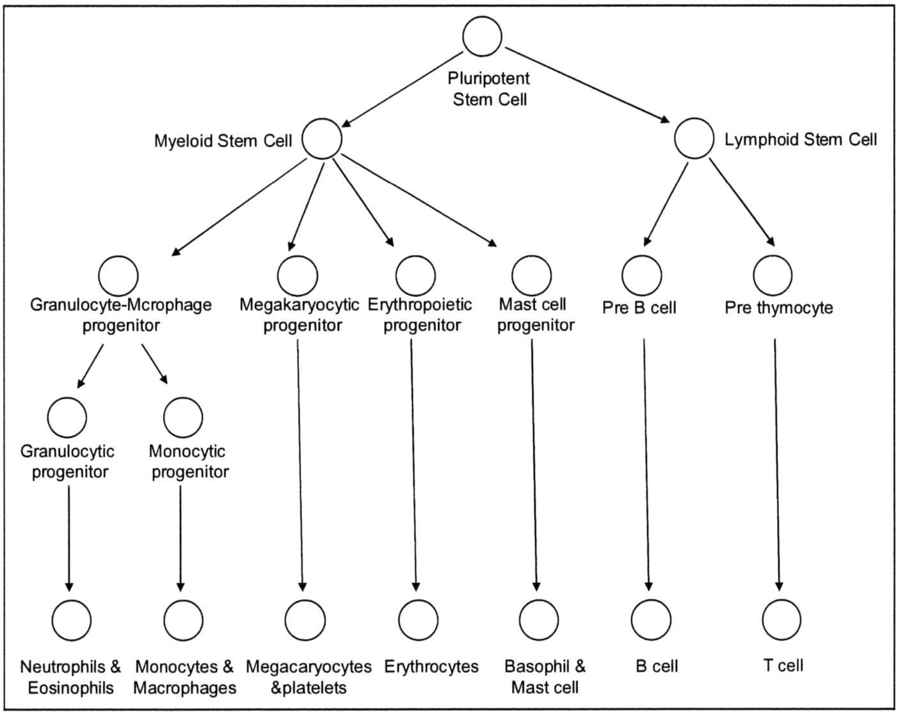

Figure 1. Diagram of Hematopoiesis in Human.

Overview of Cell Cycle Control

All growing cells undergo a cell cycle. In mammalian cells the eukaryotic cell cycle is divided into four distinct phases: G1, S, G2 and M phases. Viable cells may also remain for prolonged periods in a resting state called G0 phase. In G0 phase the cells either have irreversibly exited the cell cycle for a particular process or function, such as differentiation, or return to G1 with appropriate stimulation. G0 cells lack cell cycle regulatory molecules expressed in the cell cycle. Ki-67 antigen is a marker for distinguishing proliferating G1 from G0 cells (Schluter C 1993). The cycle of a normal cell is controlled by a group of proteins termed cyclin-dependent kinases (cdk) (Nurse P 2000, Roberts JM 1999). Cdk only becomes active when bound to a regulatory subunit called cyclin (Jeffrey PD 1995). There are several cdks and cyclins. The cell cycle progress through G2 into M phase is driven by cdc2 (cdk1) complexed with cyclin B (also termed "G2 checkpoint kinase"). Cyclin B is synthesized and accumulates in the cytoplasm, after which it associates with cdc2 as an inactive form due to the phosphorylation at Thr-14 and Tyr-25 of cdk1 by the Wee1 and Myt1 kinases (Smits VA 2001). In the late G2 phase, the phosphorylation at the two sites is removed by cdc25c phosphatase, causing a rapid rise of cdk activity followed by transition of the cells from G2 into M (Millar JB 1991). The G1 cyclin-cdk complexes (also termed "G1 checkpoint kinases") regulate the progress of the cells through G1 toward DNA replication (S phase). At least five candidate G1-phase cyclins, termed cyclins C, D1, D2, D3 and E, have been identified in mammalian cells (Inaba T 1992, Kiyokawa H 1992, Koff A 1991, Matsushime H 1991, Motokura T 1992, Xiong Y 1991). Each of these cyclins can associate with one or more of the cdk family, of which cdk2, cdk4 and/or cdk6 kinase activities have been shown to operate in G1 phase (Bates S 1994, Dou Q-P 1993, Koff A 1992, Matsuoka M 1994, Matsushime H 1992, 1994, Meyerson M 1994, Pagano M 1993, Resnitzky D 1994, Tsai L 1993, Xiong Y 1992). Cdk4 and its close relative cdk6 are driven by cylin D1, D2, and D3 and their kinase activities have been detected in mid G1, prior to the activation of any other known cdk (Matsushime H & Quelle DE 1994, Meyerson M & Harlow E 1994, Pagano M 1993). The cyclin E-bound cdk2 exhibits kinase activities at the G1/S transition (Koff A 1992, Tsai L 1993, Resnitzky D 1994, Pagano M 1993), followed by cyclinA-cdc2, and cyclin B-cdc2 kinases at the late stages of the cell division cycle (Morgan DO 1992, Motokura T & Arnold A 1993, Norbury C and Nurse P 1992, Rosenblatt J 1992). Studies of constitutive expression of cdk4 suggest that cyclin D/cdk4 may operate temporally upstream from cyclin E/cdk2, as part of mechanism that is responsible for cyclin D/cdk2 activation (Ewen ME 1993). It has been recognized that one of the major targets of cdk is Rb proteins (pRb), the product of the retinoblastoma tumor suppressor gene (Weinberg RA 1995, Dynlacht BD 1997, Inoue Y 2007), since phosphorylation level of pRb oscillates in a cell cycle- or cyclin-dependent manner. In addition, many kinase assays have shown the cdk activity on pRb phosphorylation *in vitro* (Weinberg RA 1995, Nevins J 2001). Thus, pRb acts as a signal transducer connecting the cell cycle clock with the transcriptional machinery, which is achieved by interaction with transcription factors, E2F (Cam H & Dynlacht BD 2003, Stevaux O & Dyson NJ 2002). The best understood function of E2F is its ability to regulate the G1/S transition and S-phase entry during the cell cycle progression (Frolov MV and Dyson N 2004).

Of the seven E2Fs identified to date, E2F1, E2F2, and E2F3 are considered as "activator" E2Fs, whereas, E2F4, E2F5, E2F6, and E2F7 are considered to be "repressor" E2Fs (Attwooll C 2004). pRb family contains three members: pRb, p130, and p107, which are collectively called the pocket proteins. E2F1-3 only bind to pRb, and E2F5 only binds to p130. E2F-4 can interact with all three pocket proteins. However, E2E6 and E2F7 have not been found to bind pRb family proteins (Figure 2), and the mechanisms for their action/activity are clearly different from E2F1-5.

Cdk-mediated cell cycle progression is balanced by cdk inhibitors (ckis) that have been identified. In general, these inhibitors can be divided into two families: INK and KIP/CIP. The INK family includes p16 (Serrano M 1993), p15 (Hannon GJ & Beach D 1994), p18 (Guan KL 1994) and p19 (Hirai H 1995). The members in this group associate with and inhibit cyclin D-cdk4/cdk6 kinase activity. The KIP/CIP family is composed of three members: p21 (Harper JW 1993), p27 (Hengst L 1994, Polyak K 1994, Toyoshima H & Hunter T 1994) and p57 (Matsuoka S 1995). Each molecule in this family is able to bind to and inhibit both cyclinD-cdk4/6 and cyclin E/A-cdk2 activities. This classification is based on their structure and cdk affinity. Although the first of the cdk inhibitors to receive attention was p21, p27 is one of the most widely distributed cdk inhibitors, being expressed both in proliferating and differentiating cells. In several myeloid leukemia cell lines tested, p27, but not p21, is easily detected (unpublished data). The negative role of p21 and p27 on cell cycle has been supported by several important experiments: p27 knockout mice show enhanced cell proliferation (Kiyokawa H 1996), and display increased body size and pituitary tumors (Nakayama K 1996). Oncogenic Src reduces the inhibitory activity of p27 on cyclin E-cdk2 complex (Chu I 2007). However, this widely prevailing view has been challenged by another report in which p27 or p21 could play both a positive and negative role (Cheng M 1999).

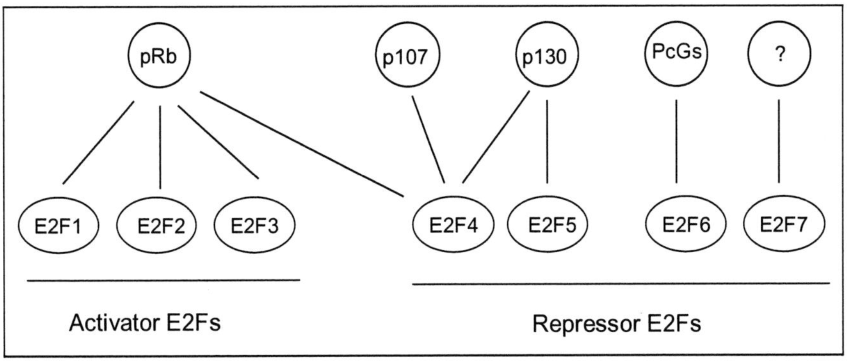

Figure 2. E2F/pRb complexes in mammals.

TGFβ Superfamily and Signal Transduction Pathway

The TGFβ superfamily contains a large number of molecules that are structurally related cytokines found in species ranging from worms and insects to mammals (Hogan BLM 1996, Massagué J 1998, Padgett RN 1998, Raftery LA & Sutherland DJ 1999, Derynck R & Zhang YE 2003), including TGFβ family, five activins, and eight BMPs (Bone Morphogenic Proteins). The TGFβ family consists of three highly homologous isoforms expressed in mammal cells, all with similar biological activities: TGFβ1, TGFβ2, and TGFβ3. As a growth inhibitor TGFβ1 has a more profound inhibitory effect on cell growth and has been isolated and purified from various human and animal tissues and cells (Assoian RK 1983, Frolik CA 1983, Roberts AB 1983). TGFβ1 is secreted as a latent precursor that requires activation into a mature form for its biologic activity (Derynck R 1985, Gentry LE 1988). Once activated, the TGFβ1 regulates cellular processes by binding to TGFβ receptors in the cell membrane. For the convenience of description, the term of TGFβ is used for TGFβ1 though the following text except where indicated otherwise.

Three membrane molecules with sizes of 53, 75, and 250 kDa have been termed TGFβ receptors type I, II, and III, respectively (Wang XF 1991, Lin HY 1992, Franzen P 1993). The type I and II receptor cDNAs have been cloned and shown to encode transmembrane serine/threonine kinases. Type II receptor contains a short extracellular domain with several cysteine residues believed to provide a binding site for TGFβ. The cytoplasmic domain contains the kinase domain and a carboxyl-terminal tail rich in serine and threonine. The kinase domain in type I receptor is distinct from the kinase domain in type II receptor. Type I receptor lacks the carboxyl-terminal tail characteristic for type II receptor and contains an additional domain of 29 amino acids rich in glycine and serine residues called the GS domain. In the absence of a ligand, type I and type II receptors exist as homodimers at the cell surface. Once bound to TGFβ, the ligand-bound type II receptor recognizes and recruits the type I receptor, forming a heterotetramer (Franzen P 1993) followed by phosphorylation of type I recepotor at serine and threonine residues in the GS domain, a region next to the kinase domain. Type III receptors consists of either β-glycan or endolin. Both proteins are involved in regulating access of ligand but have no ability to generate an intracellular signal directly (Wrana JL 1994). The expression of β-glycan is low (Klein U 2001, Jelinek DF 2003) and that of endolin is high in hematopoietic cells (Ohta M 1987).

It has been well demonstrated for years that phosphorylation of the type I receptor by the type II receptor initiates signaling from the heteromeric complex to downstream molecules, Smads. The first identified Smad member involved in TGFβ signaling is Mothers against dpp (MAD) in Drosophila cells (Raftery LA 1995). Subsequently, several Mad homologues have been reported in vertebrate cells, namely Smad 1-8 (Hahn SA 1996, Eppert K 1996, Wrana JL & Attisano L 1996, Hayashi H 1997, Chen Y 1997). Smad4 is the first one of the Smads identified in human. These Smads can be divided into three functional groups: the receptor-activated Smads (R-smads: Smads1-3, Smad5, and Smad8); the common mediator Smad (Co-Smad: Smad4); and the inhibitory Smads (I-Smads: Smad6 and Smad7) (Attisano L & Wrana JL 2002). In the absence of a signal, R-Smads (also co-Smad) are found predominantly in the cytoplasm as homomeric complexes (Lagna G 1996, Wu RY 1997). Activation of TGFβ receptors induces association of the Smad proteins with the receptors (Zhang Y 1996, Nakao A 1997). The Smads are recruited to the receptors by interaction with a membrane-associated FYVE (Fablp/YOTP/Vaclp/ETAL) domain-containing protein, SARA (Smad-anchor for receptor activation) (Tsukazaki T 1998, Wu G 2000). Type I TGFβ receptor specifically and transiently interacts with and phosphorylates the C-terminal domains of Smad2 and Smad3 proteins (Macias-Silva M 1996). Upon phosphorylation by type I receptors, the R-Smad forms heteromeric complex with Smad4. Activated Smad2/3-Smad4 complexes then translocate to the nucleus (Nakao A 1997, Macias-Silva M 1996, Yingling JM 1996), where they act as transcriptional activators by interaction with DNA-binding proteins through their MH (Mad Homology) 1 and MH2 domains (Liu F 1996). R-Smads and Co-Smads have highly conversed amino-acid sequences at their MH1 and MH2 domains. Both domains are separated by a divergent praline-rich linker region. MH1 and MH2 domains can interact with various transcription factors including ATF-2, C-Fos, C-Jun, Jub, JunD and some others (Massagué J 1998). In contrast, the inhibitory Smads, Smad 6 and Smad7, prevent R-Smad phosphorylation and their nuclear translocation (Hayashi H 1997). In some quiescent cells, Smad7 and Smad6 are localized in the nucleus and then translocated into cytoplasm in response to TGFβ and BMP stimulation respectively (Itoh F 2001, Hanyu A 2001). One of the major inhibitory mechanisms is that the inhibitory Smads recruit E3-type ubiquitin ligases to the receptor complexes, ultimately leading to degradation of the complexes (Kavsak P 2000, Ebisawa T 2001). The association of Smad8 with Smad4 has been reported to have a synergistic effect on gene transcription (Chen Y 1997). The direct evidence that Smad proteins mediate TGFβ-induced growth inhibition has been observed in a number of cell types including hematopoietic cells (Rich JN 1999, Xu J & Attisano L 2000, He W 2006).

Effect of TGFβ on Cell Proliferation of Hematopoietic Cells

TGFβ is required for maintaining hematopoietic stem/progenitor cells in quiescent states, since these cells can be released from quiescence by anti-endogenous TGFβ antibody or p21 antisense oligonucleotide (Ducos K 2000). The sensitivity of myeloid stem/progenitor cells to the inhibitory effect of TGFβ has been reported to be inversely correlated with their maturation stage (Sing GK 1988, Keller JR 1990, Ottmann OG & Pellus LM 1991, Snoeck H-W 1996). Long-term repopulating stem cells, CD34+ or CD34+/CD38- stem/progenitor cells, early myeloid and erythroid progenitors are markedly inhibited in the presence of TGFβ, whereas more differentiated myeloid progenitors forming granulocyte-macrophage colonies, and late erythroid progenitors forming CFU-e are less or not inhibited by TGFβ (Keller JR 1990, 1992, 1994, Fortunel NO 2000, Ruscetti FW 2005, Sitnicka E 1996, McNiece IK 1992, Goey H 1989, Morita N 2003, Classen S 2007). Inactivation of the TGFβ signaling pathway leads to malignant transformation of early hematopoietic cells (Fortunel NO 2000). The inhibitory or stimulatory role by TGFβ in some cases seems to correlate with the cytokines used to promote cell growth. TGFβ inhibits SCF-induced colony formation from mouse and human hematopoietic progenitors (McNiece IK 1992). In contrast, TGFβ enhances GM-CSF-induced colony formation by normal bone marrow cells and Lin^- progenitors (Keller JR 1991). Studies from *in vivo* experiments demonstrated that the major effect of TGFβ either on early or late hematopoietic cells is inhibitory. For instance, TGFβ knockout mice were shown in two studies to have a significant increase in numbers of circulating granulocytes, monocytes, and platelets (Kulkarni AB 1993, Shull MM 1992).

For leukemia cells, TGFβ1 has been shown to inhibit the growth of clonogenic blast cells from patients with acute myeloblastic leukemia (AML) (Tessier N & Hoang T 1988). Whereas, in chronic myelogenous leukemia progenitors, both inhibitory and stimulatory effects have been reported (Aglietta M 1989, Cashman JD 1992).

Based on the data described above TGFβ has both negative and positive effects on proliferation of hematopoietic cells with a major inhibitory effect being observed on early CD34+ hematopoietic cells. One possible explanation for these discordant results may be due to the complexity of the network, since cells are always exposed to multiple extracellular or intracellular signals. In the presence of growth factors, TGFβ-induced stimulation could be an indirect result. Therefore, studies on cell lines may provide more consistent results. For this purpose, leukemia cell lines have been developed to simplify the study of leukemic cell proliferation and differentiation and to allow the convenient investigation of a relatively homogenous population of neoplastic cells (Koeffler HP 1983). A TGFβ knock out hematopoietic cell line exhibited increased colony formation in response to IL-3, IL-6, and G-CSF (Kim S-J & Letterio J 2003). TGFβ has been reported to inhibit proliferation of mouse myeloid leukemic M1 cells and induce adherence of M1 myeloid leukemia cells to plastic dishes (Okabe-Kado J 1989). In the presence of dexamethasone, TGFβ1 reduced the expression of C-jun, C-fos and Jun B and blocked cell cycle progression from G1 to S phase in U937 cells (Kanatni Y 1996). In Josk-1 monoblastic leukemia cells, TGFβ1 blocked the cell cycle progression from G1 to S phase by inhibiting phosphorylation of Rb and inhibited expression of c-myc (Furukawa Y 1992). In order to determine the effect of TGFβ on more myeloid leukemia cell lines and to reach a general conclusion, we examined more than 10 human myeloid leukemia cell lines, among which MV4-11 and TF-1a are factor-independent but response cell lines; TF-1 and Mo7e are factor-dependent cell lines; and KG-1, KG-1a, K562, U937, HL60, and HEL are factor-independent and insensitive to any growth factors. Our data demonstrate that TGFβ is inhibitory for all these human myeloid leukemic cell lines, with heterogeneous inhibitory responses being observed. MV4-11 is the most and Hel is the least sensitive to TGFβ-mediated growth inhibition (Hu X & Zuckerman KS 2000). The responses are not related to the apparent differentiation stages of the cell lines (Hu X 1999). The difference in the degree of the inhibition among the cell lines could result from differences in the expression of TGFβ receptors in the different cell lines. The mechanism by which TGFβ inhibits hematopoietic cells involves in upregulation of p21 (Ducos K 2000) and p57 (Scandura JM 2004), the association of Smad4 with R-Smads in competition with TIF1 (He W 2006), and the formation of pRb pocket proteins-E2F complexes (Hu X 2000a).

TGFβ-Mediated Cell Cycle Control: Hematopoietic Cells vs. Non-Hematopoietic Cells

The effects of TGFβ on G1 cyclins and cdks are well documented. Treatment of epithelial cells with TGFβ lead to inhibition of cdk2, cdk4, cyclin A, and cyclin E. In these cases, phosphorylation of pRb was inhibited and the cell cycle was arrested at the G1 stage (Ewen ME 1993, Geng Y & Weinberg RA 1993). Cdk inhibitors play an important role in cell cycle control. TGFβ can upregulates the expression of p15, p21, and p27, and causes a redistribution of p27 with cyclin-cdk complexes. (Hannon GJ & Beach D 1994, Reynisdottir I & Massagué J 1997, Datto MB 1995). It is generally agreed that the expression of D-type cyclins, cdk2 and cdc2 in epithelial cells is not affected by TGFβ (Geng Y & Weinberg RA 1993, Fautsch MP 1995). Studies in our laboratories have extended these findings. We have detected that TGFβ not only significantly down-regulated the expression of cdk4 and cyclin A, which is consistent with previous report, but also dramatically inhibited cylin D3 and cdk2. Most important we have found that TGFβ also significantly downregulates the expression of cdc2, cyclin B1, and cdc25c in human myeloid leukemia cell line, MV4-11 (Hu X 2007). The inhibition of cdc2 is not a result of G1 arrest. Thus, our data suggested that TGFβ regulates the expression and activities of both G1 and G2 checkpoint kinases, at least in hematopoietic cells.

p27 is one of the most widely distributed cdk inhibitors. Overexpression of p27 in cultured cells arrests the cell cycle (Lee MH 1996, Porter PL 1997). The crystal structure of p27 bound to cyclin-cdk2 reveals that p27 inserts itself deep within the cdk2 catalytic site, blocking ATP access (Russo AA 1996). These data lead to a model for p27 action in which anti-proliferative stimuli up-regulate p27, followed by inhibition of cyclinD-cdk4, cyclinE-cdk2, and cyclin A-cdk2 activities. Since TGFβ significantly upregulated p27 in MV4-11 cells, we initially thought that upregulated p27 would play a negative role through formation of p27-cdk/cyclin complexes. Unexpectedly, we found that cdk2 and cyclin E are regulated in two different manners by TGFβ. TGFβ had no effect on the total quantity of cyclin E, but upregulated association of cyclin E with p27. In contrast, TGFβ downregulated the expression of cdk2, resulting in a low level of cdk2-associated p27. There were large quantities of cdk2-p27 and cdk4-p27 in proliferating cells, which were greatly decreased when cells move to G1 phase in response to TGFβ. In vitro, depletion of p27 led to a loss of cdk2 kinase activity on phosphorylation of pRb (Hu X 2001). We interpret these results to suggest that p27 may play both a positive and inhibitory effect on the cell cycle: the downregulation of cdk2 reduces the amount of the molecules available to bind cyclin E, whereas overloading of p27 on cyclin E could interfere with stable assembly of cyclin E-cdk2 complexes. However, the large quantity of p27-cdk2 and p27-cdk4 complexes in proliferating cells and the loss of cdk2 kinase activity due to depletion of p27 suggest that p27 is required for the cdk2 and cdk4 kinase activities in these human myeloid leukemia cells (Hu X 2001). The results are consistent with the observations that normal upregulation of cyclin D-cdk4 in mitogen-stimulated fibroblasts depends redundantly upon p21 and p27 (Cheng M 1999).

Control of PRB-E2F Complexes by TGFβ: Hematopoietic vs. Non-Hematopoietic Cells

It has been well demonstrated that growth inhibition by TGFβ is linked to suppression of pRb phosphorylation in MvlLu lung epithelial cells, keratinocytes, and human hematopoietic cells (Chen P 1989, Mihara K 1989, Hu X 2000). Phosphopeptide analysis demonstrates that there are multiple of distinct sites of phosphorylation on either serine or threonine residues in Rb protein but particular sites involved in growth inhibition by TGFβ1 in human hematopoietic cells were not clear. By using specific anti-phosphorylated pRb antibodies, we detected that TGFβ1 inhibited the phosphorylation of pRb at several serine and threonine residues, including Ser249/Thr252, Thr373, Ser780, Ser795, and Ser807/811, in human myeloid leukemia cell lines, MV4-11 and TF-1 (Hu X 1999). Dephosphorylated pRb obtains a binding ability with E2F transcription factors. It has been reported that underphosphorylated pRb preferentially interacts with E2F1, E2F2, and E2F3 (Weinberg R 1995), and the pRb/E2F complexes are found primarily during the G1 phase of the cell cycle (Shirodkar S 1992, Schwarz JK 1993), during which time E2F-responsive promoters are repressed. The dissociation of the pRb/E2F-repressor complex and liberation of free E2F allows for the activation of these promoters. In support of this model, reintroduction of pRb into certain cells results in a down-regulation of E2F-dependent transcription and a G1 block (Hiebert SW 1993, Qin X-Q 1995). Conversely, activation of E2F-responsive promoters is sufficient to overcome the ability of pRb to suppress cell growth. By using mutated E2F1 and pRb, Helin et al (Helin K 1993) demonstrated that E2F-1 responsive gene activation is inhibited by overexpression of wild-type pRb, but not by a mutant pRb, whereas transactivation mediated by mutants of E2F1, which do not bind to pRb, is not affected by overexpression of wild-type pRb. Loss of E2F1-3 results in an acute cell cycle arrest (Wu L 2001). These data lead to a classic model of pRb-E2F complexes as transcription repressors, whereas free-E2Fs are transcription activators (Figure 3a). This model has exposed the function of pRb as a growth inhibitor, which is consistent with a common notion that pRb is a tumor suppressor, whereas the p107 and p130 are not. Indeed, Rb gene was originally found to be mutated in patients with hereditary retinoblastoma (Friend SH 1986) and subsequently in various cancer patients, whereas p107 and p130 genes appear to be less frequently mutated in human cancers. However, studies from several labs including ours have shown that p130 and p107 also are actively involved in E2F-dependent transcription activities. For example, the major E2F complexes in quiescent fibroblasts are E2F4-p130. As cells enter the cell cycle, E2F4-p130 is replaced by E2F4-p107 (Ikeda M 1996) and E2F4-pRb (Moberg K 1996) followed by activators E2F1, 2, and 3, accompanied by induction of E2F- responsive genes (Takahashi Y 2000). In serum-deprived cells, predominantly E2F-4, p130, and a histone deacetylase (HDAC) are found to bind to corresponding E2F-responsive promoters (Rayman JB 2002, Ren B 2002, Takahashi Y 2000). Additional evidence for a role of p130 and p107 came from chromatin immnuprecipitation (ChIP) experiments, in which pRb was not detectable in the promoters containing the E2F responsive site in cycling cells but was associated with E2F4-p130 complexes or E2F4-p107 complexes during G0/G1 phase (Rayman JB 2002, Takahashi Y 2000). These data suggest that pRb may not be the primary regulator of E2F activity in control of cell cycle progression. pRb may bind other regulatory regions of DNA, directly or indirectly, for its function as a tumor suppressor. Based on these data, a modified model for E2F transcription regulation has been presented (Figure 3b). This model provides an insight into the mechanisms governing the cell cycle control by incorporation of repressor E2Fs, activator E2Fs, pRb and HDAC.

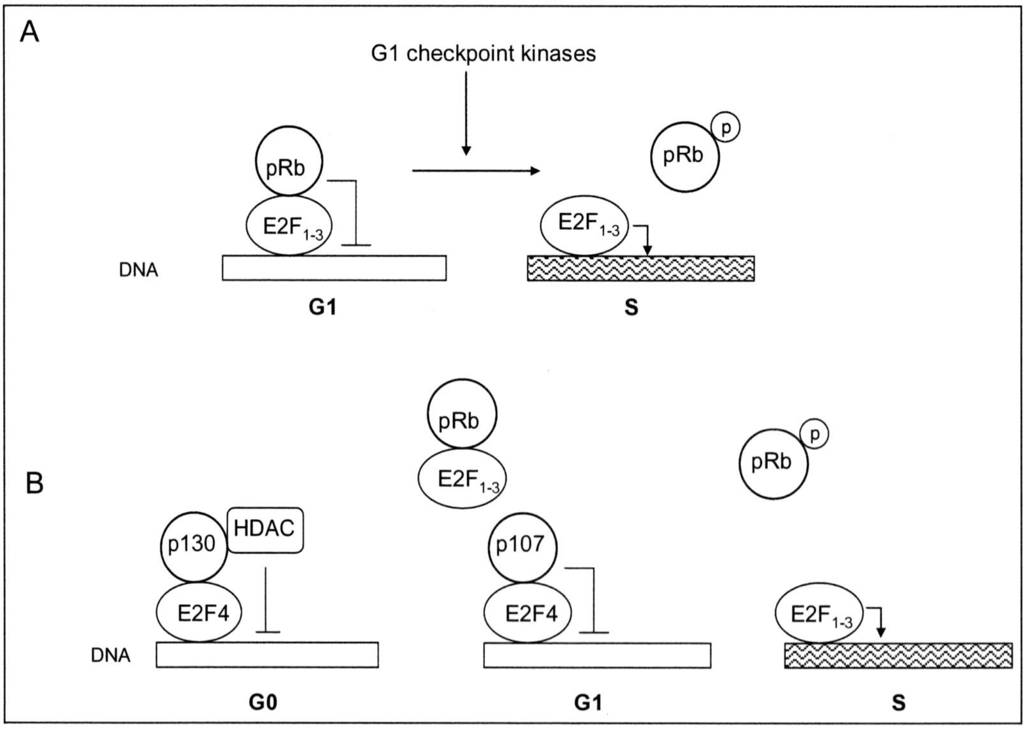

Figure 3. Models for the regulation of E2F response gene by pRb family proteins.

Since TGFβ dramatically inhibits growth of MV4-11 with up to of 90% cells in G1 phase, we wondered if the same mechanisms described above accounted for TGFβ-induced inhibition of hematopoietic cells. First, we investigated the expression of pRb, p130 and p107. In actively cycling MV4-11 cells, p130 and pRb were expressed as peptides that migrated in multiple bands in SDS-PAGE. Addition of TGFβ1 induced dephosphorylation of pRb and p130. TGFβ had no effect on p107 phosphorylation but inhibited the expression of p107 (Hu X 2000a) (Figure 4). Subsequently, we found that TGFβ-treated MV4-11 cells had very low levels of pRb-associated E2F1 and E2F3. The low levels of pRb-E2Fs resulted from a low expression of E2F1 and E2F3 in response to TGFβ (G1 cells). Unexpectedly, proliferating cells without TGFβ contained substantial quantities of pRb-E2F1 and pRb-E2F4 complexes. TGFβ had no significant effect on the expression of E2F-4, but upregulated pRb-E2F-4 and p130-E2F-4, and downregulated p107-E2F-4 complexes, as detected by immunoprecipitation and Western blotting (Hu X 200a)). Clearly, the low level of p107-E2F-4 complexes resulted from inhibition of p107 synthesis by TGFβ. Furthermore, electrophoretic gel mobility shift assay (EMSA) could not detect pRb-E2F4 DNA binding activities either in cycling cells or in G1 cells but exhibited the existence of p107-E2F4 and p130-E2F4 complexes in cycling and G0 cells, respectively, which is consistent with the data from the ChIP experiments described above. However, the difference is that E2F-4 is switched from p107 to pRb and p130 when p107 is inhibited and MV4-11 cells are arrested in G1 by TGFβ. Clearly, there is no single model of E2F regulation for all types of cells. Taken together, our dada suggest that p107 and p130, but not pRb, and the repressor E2F, but not activator E2Fs, play a critical role in regulating E2F-responsive gene expression in TGFβ-mediated cell cycle control in human myeloid leukemia cells. This conclusion is consistent with several studies that have failed to detect pRb at E2F-responsive genes, whereas p107 and p130 are readily detectable (Aslanian A 2004, Rayman JB 2002, Takahashi Y 2000, Wells J 2000). Furthermore, many cell cycle-regulated E2F targets are depressed in p107-/- and p130-/- cells but are not depressed in Rb-/- cells (Herrera RE 1996, Hurford Jr RK 1997, Mulligan GJ 1998). It is not clear if traditional activator E2Fs have any significant impact on cell cycle progression of MV4-11 cells, because pRb-E2F1 and pRb-E2F3 complexes have been found in the cycling but not in G1 phase cells. Based our data, we have presented another model for the regulation of the cell cycle by pRb family-E2F complexes in TGFβ-mediated cell cycle control of human myeloid leukemia cells (Figure 4). This model is a supplement to existed models and it partially supports the modified model illustrated in Figure 3b.

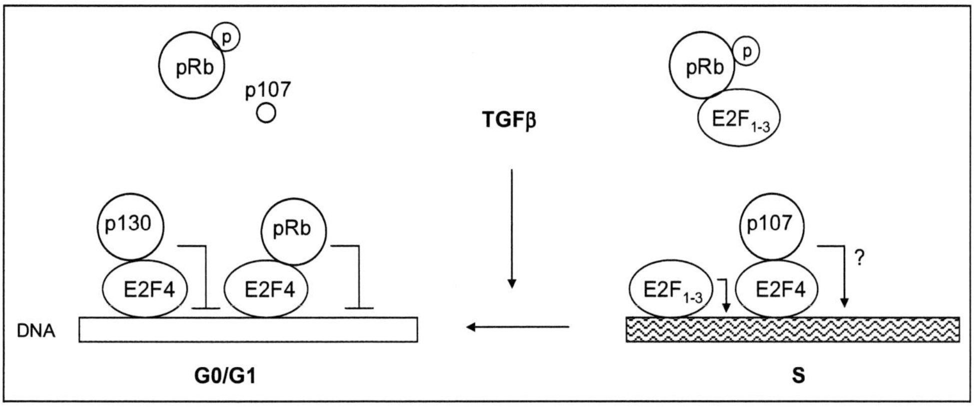

Figure 4. Schematic representation of TGFβ-mediated E2F transcription activity in MV4-11 cells.

Smad-Dependent and –Independent Pathways in Hematopoietic and Non-Hematopoietic Cells

It has been noted that there is extensive cross-talk between the Smad pathway and mitogen-activated protein kinase (MAPK) signaling. MAPK signaling is activated by receptor tyrosine kinases in response to growth stimulating or differentiation-inducing factors. When ligands bind to receptor tyrosine kinases on the cell membrane, they cause the receptor tyrosine kinases to aggregate and phosphorylate, by which the receptor is activated. Activated receptor provides a binding site for cytosol proteins with SH2 domains. The binding to the cytosolic SH2 proteins activates Ras (MEKKK) followed by activation of Raf (MEKK). The Raf is critical for activation of its downstream molecules, MEK and ERK. These three proteins (Raf, MEK, and ERK) are known as the mitogenic activated protein kinase (MAPK) signaling pathway that plays an important role in controlling cell proliferation and differentiation (Kyriakis JM 1992, Dent P 1992; Lange-Carter CA 1993). In general, the transient expression of MAPK phosphorylation is related to cell proliferation (Pages G 1993; Perkins GR 1996), whereas prolonged expression of phosphorylated MAPK is required for cell differentiation (Marshall CJ 1995; Qiu M & Green SH 1992; Hu X. 2000b). The activation of MAPK signals has a positive or negative effect on Smad pathway. It has been reported that the cells treated with hepatocyte growth factor (HGF) and/or epidermal growth factor (EGF) activate ERK in epithelial cells, resulting in activation of Smad2/Smad4 signaling (de Caestecker MP 1998). The activation of Raf enhances Smad2 phosphorylation and transcriptional activity (Brown JD 1999). In contrast, the persistent activation of MAPK pathways by Ras can inactivate Smad signaling by retention of R-Smads in the cytosol (Kretzschmar M 1999). In some types of cells, TGFβ-induced MAPK is required for Smad activation. For instance, TGFβ-induces activation of JNK in mink epithelia cells, which phosphorylates Smad3 and regulates Samd3-dependent transcription. Exactly how these MAPKs affect Smad signaling is not clear.

Although Smads play a critical role in TGFβ signal, abundant evidence demonstrates that Smad-independent pathways also exist in various cell types (Yonekura A 1999; Park HJ. 2002; Hayashida T 2003; Ji Y. 2006; Halder SK 2005). By performing immunoprecipitation/kinase assays, Engel et al observed a rapid and transient activation of JNK in mink lung epithelial cells (Mv1Lu) in response to TGFβ stimulation, which is Smad-independent (Engel ME 2002). TGFβ1 stimulates articular chondrocyte proliferation and extracellular matrix through a rapid activation of extracellular signal regulated kinase (ERK). All of these activities of TGFβ are Smad-independent. The direct evidence of Smad-independent pathway by TGFβ came from receptor mutation experiments, in which type I receptor was mutated, rendering it incapable of activating Smads but still capable of retaining kinase activity. The mutant receptor induced phosphorylation of p38 MAP kinase, which is required for TGFβ-induced apoptosis in mammary epithelial (NMuMG) cells (Yu L 2002). In these studies, the TGFβ-MAPK pathway was linked to various activities, including proliferation, apoptosis, and growth inhibition. However, the mechanisms for the linking between the SMAD–independent pathways and TGFβ receptor are poorly understood. No direct association between MAPK signaling molecules and TGFβ receptors has yet been identified.

Our previous experiments have shown that the growth of most human myeloid leukemia cell lines is inhibited by TGFβ treatment, which is linked to the inhibition of several cell cycle regulatory molecules either in their kinase activities or protein synthesis (Hu X. 2001). It is not clear if MEK-ERK signal pathway activation is induced by TGFβ in these human hematopoietic cells and whether such a signal, if there is any, contributes to TGFβ-induced cell cycle arrest and growth inhibition. Thus, we investigated the effect of TGFβ on the MEK-ERK pathway in the absence or presence of GM-CSF and PMA in three human myeloid leukemia cell lines.

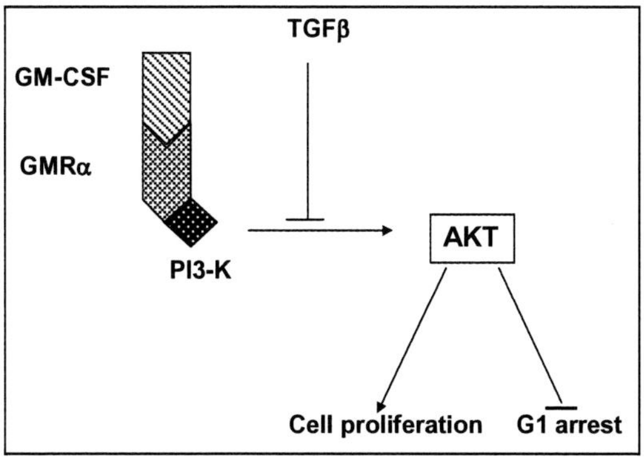

Figure 5. Diagram of TGFβ-mediated growth inhibition in the presence of GM-CSF.

GM-CSF is a growth factor that stimulates proliferation and differentiation of human myeloid hematopoietic cells via its cell surface receptors. GM-CSF receptors consist of two subunits, α and β. The α subunit (GM-CSFRα) has a short cytoplasmic tail and β subunit has a large intracytoplasmic tail. Although neither of these subunits has a tyrosine kinase activity, it is well documented that GM-CSF induced proliferation in these cells is a result of MAPK activation (Carroll MP. 1990; Kanakura Y. 1991; Okuda K. 1992). A number of studies suggest that the signal activation by GM-CSF is also mediated by phosphatidylinositol 3-kinase (PI3-k) (Al-shami A & Naccache PH 1999; Dhar-Mascareno M 2005). PMA is well known for its differentiation induction on several hematopoietic cell lines (Debbas V 2007, Huo XF 2006, Kim MS 2001, Das D 2000, Hu X 2000b). In the absence of GM-CSF or PMA, MV4-11 and TF-1 cells expressed very low level of phosphoryalted MEK and ERK. Addition of TGFβ for up to 48 hours did not increase or decrease the phosphorylation of MEK or ERK. In contrast, GM-CSF or PMA induced a rapid and transient or persistent phosphorylation of MEK and ERK, respectively. In the presence of GM-CSF or PMA, TGFβ downregulated GM-CSF- but not PMA-induced activation of MEK and ERK. In addition, GM-CSF induced a transient phosphorylation of the p85 subunit of PI-3-kinase in MV4-11 cells which was completely blocked by TGFβ (20 ng/ml) in the cells that were pretreated with TGFβ (unpublished data). Taken together our data suggest that (1) TGFβ does not activate p44/42 MAP kinase pathways in human myeloid leukemia cells whose growth is inhibited by TGFβ; (2) In the presence of a mitogen, TGFβ can turn off ERK pathway upstream from the Raf-ERK pathway via inhibition of PI3-kinase; and (3) TGFβ has no effect on PMA-stimulated differentiation-related MAPK activation. Figure 5 is a diagram of the mechanism by which TGFβ inhibits GM-CSF-induced MAPK signaling.

Conclusion

New findings in recent years have greatly altered our notions regarding the functions and signal pathway utilized by TGFβ. Traditionally, TGFβ has been recognized as a negative regulator of G1 cyclin-cdk activity. More evidence suggests that TGFβ regulates the expression and activities of both G1 and G2 checkpoint kinases, at least in hematopoietic cells. As a cdk inhibitor, p27 has both a positive and negative role in TGFβ-mediated cell cycle control. TGFβ regulates pRb phosphorylation and E2F transcription; however the mechanisms for the roles of pRb and E2F in cell cycle regulation are much more complicated than we thought previously. It is evident that there is not a single model for activation of E2F-dependent trancription. Although Smad-pathway activation is critical for TGFβ-induced signaling, Smad-independent pathways also exist in various cell types. There is extensive cross-talk between the two pathways. However, TGFβ does not activate the MEK-ERK pathway in several human myeloid leukemia cell lines whose growth is inhibited in response to TGFβ treatment, but TGFβ turns off a MAPK proliferation but not a MAPK differentiation signal. In spite of the great advances in our understanding of TGFβ signal pathways many gaps in our knowledge persist.

References

[1] Aglietta, M., Stacchini, A., Severino, A., Sanavio, F., Ferrando, M.L., Piacibello, W. (1989). Interaction of transforming growth factor-beta with hematopoietic growth factors in the regulation of human normal and leukemic myelopoiesis. *Exp Hematol, 17*, 296-299.

[2] Agrotis, A., Kalinina, N., Bobik, A. (2005). Transforming growth factor-beta, cell signaling and cardiovascular disorders. *Curr Vasc Pharmacol*, *3*, 55-61.

[3] Al-shami, A. & Naccache, P.H. (1999) Granulocyte-macrophage colony-stimulating factor-activated signaling pathways in human neutrophils. Involvement of Jak2 in the stimulation of phosphatidylinositol 3-kinase. *J Biol Chem*, *274*, 5333-5338.

[4] Aslanian, A., Iaquinta, P.J., Verona, R., Lees, J.A. (2004). Repression of the Arf tumor suppressor by E2F3 is required for normal cell cycle kinetics. *Genes Dev*, *18*, 1413-1422.

[5] Assoian, R.K., Komoriya, A., Meyers, C.A., Miller, D.M., Sporn, M.B. (1983). Transforming growth factor-beta in human platelets. Identification of a major storage site, purification, and characterization. *J Biol Chem*, *258*, 7155-7160.

[6] Attisano, L. & Wrana, J.L. (2002). Signal transduction by the TGF-β superfamily. *Science*, *296*, 1646-1647.

[7] Attwooll, C., Denchi, E.L., Helin, K. (2004). The E2F family: specific functions and overlapping interests. *EMBO J, 23*, 4709-4716.

[8] Bates, S., Bonetta, L., MacAllan, D., Parry, D., Holder, A., Dickson, C., Peters, G. (1994). Cdk6 (PLSTIRE) and cdk4 (PSK-J3) are a distinct subset of the cyclin-dependent kinases that associate with cyclin D1. *Oncogene, 9*, 71-79.

[9] Brown, J.D., DiChiara, M.R., Anderson, K.R., Gimbrone, M.A. Jr., Topper, J.N. (1999). MEKK-1, a component of the stress (stress-activated protein kinase/c-Jun N-terminal kinase) pathway, can selectively activate Smad2-mediated transcriptional activation in endothelial cells. *J Biol Chem, 274,* 8797-8805.

[10] Cam, H. & Dynlacht, B.D. (2003). Emerging roles for E2F: Beyond the G1/S transition and DNA replication. *Cancer Cell, 3*, 311-316.

[11] Carroll, M.P., Clark-lewis, I., Rapp, U.R., May, W.S. (1990) Interleukin 3 and granulocyte-macrophage colony stimulating factor mediate rapid phosphorylation and activation of cytosolic c-raf. *J Biol Chem, 265*, 19812-19817

[12] Cashman, J.D., Eaves, A.C., Eaves, C.J. (1992). Granulocyte-macrophage colony-stimulating factor modulation of the inhibitory effect of transforming growth factor-beta on normal and leukemic human hematopoietic progenitor cells. *Leukemia, 6*, 886-892.

[13] Chen, P., Scully, P.J., Shew, Y., Wang, J.Y., Lee, W.H. (1989). Phosphorylation of the retinoblastoma gene product is modulated during the cell cycle and cellula differentiation. *Cell, 58,* 1193-1198.

[14] Chen, Y., Bhushan, A., Vale, W. (1997). Smad8 mediates the signaling of the ALK-2 [corrected] receptor serine kinase. *Proc Natl Acad Sci USA, 94*, 12938-12943.

[15] Cheng, M., Olivier, P., Diehl, J.A., Fero, M., Roussel, M.F., Roberts, J.M., Sherr, C.J. (1999). The $p21^{cip1}$ and $p27^{kip1}$ `inhibitors' are essential activators of cyclin D-dependent kinases in murine fibroblasts. *EMBO J, 18,* 1571-1583.

[16] Chu, I., Sun, J., Arnaout, A., Kahn, H., Hanna, W., Narod, S., Sun, P., Tan, C.K., Hengst, L., Slingerland, J. (2007). p27 phosphorylation by Src regulates inhibition of cyclin E-Cdk2. *Cell, 128,* 281-294.

[17] Classen, S., Zander, T., Eggle, D., Chemnitz, J.M., Brors, B., Büchmann, I., Popov, A., Beyer, M., Eils, R., Debey, S., Schultze, J.L. (2007). Human resting CD4+ T cells are constitutively inhibited by TGF beta under steady-state conditions. *J Immunol, 178*, 6931-6940.

[18] Das, D., Pintucci, G., Stern, A. (2000). MAPK-dependent expression of p21(WAF) and p27(kip1) in PMA-induced differentiation of HL60 cells. *FEBS Lett, 472*, 50-52.

[19] Datto, M.B., Li, Y., Panus, J.F., Howe, D.J., Xiong, Y., Wang, X.F. (1995).Transforming growth factor beta induces the cyclin-dependent kinase inhibitor p21 through a p53-independent mechanism. *Proc Natl Acad Sci USA, 92*, 5545-5549.

[20] Debbas, V., Arai, R.J., Ferderbar, S., Schindler, F., Stern, A., Monteiro, H.P. (2007), Regulation of p21Wafl expression and TNFalpha biosynthesis by glutathione modulators in PMA induced-THP1 differentiation: involvement of JNK and ERK pathways. *Biochem Biophys Res Commun, 363,* 965-970.

[21] De Caestecker, M.P., Parks, W.T., Frank, C.J., Castagnino, P., Bottaro, D.P., Roberts, A.B., Lechleider, R.J. (1998). Smad2 transduces common signals from receptor serine-threonine and tyrosine kinases. *Genes Dev, 12,* 1587-1592.

[22] Dent, P., Hase, W., Haystead, T.A.J., Vincent, L.A., Roberts, T.M., Sturgill, T.W. (1992) Activation of mitogen-activated protein kinase kinase by v-raf in NIH3T3 cells and in vitro. *Science, 257,* 1404-1407.

[23] Derynck, R., Jarrett, J.A., Chen, E.Y., Eaton, D.H., Bell, J., Assoian, R.K., Roberts, A.B., Sporn, M.B., Goeddel, D.V. (1985). Human transforming growth factor-beta complementary DNA sequence and expression in normal and transformed cells. *Nature, 316,* 701-705

[24] Derynck, R. & Zhang, Y.E. (2003). Smad dependent and smad-independent pathways in TGF-β family signaling. *Nature, 425,* 577-584.

[25] Dhar-Mascareno, M., Pedraza, A., Golde, D.W. (2005). PI3-kinase activation by GM-CSF in endothelium is upstream of Jak/Stat pathway: role of αGMR. *BBRC, 337,* 551-556.

[26] Dou, Q-P., Levin, A.H., Zhao, S., Pardee, A.B. (1993). Cyclin E and cyclin A as candidates for the restriction point protein. *Cancer Res,* 53, 1493-1497.

[27] Ducos, K., Karterne, B., Fortunel, N., Hatzfeld. A., Monier, M.N., Hatzfeld, J. (2000). P21cip1 mRNA is controlled by endogenous transforming growth factor-beta 1 in quiescent human hematopoietic stem/progenitor cells. *J Cell Physiol,* 184, 80-85.

[28] Dynlacht, B.D. (1997). Regulation of transcription by proteins that control the cell cycle. *Nature, 389,* 149-152.

[29] Ebisawa, T., Fukuchi, M., Murakami, G., Chiba, T., Tanaka, K., Imamura, T., Miyazono, K. (2001). Smurf1 interacts with transforming growth factor-β type I receptor through Smad7 and induces receptor degradation. *J Biol Chem, 276,* 12477-12480.

[30] Engel, M.E., McDonnell, M.A., Law, B.K., Moses, H.L. (2002). Interdependent SMAD and JNK signaling in transforming growth factor-β-mediated transcription. *J Biol Chem, 274,* 37413-37420.

[31] Eppert, K., Scherer, S.W., Ozcelik, H., Pirone, R., Hoodless, P., Kim, H., Tsui, L.C., Bapat, B., Gallinger, S., Andrulis, I.L., Thomsen, G.H., Wrana, J.L., Attisano, L. (1996). MADR2 maps to 18q21 and encodes a TGFbeta-regulated MAD-related protein that is functionally mutated in colorectal carcinoma. *Cell, 86,* 543-552.

[32] Ewen, M.E., Sluss, H.K., Whitehouse, L.L., Livingston, D.M. (1993). TGF beta inhibition of Cdk4 synthesis is linked to cell cycle arrest. *Cell, 74,* 1009-1020.

[33] Fautsch, M.P., Eblen, S.T., Anders, R.A., Burnette, R.J., and Leof, E.B. (1995). *J Cell Biochem, 58*, 517-26

[34] Fortunel, N.O., Hatzfeld, A., Hatzfeld, J.A. (2000). Transforming growth factor-beta: pleiotropic role in the regulation of hematopoiesis. *Blood, 96*, 2022-2036.

[35] Franzen, P., Ten Dijke, P., Ichijo, H., Yamashita, H., Schulz, P., Heldin, C-H., Miyazono, K. (1993). Cloning of a TGFβ type I receptor that forms a heteromeric complex with the TGF-β type II receptor. *Cell, 75,* 1-20.

[36] Friend, S.H., Bernards, R., Rogelj, S., Weinberg, R.A., Rapaport, J.M., Albert, D.M., Dryja, T.P. (1986). A human DNA segment with properties of the gene that predisposes to retinoblastoma and osteosarcoma. *Nature, 323*, 643-646.

[37] Frolik, C.A., Dart, L.L., Meyers, C.A., Smith, D.M., Sporn, M.B. (1983). Purification and initial characterization of a type beta transforming growth factor from human placenta. *Proc Natl Acad Sci USA, 80*, 3676-3680.

[38] Frolov, M.V. & Dyson, N.J. (2004). Molecular mechanisms of E2F-dependent activation and pRB-mediated repression. *J Cell Science, 117*, 2173-2181.

[39] Furukawa, Y., Uenoyama, S., Ohta, M., Tsunoda, A., Griffin, J.D., Saito, M. (1992). Transforming growth factor-β inhibits phosphorylation of the retinoblastoma susceptibility gene product in human monocytic leukemia cell line JOSK-I. *J Bio Chem, 267*, 17121-17127.

[40] Fortunel, N. O., Hatzfeld, A. & Hatzfeld, J. A. (2000). Transforming growth factor-beta: pleiotropic role in the regulation of hematopoiesis *Blood, 96*, 2022-2036.

[41] Geng, Y. & Weinberg, R.A. (1993).Transforming growth factor beta effects on expression of G1 cyclins and cyclin-dependent protein kinases. *Proc Natl Acad Sci USA, 90*, 10315-10319.

[42] Gentry, L.E., Liobin, M.N., Purchio, A.F., Marquardt, H. (1988). Molecular events in the processing of recombinant type 1 pre-pro-transforming growth factor beta to the mature polypeptide. *Mol Cell Biol, 8,* 4162-4168.

[43] Goey, H., Keller, J.R., Back, T., Longo, D.L., Ruscetti, F.W., Wiltrout, R.H. (1989). Inhibition of early murine hematopoietic progenitor cell proliferation after in vivo localregional administration of transforming growth factor-β1. *J Immunol, 143,* 877-880.

[44] Guan, K.L., Jenkins, C.W., Li, Y., Nichols, M.A., Okeefe, C.L., Matera, A.G., Xiong, Y. (1994). Growth suppression by p18, a p161NK4/MTS1- and p141NK4B/MTS2-related CDK6 inhibitor, correlates with wild-type pRb function. *Gene & Dev, 8,* 2939-2952.

[45] Guasch, G., Schober, M., Pasolli, H.A., Conn, E.B., Polak, L., Fuchs, E. (2007). Loss of TGFbeta signaling destabilizes homeostasis and promotes squamous cell carcinomas in stratified epithelia. *Cancer Cell, 12*, 313-327.

[46] Hahn, S.A., Schutte, M., Hoque, A.T., Moskaluk, C.A., da Costa, L.T., Rozenblum, E., Weinstein, C.L., Fischer, A., Yeo, C.J., Hruban, R.H., Kern, S.E. (1996). DPC4, a candidate tumor suppressor gene at human chromosome 18q21.1. *Science, 271,* 350-353.

[47] Halder, S.K., Beauchamp, R.D., Datta, P.K. (2005). Smad7 induces tumorigenicity by blocking TGF-β-induced growth inhibition and apoptosis. *Exp Cell Res, 307*, 231-246.

[48] Hannon, G.J. & Beach, D. (1994). p15INK4B is a potential effector of TGF-β-induced cell cycle arrest. *Nature, 371*, 257-261.

[49] Hanyu, A., Ishidou, Y., Ebisawa, T., Shimanuki, T., Imamura, T., Miyazono, K. (2001). The N-domain of Smad7 is essential for specific inhibition of transforming growth factor-beta signaling. *J Cell Biol, 155*, 1017-1027.

[50] Harper, J.W., Admi, G.R., Wei, N., Keymarsi, K., Elledge, S.J. (1993). The p21 cdk-interacting protein cip is a potent inhibitor of G1 cyclin-dependent kinases. *Cell, 75*, 805-816.

[51] Hayashi, H., Abdollah, S., Qiu, Y., Cai, J. Xu, Y.Y., Grinnell, B.W., Richardson, M.A., Topper, J.N., Gimbrone, MA. Jr. Wrana J, Falb. D (1997) The MAD-related protein Smad7 associates with the TGFβ receptor and functions as an antagonist of TGFβ signaling. *Cell, 89*, 1165-1173.

[52] Hayashida, T., Decaestecker, M., Schnaper, H.W. (2003) Cross-talk between ERK MAP kinase and Smad signaling pathways enhances TGF-beta-dependent responses in human mesangial cells. *FASEB J, 17*, 1576-1578.

[53] He, W., Dorn, D.C., Erdjument-Bromage, H., Tempst, P., Moore, M.A., Massagué, J.(2006). Hematopoiesis controlled by distinct TIF1gamma and Smad4 branches of the TGFbeta pathway. *Cell, 125*, 929-941.

[54] Helin, K., Harlow, E., and Fattaey, A. (1993). Inhibtion of E2F-1 transactivation by direct binding of the retinoblastoma gene. *Mol cell Biol, 13*, 6501-6508.

[55] Hengst, L., Dulic, V., Slingefland, J.M., Lees, E., Reed, S.I. (1994). A cell cycle-regulated inhibitor of cyclin-dependent kinases. *Proc Natl Acad Sci USA, 91*, 5291-5295.

[56] Herrera, R.E., Sah, V.P., Williams, B.O., Makela, T.P., Weinberg, R.A., Jacks, T. (1996). Altered cell cycle kinetics, gene expression, and G1 restriction point regulation in Rb-deficient fibroblasts. *Mol. Cell Biol, 16*, 2402-2407.

[57] Hiebert, S.W. (1993). Regions of the retinoblastoma gene product required for its interaction with the E2F transcription factor are necessary for E2F promoter repression and pRb-mediated growth suppression. *Mol Cell Biol, 13*, 3384-3391.

[58] Hinshelwood, R.A., Huschtscha, L.I., Melki, J., Stirzaker, C., Abdipranoto, A., Vissel, B., Ravasi, T., Wells, C.A., Hume, D.A., Reddel, R.R., Clark, S.J. (2007). Concordant epigenetic silencing of transforming growth factor-beta signaling pathway genes occurs early in breast carcinogenesis. *Cancer Res, 67*,11517-27

[59] Hirai, H., Roussei, M.F., Kato, J.Y., Ashmum, R.A., Sherr, C.J. (1995). Novel INK4 proteins, p19 and p18, are specific inhibitors of the cyclin D-dependent kinases cdk4 and cdk6. *Mol Cell Biol, 15,* 2672-2681.

[60] Hogan, B.L.M. (1996). Bone morphogenetic proteins: multifunctional regulators of vertebrate development. *Genes Dev, 10***,** 1580–1594.

[61] Hu, X., Cress, D.W., Zhong, Q., Zuckerman, K.S. (2000a). Phosphorylation of pRB family and formation of pRb family/E2F-4 complexes are differentially regulated in proliferating *cells and cells arrested in G1-phase by TGFβ1. Biochem Biophys Res Commun, 276,* 930-939.

[62] Hu, X., Cui, D., Moscinski, L.C., Zhang, X., Maccachero, V., Zuckerman, K.S. (2007). TGFbeta regulates the expression and activities of G2 checkpoint kinases in human myeloid leukemia cells. *Cytokine, 37,* 155-162.

[63] Hu, X., Moscinski, L.C., Valkov, N.I., Fisher, A.B., Hill, B.J., Zuckerman, K.S. (2000b). Prolonged activation of the mitogen-activated protein kinase pathway is required for macrophage-like differentiation of a human myeloid leukemic cell line. *Cell Growth Differ. 11*,191-200.

[64] Hu, X., Moscinski, L.C., Zuckerman, K.S. (1999). Transforming growth factor-β (TGFβ) inhibits growth of more differentiated myeloid leukemia cells and retinoblastoma protein phosphorylation at Ser795. *Exp Hematol, 27*, 605-614.

[65] Hu, X., Zhang, X., Zhong, Q., Fisher, A.B., Bryington, M., Zuckerman. K.S. (2001). Differential effects of transforming growth factor on cell cycle regulatory molecules in human myeloid leukemia cells. *Oncogene, 20*, 6840-6850.

[66] Hu, X. & Zuckerman KS (2000). Cell cycle and transcriptional control of human myeloid leukemic cells by transforming growth factor beta. *Leuk Lymphoma, 38*, 235-246.

[67] Huo, X.F., Yu, J., Peng, H., Du, Z.W., Liu, X.L., Ma, Y.N., Zhang, X., Zhang, Y., Zhao, H.L., Zhang, J.W. (2006). Differential expression changes in K562 cells during the hemin-induced erythroid differentiation and the phorbol myristate acetate (PMA)-induced megakaryocytic differentiation. *Mol Cell Biochem, 292,*155-167.

[68] Hurford, R.K. Jr., Cobrinik, D., Lee, M.H., Dyson, N. (1997). pRB and p107/p130 are required for the regulated expression of different sets of E2F responsive genes. *Genes Dev, 11,* 1447-1463.

[69] Ikeda, M.A., Jakoi, L., Nevins, J.R. (1996). A unique role for the Rb protein in controlling E2F accumulation during cell growth and differentiation. *Proc Natl Acad Sci U S A, 93,* 3215-3220.

[70] Inaba, T., Matsushime, H., Valentine, M., Roussel, M.F., Sherr, C.J., Look, A.T. (1992). Genomic organization, chromosomal localization, and independent expression of human cyclin D genes. *Genomics, 13,* 565-574.

[71] Inoue, Y., Kitagawa, M., Taya, Y. (2007). Phosphorylation of pRB at Ser612 by Chk1/2 leads to a complex between pRB and E2F-1 after DNA damage. *EMBO J. 26,* 2083-2093.

[72] Itoh, F. Asao, H., Sugamura, K., Heldin, C.H., ten Dijke, P., Itoh, S. (2001). Promoting bone morphogenetic protein signaling through negative regulation of inhibitory Smads. *EMBO J, 20,* 4132-4142.

[73] Jeffrey, P.D., Russo, A.A., Polyak, K., Gibbs, E., Hurwitz, J., Massagué, J., Pavletich, N.P. (1995). Mechanism of CDK activation revealed by the structure of a cyclinA-CDK2 complex. *Nature, 376,* 313-320

[74] Jelinek, D.F., Tschumper, R.C., Stolovitzky, G.A. Iturria, S.J., Tu, Y., Lepre, J., Shah, N., Kay, N.E. (2003). Identification of a global gene expression signature of B-chronic lymphocytic leukemia. *Mol Cancer Res, 1,* 346-361.

[75] Ji, Y., Lee, H.J., Goodman, C., Uskokovic, M., Liby, K., Sporn, M., Suh, N. (2006). The synthetic triterpenoid CDDO-imidazolide induces monocytic differentiation by activating the Smad and ERK signaling pathways in HL60 leukemia cells. *Mol Cancer Ther, 5,* 1452-1458.

[76] Kanakura Y, Druker B, Wood K, Mamon H, Okuda K, Roberts T, Griffin JD (1991) GMCSF and IL-3 induce rapid phosphorylation and activation of the proto-oncogene Raf-1 in a human factor-dependent myeloid cell line. *Blood, 77,* 243-248

[77] Kanatni, Y., Kasukabe, T., Okabe-Kado, J., Hayashi, S.I., Yamamoto-Yamaguchi, Y., Motoyoshi, K., Nagata, N., Honma, Y. (1996). Transforming growth factor β and dexamethasone cooperatively enhance c-jun gene expression and inhibit the growth of human monocytoid leukemia cells. *Cell Growth & Differ, 7,* 187-196.

[78] Kavsak, P., Rasmussen, R.K., Causing, C.G., Bonni, S., Zhu, H., Thomsen, G.H., Wrana, J.L. (2000). Smad7 binds to Smurf2 to form an E3 ubiquitin ligase that targets the TGFβ receptor for degradation. *Mol. Cell, 6,* 1365-1375.

[79] Kehrl, J.H., Roberts, A.B., Wakefield, L.M., Jakowlew, S., Sporn, M.B., Fauci, A.S. (1986). Transforming growth factor-beta is an important immunomodulatory protein for human B lymphocytes. *J Immunol, 137,* 3855-3860.

[80] Keller, J.R., Bartelmez, S.H., Sitnicka, E., Ruscetti, F.W., Ortiz, M., Gooya, J.M., Jacobsen, S.E.W. (1994). Distinct and overlapping direct effects of macrophage

inflammatory protein-1α and transforming growth factor β on hematopoietic progenitor/stem cell growth. *Blood, 84*, 2175-2181.

[81] Keller, J.R., Jacobsen, S.E.W., Dubois, C.M., Hestda, I.K., Ruscetti, F.W. (1992). Transforming growth factor-β: A bidirectional regulator of hematopoietic cell growth. *Int. J Cell Cloning, 10*:2-11.

[82] Keller, J.R., Jacobsen, S.E.W., Sill, K., Ellingsworth, L., Ruscetti, F.W. (1991). Stimulation of granulopoiesis by transforming growth factor β: Synergy with granulocyte-macrophage colony-stimulating factor. *Proc Natl Acad Sci USA, 88:* 7190-7194.

[83] Keller, J.R., McNiece, I.M., Sill, K.T., Ellingsworth, L.R., Quesenberry, P.J., Sing, G.K., Ruscetti, F.W. (1990). Transforming growth factor-β directly regulates primitive murine hematopoietic cell proliferation. *Blood, 75*, 596-602.

[84] Kim, M.S., Lim, W.K., Cha, J.G., An, N.H., Yoo, S.J., Park, J.H., Kim, H.M., Lee, Y.M. (2001). The activation of PI 3-K and PKC zeta in PMA-induced differentiation of HL-60 cells. *Cancer Lett, 171*, 79-85.

[85] Kim, S-J., Letterio, J. (2003). Transforming growth factor-β signaling in normal and malignant hematopoiesis. *Leukemia, 17*, 1731-1737.

[86] Kiyokawa, H., Busquets, X., Powell, C.T., Ngo, L., Rifkind, R.A., Marks, P.A. (1992). Cloning of a D-type cyclin from murine erythroleukemia cells. *Proc Natl. Acad. Sci USA, 89,* 2444-2447.

[87] Kiyokawa, H., Kineman, R.D., Manova-Todorova, K.O., Soares, V.C., Hoffman, E.S., Ono, M., Khanam, D., Hayday, A.C., Frohman, L.A. (1996). Enhanced growth of mice lacking the cyclin-dependent kinase inhibitor function of p27 (Kip1). *Cell, 85,* 721-732.

[88] Klein, U., Tu, Y., Stolovitzky, G.A., Mattioli, M., Cattoretti, G., Husson, H., Freedman, A., Inghirami, G., Cro, L., Baldini, L., Neri, A., Califano, A., Dalla-Favera, R. (2001). Gene expression profiling of B cell chronic lymphocytic leukemia reveals a homogeneous phenotype related to memory B cells. *J Exp Med, 194*, 1625-1638.

[89] Knabbe, C., Lippman, M.E., Wakefield, L.M., Flanders, K.C., Kasid, A. (1987). Evidence that transforming growth factor-beta is a hormonally regulated negative growth factor in human breast cancer cells. *Cell, 48*, 417-428.

[90] Koeffler, H.P. (1983). Induction of differentiation of human acute myelogenous leukemia cells: therapeutic implications. *Blood, 62,* 709-721.

[91] Koff, A., Cross, F., Fisher, A., Schumacher, J., Leguellec, K., Philippe, M., Roberts, J.M. (1991). Human cyclin E, a new cylin that interacts with two members of the CDC2 gene family. *Cell, 66*, 1217-1228.

[92] Koff, A., Giordano, A., Desai, D., Yamashita, K., Harper, J.W., Elledge, S., Nishimoto, T., Morgan, D.O., Franza, B.R., Roberts, J.M. (1992). Formation and activation of a cyclin E-cdk2 complex during the G1 phase of the human cell cycle. *Science, 257*, 1689-1694.

[93] Kretzschmar, M., Doody, J., Timokhina, I., Massagué, J. (1999). A mechanism of repression of TGF-β/Smad signaling by oncogenic Ras. *Genes Dev, 13*, 804-816.

[94] Kulkarni, A.B., Huh, C.G., Becker, D., Geiser, A., Lyght, M., Flanders, K.C., Roberts, A.B., Sporn, M.B., Ward, J.M., Karlsson, S. (1993). Transforming growth factor beta 1

null mutation in mice causes excessive inflammatory response and early death. *Proc Natl Acad Sci USA, 90,* 770-774.

[95] Kyriakis, J.M., App, H., Zhang, X., Banerjee, P., Brautigan, D.L., Rapp, U.R., Avruch, J. (1992). Raf-1 activates MAP kinase-kinase. *Nature, 358*, 417-421.

[96] Lagna, G., Hata, A., Hemmati-Brivanlou, A., Massague, J. (1996). Partnership between DPC4 and SMAD proteins in TGF-β signaling pathways. *Nature, 383*, 832-836.

[97] Lange-Carter, C.A., Pleiman, C.M., Gardner, A.M., Blumer, K.J., Johnson, G.L. (1993) A divergence in the MAP kinase regulatory network defined by MEK kinase and Raf. *Science, 260,* 315-319.

[98] Lee, M.H., Nikolic, M., Baptista, C.A., Lai, E., Tsai, L.H. and Massague, J. (1996). The brain-specific activator p35 allows cdk5 to escape inhibition by p27kip1 in neurons. *Proc Natl Acad Sci USA, 93*, 3259-3263.

[99] Lin, H.Y., Wang, X.F., Ng-eaton, E., Downward, J., Lodish, H.F. (1992). Expression cloning of the TGF-β type II receptor, a functional transmembrane serine/threonine kinase. *Cell, 68*, 775-785.

[100] Liu, F., Hata, A., Baker, J.C., Doody, J., Carcamo, J., Harland, R.M., Massague, J. (1996). A human Mad protein acting as a BMP-regulated transcriptional activator. *Nature, 381*, 620-623.

[101] Macias-Silva, M., Abdollah, S., Hoodless, P.A., Pirone, R., Attisano, L., Wrana, J.L. (1996). MADR2 is a substrate of the TGF-β receptor and its phosphorylation is required for nuclear accumulation and signaling. *Cell, 87*, 1215-1224.

[102] Marshall, C.J. (1995) Specificity of receptor tyrosine kinase signaling: transient versus sustained extracellular signal-regulated kinase activation. *Cell, 80*, 179-185.

[103] Massagué. J. (1998). TGFβ signal transduction. *Annu Rev Biochem, 67,* 753–791.

[104] Massagué, J. (2000). How cells read TGF-β signals. *Nature Rev Mol Cell Biol, 1,* 169-178.

[105] Matsuoka, M., Kato, J.Y., Fisher, R.P., Morgan, D.O., Sherr, C.J. (1994). Activation of cyclin-dependent kinase 4 (cdk4) by mouse MO15-associated kinase. Mol. *Cell Bio, 14*, 7265-7275.

[106] Matsuoka, S., Edwards, M.C., Bai, C., Parker, S., Zhang, P., Baldini, A., Harpen, J.W., Elledge, S.J. (1995). P57KIP2, a structurally distinct member of the p21CIP1 cdk inhibitor family, is a candidate tumor suppressor gene. *Gene & Dev, 9,* 650-662.

[107] Matsushime, H., Ewen, M.E., Strom, D.K., Kato, J.Y., Hanks, S.K., Roussel, M.F., Shern, C.J. (1992). Identification and properties of an atypical catalytic subunit (p34PSK-J3/cdk4) for mammalian D type G cyclins. *Cell, 71*, 323-334.

[108] Matsushime, H., Quelle, D.E., Shurtletf, S.A., Shibuya, M., Sherr, C. J., Kato, J.Y. (1994). D-type cyclin-dependent kinase activity in mammalian cells. *Mol. Cell. Biol, 14*, 2066-2076.

[109] Matsushime, H., Roussel, M.F., Ashmun, R.A., Sherr, C.J. (1991). Colony-stimulating factor 1 regulates novel cyclins during the G1 phase of the cell cycle. *Cell, 65,* 701-713.

[110] McNiece, I.K., Bertoncello, I., Keller, J.R., Ruscetti, F.W., Hartley, C.A., Zsebo, K.M. (1992). Transforming growth factor β inhibits the action of stem cell factor on mouse and human hematopoietic progenitors. *Int. J. Cell Cloning, 10,* 80-86.

[111] Meyerson, M. & Harlow, E. (1994). Identification of G1 kinase activity for cdk6, a novel cyclin D partner. *Mol Cell Biol, 14*, 2077-2086.

[112] Mihara, K., X.R. Cao, A. Yen, S. Chandler, B. Driscoll, AL. Murphree, A. Tang, and YK Fung. (1989). Cell cycle-dependent regulation of phosphorylation of the human retinoblastoma gene product. *Science, 246*, 1300-1303.

[113] Millar, J.B., McGowan, C.H., Lenaers, G., Jones, R., Russell, P. (1991). p80cdc25 mitotic inducer is the tyrosine phosphatase that activates p34cdc2 kinase in fission yeast. *EMBO J, 10*, 4301-4309.

[114] Moberg, K., Starz, M.A., Lees, J.A. (1996). E2F-4 switches from p130 to p107 and pRb in response to cell cycle reentry. *Mol Cell Biol, 16*, 1436-1449.

[115] Morgan, D.O. (1992). Cell cycle control in normal and neoplastic cells. *Curr Opin Genet Dev, 2*, 33-37.

[116] Morita, N., Yamamoto, M., Tanizawa, T. (2003). Correlation of c-kit expression and cell cycle regulation by transforming growth factor-beta in CD34+ CD38- human bone marrow cells. *Eur J Haematol, 71*, 351-358.

[117] Motokura, T., & Arnold, A. (1993). Cyclins and oncogenesis. Biophys. Acta 1155: 63-78.

[118] Motokura, T., Bloom, T., Kim, H.G., Juppner, H., Ruderman, J.V., Kronenberg, H.M., Arnold, A. (1992). Motokura T, Keyomarsi K, Kronenberg HM, and Arnold A. Cloning and characterization of human cyclin D3, a cDNA closely related in sequence to the PRAD1/cyclin D1 proto-oncogene. *J Biol Chem, 267*, 20412-20415.

[119] Mulligan, G.J., Wong, J., Jacks, T. (1998). p130 is dispensable in peripheral T lymphocytes: evidence for functional compensation by p107 and pRB. *Mol Cell Biol, 18*, 206-220.

[120] Nakao, A., Roijer, E., Imamura, T., Souchelnytskyi, S., Stenman, G., Heldin, C.H., ten Dijke, P. (1997). Identification of Smad2, a human Mad-related protein in the transforming growth factor beta signaling pathway. *J Biol Chem, 272*, 2896-2900.

[121] Nakayama, K., Ishida, N., Shirane, M., Inomata, A., Inoue, T., Shishido, N., Horii, I., Loh, D.Y., Nakayama, K (1996). Mice lacking p27(Kip1) display increased body size, multiple organ hyperplasia, retinal dysplasia, and pituitary tumors. *Cell, 85,* 707-720.

[122] Nevins, J.R. (2001). The Rb/E2F pathway and cancer. *Human Molecular Genetics, 10*, 699-703.

[123] Norbury, C. & Nurse, P. (1992). Animal cell cycles and their control. *Annu. Rev. Biochem, 61*, 441-470.

[124] Nurse, P. (2000). A long twentieth century of the cell cycle and beyond. Cell, 100,71-78.

[125] Ohta, M., Greenberger, J.S., Anklesaria, P., Bassols, A., Massague, J. (1987). Two forms of transforming growth factor-beta distinguished by multipotential haematopoietic progenitor cells. *Nature, 329*, 539-541.

[126] Okabe-Kado, J., Honma, Y., Hayashi, M., Hozumi, M. (1989). Inhibitory action of transforming growth factor-beta on induction of differentiation of myeloid leukemia cells. *Jpn J Cancer Res, 80,* 228-232.

[127] Okuda K, Sanghera JS, Pelech SL, Kanakura Y, Hallek M, Griffin JD, Druker BJ (1992). Granulocyte-macrophage colony-stimulating factor, interleukin-3, and steel

factor induce rapid tyrosine phosphorylation of p42 and p44 MAP kinase. *Blood, 79*, 2880-2887.

[128] Ottmann, O.G. & Pelus, L.M. (1991). Differential proliferative effects of transforming growth factor-β on human hematopoietic progenitor cells. *J Immunol, 140*, 2661-2665.

[129] Padgett, R.W., Das, P., Krishna, S.(1998). TGF-beta signaling, Smads, and tumor suppressors. *Bioessays, 20*, 382–390

[130] Pagano, M., Pepperkok, R., Lukas, J., Baldin, V., Ansorge, W., Bartek, J., Draetta, G. (1993). Regulation of the cell cycle by the cdk2 protein kinase in cultured human fibroblasts. *J Cell Biol, 121*, 101-111.

[131] Pages, G., Lenormand, P., Allemain, G.L., Chambard, J.C., Meloche, S., Pouyssegur, J. (1993) Mitogen-activated protein kinases $p42^{mapk}$ and $p44^{mapk}$ are required for fibroblast proliferation. *Proc. Natl. Acad. Sci. USA, 90*, 8319-8323.

[132] Park, H.J., Kim, B.C., Kim, S.J., Choi, K.S. (2002) Role of MAP kinases and their cross-talk in TGF-beta1-induced apoptosis in FaO rat hepatoma cell line. *Hepatology, 35,*1360-1371.

[133] Perkins, G.R., Marshall, C.J., Collins, M.K.L. (1996) The role of MAP kinase kinase in Interleukin -3 stimulation of proliferation. *Blood, 87*, 3669-3675.

[134] Polyak, K., Lee, M.H., Bromage, H.E., Koff, A., Roberts, J.M., Tempst, P. Massagué, J. (1994) Cloning of p27Kip 1, a cyclin-dependent kinase inhibitor and a potential mediator of extracellular antimitogenic signals. *Cell, 78,* 59-66.

[135] Porter, P.L., Malone, K.E., Heagerty, P.J., Alexander, G.M., Gatli, L.A., Firpo, E.J., Daling J.R,. and Roberts, J.M. (1997). Expression of cell-cycle regulators p27Kip1 and cyclin E, alone and in combination, correlated with survival in young breast cancer pathents. *Nat Med, 3*, 222-225.

[136] Qin, X-Q., Livingston, D.M., Ewen, M., Sellers, W.R., Arany, Z., Kaelin, W.G. Jr. (1995). The transcription factor E2F1 is a downstream target of Rb action. *Mol Biol Cell, 15*, 742-755.

[137] Qiu, M. & Green, S.H. (1992) PC12 cell neuronal differentiation is associated with prolonged p21ras activity and consequent prolonged ERK activity. *Neuron, 9*, 705-717.

[138] Raftery, L.A. & Sutherland, D.J. (1999). TGF-beta family signal transduction in Drosophila development: from Mad to Smads. *Dev Biol, 210*, 251–268.

[139] Raftery, L.A., Twombly, V., Wharton, K., Gelbart, W.M. (1995). Genetic screens to identify elements of the decapentaplegic signaling pathway in Drosophila. *Genetics, 139*, 241-254.

[140] Rayman, J.B., Takahashi, Y., Indjeian, V.B., Dannenberg, J.H., Catchpoles, S., Watson, R.J., te Riele, H., Dynlacht, B.D. (2002). E2F mediates cell cycle-dependent transcriptional repression in vivo by recruitment of an HDAC1/mSin3B corepressor complex. *Genes Dev, 16*, 933-947.

[141] Ren, B., Cam, H., Takahashi, Y., Volkert, T., Terragni, J., Young, R., Dynlacht, B. (2002). E2F integrates cell cycle progression with DNA repair, replication, and G2/M checkpoints. *Genes Dev, 16*, 245-256.

[142] Resnitzky, D., Gossen, M., Bujard, H., Reed, S.I. (1994). Acceleration of the G1/S phase transition by expression of cyclins D1 and E with an inducible system. *Mol Cell Biol, 14*, 1669-1679.

[143] Reynisdottir, I. & Massague, J. (1997).The subcellular locations of p15ink4b and p27kip1 coordinate their inhibitory interactions with Cdk4 and cdk2. *Genes Dev, 11*, 492-503.

[144] Rich, J.N., Zhang, M., Datto, M.B., Bigner, D.D., Wang, X.F. (1999). Transforming growth factor-beta-mediated p15(INK4B) induction and growth inhibition in astrocytes is SMAD3-dependent and a pathway prominently altered in human glioma cell lines. *J Biol Chem, 274*, 35053-35058.

[145] Roberts, A.B., Anzano, M.A., Meyers, C.A., Wideman, J., Blacher, R., Pan, Y.C., Stein, S., Lehrman, S.R., Smith, J.M., Lamb, L.C. (1983). Purification and properties of a type beta transforming growth factor from bovine kidney. *Biochemistry, 22*, 5692-5698.

[146] Roberts, J.M. (1999). Evolving ideas about cyclins. *Cell, 97*,129-132.

[147] Rosenblatt, J., Gu, Y., Morgan, D.O. (1992). Human cyclin-dependent kinase 2 is activated during the S and G2 phases of the cell cycle and associates with cyclin A. *Proc Natl. Acad. Sci. USA, 89*, 2824-2828.

[148] Ruscetti, F.W., Akel, S., Bartelmez, S.H. (2005). Autocrine transforming growth factor-beta regulation of hematopoiesis: many outcomes that depend on the context. *Oncogene, 24,* 5751-5763.

[149] Russo, A.A., Jeffrey, P.D., Patten, A.K., Massague, J., and Pavletich, N.P. (1996). Crystal structure of the p27Kip1 cyclin-dependent-kinase inhibitor bound to the cyclin A-cdk2 complex. *Nature, 382,* 325-331.

[150] Scandura, J.M., Boccuni, P., Massagué, J., Nimer, S.D. (2004). Transforming growth factor beta-induced cell cycle arrest of human hematopoietic cells requires p57KIP2 up-regulation. *Proc Natl Acad Sci U S A, 101*, 15231-15236.

[151] Schluter, C., Duchrow, M., Wohlenberg, C., Becker, M.H., Key, G., Flad, H.D., Gerdes, J. (1993). The cell proliferation-associated antigen of antibody Ki-67: a very large, ubiquitous nuclear protein with numerous repeated elements, representing a new kind of cell cycle-maintaining proteins. *J Cell Biol, 123*, 513-522.

[152] Schwarz, J.K., Devoto, S.H., Smith, E.J., Chellappan, S.P., Jakoi, L., Nevins, J.R. (1993). Interactions of the p107 and Rb proteins with E2F during the cell proliferation response. *EMBO J, 12*, 1013-1020.

[153] Serrano, M., Hannon, G.J., and Beach, D. (1993). A new regulatory motif in cell-cycle control causing specific inhibition of cyclin D/Cdk4. *Nature, 366*, 704-707.

[154] Shirodkar, S., Ewen, M., DeCaprio, J.A., Morgan, J., Livingston, D.M., Chittenden, T. (1992). The transcription factor E2F interacts with the retinoblastoma product and a p107-cyclin A complex in a cell cycle-regulated manner. *Cell, 68*, 157-166.

[155] Shull, M.M., Ormsby, I., Kier, A.B., Pawlowski, S., Diebold, R.J., Yin, M. Allen R, Sidman C, Proetzel G, Calvin D., Annunziata N., Doetschman, T. (1992). Targeted disruption of the mouse transforming growth factor-β1 gene results in multifocal inflammatory disease. *Nature, 359*, 693-699.

[156] Sing, G.K., Keller, J.R., Ellingsworth, L.R., Ruscetti, F.W. (1988). Transforming growth factor-β selectively inhibits normal and leukemic human bone marrow cell growth in vitro. *Blood, 72,* 1504-1511.

[157] Sitnicka, E., Ruscetti, F.W., Priestley, G.V., Wolf, N.S., Bartelmez, S.H. (1996) Transforming growth factor β1 directly and reversibly inhibits the initial cell divisions of long-term repopulating hematopoietic stem cells. *Blood, 88*, 82-88

[158] Smits, V.A. & Medema, R.H. (2001). Checking out the G2/M transition. *Biochim Biophys Acta, 1519*, 1-12.

[159] Snoeck, H-W., Weekx, S., Moulijn, A., Lardon, F., Lenjou, M., Nys, F., Van Ranst, P.C.F., Van Bockstaele, D.R., Berneman, Z.N. (1996). Tumor necrosis factor αis a potent synergistic factor for the proliferation of primitive human hematopoietic progenitor cells and induces resistance to transforming growth factor β but not to interferon γ. *J Exp Med, 183,* 705-710.

[160] Stevaux, O. & Dyson, N.J. (2002. A revised picture of the E2F transcriptional network and RB function. *Curr. Opin Cell Biol, 14*, 684-691.

[161] Takahashi, Y., Rayman, J.B., Dynlacht, B.D. (2000). Analysis of promoter binding by the E2F and pRB families in vivo: distinct E2F proteins mediate activation and repression. *Genes Dev, 14*, 804-816.

[162] Tessier, N. & Hoang, T. (1988). Transforming growth factor beta inhibits the proliferation of the blast cells of acute myeloblastic leukemia. *Blood, 72*, 159-164

[163] Toyoshima, H. & Hunter, T. (1994) p27, a novel inhibitor of G1 cyclin-cdk protein kinase activity, is related to p21. *Cell, 78*, 67-74.

[164] Tsai, L., Lees, E., Faha, B., Harlow, E., Riabowol, K. (1993). The cdk2 kinase is required for the G1-to-S transition in mammalian cells. *Oncogene, 8,* 1593-1602.

[165] Tsukazaki, T., Chiang, T.A., Davison, A.F., Attisano, L., Wrana, J.L. (1998). SARA, a FYVE domain protein that recruits Smad2 to the TGFbeta receptor. *Cell, 95*, 779-791.

[166] Wang, X.F., Lin, H.Y., Ng-eaton, E., Downward, J., Lodish, H.F., Weinberg, R.A. (1991). Expression cloning and characterization of the TGF-β type III receptor. *Cell, 67,* 797-805.

[167] Weinberg, R.A. (1995). The retinoblastoma protein and cell cycle control. *Cell, 81*, 323-330.

[168] Wells, J., Boyd, K.E., Fry, C.J., Bartley, S.M., Farnham, P.J. (2000). Target gene specificity of E2F and pocket protein family members in living cells. *Mol Cell Biol, 20,* 5797-5807.

[169] Wrana, J.L. & Attisano, L. (1996). MAD-related proteins in TGF-β signaling. *Trends Genet, 12,* 493-496.

[170] Wrana, J.L., Attisano, L., Wieser, R., Venture, F., Massague, J. (1994) Mechanism of activation of the TGF-β receptor. *Nature, 370*, 341-347.

[171] Wu, G., Chen, Y.G., Ozdamar, B., Gyuricza, C.A., Chong, P.A., Wrana, J.L., Massagué, J., Shi, Y. (2000). Structural basis of Smad2 recognition by the Smad anchor for receptor activation. *Science, 287*, 92-97.

[172] Wu, L., Timmers, C., Maiti, B., Saabedra, H.I., Sang, L., Chong, G.T., Nuckolls, F., Giangrande, P., Wright, F.A., Field, S.J., Greenberg, M.E., Orkin, S., Nevins, J.R., Robinson, M.L., Leone, G. (2001). The E2F1-3 transcription factors are essential for cellular proliferation. *Nature, 414*, 457-462.

[173] Wu, R.Y., Zhang, Y., Feng, X.H., Derynck, R. (1997). heteromeric and homomeric interactions correlate with signaling activity and functional cooperatively of Smad3 and Smad4DPC4. *Mol. Cell Biol, 17*, 2521-2528.

[174] Xiong, Y., Connolly, T., Futcher, B., Beach, D. (1991). Human D-type cyclin. *Cell, 65*, 691-699.

[175] Xiong, Y., Zhang, H., Beach, D. (1992). D type cyclins associate with multiple potein kinases and the DNA replication and repair factor PCNA. *Cell, 71*, 505-514.

[176] Xu, J. & Attisano, L. (2000). Mutations in the tumor suppressors Smad2 and Smad4 inactivate transforming growth factor beta signaling by targeting Smads to the ubiquitin-proteasome pathway. *Proc Natl Acad Sci U S A, 97*, 4820-4825.

[177] Yingling, J.M., Das, P., Savage, C., Zhang, M., Padgett, R.W., Wang, X.F. (1996). Mammalian dwarfins are phosphorylated in response to TGF-β and are implicated in control of cell growth. *Proc Natl Acad Sci USA, 93*, 8940-8944.

[178] Yonekura, A., Osaki, M., Hirota, Y., Tsukazaki, T., Miyazaki, Y., Matsumoto, T., Ohtsuru, A., Namba, H., Shindo, H., Yamashita, S. (1999) Transforming growth factor-beta stimulates articular chondrocyte cell growth through p44/42 MAP kinase (ERK) activation. *Endocr J, 46*, 545-553.

[179] Yu, L., Hebert, M.C., Zhang, Y.E. (2002). TGF-β receptor-activated p38 MAP kinase mediates Smad-independent TGF-β responses. *EMBO J, 21*, 3749-3759.

[180] Zhang, Y., Feng, X.H., Wu, R.Y., Derynck, R. (1996). Receptor-associated Mad homologues synergize as effectors of the TGF-β response. *Nature, 383,* 168-172.

In: Cell Cycle Control: New Research
Editors: N. H. Leroy, N. T. Fournier

ISBN: 978-1-60456-796-0

Chapter V

New Directions for CDKs and HDACs Inhibition in Cancer Therapy

Graziano Lolli and Stefania Di Marco
Istituto di Ricerche di Biologia Molecolare "P. Angeletti",
Via Pontina Km 30.600, 00040 Pomezia, Rome, Italy

Abstract

Cell cycle is dependent on correct cycling and activation of Cyclin Dependent Kinases (CDKs)/Cyclin pairs. Their protein levels and regulation mechanisms are altered in transformed cells and CDKs are considered attractive targets for cancer therapy. Data from knockout and siRNA experiments have shown a redundancy in CDKs functions. Recent clinical trials and preclinical studies have identified the possibility to use inhibitors specific for subclasses of cell-cycle CDKs (CDK1/2 or CDK4/6) as single agents for a few specific cancer types that show a more pronounced dependency on CDKs functions. Pan-CDK inhibitors are giving, instead, more promising results. They inhibit both cell-cycle CDKs and transcriptional CDKs by combining cytostatic and apoptotic effects in cancer cells. Apoptosis is induced through depletion of anti-apoptotic proteins, sensitive to inhibition of transcription because encoded by mRNAs with short half-lifes and rapid turnover. Recently, new crystal structures of transcriptional CDKs (CDK7 and CDK9) have provided molecular basis for targeting this subclass. Combinations of pan-CDK inhibitors with traditional chemotherapic agents have shown a synergistic cytotoxic effect in a variety of cancers and the restoration of susceptibility in cancers that developed resistance to traditional therapies. Flavopiridol, a pan-CDK inhibitor, is also being used in clinical trials in combination with Histone deacetylase (HDAC) inhibitors, which are also being developed as anti-neoplastic drugs. Evidence has been reported of cross-talk between CDKs and HDACs that indicates the existence of a complex cross-regulation mechanism. This cross-talk between CDKs and HDACs mediated by Retinoblastoma (pRb) and/or CDK inhibitors (CKI) is of particular pharmacological relevance. It engenders the expectation that the combined pharmacological inhibition of CDKs and HDACs might have a higher therapeutic profile than the single treatment.

Introduction

Cyclin-Dependent Kinases

Cyclin Dependent Kinases (CDKs) regulate the cell division cycle, apoptosis, transcription and differentiation in addition to functions in the nervous system. These kinases are activated by the association with their regulatory partners (cyclins) and by phosphorylation of a specific Thr residue. Thirteen CDKs and 25 cyclin-box-containing proteins are encoded in the human genome [1]. Among them, those known to interact in a complex are presented in table 1 with their respective functions. Inhibitors of CDKs are currently being evaluated as possible therapeutic agents for different diseases (table 1).

Table 1. Involvement of CDK/cyclin complexes in cellular functions and in diseases

CDK	Cyclin	Function	Diseases targeted
CDK1	Cyclin A Cyclin B Ringo	Regulates G2/M transition and progression of M phase	Cancer
CDK2	Cyclin E Cyclin A	Regulates G1/S transition and progression of S phase	Cancer Viral infections
CDK3	Cyclin E	Regulates G1/S transition	
CDK4	Cyclin D	Regulates progression through G1 phase	Cancer
CDK5	p35, p25 p39, p29	Involved in several processes in the nervous system (neurone migration, neurotransmission)	Alzheimer's disease Amyotrophic lateral sclerosis
CDK6	Cyclin D	Regulates progression through G1 phase	Cancer
CDK7	Cyclin H	Activates CDK1, CDK2, CDK4, CDK6 Regulates initiation of transcription	Cancer Viral infections
CDK8	Cyclin C	Regulates transcription	
CDK9	Cyclin T Cyclin K	Regulates elongation phase of transcription Involved in HIV-TAT dependent transcription Involved in signal transduction and differentiation	Cancer HIV Cardiac hypertrophy
CDK11	Cyclin L	Involved in transcription	

Only known complexes are presented. Other potential CDKs include the PCTAIREs, PFTAIRE, PITAIRE, KKIALRE, PISSLRE and NKIAMRE (named after the sequence of their PSTAIRE motif, a conserved motif located in the cyclin-binding domain). 'Orphan' cyclins include cyclins F, G, I, M, O and P.

Role of CDKs in Cell Cycle

CDKs in association with their respective cyclins regulate cell cycle progression. While CDKs expression is constant during cell cycle, specific cyclins show periodic synthesis and degradation in synchrony with the different phases of cell cycle (figure 1). Since CDKs are inactive without their respective cyclins, the temporary programmed activity of CDKs is dictated by the temporary programmed expression and destruction of cyclins. Specific CDKs are then turned on and off as required by the different cell cycle phases (figure 1).

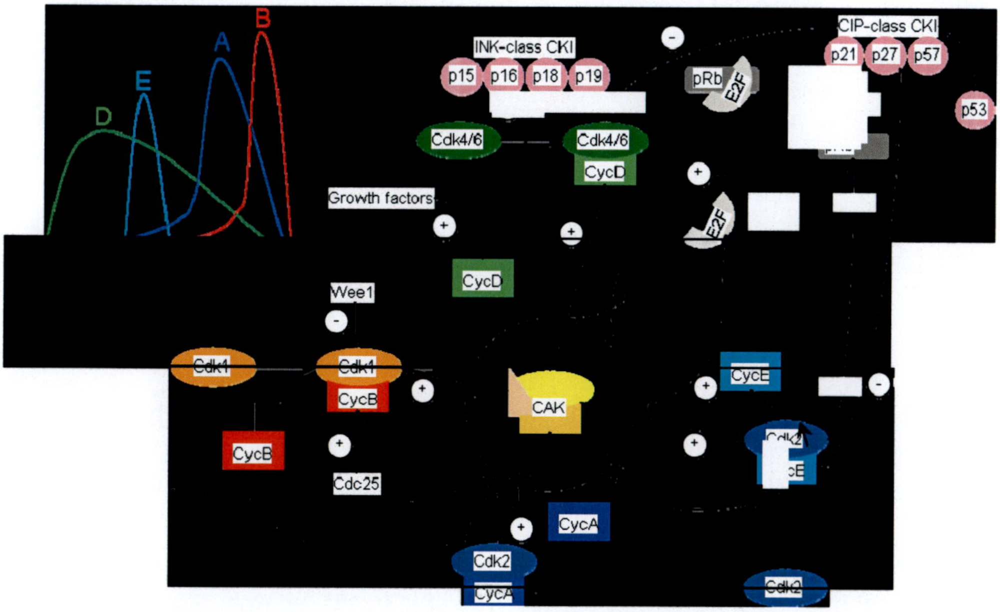

Figure 1. Regulation of cell cycle by CDK/Cyclin complexes. Left: Periodic changes in the expression of the different cyclins during cell cycle progression. Right: Regulation of cell cycle by activation and inhibition of CDKs. The activities of CDKs are influenced not only by cyclins but also by phosphorylation (CAK and Wee1), dephosphorylation (Cdc25), ubiquitin-dependent proteolysis and induction of CDIs by the tumor suppressor protein p53.

Our present state of understanding of the roles of the CDKs can be summarised briefly as follows. G0 quiescent cells enter G1 phase following stimulation from growth factors. Mitogen-induced Ras signalling induces the expression of D-type cyclins that activate CDK4 and CDK6. The major substrate of CDK4 and CDK6 is pRB. In quiescent and early G1 cells, pRB is hypo-phosphorylated and acts as a transcriptional repressor binding and inactivating the transcription factor E2F [2-3]. pRB additionally recruits chromatin remodelling enzymes like histone deacetylases (HDACs) to promoters [4]. pRB phosphorylation by CDK4 and CDK6 abolishes HDAC and E2F binding. This allows the synthesis of E2F-responsive genes including several proteins required for DNA synthesis and Cyclin E (CycE) [5-6].

CDK2/CycE activity increases rapidly at the end of G1 phase. This is due to CycE synthesis but also to additional mechanisms. D-type cyclin/CDK complexes sequester the CIP/KIP inhibitors of CDK2 [7-8]. At least one CDK2 inhibitor, p27, is phosphorylated by CDK2/CycE and targeted for degradation [9]. This generates a positive feedback loop for CDK2/CycE activity. CDK2/CycE targets pRB at different phosphorylation sites from D-type cyclin/CDK complexes [10-11]. These phosphorylations are required for the complete inactivation of pRB and abrogation of the mitogen dependence for cell cycle progression. DNA replication is also initiated. Activation of Cdc45 by phosphorylation allows loading of DNA polymerases on replication origins [12]. Cyclins D are phosphorylated and targeted for degradation by SCF (Skp1/cullin/F-box protein) complexes. Destruction of CycE is also dictated by CDK2/CycE that auto-phosphorylates on CycE [13].

Cyclin A (CycA) is an E2F-dependent gene, expressed following pRB phosphorylation by CDK2/CycE [14-15]. CycA synthesis and CycE degradation redistribute CDK2 to CycA, possibly changing its substrate specificity. However a difference in substrate specificity between CDK2/CycE and CDK2/CycA has not been definitively established. Cdc6 is phosphorylated and targeted for nuclear export [16]. Since Cdc6 is required for initiation of DNA replication, its nuclear export prevents DNA re-replication. E2F transcription factors are also inhibited by phosphorylation and by CycA binding [12, 17]. Failure to inhibit E2F leads to apoptosis [18].

CDK1 associates with CycA and Cyclin B (CycB) and is responsible for the G2/M transition and for the progression of M-phase [19]. CycA and CycB association with CDK1 does not generate an active kinase. CDK1 is inactivated by Wee1-mediated phosphorylation of Tyr15 and Myt1-mediated phosphorylation of Thr14 [20-21] that retard entry into mitosis. In case of DNA damage, ATM and ATR kinases are activated [22]. They phosphorylate Chk1 and Chk2 that in turn phosphorylate a number of substrates ensuring retention of cells in the G2 phase [23]. Inhibitory phosphorylations on CDK1 are not removed, CycB is retained in the cytoplasm and Cdc25C is phosphorylated and inhibited [24-25]. In the case of extensive DNA damage, cells undergo apoptosis [26]. In absence of DNA damage or after DNA has been repaired, cells can enter mitosis. The first event is phosphorylation and activation of Cdc25C by polo-like kinases [27-28]. Cdc25 phosphatases comprise three isoforms: Cdc25A and Cdc25B are expressed during G1 and at the G1/S boundary, while Cdc25C is expressed only during mitosis [29]. Cdc25C dephosphorylates and activates CDK1 that in turn phosphorylates Cdc25C for its full activation [30]. This feedback loop ensures complete activation of CDK1. CDK1/CycB is then imported in the nucleus where it phosphorylates nuclear lamins, leading to nuclear membrane breakdown [31-32]. CDK1/CycB also phosphorylates several Anaphase Promoting Complex (APC) subunits [33]. APC activation is required for progression through and exit from M-phase [34-35]. Its degradation of Pds1 results in proteolysis of cohesins, allowing separation of sister chromatids, and activation of the Cdc14 phosphatase, which in turn allows degradation of CycB and accumulation of CKIs (CDK inhibitors) as cells exit M-phase and enter G1.

A special role in cell cycle is covered by CDK7. In association with Cyclin H (CycH) and Menage-a-Trois-1 (MAT1), it forms the CDK-activating kinase (CAK). CAK phosphorylates CDK1, CDK2, CDK4 and CDK6 on the activation segment threonine, activating all of them [36-42]. Its activity does not fluctuate during cell cycle [43].

Role of CDKs in Transcription

Phosphorylation is a key mechanism of regulation in transcription. The primary target of regulating phosphorylations is the carboxy-terminal domain (CTD) of RNA Pololymerase (Pol) II. The CTD consists of heptapeptide repeats with a consensus sequence YSPTSPS that has been conserved through evolution. However, the number of repeats varies among species: 26 repeats in *Saccharomyces cerevisiae*, 42 repeats in *Drosophila melanogaster* and 52 repeats in human and mouse [44-45]. Depending on its phosphorylation state, it recruits different pools of transcription factors and mRNA-processing enzymes [46]. When unphosphorylated, it recruits transcription factors required for the assembly of the inactive pre-initiation complex (PIC) on the promoter region. After CTD has been phosphorylated on

Ser5, RNA Pol II becomes competent for initiation of transcription, but it is unable to transcribe the entire mRNA and it pauses after synthesis of a 20-30 bases short RNA [47-48]. Phosphorylation on Ser2 allows the transcription machinery to escape the promoter and to recruit a new pool of transcription factors required for elongation [49]. At the same time, enzymes required for mRNA splicing and maturation, are recruited by the CTD [50].

CDKs play a pivotal role in this process (figure 2). CDK7/Cyclin H together with other eight subunits forms the basal transcription factor TFIIH that is responsible for CTD phosphorylation on Ser5 and regulates transition of RNA Pol II from the PIC to the initiation competent complex [51-52]. Phosphorylation by CDK7 of CTD Ser5 and of transcription initiation factors like TBP and TFIIE disrupts the interactions stabilising the PIC and results in an active RNA Pol II and a scaffold complex that remains at the promoter. Phospho-Ser5 CTD is recognised by the capping enzyme [53-54]. The cap structure is added shortly after transcription initiation, when nascent RNA is about 25-30 nucleotides in length. Phospho-Ser5 CTD is also an allosteric regulator of the capping enzyme that is activated following CTD binding [55]. Another allosteric activator of the capping enzyme is Spt5 [56]. Spt5 in complex with Spt4 forms the DSIF complex that together with NELF inhibits transcription elongation. CDK9/Cyclin T complex, originally called Positive Transcription Elongation Factor b (P-TEFb), regulates the next transition from initiation to elongation, phosphorylating the CTD on Ser2 [57-58]. Phosphorylation of the elongation inhibitors DSIF and NELF by CDK9 removes the transcriptional block that stops RNA Pol II after synthesis of ~30 nucleotides long RNAs. Phosphorylation of CTD Ser2 by CDK9 allows recognition of RNA Pol II by elongation factors like Elongator and the assembly of the elongation-competent RNA Pol II [59]. Moreover the Ser2-Ser5 phosphorylated CTD is recognised by splicing factors and is required for efficient assembly of the spliceosome and for efficient splicing reactions on the elongating RNA. Finally polyadenylation, cleavage and transcription termination require interactions of the Cleavage/Polyadenylation Specificity Factor (CPSF) and of the Cleavage-stimulation Factor (CstF) with CTD. While CDK7 and CDK9 activate transcription, two other CDKs have been shown to have inhibitory effects on transcription. CDK8/Cyclin C phosphorylates Cyclin H and this leads to inhibition of the TFIIH CTD kinase and transcription activities [60]. Moreover and in contrast to TFIIH, CDK8 is capable of efficiently phosphorylating the CTD prior to formation of the PIC on DNA thus inhibiting the assembly of the transcription machinery at the promoter [61]. CDK1 phosphorylates CDK7 and two other subunits of TFIIH, p36 and p62. This results in reduced TFIIH CTD kinase and transcription activities and has been correlated with silencing of transcription during mitosis [62].

Inhibition of CDKs as Therapeutic Strategy for Cancer

Cancer is a highly heterogeneous disease. However a direct or indirect deregulation of cell cycle underlies the majority of cancers. These deregulations include a decreased dependence on growth factors, insensitivity to anti-growth signals, unlimited proliferative potential, evasion of apoptosis, sustained angiogenesis and the ability to metastasise and

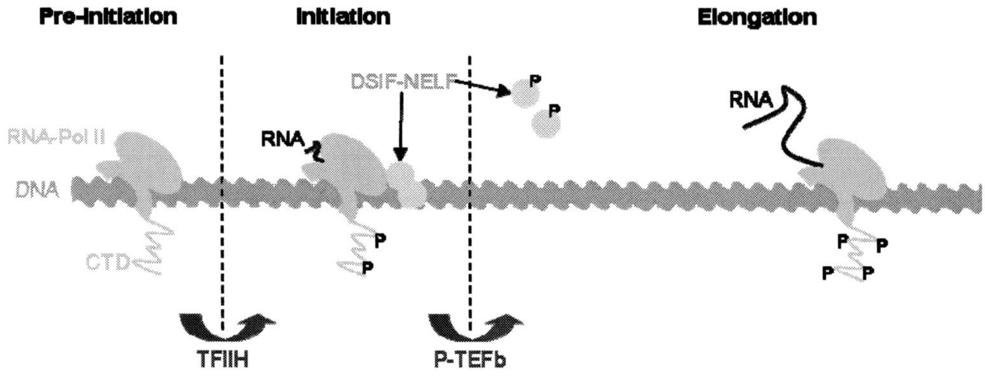

Figure 2. Transcriptional activation by CDK complexes. As part of TFIIH, CDK7 regulates progression of transcription from the pre-initiation to the initiation stage through phosphorylation of RNA-PolII CTD at Ser5. Transcription is blocked after synthesis of short RNAs by DSIF and NELF. CDK9, in the P-TEFb complex, removes this block by phosphorylating both DSIF and NELF and activates transcription elongation through further phosphorylation of CTD at Ser2.

invade tissue [63-64]. Uncontrolled cellular proliferation results from mutations in two families of genes. The first family, termed proto-oncogenes, encodes products that drive cellular proliferation. Up-regulated activity of proto-oncogenes can occur via over-expression of the gene or via mutations that yield over-active gene products. These processes convert proto-oncogenes to oncogenes. The second family comprises inhibitors of proliferation that are termed tumour suppressors. Their genetic alteration leads to down-regulation of their function.

Oncogenes can determine a reduced dependency on growth factors via over-expression of the receptors, or constitutive activity of associated enzymes (like tyrosine kinases of the Src family) [65] or of down-stream component of the signalling pathway (like G-proteins of the Ras family) [66].

Mutations in tumour suppressors have been found to be the most diffuse alterations in human cancers [67-68]. pRB represses the expression of factors required for cell cycle progression. Cells with inactive pRB constitutively express those factors [69]. p53 prevents cells from replicating damaged DNA and induces apoptosis in these cells [70-71]. Inactivation of p53 primarily results in the disengagement of apoptotic signalling from the apoptotic machinery. Abnormalities in the pRb pathway are observed in ~90% of human cancers [72]. These may arise through inhibition of pRb itself or through deregulation of the proteins that phosphorylate pRb, namely CDK4 and CDK2. CDK4 and CDK2 are deregulated in several cancer types (table 2) [73]. Deregulation of CDK4 is due to increased levels of Cyclins D and CDK4 [74-75] or to deletion or reduced expression of CKI of the INK4 family [76]. CDK2 activity is deregulated by increased levels of CycE [77], sometimes arising from defects in the proteins involved in targeting CycE for degradation, or decreased levels of p27 [9, 78].

CDKs have been considered for long time attractive targets for cancer therapy and all major pharmaceutical companies started research programs devoted at the development of CDKs inhibitors as anti-neoplastic drugs. Potent inhibitors have been designed for CDK4 and CDK2 and several are in clinical trials [79-80]. However, following the observations that ablation of single CDKs does not affect the ability of cells to cycle and indeed $CDK2^{-/-}$, $CDK4^{-/-}$ and $CDK6^{-/-}$ mice are viable [81-82], the logic of this approach has been questioned [83-84]. This seems to indicate that a suitable therapy should be based on combined inhibition of more than one CDK. $CDK2^{-/-}CDK6^{-/-}$ double knockouts mice are viable while $CDK4^{-/-}CDK6^{-/-}$ and $CDK2^{-/-}CDK4^{-/-}$ knockouts are embryonic lethal [85-87]. However $CDK4^{-/-}CDK6^{-/-}$ mice die due to very circumscribed abnormalities while a pronounced impairment of cell cycle progression is observed in $CDK2^{-/-}CDK4^{-/-}$ mice. Combined depletion of CDK1 and CDK2 induced apoptosis in U2OS and a strong G2/M block in NCI-H1299 cells [88]. The actual picture shows that CDK1 can bind CycE and substitute CDK2 [89], while CDK2 can compensate for CDK4 loss [85] that instead cannot be completely compensated by CDK6 or CDK1.

Table 2. Deregulation of CDK2 and CDK4 in human cancers

Protein	Alteration	Representative tumour
CDK4	Amplification, mutation	Sarcomas, melanoma
CDK2	Amplification	Hepatocellular carcinoma
Cyclin D	Amplification, translocation, overexpression	Breast and prostate cancers, parathyroid adenoma, multiple myeloma
Cyclin E	Overexpression	Breast, ovary and gastric carcinomas
INK4A	Deletion, mutation, methylation	Melanoma, lung cancer, pancreatic carcinoma
p27	Reduced protein levels/degradation	Colon, breast and prostate cancers

Structural Determinants of CDKs Function and Regulation

A number of CDK structures have been solved elucidating the molecular mechanisms of CDK activation by cyclin binding and activation segment phosphorylation and of CDK inactivation by inhibitor binding.

CDK2 structure shows the classical kinase bi-lobate architecture consisting of a smaller N-terminal lobe and a larger C-terminal lobe [90] linked by the hinge region (figure 3A). The N-terminal lobe is formed by five anti-parallel β-strands plus the αC helix. The C-terminal lobe contains a four helical bundle, a small β-ribbon and two additional α-helices. ATP and substrates bind at the active site in between the two lobes [91-92]. In monomeric CDK2 the activation segment interacts with the glycine loop and is stabilised in a conformation that extends across the entrance of CDK2 active site. Substrate entry is then blocked and the kinase is inactive. ATP is also misaligned with residues critical for the phosphotransfer catalysis (figure 3C).

Cyclins are characterised by their cyclin-box fold (CBF) comprising five α-helices [93-94]. Regions outside the CBF are responsible for cellular localisation and degradation via destruction boxes [95-96]. Despite their low homology, all cyclins contain very similar CBFs. CycA contains two CBFs joined by a flexible linker and binds CDK2 via its N-terminal α-helix and the α1-α2 and α4-α5 loops. The interaction involves only the N-terminal CBF. The binding event does not alter CycA structure [93-94]. CDK2 instead undergoes a profound structural rearrangement (figure 3B). The C-helix forms the centre of the interface with CycA that extends on both CDK2 lobes via a network of hydrogen bonds and electrostatic interactions. Upon cyclin binding, the C-helix moves towards the active site and rotates approximately 90° along its helical axis. In order to accommodate the C-helix, the N-terminal lobe swings away from the C-terminal lobe, pivoting around the hinge region. The C-helix movement removes Glu51 from the solvent inserting it into the active site where it interacts with Lys33. Asp145 no longer interacts with Lys33 as observed in monomeric CDK2 and is free to bind the Mg2+ ion, essential for catalysis. These changes collectively results in correct orientation of ATP for phosphotransfer (figure 3D).

The other relevant structural change regards the activation segment. The L_{12} helix that packs against the αC helix in monomeric CDK2 occluding the active site entrance, melts as a consequence of cyclin binding. The activation segment reorients moving away from the active site and leaving free access to it (figure 3B).

CDK2/CycA3 possesses only 1% of the activity of pThr160-CDK2/CycA3 [97]. Phosphorylation of Thr160 fully destabilises its interaction with the glycine loop that is characteristic of the inactive conformation. pThr160 instead interacts with Arg50, Arg126 and Arg150 stabilising the open active conformation. Furthermore, Val164 adopts an unusual left-handed conformation that confers proline-directed specificity to CDK2.

The structure of CDK2/CycA in complex with the inhibitor p27 shows how the inhibitor protein exerts its function (figure 3E) [98]. The N-terminal region of p27 interacts with CycA hydrophobic pocket (also known as recruitment site because involved in CDKs substrate recognition and binding), while residues 53-93 make extensive contacts with CDK2 in two distinct regions. The N-terminal lobe of CDK2 is rearranged (amino acids 1-35) with several hydrophobic residues becoming exposed and interacting with hydrophobic residues from p27. The other interacting region is the catalytic cleft, occupied now by p27.

INK inhibitors bind CDK4 and CDK6 either in the absence or in the presence of cyclins. They induce a rotation between CDKs N-terminal and C-terminal lobes reducing interactions with cyclins and, therefore their binding affinities. Residues involved in the phosphotransfer are misaligned and the activation segment assumes a β-hairpin structure moving to a position opposite to the one observed in active CDKs (figure 3F) [99].

Design of CDKs Inhibitors

Protein kinases are targets for treatment of several diseases, including cancer, diabetes and inflammation. The most exploited approach for kinase inhibition is the use of small-molecule ATP antagonists. For most kinases the K_m for ATP is in the micromolar range with a cellular concentration of ATP in the millimolar range. In order to be effective, ATP-

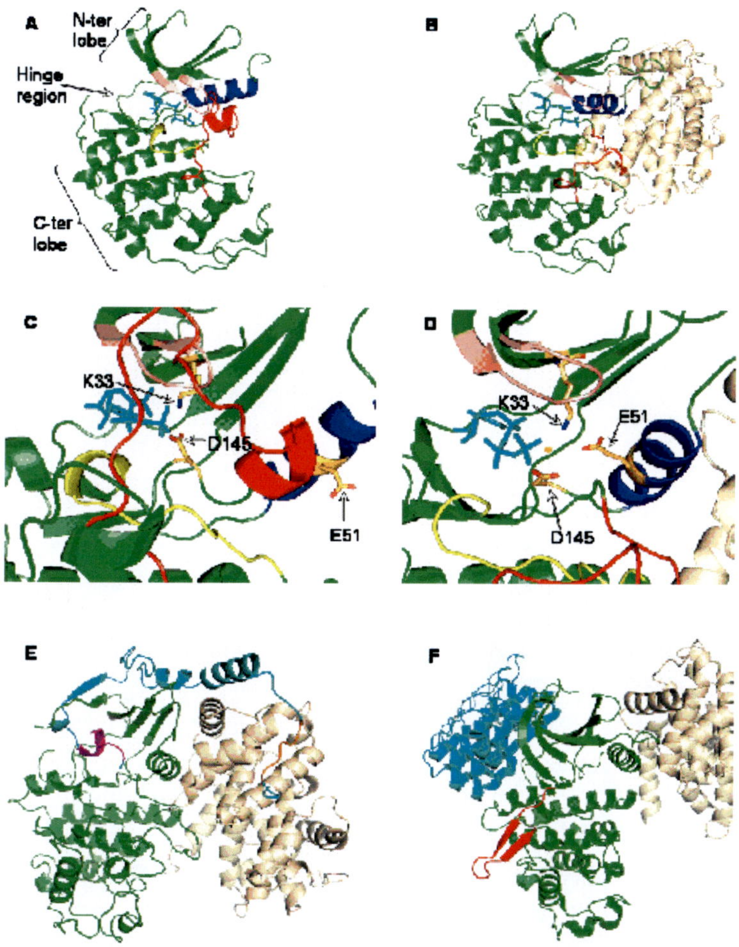

Figure 3. Structural rearrangement of CDK2 upon cyclin or CKI binding. A: Structure of CDK2. C-helix is shown in blue, activation segment in red, catalytic loop in yellow, glycine loop in salmon and ATP in cyan. The activation segment closes active site entrance. B: Structure of CDK2/CycA. CycA is shown in wheat. CDK2 is coloured as described in A. Activation segment rearranges leaving free entrance to the active site. C: In monomeric CDK2, D145 interacts with K33 and ATP is misaligned for phosphotransfer. D: In the CDK2/CycA complex, rearrangement of the C-helix creates a new interaction between E51 and K33. D145 is now free to bind the Mg ion required for catalysis. E: Structure of CDK2/CycA/p27. p27 is shown in cyan. Its region interacting with CycA hydrophobic pocket is in orange. p27 N-terminal region occupying catalytic cleft is shown in magenta. F: Structure of CDK6/CycK/p18. CycK is a D-type cyclin encoded by the Kaposi's sarcoma-associated herpes virus. Colour code is the same as E. CDK6 activation segment is shown in red.

competitive inhibitors must bind their targets with a significantly higher affinity than ATP. Target selectivity is another requisite for such inhibitors. They must discriminate their targets among the 518 kinases encoded in the human genome [100]. Other non-kinase nucleotide-binding proteins could provide additional difficulties for the development of specific inhibitors [101].

In the last few years, both potency and selectivity have been achieved for a number of inhibitors (reviewed by Noble *et al.* [102]). The most successful case is the inhibition of the

Bcr-Abl fusion protein by Gleevec. The Breakpoint Cluster Region (Bcr) protein substitutes part of the N-terminal region of the Abelson Leukemia virus (Abl) kinase as a consequence of a chromosomal translocation. This constitutively active kinase is responsible for the development of the rare cancer Chronic Myeloid Leukemia (CML). Treatment with Gleevec results in remission in nearly 100% of patients in the early stages of the disease [103].

Structures of kinases in complex with small molecule inhibitors demonstrate that, these compounds mainly target four pockets around the ATP binding site (figure 4A). We examine each of these pockets for differences that might allow selectivity.

The "p38 pocket" represents a pocket above the adenine, which in p38 and other kinases such as Abl and RAF is guarded by a small residue (usually threonine). In all CDKs the pocket is blocked by a phenylalanine (Phe80 in CDK2). This pocket allows discrimination between CDKs and other kinases, but offers no scope for additional groups in selectivity among CDKs. The nature of the residues that line the ribose pocket is similar in all CDKs suggesting that there would be little scope for discrimination for different groups bound to this pocket. The hinge regions do have different sequences. However, CDKs inhibitors interact with this region via two or three hydrogen bonds to main chain atoms that are in similar conformations in all known CDKs structures [102].

Major differences among CDKs are in the CDK2 "Lys89 pocket" or “hydrophobic pocket” (figure 4B) [104]. This pocket has been targeted successfully by CDK2 inhibitors such as purvanalol B [105] and the 6-cyclohexylmethoxy-2-(4’-sulfamoylanilino)-purine compound named NU6102 [106]. Both compounds do not inhibit CDK4 and CDK6. In CDK1 and CDK5, residues defining this pocket are essentially conserved compared to CDK2 (figure 4C) so that CDK2 inhibitors targeting this pocket also inhibit CDK1 and CDK5 with similar potency. With some inhibitors, Lys89 of CDK2 makes a direct interaction (e.g. the carboxylate moiety of purvalanol B), while with others like NU6102, Lys89 makes no interaction. However this residue appears to play a role in the electrostatic and steric repulsion of inhibitors that carry a positively charged group in this region [107]. For this reason such inhibitors are weakly active against CDK2. In CDK4 and CDK6, Lys89 of CDK2 is replaced by a threonine (figure 4D). Therefore, the changes in this pocket have been exploited for the development of specific CDK4 inhibitors [107-111]. To this purpose, Ikuta *et al.* [112] engineered several mutations into CDK2, Phe82His, Leu83Val and Lys89Thr, to simulate corresponding residues in CDK4. They showed that these mutants have additional space in the “hydrophobic pocket” that could accommodate larger substituents and bind CDK4-specific-inhibitors with increased selectivity. In CDK7 the residue corresponding to Lys89 at the bottom of the pocket is Val100. This change, together with other changes in the region such as Thr96 (in place of Gln85 in CDK2), generates an even larger pocket of non-polar character (figure 4E) [113]. Similarly in CDK9, additional space with respect to CDK2 is provided by the substitution of CDK2 Lys89 with CDK9 Gly112 (figure 4F) [114].

The different nature of the “hydrophobic pocket” in CDKs allowed generation of inhibitors specific for subclasses of CDKs (CDK1/2 or CDK4/6) (figure 4G). This pocket also offers additional space for the generation of specific CDK7 or CDK9 inhibitors. Inhibitors with limited occupancy of the “hydrophobic pocket” have also been developed.

Those inhibitors, named pan-CDK inhibitors, are less selective and bind all CDKs although with different affinities.

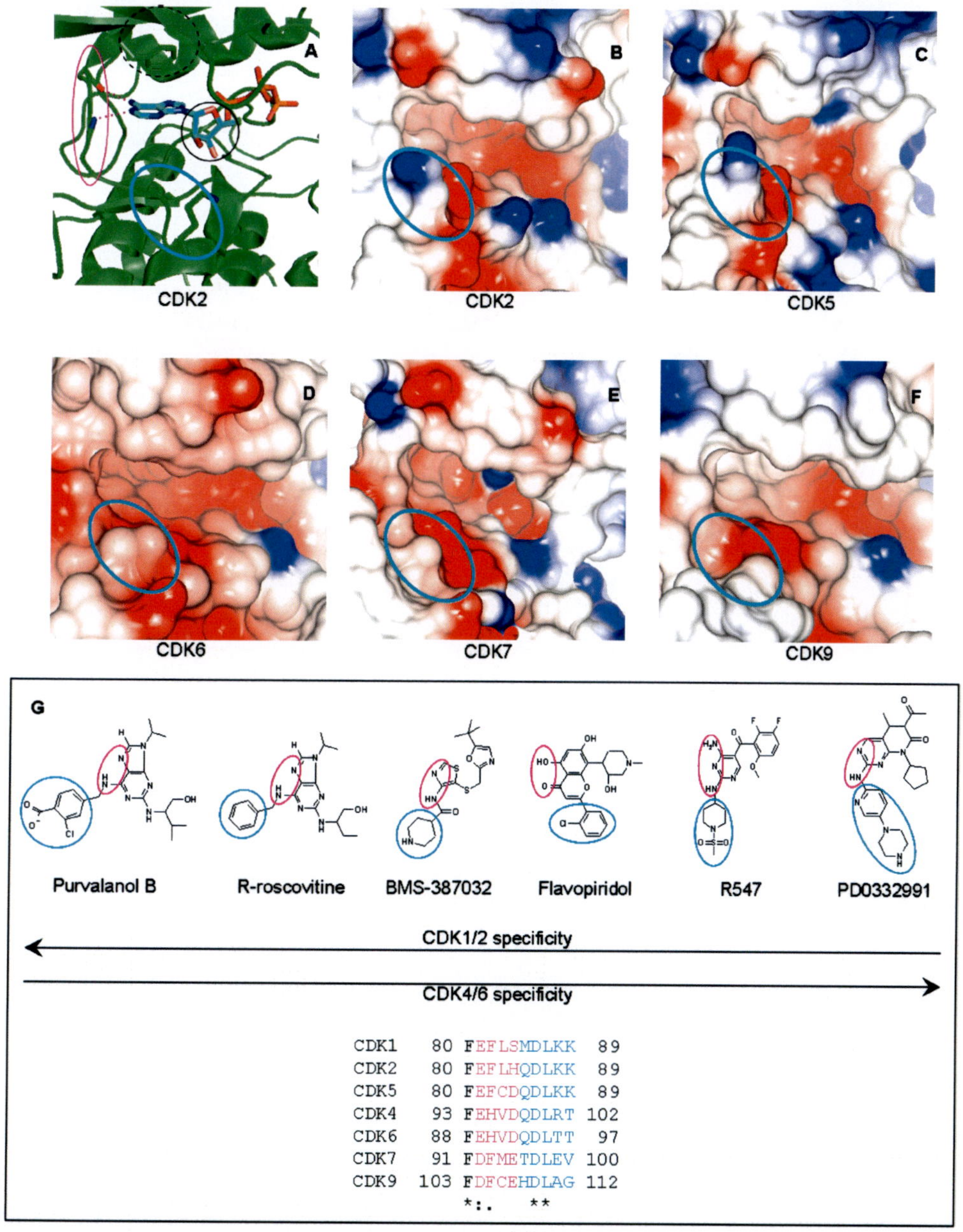

Figure 4. Structural rearrangement of CDK2 upon cyclin or CKI binding. A: ATP bound to CDK2. Ribose pocket is circled in black. p38 pocket, guarded by Phe80, is enclosed in the dashed circle. Hinge region with main chain atoms interacting with the adenine ring is circled in magenta. Lys89 pocket is circled in cyan. B-C-D-E-F: Electrostatic surface representation for ATP pockets of CDK2, CDK5, CDK6, CDK7 and CDK9. Lys89 pockets are circled in cyan. G: Representative CDKs inhibitors. Regions interacting with the hinge region (magenta in the sequence alignment) are circled in magenta. Groups occupying the Lys89 pockets (cyan in the sequence alignment) are highlighted in cyan.

Specific CDKs Inhibitors

Inactivation of CDKs in mice highlighted a widespread compensation mechanism for those kinases. However, recent reports identified a few cancer types with a more pronounced dependency on CDKs functions.

Breast cancer driven by the ErbB-2 oncogene is critically dependent on CDK4/CycD1 function. Mice lacking CycD1 are resistant to ErbB-2-driven mammary neoplasia [115]. CDK4 ablation resulted in complete protection against breast cancer in transgenic mice with enhanced expression of ErbB-2 (MMTV-ErbB-2) [116-117]. After about 40 weeks from birth, all MMTV-ErbB-2/CDK4$^{+/+}$ mice developed mammary carcinoma, while no tumor formation was observed in MMTV-ErbB-2/CDK4$^{-/-}$ animals. Mice expressing kinase deficient CDK4/CycD1 complexes (through point mutation on CycD1, abolishing CDK4 activation), like CycD1-null or CDK4-null animals, were fully refractory to mammary carcinoma triggered by ErbB-2 [118]. Tumorigenic potential of ErbB-2-driven breast cancer cells is also blocked by silencing of CDK4 or CycD1 through specific shRNAs or by replacement of active CDK4 with its kinase-dead mutant [116]. In addition, CDK4-null mice develop normal mammary glands that can undergo full expansion during pregnancy (as observed also in mice expressing kinase deficient CDK4/CycD1 complexes). There is then a differential requirement for CDK4/CycD1 function in mammary neoplasia *versus* normal development.

The most relevant CDK4/6 specific inhibitor is PD0332991 (Pfizer, figure 4G). It has an IC_{50} of ~10 nM for several CDK4/6-Cyclins D complexes, and no activity on CDK1, CDK2 and CDK5. In breast and colon carcinoma cells, PD0332991 caused cell cycle arrest at G1 phase and a reduction in pRb phosphorylation. Oral administration of this drug to mice bearing human tumor xenografts halted tumor growth and in some cases caused tumor regression [119]. A very rapid regression was observed in mice bearing SF295 (glioblastoma) or Colo-205 (colon carcinoma) xenografts. Importantly, no adverse effects were observed on animals' health. PD0332991 is currently in phase I clinical trial for the treatment of solid tumors and of non-Hodgkin lymphoma. Another phase I clinical trial is starting in patients with mantle-cell lymphoma, while a phase I/II trial is planned for the treatment of multiple myeloma in combination with bortezomib (proteasome inhibitor) (http://clinicaltrials.gov).

Compensation of CDK2 function by other CDKs in knockout cells and mice suggests that highly selective inhibition of CDK2 may not be useful therapeutically. One exception could be malignant melanoma where CDK2 levels are under strict control of the melanocyte transcription factor MITF (microphthalmia-associated transcription factor). Melanoma cells critically require CDK2 for proliferation, as demonstrated by CDK2 shRNAs that are able to block cell proliferation in melanoma cell lines [120].

CDK1 and CDK2 have however very similar ATP pockets and CDK2 inhibitors developed so far also inhibit CDK1. They could be therapeutically valid for a variety of cancers. The inhibitor R-roscovitine (Cyclacel Pharmaceuticals, figure 4G) inhibits CDK1 and CDK2 with IC_{50} values of 650 nM and 700 nM, respectively [121], but it has also an inhibitory effect on CDK7 and CDK9 [122]. Instead, it is inactive against CDK4 and CDK6 (IC_{50} > 100 μM) [121]. Although not very potent, R-roscovitine has excellent pharmacokinetic properties. Phase I clinical trials have shown that the drug is well tolerated with an excellent oral bioavailability. High plasma concentrations were reached and none of the side effects, typically associated with cancer chemotherapy, such as reduction of blood cell counts and loss of hair, was observed [123]. Dose-limiting-toxicities (DLTs) had as side effects nausea and vomiting, hypokalaemia, elevated creatinine and skin rash. These side effects were reversible indicating that R-roscovitine does not induce a permanent damage. A total of 70 patients were analysed for antitumor activity. One partial response in a patient with hepatocellular carcinoma and in ten patients, a stable disease lasting for more than four months have been reported; two of these patients had non-small-cell lung cancer (NSCLC), which remained stable for more than one year [124-125]. The efficacy of R-roscovitine is currently tested in phase II clinical trial to threat NSCLC (http://clinicaltrials.gov).

Pan-CDKs Inhibitors

Flavopiridol [Aventis and National Cancer Institute (NCI)] inhibits CDK1, CDK2 and CDK4 with K_i values between 40 and 70 nM, CDK7 with $K_i = 300$ nM and CDK9 ($K_i = 3$ nM). In preclinical studies, it induced growth inhibition or apoptosis in several tumor cell lines as well as human tumor xenografts (lung, colon, ovary, brain and breast) in nude mice [126]. Apoptosis or growth inhibition was observed in more than 60 tumor cell lines in the NCI *in vitro* screening panel [127-128]. In initial phase I and II clinical trials only low response rates were achieved, with stabilization of disease observed in patients with metastatic androgen-independent prostate cancer (18% of treated patients), endometrial carcinoma (22%), and metastatic malignant melanoma (44%) [129-131]. In 34 patients with advanced renal cell carcinoma, one complete response, three partial responses and 14 stable diseases (41%) were observed [132]. Associated toxicities included neutropenia, vomiting, diarrhea and proinflammatory syndrome (fever, fatigue and local tumor pain). Recent data indicate that the proapoptotic effect of flavopiridol may be dependent on the dose and the schedule, because flavopiridol is highly bound (>90%) to human plasma proteins [133]. Effective drug concentrations may not have been achieved in previous clinical trials. A phase I study in refractory chronic lymphocytic leukemia (CLL) patients used a 30-minute bolus dose followed by a 4-hour infusion, pharmacokinetically modeled to achieve and sustain micromolar concentrations for several hours. About 50% of the treated patients achieved partial durable responses [134]. Furthermore, some of the responding patients with enlarged lymph nodes showed a 50% reduction after the first treatment. Such cell killing as to induce tumor lysis syndrome (TLS), was observed in many patients and TLS was in fact the DLT. Considering the improvement in response rate, several phase I/II trials of bolus/infusion flavopiridol in patients with lymphoma, acute leukemia and solid tumors have started recently (http://clinicaltrials.gov).

BMS-387032 (Bristol-Myers Squibb and Sunesis Pharmaceuticals) is slightly more selective than flavopiridol with primary targets being CDK2, CDK7 and CDK9 ($IC_{50} = 48$ nM, 62 nM and 4 nM, respectively) and less potent inhibitory effect on CDK1 and CDK4 ($IC_{50} = 480$ nM and 925 nM, respectively) [135]. Its *in vivo* activity was confirmed in two murine (P388 leukaemia and Br-CycE breast carcinoma) and three human (A2780 ovarian carcinoma, A431 squamous cell carcinoma and Colo205 colon carcinoma) tumor models [136]. BMS-387032 is currently evaluated in phase I trials (http://clinicaltrials.gov).

R547 (Hoffmann-La Roche) is a very potent inhibitor of CDK1, CDK2 and CDK4 (K_i = 2 nM, 3 nM and 1 nM, respectively) with less potent effect on CDK7 (IC_{50} = 171 nM). It showed antiproliferative activity in all cell lines tested and in six human tumor xenografts [137] and a phase I trial in patients with advanced solid tumors is now starting (http://clinicaltrials.gov).

P276-00 (Nicholas Piramal India Limited – unreleased structure) inhibits CDK1, CDK2, CDK4 and CDK9 (IC_{50} = 79 nM, 224 nM, 63 nM and 10 nM, respectively) with significant selectivity over CDK7 (IC_{50} = 2.87 μM) [138]. When tested against a panel of 16 cisplatin-sensitive and cisplatin-resistant cell lines, the antiproliferative potential of P276-00 was found to be approximately 30-fold higher than cisplatin. In 22 human xenografts, P276-00 was approximately 26-fold more potent than cisplatin, and further, it was also found to be active against cisplatin-resistant tumors of central nervous system, melanoma, prostate, and renal cancers [139]. It is now starting phase I trials in patients with advanced refractory neoplasms and refractory multiple myeloma (http://clinicaltrials.gov).

Histone Deacetylases

Core histones have long N-terminal extensions that have been known for decades to undergo extensive post-translational modifications such as acetylation, methylation and phosphorylation, but also ubiquitinylation, sumoylation and ADP-ribosylation [140-144]. Acetylation occurs at the ε-amino groups of lysine residues present within the N-terminal extensions of the core histones. In all documented cases, the acetylation status of target lysines is tightly controlled by the balance of two counteracting enzymatic activities: those of histone acetyl transferases (HAT) and histone deacetylases (HDACs) [145-146]. Histone deacetylation represses transcription by different mechanisms. This process increases the charge density on the N-termini of the core histones thereby strengthening histone tail-DNA interactions and blocking access of the transcriptional machinery to the DNA template. In addition, histone modifications are specifically recognized by several chromatin-interacting proteins. As a result, there is a progressive decrease in the accessibility of DNA to the transcription machinery and hence an increasing transcriptional silencing. In addition to histones, an increasing number of non-histone proteins also undergo lysine acetylation: many of these proteins are transcription factors, but also heat shock proteins and structural proteins have been shown to be acetylated. Generally, acetylation has profound influences on either metabolic stability or on the biological function of the modified protein and can be regarded akin to phosphorylation in terms of being a widely used post-translational control mechanism of protein function.

HATs utilize acetyl-CoA as a cofactor in the acetylation reaction. About 30 different HATs were identified that are grouped in five different families [147]. In the eukaryotic cell, HATs may reside either in the nucleus or in the cytosol. HDACs catalyze the inverse reaction by removing the ε-amino group from the acetyl lysine residue.

Eukaryotic HDACs have been classified into four groups based on phylogenetic analysis [148]. Class I HDACs (homologous to yeast Rpd3p), comprise HDAC1, 2, 3 and 8. Class II HDACs (homologous to yeast Hda1p) comprise HDAC4, 5, 6, 7, 8, 9 and 10 and are divided into two subclasses based on sequence homology and domain organization [149]. Subclass IIa [143, 144, 146, 148] contains one catalytic domain, while subclass IIb [145, 149] contains two catalytic domains, although the duplication is partial in the case of HDAC10. HDAC11 is as distinct from class I and class II HDACs as either class is from the other, therefore this protein has been placed in its own class, class IV [148], because the name class III is already used to refer to the unrelated sirtuin deacetylases [150]. HDACs 1-11 are metalloenzymes, which require zinc for deacetylation. Class I-IV enzymes are 350-500 amino acid residues in length, while class II enzymes are about 1000 amino acid residues long, but they all have homologous catalytic sites and are considered to go through similar reaction mechanism, the Zn-catalyzed hydrolysis of the acetyl-lysine amide bond.

Cancer has traditionally been considered a disease of genetic defects that result in the loss of function of tumor-suppressor genes and/or gain of function or hyperactivation of oncogenes. However, there are today many evidences that gene expression governed by epigenetic modifications is also crucial to the onset and progression of cancer and HDACs are promising targets for therapeutic intervention aimed to reverse aberrant epigenetic states associated with cancers. Remarkably, imbalance of histone and non-histone protein acetylation is common in human cancer. HDACs are overexpressed, aberrantly recruited by oncoproteins or mutated in malignant cells [151].

HDAC Inhibitors

To date there are five distinct classes of HDAC inhibitors: short chain fatty acids like butyric acid, hydroxamic acids, electrophilic ketones, aminobenzamides exemplified by MS-275, and natural cyclic peptides like Romidepsin and Apicidin. Several HDAC inhibitors from these chemical classes have already advanced into clinical trials and recently Vorinostat, a hydroxamic acid compound, was approved by the FDA for the treatment of advanced cutaneous T-cell lymphoma (reviewed in Gallinari *et al.*, [152]).

The first inhibitors to be identified were the short chain fatty acids like butyric acid (BA) and valproic acid [153-156]. These agents are the least potent of the various classes of HDAC inhibitors with IC_{50}s in the millimolar range. Indeed in 1975 BA was shown to induce the differentiation in Friend Leukemia cells and then three years later to cause a dramatic and reversible increase in histone acetylation [153]. In a child with acute myelogenous leukaemia, daily treatment with BA, caused the reduction and then the disappearance of the myeloblasts. However, more extensive clinical trials failed to reproduce these findings. In an effort to obtain higher plasma levels, several prodrugs of BA have been developed, the most studied of these is AN9/Pivanex, which entered phase II clinical trials [152]. Pivanex is highly lipophilic and is able to cross cell membranes and it has been shown to affect the growth, differentiation and apoptosis of various cancer cell lines faster and at lower concentrations than BA.

TSA a fungistatic antibiotic was the first of a series of hydroxamic acid inhibitors to be identified and has become one of the main research tools for probing the function of histone acetylation [157]. Like BA, interest in this natural product raised due to the demonstration that it was able to cause induction of Friend leukemia cell differentiation, as well as G1 and G2/M cell cycle arrest. Poor pharmacokinetics precluded any in vivo work with this inhibitor, nevertheless these finding stimulated the development of structurally related inhibitors.

The hydroxamic acid compound, vorinostat (SAHA) is the most clinically advanced inhibitor [152]. Vorinostat was identified in the mid 1990's as an agent capable of inducing differentiation of murine erythroleukemia cells and subsequently has been shown to induce differentiation, cell growth arrest and/or apoptosis in a wide variety of cell lines at low micromolar concentrations (reviewed in [152]). Vorinostat is able to inhibit both class I and II HDACs at nanomolar concentrations and like TSA has been shown to bind at the active site of the enzyme (reviewed in [152]). Vorinostat has been demonstrated to have anti-proliferative effects against a range of cultured transformed cells and these effects in part are due to altered gene expression, where this compound causes both increased and decreased expression of 2-10% of expressed genes. In vivo, this compound inhibits tumor growth with little toxicity in a wide range of animal models for solid tumors (including breast, prostate, lung and gastric cancers) and hematological malignancies. Clinical studies have been initiated against diverse tumors types and meaningful clinical responses have been seen in patients with T-cell and B-cell lymphomas, mesothelioma, and laryngeal or thyroid cancers. Stable disease has also been achieved in patients with renal-cell carcinoma (RCC), urothelial cancer or Hodgkin's Lymphoma. Vorinostat has also been shown to be synergistic or additive with a wide range of chemotherapeutics and clinical combination studies are underway (see http://www.cancer.gov/clinicaltrials, for an updated list). Further examples of hydroxamic acids inhibitors include LAQ824 and LBH589, PXD-101, and CRA-24781 (reviewed in [152]).

HDACs and CDKs

Evidence has been reported of a cross-talk between CDKs and HDACs that indicates the existence of a complex cross-regulation mechanism. HDACs bind to pRb and are recruited to the E2F promoter on DNA, where they inhibit expression of cyclins and other cell-cycle promoting genes. pRb phosphorylation by CDK4/CyclinD, CDK6/CyclinD and CDK2/CyclinE determines the release of HDACs from pRb and of pRb from E2F, therefore promoting cell cycle progression [158-159]. However, HDACs also regulate the expression of CDK inhibitors (CKI) and HDAC inhibitors induce CKI expression, which may contribute to their cell-cycle inhibitory effects [160]. This cross-talk between CDKs and HDAC mediated by pRb and/or CKI is of particular pharmacological relevance. It engenders the expectation that the combined pharmacological inhibition of CDKs and HDACs might have a higher therapeutic profile than single treatments [161]. Direct interactions between CDK/Cyclin pairs and HDACs have been reported. CyclinD1 binds in vitro and in vivo to HDAC1, HDAC2, HDAC3 and HDAC5 [162] by recruiting them to the Peroxisome Proliferator Activated Receptor (PPAR) response element, thus blocking PPAR-mediated adipogenesis. Cdk2/CyclinE binds and phosphorylates HDAC1 [163]. HDAC1 then dissociates from the Myc/Mad complex and after that, the transcription of cell-cycle promoting factors is activated from Myc promoters. The CDK CAK recruits HDAC2 to the estrogen receptor (ER) [164]. This recruitment, mediated by the Metastasis-associated-protein 1 (MTA1), was found to be responsible for the loss of ER responses, which is generally associated with the progression of breast cancer to more invasive phenotypes.

CDKs, like CAK and PTEF-b, activate many transcription factors, such as nuclear hormone receptors (estrogen, androgen and retinoic acid receptors), NF-kB, MyoD, c-Myc and CIITA [165-168]. Expression of genes by their respective promoters directs cells towards their fate (proliferation, differentiation, survival or apoptosis). The majority of these transcription factors are inhibited by HDACs [169]. Interactions between CDK/Cyclin pairs and HDACs could be direct, as demonstrated for CyclinD in the case of PPAR-activated transcription or indirect as demonstrated for CAK in the case of ER activated transcription.

Except for HDAC3, one or more putative CDKs phosphorylation sites can be predicted for class I HDACs. In particular, the CDK2/CyclinE phosphorylation site identified in HDAC1 [163] is also conserved in HDAC2 and HDAC8 (SP in HDAC1, SP and TP in HDAC2, TP in HDAC8, figure 5). Phosphorylation for many CDK2 substrates is enhanced by their recruitment from CyclinE or CyclinA [92, 170-171]. These cyclins contain a docking site that specifically recognises the consensus sequence RXL or KXL, efficiently increasing the local concentration of substrate and consequently its phosphorylation by the associated CDK2. HDAC1, 2 and 3 but not HDAC8 contain such sequence (figure 5). It is thus feasible that CKD2/CyclinE takes advantage of these docking sites, and prefers HDAC1 and HDAC2 as substrates over HDAC8. Since HDAC3 contains no CDK phosphorylation sites, but contains a KXL motif, it could compete for binding to the Cyclins with HDAC1 and HDAC2, reducing their recruitment to the active CDK2.

HDAC1	341 FKLHISPSN-MTNQNTNEYLEKIKQRLFENLRMLPHA 376
HDAC2	342 FKLHISPSN-MTNQNTPEYMEKIKQRLFENLRMLPHA 377
HDAC3	336 FTLHPDVSTRIENQNSRQYLDQIRQTIFENLKMLNHA 372
HDAC8	344 YVLEITPSC-RPDRNEPHRIQQILNYIKGNLKHVV—377

Figure 5. Putative HDAC1, 2, 3 and 8 phosphorylation sites and docking segments (underlined).

Combined Treatments

Cell cycle-mediated drug resistance indicates the non-respondance to a chemotherapeutic agent due to a cell cycle block [172]. Cells are blocked in a phase in which the chemotherapeutic agent is inactive or slightly active. This resistance limits the efficacy of standard cytotoxic drugs and can be overcome by the appropriately scheduled administration of CDKs inhibitors.

Treatment of cancer cells or human xenografts with flavopiridol administered before or concomitantly with paclitaxel or docetaxel did not show any better effect than the taxanes alone. CDKs inhibitors like flavopiridol block or slow the mitotic entry and antagonize spindle agents such as taxanes. Inversion of the treatment sequence resulted in increased apoptosis and reduction of tumor volume [173-174]. This can be explained through CDK1/CycB inhibition by flavopiridol that accelerates the exit from taxane-induced M-phase arrest and leads to post-replicative cell death. Mitotic arrest through microtubule stabilization also results in elevated CDK1/CycB activity that phosphorylates and stabilizes the apoptotic inhibitor Survivin. Suppression of this activity with CDKs inhibitors abrogates CDK1 survival checkpoint and enhances apoptosis [175-176]. In clinical trial, sequential combination of paclitaxel followed by escalating doses of flavopiridol showed clinical activity in patients with chemotherapy refractory malignancies, including patients previously treated with paclitaxel [177]. Recently and considering that flavopiridol is highly protein-bound, its schedule with taxanes has been changed in order to achieve higher flavopiridol peak dose. One complete response in a patient with pancreatic carcinoma was observed, four partial responses in patients with pancreatic (1), breast (2), and ovarian (1) cancers and stable diseases in ten patients, over a total of 27 patients [178]. Similar results were obtained by combining flavopiridol and S-phase chemotherapic agents such as antimetabolites (fluorouracil, capecitabine, cytarabine, gemcitabine, fludarabine and methotrexate), topoisomerase inhibitors (irinotecan, topotecan, etoposide, mitoxantrone and the DNA-intercalating agent doxorubicin) and to some extent the platins (cisplatin, carboplatin and oxaliplatin). Treatment with these agents is generally associated with an induction of p21 and activation of the p53-dependent checkpoint that results in G1 and G2 arrest after DNA damage, rendering cells resistant to S-phase agents. Flavopiridol followed by S-phase agents did not have a better effect on tumor cell lines than the single drugs. A better antitumor activity has been demonstrated in several cancer cell lines and human xenografts [179-181], when the sequence is inverted, S-phase agents and then flavopiridol. In clinical trials, treatment with irinotecan followed by flavopiridol resulted in partial responses in patients with gastric, esophagus and colorectal cancer, stable disease for 36% of all enrolled patients and 52% of patients with previously treated colorectal cancer [182]. Clinical benefit appeared to be most pronounced in patients with p53-wild type tumors. High response rates (90%) were also obtained by combining flavopiridol with fludarabine and rituximab [183]. Other ongoing or starting clinical trials include combinations of flavopiridol with fluorouracil and irinotecan, with fluorouracil and oxaliplatin, with gemcitabine, with gemcitabine and irinotecan, with cytarabine and mitoxantrone, with doxorubicin, with irinotecan and cisplatin, with cisplatin/carboplatin, and with cisplatin and paclitaxel (http://clinicaltrials.gov).

Combining HDAC Inhibitors with Kinase Inhibitors

Flavopiridol functions in synergy with the weak HDAC inhibitor sodium butyrate (a short chain fatty acid) to kill cultured and primary tumor cells [184]. These authors presented evidence that attenuation of p21 induction by flavopiridol played a role in promoting butyrate-induced apoptosis in human leukemia cells. Considering that p21 plays a major role in regulating the apoptotic response of leukemic as well as other neoplastic cells to HDAC inhibitors [152], the concept of combining these agents with flavopiridol, which transcriptionally represses the expression of p21 and other antiapoptotic proteins seems to be a rational one.

Antileukemic synergism between HDAC inhibitors and CDK inhibitors has been observed also using the potent HDAC inhibitor LAQ824 (a cinnamic acid) and the CDK inhibitor roscovitine [185], an inhibitor that is less pleiotropic than flavopiridol. The mechanisms underlying synergism between LAQ824 and roscovitine differ from those involved in flavopiridol/butyrate interactions, including the lack of transcriptional repression of p21 and the involvement of oxidative damage.

As single agents, HDAC inhibitors can decrease the mRNA expression of BCR-ABL kinase and promote the ubiquitinylation and degradation of BCR-ABL by inhibiting HDAC6 [186]. Therefore, HDAC inhibitors were tested in combination with imatinib (Gleevec; Novartis), an ABL kinase inhibitor that can kill BCR-ABL-positive chronic myeloid leukemia cells (CML), and has an impressive clinical activity [187]. All HDAC inhibitors tested so far, like the hydroxamic acids vorinostat and LAQ824, the natural product apicidin and the cyclic peptide, depsipeptide, function synergistically with imatinib to kill CML cells expressing BCR-ABL kinase, [188, review]. Notably, the combination imatinib/HDAC inhibitor, is effective against imatinib-refractory CML [188].

Clinical studies are ongoing using the HDAC inhibitor vorinostat in combination with the CDK inhibitor flavopiridol: a phase I trial is underway to study the side effects and best dose of vorinostat when given together with flavopiridol in treating patients with advanced solid tumors; a phase I trial is studying the side effects and best dose of flavopiridol when given together with vorinostat in treating patients with relapsed or refractory acute leukemia or chronic myelogenous leukemia or refractory anemia (http://clinicaltrials.gov).

New Directions

Until very recently, CDK7 has never been considered a suitable target for cancer chemotherapy. Due to its central role in cell cycle as activator of CDK1, CDK2, CDK4 and CDK6, its inhibition should be associated with a complete inactivation of all cell cycle CDKs, with increased probability of unwanted side effects. For the same reason, CDKs drug development programs aimed to develop inhibitors as much selective as possible for a single CDK. With the emerging CDKs redundancy, this approach has been reconsidered. Other concerns regarding possible toxicity of CDK7 inhibitors could be raised considering its role in transcription. Several recent reports encourage however the search for such inhibitors [189-190]. In *Drosophila melanogaster*, CDK7 mutants showed death before or during pupation caused by severe mitotic defects [191]. Interestingly, defects in transcription were not detectable. In mice, MAT1 (a component of the CAK complex) disruption severely affected mitosis and led to early embryonic lethality [192]. No decrease in transcription was observed. G1 arrest following MAT1 abrogation was also observed in human osteosarcoma MG-63 and U-2 OS cells [193]. This block has been correlated with the observed reduction in pRb phosphorylation. CDK7 appears then as an essential component both for G1 progression (when CDK2 and CDK4 action is required) and during the M phase (when CDK2 and CDK1 activities are needed), while its role in transcription can probably be compensated by CDK9. Selectivity against CDK1, CDK2, CDK4 and CDK6 is not required because, in presence of a CDK7 inhibitor, they will not be activated.

Flavopiridol-induced apoptosis has been ascribed to CDK9 inhibition and to its role in transcription. CDK9 has then emerged as a relevant target: apoptosis is induced through depletion of anti-apoptotic proteins (such as Bcl-2, Mcl-1 and XIAP), sensitive to inhibition of transcription because encoded by mRNAs with short half-lifes and rapid turnover. Other gene classes encoding labile mRNAs include cell cycle regulators (D-type cyclins, c-myc and mitotic regulatory kinases), NF-kB-responsive gene transcripts and those contributing to the p53 pathway (such as p21). Inhibition of CDK9 could be particularly beneficial in tumors that are dependent on D-type cyclins, such as ErbB-2-positive breast cancer. This rationale is currently being evaluated in clinical trials for the flavopiridol-herceptin (antibody anti-ErbB-2 receptor) combined therapy. Flavopiridol also potentiates Gleevec-mediated apoptosis in BCR-ABL-positive leukemia cells, through reduction of Mcl-1 levels [194] (phase I trial ongoing). In preclinical studies, synergism has been observed for the combination of flavopiridol with bortezomib, a proteasome inhibitor preventing degradation of pro-apototic factors [195] (phase I trial planned). CDK9 could also have a direct role in some malignant transformation. CDK9 is strongly overexpressed in follicular lymphoma and diffuse large B-cell lymphomas and in the cell lines of classical Hodgkin's lymphomas, Burkitt's lymphomas and anaplastic large cell lymphoma [196]. More than half of CDK9 is inactive in cells due to its association with the inhibitory protein HEXIM1 and with 7SK-RNA. HEXIM1, also called estrogen down-regulated gene 1 (EDG1), was found to be a growth inhibitor that is down-regulated in breast cancer in response to estrogen [197]. HEXIM1 expression was found to be significantly higher in normal breast epithelial tissue as compared with adjacent breast epithelial tumor tissue, suggesting a role for P-TEFb in estrogen-mediated initiation and development of breast cancer. HEXIM1 has also been identified as a potential indicator of ovarian cancer, because HEXIM1 antibodies were elevated in the serum of ovarian cancer patients compared to normal controls [198], hence down-regulating HEXIM1 levels. A CDK9-specific inhibitor could be particularly beneficial in these types of cancers, eventually showing reduced toxicity in respect to flavopiridol.

Conclusion

CDKs and HDACs are sensible targets for cancer therapy. Their specific inhibitors are showing promising anti-tumor activity in clinical trials for different subsets of cancer types. Classic cytotoxic agents are synergistic with these inhibitors and combination therapies seem to be most effective in terms of response rates. Combinations of CDKs or HDACs inhibitors with chemotherapeutic agents targeting different oncogenes have shown encouraging results as in the case of imatinib-refractory CML. Finally, direct combination of CDKs and HDACs inhibitors should interfere with the cross-talk between the two classes of proteins with strong synergistic effects. This therapeutic approach could be of high impact and is currently tested in clinical trials.

Acknowledgments

GL is supported by a FEBS Long-Term Fellowship.

References

[1] Knockaert, M., Greengard, P., Meijer, L. (2002). "Pharmacological inhibitors of cyclin-dependent kinases." *Trends Pharm. Sci.* 23(9): 417-425.

[2] Bagchi, S., Weinmann, R., Raychaudhuri, P. (1991). "The retinoblastoma protein copurifies with E2F-I, an E1A-regulated inhibitor of the transcription factor E2F." *Cell.* 65(6): 1063-72.

[3] Dyson, N. (1998). "The regulation of E2F by pRB-family proteins." *Genes Dev.* 12(15): 2245-62.

[4] Harbour, J. W., Dean, D.C. (2000). "Rb function in cell-cycle regulation and apoptosis." *Nat. Cell. Biol.* 2(4): E65-7.

[5] Harbour, J. W., Luo, R. X., Dei Santi, A., Postigo, A. A., Dean, D. C. (1999). "CDK phosphorylation triggers sequential intramolecular interactions that progressively block Rb functions as cells move through G1." *Cell.* 98(6): 859-69.

[6] Sherr, C. J., Roberts, J. M. (1999). "CDK inhibitors: positive and negative regulators of G1-phase progression." *Genes Dev.* 13(12): 1501-12.

[7] LaBaer, J., Garrett, M. D., Stevenson, L. F., Slingerland, J. M., Sandhu, C., Chou, H. S., Fattaey, A., Harlow, E. (1997). "New functional activities for the p21 family of CDK inhibitors." *Genes Dev.* 11(7): 847-62.

[8] Planas-Silva, M. D., Weinberg, R. A. (1997). "Estrogen-dependent cyclin E-CDK2 activation through p21 redistribution." *Mol. Cell. Biol.* 17(7): 4059-69.

[9] Pagano, M., Tam, S. W., Theodoras, A. M., Beer-Romero, P., Del Sal, G., Chau, V., Yew, P. R., Draetta, G. F., Rolfe, M. (1995). "Role of the ubiquitin-proteasome pathway in regulating abundance of the cyclin-dependent kinase inhibitor p27." *Science.* 269(5224): 682-5.

[10] Knudsen, E. S., Wang, J. Y. (1996). "Differential regulation of retinoblastoma protein function by specific CDK phosphorylation sites." *J. Biol. Chem.* 271(14): 8313-20.

[11] Lundberg, A. S., Weinberg, R. A. (1998). "Functional inactivation of the retinoblastoma protein requires sequential modification by at least two distinct cyclin-CDK complexes." *Mol. Cell. Biol.* 18(2): 753-61.

[12] Arata, Y., Fujita, M., Ohtani, K., Kijima, S., Kato, J. Y. (2000). "CDK2-dependent and -independent pathways in E2F-mediated S phase induction." *J. Biol. Chem.* 275(9): 6337-45.

[13] Clurman, B. E., Sheaff, R. J., Thress, K., Groudine, M., Roberts, J. M. (1996). "Turnover of cyclin E by the ubiquitin-proteasome pathway is regulated by CDK2 binding and cyclin phosphorylation." *Genes Dev.* 10(16): 1979-90.

[14] Pagano, M., Draetta, G., Jansen-Durr, P. (1992). "Association of CDK2 kinase with the transcription factor E2F during S phase." *Science.* 255(5048): 1144-7.

[15] Helin, K. (1998). "Regulation of cell proliferation by the E2F transcription factors." *Curr. Opin. Genet. Dev.* 8(1): 28-35.

[16] Jiang, W., Wells, N. J., Hunter, T. (1999). "Multistep regulation of DNA replication by CDK phosphorylation of HsCdc6." *Proc. Natl. Acad. Sci. USA*. 96(11): 6193-8.

[17] Krek, W., Ewen, M. E., Shirodkar, S., Arany, Z., Kaelin, W. G. Jr, Livingston, D. M. (1994). "Negative regulation of the growth-promoting transcription factor E2F-1 by a stably bound cyclin A-dependent protein kinase." *Cell*. 78(1): 161-72.

[18] Krek, W., Xu, G., Livingston, D. M. (1995). "Cyclin A-kinase regulation of E2F-1 DNA binding function underlies suppression of an S phase checkpoint." *Cell.* 83(7): 1149-58.

[19] van den Heuvel, S., Harlow, E. (1993). "Distinct roles for cyclin-dependent kinases in cell cycle control." *Science*. 262(5142): 2050-4.

[20] Parker, L. L., Piwnica-Worms, H. (1992). "Inactivation of the p34cdc2-cyclin B complex by the human WEE1 tyrosine kinase." *Science*. 257(5078): 1955-7.

[21] Mueller, P. R., Coleman, T. R., Kumagai, A., Dunphy, W. G. (1995). "Mytl: a membrane-associated inhibitory kinase that phosphorylates Cdc2 on both threonine-14 and tyrosine-15." *Science*. 270(5233): 86-90.

[22] Durocher, D., Jackson, S. P. (2001). "DNA-PK, ATM and ATR as sensors of DNA damage: variations on a theme?" *Curr. Opin. Cell. Biol.* 13(2): 225-31.

[23] Zhou, B. B., Elledge, S. J. (2000). "The DNA damage response: putting checkpoints in perspective." *Nature*. 408(6811): 433-9.

[24] Furnari, B., Blasina, A., Boddy, M. N., McGowan, C. H., Russell, P. (1999). "Cdc25 inhibited in vivo and in vitro by checkpoint kinases Cds1 and Chk1." *Mol. Biol. Cell.* 10(4): 833-45.

[25] Lee, J., Kumagai, A., Dunphy, W. G. (2001). "Positive regulation of Wee1 by Chk1 and 14-3-3 proteins." *Mol. Biol. Cell.* 12(3): 551-63.

[26] Hickman, E. S., Moroni, M. C., Helin, K. (2002). "The role of p53 and pRB in apoptosis and cancer." *Curr. Opin. Genet. Dev.* 12(1): 60-6.

[27] Glover, D. M., Hagan, I. M., Tavares, A. A. (1998). "Polo-like kinases: a team that plays throughout mitosis." *Genes Dev.* 12(24): 3777-87.

[28] Nigg, E. A. (1998). "Polo-like kinases: positive regulators of cell division from start to finish." *Curr. Opin. Cell. Biol.* 10(6): 776-83.

[29] Nilsson, I., Hoffmann, I. (2000). "Cell cycle regulation by the Cdc25 phosphatase family." *Prog. Cell Cycle Res.* 4: 107-14.

[30] Hoffmann, I., Clarke, P. R., Marcote, M. J., Karsenti, E., Draetta, G. (1993). "Phosphorylation and activation of human cdc25-C by cdc2--cyclin B and its involvement in the self-amplification of MPF at mitosis." *EMBO J.* 12(1): 53-63.

[31] Peter, M., Nakagawa, J., Doree, M., Labbe, J. C., Nigg, E. A. (1990). "In vitro disassembly of the nuclear lamina and M phase-specific phosphorylation of lamins by cdc2 kinase." *Cell.* 61(4): 591-602.

[32] Li, J., Meyer, A. N., Donoghue, D. J. (1997). "Nuclear localization of cyclin B1 mediates its biological activity and is regulated by phosphorylation." *Proc. Natl. Acad. Sci. USA*. 94(2): 502-7.

[33] Lahav-Baratz, S., Sudakin, V., Ruderman, J. V., Hershko, A. (1995). "Reversible phosphorylation controls the activity of cyclosome-associated cyclin-ubiquitin ligase." *Proc. Natl. Acad. Sci. USA.* 92(20): 9303-7.

[34] Visintin, R., Craig, K., Hwang, E. S., Prinz, S., Tyers, M., Amon, A. (1998). "The phosphatase Cdc14 triggers mitotic exit by reversal of CDK-dependent phosphorylation." *Mol. Cell.* 2(6): 709-18.

[35] Shirayama, M., Toth, A., Galova, M., Nasmyth, K. (1999). "APC(Cdc20) promotes exit from mitosis by destroying the anaphase inhibitor Pds1 and cyclin Clb5." *Nature.* 402(6758): 203-7.

[36] Solomon, M. J., Lee, T., Kirschner, M. W. (1992). "Role of phosphorylation in p34cdc2 activation: identification of an activating kinase." *Mol. Biol. Cell.* 3: 13-27.

[37] Solomon, M. J., Harper, J. W., Shuttleworth, J. (1993). "CAK, the p34cdc2 activating kinase, contains a protein identical or closely related to p40MO15." *EMBO J.* 12: 3133-42.

[38] Fesquet, D., Labbe', J.-C., Derancourt, J., Capony, J.-P., Galas, S., Girard, F., Lorca, T., Shuttleworth, J., Doree, M., Cavadore, J. C. (1993). " The MO15 gene encodes the catalytic subunit of a protein kinase that activates cdc2 and other cyclin-dependent kinases (CDKs) through phosphorylation of Thr161 and its homologues." *EMBO J.* 12: 3111-21.

[39] Poon, R. Y. C., Yamashita, K., Adamczewski, J. P., Hunt, T., Shuttleworth, J. (1993). "The cdc2-related protein p40MO15 is the catalytic subunit of a protein kinase that can activate p33CDK2 and p34cdc2." *EMBO J.* 12: 3123-32.

[40] Fisher, R. P., Morgan, D. O. (1994). "A novel cyclin associates with MO15/CDK7 to form the CDK-activating kinase." *Cell.* 78: 713-24.

[41] Kato, J.-Y., Matsuoka, M., Storm, D. K., Sherr, C. J. (1994). "Regulation of cyclin D-dependent kinase 4 (CDK4) by CDK4-activating kinase." *Mol. Cell. Biol.* 14: 2713-21.

[42] Matsuoka, M., Kato, J.-Y., Fisher, R. P., Morgan, D. O., Sherr, C. J. (1994). "Activation of cyclin-dependent kinase 4 (CDK4) by mouse MO15-associated kinase." *Mol. Cell. Biol.* 14: 7265-75.

[43] Tassan, J.-P., Schultz, S., Bartek, J., Nigg, E. A. (1994). "Cell cycle analysis of the activity, subcellular localization and subunit composition of human CAK (CDK-activating kinase)." *J. Cell. Biol.* 127: 467-78.

[44] Allison, L. A., Moyle, M., Shales, M., Ingles, C. J. (1985). "Extensive homology among the largest subunits of eukaryotic and prokaryotic RNA polymerases." *Cell.* 42(2): 599-610.

[45] Corden, J. L., Cadena, D. L., Ahearn, J. M., Dahmus, M. E. (1985). "A unique structure at the carboxyl terminus of the largest subunit of eukaryotic RNA polymerase II." *Proc. Natl. Acad. Sci. USA.* 82(23): 7934-38.

[46] Fong, N., Bentley, D. L. (2001). "Capping, splicing, and 3' processing are independently stimulated by RNA polymerase II: different functions for different segments of the CTD." *Genes Dev.* 15(14): 1783-95.

[47] Kobor, M. S., Greenblatt, J. (2002). "Regulation of transcription elongation by phosphorylation." *Biochim. Biophys. Acta.* 1577(2): 261-275.

[48] Oelgeschlager, T. (2002). "Regulation of RNA polymerase II activity by CTD phosphorylation and cell cycle control." *J. Cell. Physiol.* 190(2): 160-9.

[49] Shilatifard, A., Conaway, R. C., Conaway, J. W. (2003). "The RNA polymerase II elongation complex." *Annu. Rev. Biochem.* 72(693-715).

[50] Corden, J. L., Patturajan, M. (1997). "A CTD function linking transcription to splicing." *Trends Biochem. Sci.* 22(11): 413-6.

[51] Trigon, S., Serizawa, H., Conaway, J. W., Conaway, R. C., Jackson S. P., Morange, M. (1998). "Characterization of the residues phosphorylated in vitro by different C-terminal domain kinases." *J. Biol. Chem.* 273(12): 6769-75.

[52] Rickert, P., Corden, J. L., Lees, E. (1999). "Cyclin C/CDK8 and cyclin H/CDK7/p36 are biochemically distinct CTD kinases." *Oncogene*. 18(4): 1093-1102.

[53] Yue, Z., Maldonado, E., Pillutla, R., Cho, H., Reinberg, D., Shatkin, A. J. (1997). "Mammalian capping enzyme complements mutant Saccharomyces cerevisiae lacking mRNA guanylyltransferase and selectively binds the elongating form of RNA polymerase II." *Proc. Natl. Acad. Sci. USA*. 94(24): 12898-903.

[54] McCracken, S., Fong, N., Rosonina, E., Yankulov, K., Brothers, G., Siderovski, D., Hessel, A., Foster, S., Shuman, S., Bentley, D. L. (1997). "5'-Capping enzymes are targeted to pre-mRNA by binding to the phosphorylated carboxy-terminal domain of RNA polymerase II." *Genes Dev.* 11(24): 3306-18.

[55] Ho, C. K., Shuman, S. (1999). "Distinct roles for CTD Ser-2 and Ser-5 phosphorylation in the recruitment and allosteric activation of mammalian mRNA capping enzyme." *Mol. Cell.* 3(3): 405-11.

[56] Wen, Y., Shatkin, A. J. (1999). "Transcription elongation factor hSPT5 stimulates mRNA capping." *Genes Dev.* 13(14): 1774-9.

[57] Zhou, M., Halanski, M. A., Radonovich, M. F., Kashanchi, F., Peng, J., Price, D. H., Brady, J. N. (2000). "Tat modifies the activity of CDK9 to phosphorylate serine 5 of the RNA polymerase II carboxyl-terminal domain during human immunodeficiency virus type 1 transcription." *Mol. Cell. Biol.* 20(14): 5077-86.

[58] Ramanathan, Y., Rajpara, S. M., Reza, S. M., Lees, E., Shuman, S., Mathews, M. B., Pe'ery, T. (2001). "Three RNA polymerase II carboxyl-terminal domain kinases display distinct substrate preferences." *J. Biol. Chem.* 276(14): 10913-20.

[59] Otero, G., Fellows, J., Li, Y., de Bizemont, T., Dirac, A. M., Gustafsson, C. M., Erdjument-Bromage, H., Tempst, P., Svejstrup, J. Q. (1999). "Elongator, a multisubunit component of a novel RNA polymerase II holoenzyme for transcriptional elongation." *Mol. Cell.* 3(1): 109-18.

[60] Akoulitchev, S., Chuikov, S., Reinberg, D. (2000). "TFIIH is negatively regulated by cdk8-containing mediator complexes." *Nature*. 407(6800): 102-6.

[61] Hengartner, C. J., Myer, V. E., Liao, S. M., Wilson, C. J., Koh, S. S., Young, R. A. (1998). "Temporal regulation of RNA polymerase II by Srb10 and Kin28 cyclin-dependent kinases." *Mol. Cell.* 2(1): 43-53.

[62] Long, J. J., Leresche, A., Kriwacki, R. W., Gottesfeld, J. M. (1998). "Repression of TFIIH transcriptional activity and TFIIH-associated cdk7 kinase activity at mitosis." *Mol. Cell. Biol.* 18(3): 1467-76.

[63] Weinberg, R. A. (1995). "The retinoblastoma protein and cell cycle control." *Cell.* 81(3): 323-30.

[64] Hanahan, D., Weinberg, R. A. (2000). "The hallmarks of cancer." Cell 100(1): 57-70.

[65] Blume-Jensen, P., Hunter, T. (2001). "Oncogenic kinase signalling." *Nature.* 411(6835): 355-65.

[66] Malumbres, M., Hunt, S. L., Sotillo, R., Martin, J., Odajima, J., Martin, A., Dubus, P., Ortega, S., Barbacid, M. (2003). "Driving the cell cycle to cancer." *Adv. Exp. Med. Biol.* 532: 1-11.

[67] Hollstein, M., Rice, K., Greenblatt, M. S., Soussi, T., Fuchs, R., Sorlie, T., Hovig, E., Smith-Sorensen, B., Montesano, R., Harris, C. C. (1994). "Database of p53 gene somatic mutations in human tumors and cell lines." *Nucleic Acids Res.* 22(17): 3551-5.

[68] Hollstein, M., Sidransky, D., Vogelstein, B., Harris, C. C. (1991). "p53 mutations in human cancers." *Science.* 253(5015): 49-53.

[69] Huang, H. J., Yee, J. K., Shew, J. Y., Chen, P. L., Bookstein, R., Friedmann, T., Lee, E. Y., Lee, W. H. (1988). "Suppression of the neoplastic phenotype by replacement of the RB gene in human cancer cells." *Science.* 242(4885): 1563-6.

[70] Xu, Y., Baltimore, D. (1996). "Dual roles of ATM in the cellular response to radiation and in cell growth control." *Genes Dev.* 10(19): 2401-10.

[71] Adams, J. M., Cory, S. (1998). "The Bcl-2 protein family: arbiters of cell survival." *Science.* 281(5381): 1322-6.

[72] Mani, S., Wang, C., Wu, K., Francis, R., Pestell, R. (2000). "Cyclin-dependent kinase inhibitors: novel anticancer agents." *Expert Opin. Investig. Drugs.* 9(8): 1849-70.

[73] Malumbres, M., Carnero, A. (2003). "Cell cycle deregulation: a common motif in cancer." *Prog. Cell Cycle Res.* 5: 5-18.

[74] Donnellan, R., Chetty, R. (1998). "Cyclin D1 and human neoplasia." *Mol. Pathol.* 51(1): 1-7.

[75] Wei, G., Lonardo, F., Ueda, T., Kim, T., Huvos, A. G., Healey, J. H., Ladanyi, M. (1999). "CDK4 gene amplification in osteosarcoma: reciprocal relationship with INK4A gene alterations and mapping of 12q13 amplicons." *Int. J. Cancer.* 80(2): 199-204.

[76] Ruas, M., Peters, G. (1998). "The p16INK4a/CDKN2A tumor suppressor and its relatives." *Biochim. Biophys. Acta.* 1378(2): F115-77.

[77] Donnellan, R., Chetty, R. (1999). "Cyclin E in human cancers." *FASEB J.* 13(8): 773-80.

[78] Catzavelos, C., Bhattacharya, N., Ung, Y. C., Wilson, J. A., Roncari, L., Sandhu, C., Shaw, P., Yeger, H., Morava-Protzner, I., Kapusta, L., Franssen, E., Pritchard, K. I., Slingerland, J. M. (1997). "Decreased levels of the cell-cycle inhibitor p27Kip1 protein: prognostic implications in primary breast cancer." *Nat. Med.* 3(2): 227-30.

[79] Dai, Y., Grant, S. (2003). "Cyclin-dependent kinase inhibitors." *Curr. Opin. Pharmacol.* 3(4): 362-70.

[80] Senderowicz, A. M. (2003). "Small-molecule cyclin-dependent kinase modulators." *Oncogene.* 22(42): 6609-20.

[81] Berthet, C., Aleem, E., Coppola, V., Tessarollo, L., Kaldis, P. (2003). "Cdk2 knockout mice are viable." *Curr. Biol.* 13(20): 1775-85.

[82] Ortega, S., Prieto, I., Odajima, J., Martin, A., Dubus, P., Sotillo, R., Barbero, J. L., Malumbres, M., Barbacid, M. (2003). "Cyclin-dependent kinase 2 is essential for meiosis but not for mitotic cell division in mice." *Nat. Genet.* 35(1): 25-31.

[83] Tetsu, O., McCormick, F. (2003). "Proliferation of cancer cells despite CDK2 inhibition." *Cancer Cell.* 3(3): 233-45.

[84] Aleem, E., Berthet, C., Kaldis, P. (2004). "Cdk2 as a master of S phase entry: fact or fake?" *Cell Cycle.* 3(1): 35-7.

[85] Malumbres, M., Sotillo, R., Santamaría, D., Galán, J., Cerezo, A., Ortega, S., Dubus, P., Barbacid, M. (2004). "Mammalian cells cycle without the D-type cyclin-dependent kinases Cdk4 and Cdk6." *Cell.* 118(4): 493-504.

[86] Berthet, C., Kaldis, P. (2006) "Cdk2 and Cdk4 cooperatively control the expression of Cdc2." *Cell Div.* 1:10.

[87] Berthet, C., Kaldis, P. (2007) "Cell-specific responses to loss of cyclin-dependent kinases." *Oncogene.* 26(31):4469-77.

[88] Cai, D., Latham, V. M. Jr, Zhang, X., Shapiro, G. I. (2006) "Combined depletion of cell cycle and transcriptional cyclin-dependent kinase activities induces apoptosis in cancer cells." *Cancer Res.* 66(18):9270-80.

[89] Aleem, E., Kiyokawa, H., Kaldis, P. (2005) "Cdc2-cyclin E complexes regulate the G1/S phase transition." *Nat. Cell. Biol.* 7(8):831-6.

[90] De Bondt, H. L., Rosenblatt, J., Jancarik, J., Jones, H. D., Morgan, D. O., Kim, S. H. (1993). "Crystal structure of cyclin-dependent kinase 2." *Nature.* 363(6430): 595-602.

[91] Schulze-Gahmen, U., De Bondt, H. L., Kim, S. H. (1996). "High-resolution crystal structures of human cyclin-dependent kinase 2 with and without ATP: bound waters and natural ligand as guides for inhibitor design." *J. Med. Chem.* 39(23): 4540-6.

[92] Brown, N. R., Noble, M. E., Endicott, J. A., Johnson, L. N. (1999). "The structural basis for specificity of substrate and recruitment peptides for cyclin-dependent kinases." *Nat. Cell. Biol.* 1(7): 438-43.

[93] Brown, N. R., Noble, M. E., Endicott, J. A., Garman, E. F., Wakatsuki, S., Mitchell, E., Rasmussen, B., Hunt, T., Johnson, L. N. (1995). "The crystal structure of cyclin A." *Structure.* 3(11): 1235-47.

[94] Jeffrey, P. D., Russo, A. A., Polyak, K., Gibbs, E., Hurwitz, J., Massague, J., Pavletich, N. P. (1995). "Mechanism of CDK activation revealed by the structure of a cyclinA-CDK2 complex." *Nature.* 376(6538): 313-20.

[95] Glotzer, M., Murray, A. W., Kirschner, M. W. (1991). "Cyclin is degraded by the ubiquitin pathway." *Nature.* 349(6305): 132-8.

[96] Kobayashi, H., Stewart, E., Poon, R., Adamczewski, J. P., Gannon, J., Hunt, T. (1992). "Identification of the domains in cyclin A required for binding to, and activation of, p34cdc2 and p32cdk2 protein kinase subunits." *Mol. Biol. Cell.* 3(11): 1279-94.

[97] Connell-Crowley, L., Solomon, M. J., Wei, N., Harper, J. W. (1993). "Phosphorylation independent activation of human cyclin-dependent kinase 2 by cyclin A in vitro." *Mol. Biol. Cell.* 4(1): 79-92.

[98] Russo, A. A., Jeffrey, P. D., Patten, A. K., Massague, J., Pavletich, N. P. (1996). "Crystal structure of the p27Kipl cyclin-dependent-kinase inhibitor bound to the cyclin A-Cdk2 complex." *Nature.* 382(6589): 325-31.

[99] Jeffrey, P.D., Tong, L., Pavletich, N.P. (2000). "Structural basis of inhibition of CDK-cyclin complexes by INK4 inhibitors." *Genes Dev.* 14(24): 3115-25.

[100] Manning, G., Whyte, D. B., Martinez, R., Hunter, T., Sudarsanam, S. (2002). "The protein kinase complement of the human genome." *Science.* 298(5600): 1912-34.

[101] Lolli, G., Thaler, F., Valsasina, B., Roletto, F., Knapp, S., Uggeri, M., Bachi, A., Matafora, V., Storici, P., Stewart, A., Kalisz, H. M., Isacchi, A. (2003). "Inhibitor affinity chromatography: profiling the specific reactivity of the proteome with immobilized molecules." *Proteomics.* 3(7): 1287-98.

[102] Noble, M. E., Endicott, J. A., Johnson, L. N. (2004). "Protein kinase inhibitors: insights into drug design from structure." *Science.* 303(5665): 1800-5.

[103] Druker, B. J., Talpaz, M., Resta, D. J., Peng, B., Buchdunger, E., Ford, J. M., Lydon, N. B., Kantarjian, H., Capdeville, R., Ohno-Jones, S., Sawyers, C. L. (2001). "Efficacy and safety of a specific inhibitor of the BCR-ABL tyrosine kinase in chronic myeloid leukemia." *N. Engl. J. Med.* 344(14):1031-7.

[104] Fabbro, D., Ruetz, S., Buchdunger, E., Cowan-Jacob, S. W., Fendrich, G., Liebetanz, J., Mestan, J., O'Reilly, T., Traxler, P., Chaudhuri, B., Fretz, H., Zimmermann, J., Meyer, T., Caravatti, G., Furet, P., Manley, P.W. (2002). "Protein kinases as targets for anticancer agents: from inhibitors to useful drugs." *Pharmacol. Ther.* 93: 79-98.

[105] Gray, N. S., Wodicka, L., Thunnissen, A. M., Norman, T. C., Kwon, S., Espinoza, F. H., Morgan, D. O., Barnes, G., LeClerc, S., Meijer, L., Kim, S. H., Lockhart, D. J., Schultz, P. G. (1998). "Exploiting chemical libraries, structure, and genomics in the search for kinase inhibitors." *Science.* 281: 533-38.

[106] Davies, T. G., Bentley, J., Arris, C. E., Boyle, F. T., Curtin, N. J., Endicott, J. A., Gibson, A. E., Golding, B. T., Griffin, R. J., Hardcastle, I. R., Jewsbury, P., Johnson, L. N., Mesguiche, V., Newell, D. R., Noble, M. E. M., Tucker, J. A., Wang, L., Whitfield, H. J. (2002). "Structure-based design of a potent purine-based cyclin dependent kinase inhibitor." *Nat. Struct. Biol.* 9: 745-49.

[107] Honma, T., Yoshizumi, T., Hashimoto, N., Hayashi, K., Kawanishi, N., Fukasawa, K., Takaki, T., Ikeura, C., Ikuta, M., Suzuki-Takahashi, I., Hayama, T., Nishimura, S. and Morishima, H. (2001). "A novel approach for the development of selective Cdk4 inhibitors: library design based on locations of Cdk4 specific amino acid residues." *J. Med. Chem.* 44: 4628-40.

[108] Honma, T., Hayashi, K., Aoyama, T., Hashimoto, N., Machida, T., Fukasawa, K., Iwama, T., Ikeura, C., Ikuta, M., Suzuki-Takahashi, I., Iwasawa, Y., Hayama, T., Nishimura, S., Morishima, H. (2001). "Structure-based generation of a new class of potent Cdk4 inhibitors: new de novo design strategy and library design." *J. Med. Chem.* 44: 4615-27.

[109] Soni, R., O'Reilly, T., Furet, P., Muller, L., Stephan, C., Zumstein-Mecker, S., Fretz, H., Fabbro, D., Chaudhuri, B. (2001). "Selective in vivo and in vitro effects of a small molecule inhibitor of cyclin-dependent kinase 4." *J. Natl. Cancer Inst.* 93: 436-46.

[110] Beattie, J. F., Breault, G. A., Ellston, R. P., Green, S., Jewsbury, P. J., Midgley, C. J., Naven, R. T., Minshull, C. A., Pauptit, R. A., Tucker, J. A., Pease, J. E. (2003). "Cyclin-dependent kinase 4 inhibitors as a treatment for cancer. Part 1: identification and optimisation of substituted 4,6-bis anilino pyrimidines." *Bioorg. Med. Chem. Lett.* 13: 2955-60.

[111] Breault, G. A., Ellston, R. P., Green, S., James, S. R., Jewsbury, P. J., Midgley, C. J., Pauptit, R. A., Minshull, C. A., Tucker, J. A., Pease, J. E. (2003). "Cyclin-dependent kinase 4 inhibitors as a treatment for cancer. Part 2: identification and optimisation of substituted 2,4-bis anilino pyrimidines." *Bioorg. Med. Chem. Lett.* 13: 2961-66.

[112] Ikuta, M., Kamata, K., Fukasawa, K., Honma, T., Machida, T., Hirai, H., Suzuki-Takahashi, I., Hayama, T., Nishimura, S. (2001). "Crystallographic approach to identification of cyclin-dependent kinase 4 (CDK4)-specific inhibitors by using CDK4 mimic CDK2 protein." *J. Biol. Chem.* 276: 27548-54.

[113] Lolli, G., Lowe, E. D., Brown, N. R., Johnson, L. N. (2004). "The crystal structure of human CDK7 and its protein recognition properties." *Structure.* 12(11):2067-79.

[114] Baumli, S., Lolli, G., Lowe, E. D., Troiani, S., Rusconi, L., Bullock, A. N., Debreczeni, J. E., Knapp, S., Johnson, L.N. (2008). "The structure of P-TEFb (CDK9/cyclin T), its complex with flavopiridol and regulation by phosphorylation." *EMBO J.* In press.

[115] Yu, Q., Geng, Y., Sicinski, P. (2001). "Specific protection against breast cancers by cyclin D1 ablation." *Nature.* 411(6841):1017-21.

[116] Yu, Q., Sicinska, E., Geng, Y., Ahnström, M., Zagozdzon, A., Kong, Y., Gardner, H., Kiyokawa, H., Harris, L. N., Stål, O., Sicinski, P. (2006). "Requirement for CDK4 kinase function in breast cancer." *Cancer Cell.* 9(1):23-32.

[117] Reddy, H. K., Mettus, R. V., Rane, S. G., Graña, X., Litvin, J., Reddy, E. P. (2005). "Cyclin-dependent kinase 4 expression is essential for neu-induced breast tumorigenesis." *Cancer Res.* 65(22):10174-8.

[118] Landis, M. W., Pawlyk, B. S., Li, T., Sicinski, P., Hinds, P. W. (2006). "Cyclin D1-dependent kinase activity in murine development and mammary tumorigenesis." *Cancer Cell.* 9(1):13-22.

[119] Fry, D. W., Harvey, P. J., Keller, P. R., Elliott, W. L., Meade, M., Trachet, E., Albassam, M., Zheng, X., Leopold, W. R., Pryer, N. K., Toogood, P. L. (2004). "Specific inhibition of cyclin-dependent kinase 4/6 by PD 0332991 and associated antitumor activity in human tumor xenografts." *Mol. Cancer Ther.* 3(11):1427-38.

[120] Du, J., Widlund, H. R., Horstmann, M. A., Ramaswamy, S., Ross, K., Huber, W. E., Nishimura, E. K., Golub, T. R., Fisher, D. E. (2004). "Critical role of CDK2 for melanoma growth linked to its melanocyte-specific transcriptional regulation by MITF." *Cancer Cell.* 6(6):565-76.

[121] Meijer, L., Borgne, A., Mulner, O., Chong, J. P., Blow, J. J., Inagaki, N., Inagaki, M., Delcros, J. G., Moulinoux, J. P. (1997). "Biochemical and cellular effects of roscovitine, a potent and selective inhibitor of the cyclin-dependent kinases cdc2, cdk2 and cdk5." *Eur. J. Biochem.* 243(1-2):527-36.

[122] Bach, S., Knockaert, M., Reinhardt, J., Lozach, O., Schmitt, S., Baratte, B., Koken, M., Coburn, S. P., Tang, L., Jiang, T., Liang, D. C., Galons, H., Dierick, J. F., Pinna,

L. A., Meggio, F., Totzke, F., Schächtele, C., Lerman, A. S., Carnero, A., Wan, Y., Gray, N., Meijer, L. (2005). "Roscovitine targets, protein kinases and pyridoxal kinase." *J. Biol. Chem.* 280(35):31208-19.

[123] Clough, J. (2002). "CDK inhibitor shows promise for inflammatory kidney disease." *Drug Discov. Today*. 7(15): 789-90.

[124] White, J. D., Cassidy, J., Twelves, C., Benson, C., Pacey, S., Judson, I., McGrath, H., Rose, F., Frenz L. (2004). "A phase I trial of the oral cyclin dependent kinase inhibitor CYC202 in patients with advanced malignancy" *J. Clin. Oncol.* 22S: 3042.

[125] Pierga, J.-Y., Faivre, S., Vera, K., Laurence, V., Delbaldo, C., Bekradda, M., Armand, J.-P., Gianella-Borradori, A., Dieras, V., Raymond E. (2003). "A phase I and pharmacokinetic (PK) trial of CYC202, a novel oral cyclin-dependent kinase (CDK) inhibitor, in patients (pts) with advanced solid tumors." *Proc. Am. Soc. Clin. Oncol.* 22: abstr. 840.

[126] Czech, J., Hoffman, D., Naik, R., Sedlacek, H. H. (1995). "Antitumor activity of flavone L86-8275." *Int. J. Oncol.* 6: 31-36.

[127] Arguello, F., Alexander, M., Sterry, J. A., Tudor, G., Smith, E. M., Kalavar, N. T., Greene, J. F. Jr, Koss, W., Morgan, C. D., Stinson, S. F., Siford, T. J., Alvord, W. G., Klabansky, R. L., Sausville, E. A. (1998). "Flavopiridol induces apoptosis of normal lymphoid cells, causes immunosuppression, and has potent antitumor activity In vivo against human leukemia and lymphoma xenografts." *Blood.* 91(7): 2482-90.

[128] Patel, V., Senderowicz, A. M., Pinto, D. Jr, Igishi, T., Raffeld, M., Quintanilla-Martinez, L., Ensley, J. F., Sausville, E. A., Gutkind, J. S. (1998). "Flavopiridol, a novel cyclin-dependent kinase inhibitor, suppresses the growth of head and neck squamous cell carcinomas by inducing apoptosis." *J. Clin. Invest.* 102(9): 1674-81.

[129] Liu, G., Gandara, D. R., Lara, P. N. Jr, Raghavan, D., Doroshow, J. H., Twardowski, P., Kantoff, P., Oh, W., Kim, K., Wilding, G. (2004). "A Phase II trial of flavopiridol (NSC #649890) in patients with previously untreated metastatic androgen-independent prostate cancer." *Clin. Cancer Res.* 10(3):924-8.

[130] Grendys, E. C. Jr, Blessing, J. A., Burger, R., Hoffman, J. (2005). "A phase II evaluation of flavopiridol as second-line chemotherapy of endometrial carcinoma: a Gynecologic Oncology Group study." *Gynecol. Oncol.* 98(2):249-53.

[131] Burdette-Radoux, S., Tozer, R. G., Lohmann, R. C., Quirt, I., Ernst, D. S., Walsh, W., Wainman, N., Colevas, A. D., Eisenhauer, E. A. (2004). "Phase II trial of flavopiridol, a cyclin dependent kinase inhibitor, in untreated metastatic malignant melanoma." *Invest. New Drugs*. 22(3):315-22.

[132] Van Veldhuizen, P. J., Faulkner, J. R., Lara, P. N. Jr, Gumerlock, P. H., Goodwin, J. W., Dakhil, S. R., Gross, H. M., Flanigan, R. C., Crawford, E. D. (2005). "A phase II study of flavopiridol in patients with advanced renal cell carcinoma: results of Southwest Oncology Group Trial 0109." *Cancer Chemother. Pharmacol.* 56(1):39-45.

[133] Rudek, M. A., Bauer, K. S. Jr, Lush, R. M. 3rd, Stinson, S. F., Senderowicz, A. M., Headlee, D. J., Arbuck, S. G., Cox, M. C., Murgo, A. J., Sausville, E. A., Figg, W. D. (2003). "Clinical pharmacology of flavopiridol following a 72-hour continuous infusion." *Ann. Pharmacother.* 37(10):1369-74.

[134] Byrd, J. C., Lin, T. S., Dalton, J. T., Wu, D., Phelps, M. A., Fischer, B., Moran, M., Blum, K. A., Rovin, B., Brooker-McEldowney, M., Broering, S., Schaaf, L. J., Johnson, A. J., Lucas, D. M., Heerema, N. A., Lozanski, G., Young, D. C., Suarez, J. R., Colevas, A. D., Grever, M. R. (2007). "Flavopiridol administered using a pharmacologically derived schedule is associated with marked clinical efficacy in refractory, genetically high-risk chronic lymphocytic leukemia." *Blood.* 109(2):399-404.

[135] Misra, R. N., Xiao, H. Y., Kim, K. S., Lu, S., Han, W. C., Barbosa, S. A., Hunt, J. T., Rawlins, D. B., Shan, W., Ahmed, S. Z., Qian, L., Chen, B. C., Zhao, R., Bednarz, M. S., Kellar, K. A., Mulheron, J. G., Batorsky, R., Roongta, U., Kamath, A., Marathe, P., Ranadive, S. A., Sack, J. S., Tokarski, J. S., Pavletich, N. P., Lee, F. Y., Webster, K. R., Kimball, S. D. (2004). "N-(cycloalkylamino)acyl-2-aminothiazole inhibitors of cyclin-dependent kinase 2. N-[5-[[[5-(1,1-dimethylethyl)-2-oxazolyl]methyl]thio]-2-thiazolyl]-4-piperidinecarboxamide (BMS-387032), a highly efficacious and selective antitumor agent." *J. Med. Chem.* 47(7):1719-28.

[136] Heath, E. I., Bible, K., Martell, R. E., Adelman, D. C., Lorusso, P. M. (2008). "A phase 1 study of SNS-032 (formerly BMS-387032), a potent inhibitor of cyclin-dependent kinases 2, 7 and 9 administered as a single oral dose and weekly infusion in patients with metastatic refractory solid tumors." *Invest. New Drugs.* 26(1):59-65.

[137] DePinto, W., Chu, X. J., Yin, X., Smith, M., Packman, K., Goelzer, P., Lovey, A., Chen, Y., Qian, H., Hamid, R., Xiang, Q., Tovar, C., Blain, R., Nevins, T., Higgins, B., Luistro, L., Kolinsky, K., Felix, B., Hussain, S., Heimbrook, D. (2006). "In vitro and in vivo activity of R547: a potent and selective cyclin-dependent kinase inhibitor currently in phase I clinical trials." *Mol. Cancer Ther.* 5(11):2644-58.

[138] Joshi, K. S., Rathos, M. J., Joshi, R. D., Sivakumar, M., Mascarenhas, M., Kamble, S., Lal, B., Sharma, S. (2007). "In vitro antitumor properties of a novel cyclin-dependent kinase inhibitor, P276-00." *Mol. Cancer Ther.* 6(3):918-25.

[139] Joshi, K. S., Rathos, M. J., Mahajan, P., Wagh, V., Shenoy, S., Bhatia, D., Chile, S., Sivakumar, M., Maier, A., Fiebig, H. H., Sharma, S. (2007). "P276-00, a novel cyclin-dependent inhibitor induces G1-G2 arrest, shows antitumor activity on cisplatin-resistant cells and significant in vivo efficacy in tumor models." *Mol. Cancer Ther.* 6(3):926-34.

[140] Jenuwein, T., Allis, C. D. (2001). "Translating the histone code." *Science.* 293: 1074–80.

[141] Lachner, M., Jenuwein, T. (2002). "The many faces of histone lysine methylation." *Curr. Opin. Cell. Biol.* 14(3):286-98.

[142] Nowak, S. J., Corces, V. G. (2004). "Phosphorylation of histone H3: a balancing act between chromosome condensation and transcriptional activation." *Trends Genet.* 20(4):214-20.

[143] Jason, L. J., Moore, S. C., Lewis, J. D., Lindsey, G., Ausió, J. (2002). "Histone ubiquitination: a tagging tail unfolds?" *Bioessays*. 24(2):166-74.

[144] Nathan, D., Sterner, D. E. Berger, S. L. "Histone modifications: Now summoning sumoylation." *Proc. Natl. Acad. Sci. USA*. 100(23):13118-20.

[145] Roth, S. Y., Denu, J. M., Allis, D. C. (2001). "Histone Acetyltransfereases." *Ann. Rev. Biochem.* 70:81-120.

[146] Marks, P. A., Miller, T., Richon, V. M. (2003). "Histone Deacetylases." *Curr. Opin. Pharmacol.* 3:344-351.

[147] Carrozza, M. J., Utley, R. T., Workman, J. L., Cote, J. (2003). "The diverse functions of histone acetyltransferase complexes." *Trends in Genetics* 19:321-29.

[148] Gregoretti, I. V., Lee, Y. M., Goodson, H.V. (2004). "Molecular evolution of the histone deacetylase family: functional implications of phylogenetic analysis." *J. Mol. Biol.* 338:17-31.

[149] Verdin, E., Dequiedt, F., Kasler, H. G. (2003). "Class II histone deacetylases: versatile regulators." *Trends Genet.* 19(5):286-93.

[150] Blander, G. Guarente, L. (2004). "The Sir2 family of protein deacetylases." *Ann. Rev. Biochem.* 73:417-35.

[151] Minucci, S., Pelicci, P. G. (2006). "Histone deacetylase inhibitors and the promise of epigenetic (and more) treatments for cancer." *Nat. Rev. Cancer* 6(1):38-51.

[152] Gallinari, P., Di Marco, S., Jones, P., Pallaoro, M., Steinkühler, C. (2007). "HDACs, Histone deacetylation and gene transcription: from molecular biology to cancer therapeutics." *Cell Research* 17:195-211.

[153] Leder, A., Leder, P. (1975). "Butyric acid, a potent inducer of erythroid differentiation in cultured erythroleukemic cells." *Cell* 5:319-22.

[154] Riggs, M. G., Whittaker, R. G., Neumann, J. R., Ingram, V. M. (1977). "Butyrate causes histone modification in HeLa and Friend erythroleukemia cells." *Nature* 268:462-4.

[155] Novogrodsky, A., Dvir, A., Ravid, A., Shkolnik, T., Stenzel, K. H., Rubin, A. L., Zaizov, R. (1983). "Effect of polar organic compounds on leukemic cells." *Cancer* 51:9-14.

[156] Rephaeli, A., Zhuk, R., Nudelman, A. (2000). "Prodrugs of butyric acid from bench to bedside: synthetic design, mechanisms of action, and clinical applications." *Drug Dev. Res.* 50:379-91.

[157] Yoshida, M., Kijima, M., Akita, M., Beppu, T. (1990). "Potent and specific inhibition of mammalian histone deacetylase both in vivo and in vitro by trichostatin A." *J. Biol. Chem.* 265(28):17174-9.

[158] Brehm, A., Miska, E. A., McCance, D. J., Reid, J. L., Bannister, A. J., Kouzarides, T. (1998). "Retinoblastoma protein recruits histone deacetylase to repress transcription." *Nature* 391(6667):597-601.

[159] Magnaghi-Jaulin, L., Groisman, R., Naguibneva, I., Robin, P., Lorain, S., Le Villain, J. P., Troalen, F., Trouche, D., Harel-Bellan, A. (1998). "Retinoblastoma protein represses transcription by recruiting a histone deacetylase." *Nature* 391(6667):601-5.

[160] Lagger, G., O'Carroll, D., Rembold, M., Khier, H., Tischler, J., Weitzer, G., Schuettengruber, B., Hauser, C., Brunmeir, R., Jenuwein, T., Seiser, C. (2002). "Essential function of histone deacetylase 1 in proliferation control and CDK inhibitor repression." *EMBO J.* 21(11):2672-81.

[161] Dasmahapatra, G., Almenara, J., Grant, S. (2006). "Flavopiridol and histone deacetylase inhibitors promote mitochondrial injury and cell death in human leukemia cells overexpressing BCL-2." *Mol. Pharmacol.* 69(1):288-98.

[162] Fu, M., Wang, C., Rao, M., Wu, X., Bouras, T., Zhang, X., Li, Z., Jiao, X., Yang, J., Li, A., Perkins, N. D., Thimmapaya, B., Kung, A. L., Munoz, A., Giordano, A., Lisanti, M. P., Pestell, R. G. (2005). "Cyclin D1 represses p300 transactivation through a cyclin-dependent kinase-independent mechanism." *J. Biol. Chem.* 280(33):29728-42.

[163] Rottmann, S., Menkel, A. R., Bouchard, C., Mertsching, J., Loidl, P., Kremmer, E., Eilers, M., Luscher-Firzlaff, J., Lilischkis, R., Luscher, B. (2005). "Mad1 function in cell proliferation and transcriptional repression is antagonized by cyclin E/CDK2." *J. Biol. Chem.* 280(16):15489-92.

[164] Talukder, A. H., Mishra, S. K., Mandal, M., Balasenthil, S., Mehta, S., Sahin, A. A., Barnes, C. J., Kumar, R. (2003). "MTA1 interacts with MAT1, a cyclin-dependent kinase-activating kinase complex ring finger factor, and regulates estrogen receptor transactivation functions." *J. Biol. Chem.* 278(13):11676-85.

[165] Chen, D., Riedl, T., Washbrook, E., Pace, P. E., Coombes, R. C., Egly, J. M., Ali, S. (2000). "Activation of estrogen receptor alpha by S118 phosphorylation involves a ligand-dependent interaction with TFIIH and participation of CDK7." *Mol. Cell* 6(1):127-37.

[166] Kanazawa, S., Soucek, L., Evan, G., Okamoto, T., Peterlin, B. M. (2003). "c-Myc recruits P-TEFb for transcription, cellular proliferation and apoptosis." *Oncogene* 22:5707-11.

[167] Rochette-Egly, C., Adam, S., Rossignol, M., Egly, J.-M., Chambon, P. (1997). "Stimulation of RARa activation function AF-1 through binding to the general transcription factor TFIIH and phosphorylation by CDK7." *Cell* 90:97-107.

[168] Simone, C., Stiegler, P., Bagella, L., Pucci, B., Bellan, C., De Falco, G., De Luca, A., Guanti, G., Puri, P. L., Giordano, A. (2002). "Activation of MyoD-dependent transcription by cdk9/cyclin T2." *Oncogene* 21:4137-48.

[169] Fu, M., Wang, C., Zhang, X., Pestell, R. G. (2004). "Acetylation of nuclear receptors in cellular growth and apoptosis." *Biochem. Pharmacol.* 68(6):1199-208.

[170] Kelly, B. L., Wolfe, K. G., Roberts, J. M. (1998). "Identification of a substrate-targeting domain in cyclin E necessary for phosphorylation of the retinoblastoma protein." *Proc. Natl. Acad. Sci. USA* 95(5):2535-40.

[171] Schulman, B. A., Lindstrom, D. L., Harlow, E. (1998). "Substrate recruitment to cyclin-dependent kinase 2 by a multipurpose docking site on cyclin A." *Proc. Natl. Acad. Sci. USA* 95(18):10453-8.

[172] Shah, M. A., Schwartz, G. K. (2001). "Cell cycle-mediated drug resistance: an emerging concept in cancer therapy." Clin. *Cancer Res.* 7(8):2168-81.

[173] Motwani, M., Delohery, T. M., Schwartz, G. K. (1999). "Sequential dependent enhancement of caspase activation and apoptosis by flavopiridol on paclitaxel-treated human gastric and breast cancer cells." Clin. *Cancer Res.* 5(7):1876-83.

[174] Motwani, M., Rizzo, C., Sirotnak, F., She, Y., Schwartz, G. K. (2003). "Flavopiridol enhances the effect of docetaxel in vitro and in vivo in human gastric cancer cells." *Mol. Cancer Ther.* 2(6):549-55.

[175] Wall, N. R., O'Connor, D. S., Plescia, J., Pommier, Y., Altieri, D. C. (2003). "Suppression of survivin phosphorylation on Thr34 by flavopiridol enhances tumor cell apoptosis." *Cancer Res.* 63(1):230-5.

[176] O'Connor, D. S., Wall, N. R., Porter, A. C., Altieri, D. C. (2002). "A p34(cdc2) survival checkpoint in cancer." *Cancer Cell.* 2(1):43-54.

[177] Schwartz, G. K., O'Reilly, E., Ilson, D., Saltz, L., Sharma, S., Tong, W., Maslak, P., Stoltz, M., Eden, L., Perkins, P., Endres, S., Barazzoul, J., Spriggs, D., Kelsen, D. (2002). "Phase I study of the cyclin-dependent kinase inhibitor flavopiridol in combination with paclitaxel in patients with advanced solid tumors." *J. Clin. Oncol.* 20(8):2157-70.

[178] Fornier, M. N., Rathkopf, D., Shah, M., Patil, S., O'Reilly, E., Tse, A. N., Hudis, C., Lefkowitz, R., Kelsen, D. P., Schwartz, G. K. (2007). "Phase I dose-finding study of weekly docetaxel followed by flavopiridol for patients with advanced solid tumors." *Clin. Cancer Res.* 13(19):5841-6.

[179] Jung, C. P., Motwani, M. V., Schwartz, G. K. (2001). "Flavopiridol increases sensitization to gemcitabine in human gastrointestinal cancer cell lines and correlates with down-regulation of ribonucleotide reductase M2 subunit." *Clin. Cancer Res.* 7(8):2527-36.

[180] Motwani, M., Jung, C., Sirotnak, F. M., She, Y., Shah, M. A., Gonen, M., Schwartz, G. K. (2001). "Augmentation of apoptosis and tumor regression by flavopiridol in the presence of CPT-11 in Hct116 colon cancer monolayers and xenografts." *Clin. Cancer Res.* 7(12):4209-19.

[181] Ali, S., El-Rayes, B. F., Aranha, O., Sarkar, F. H., Philip, P. A. (2005). "Sequence dependent potentiation of gemcitabine by flavopiridol in human breast cancer cells." *Breast Cancer Res. Treat.* 90(1):25-31.

[182] Shah, M. A., Kortmansky, J., Motwani, M., Drobnjak, M., Gonen, M., Yi, S., Weyerbacher, A., Cordon-Cardo, C., Lefkowitz, R., Brenner, B., O'Reilly, E., Saltz, L., Tong, W., Kelsen, D. P., Schwartz, G. K. (2005). "A phase I clinical trial of the sequential combination of irinotecan followed by flavopiridol." *Clin. Cancer Res.* 11(10):3836-45.

[183] Lin, T. S., Fischer, B., Moran, M. E., Buckner, M. M., Shank, R., Kraut, E. H., Porcu, P., Farag, S. S., Eisenbeis, . F., Colevas, D., Grever, M. R., Byrd, J. C. (2005). "Flavopiridol, Fludarabine and Rituximab Is a Highly Active Regimen in Indolent B-Cell Lymphoproliferative Disorders Including Mantle Cell Lymphoma." *Blood* (ASH Annual Meeting Abstracts) 106:944.

[184] Rosato, R., R., Almenara, J. A., Yu, C., Grant, S. (2004). "Evidence of a functional role for p21WAF1/CIP1 down-regulation in synergistic antileukemic interactions between the histone deacetylase inhibitor sodium butyrate and flavopiridol." *Mol Pharmacol.* 65(3):571-81.

[185] Rosato, R. R., Almenara, J. A., Maggio, S. C., Atadja, P., Craig, R., Vrana, J., Dent, P., Grant, S. (2005). "Potentiation of the lethality of the histone deacetylase inhibitor

LAQ824 by the cyclin-dependent kinase inhibitor roscovitine in human leukemia cells." *Mol Cancer Ther.* 4(11):1772-85.

[186] Nimmanapalli, R., Fuino, L., Bali, P., Gasparetto, M., Glozak, M., Tao, J., Moscinski, L., Smith, C., Wu, J., Jove, R., Atadja, P., Bhalla, K. (2003). "Histone deacetylase inhibitor LAQ824 both lowers expression and promotes proteasomal degradation of Bcr-Abl and induces apoptosis of imatinib mesylate-sensitive or -refractory chronic myelogenous leukemia-blast crisis cells." *Cancer Res.* 63(16):5126-35.

[187] Deininger, M., Buchdunger, E., Druker, B. J. (2005). "The development of imatinib as a therapeutic agent for chronic myeloid leukemia." *Blood* 105(7):2640-53.

[188] Bolden, J. E., Peart, M. J., Johnstone, R. W. (2006). "Anticancer activities of histone deacetylase inhibitors." *Nat Rev Drug Discov.* 5(9):769-84.

[189] Lolli, G., Johnson, L. N. (2005). "CAK-Cyclin-dependent Activating Kinase: a key kinase in cell cycle control and a target for drugs?" *Cell Cycle* 4(4):572-7.

[190] Fisher, R. P. (2005). "Secrets of a double agent: CDK7 in cell-cycle control and transcription." J. *Cell. Sci.* 118(Pt 22):5171-80.

[191] Larochelle, S., Pandur, J., Fisher, R.P., Salz, H.K., Suter, B. (1998). "Cdk7 is essential for mitosis and for in vivo Cdk-activating kinase activity." *Genes Dev.* 12: 370-381.

[192] Rossi, D. J., Londesborough, A., Korsisaari, N., Pihlak, A., Lehtonen, E., Henkemeyer, M., Makela, T.P. (2001). "Inability to enter S phase and defective RNA polymerase II CTD phosphorylation in mice lacking Mat1." *EMBO J.* 20(11): 2844-56.

[193] Wu, L., Chen, P., Shum, C. H., Chen, C., Barsky, L. W., Weinberg, K. I., Jong, A., Triche, T. J. (2001). "MAT1-modulated CAK activity regulates cell cycle G1 exit." *Mol. Cell. Biol.* 21(1): 260-70.

[194] Yu, C., Krystal, G., Dent, P., Grant, S. (2002). "Flavopiridol potentiates STI571-induced mitochondrial damage and apoptosis in BCR-ABL-positive human leukemia cells." *Clin. Cancer Res.* 8(9):2976-84.

[195] Dai, Y., Rahmani, M., Grant, S. (2003). "Proteasome inhibitors potentiate leukemic cell apoptosis induced by the cyclin-dependent kinase inhibitor flavopiridol through a SAPK/JNK- and NF-kappaB-dependent process." *Oncogene* 22(46):7108-22.

[196] Bellan, C., De Falco, G., Lazzi, S., Micheli, P., Vicidomini, S., Schurfeld, K., Amato, T., Palumbo, A., Bagella, L., Sabattini, E., Bartolommei, S., Hummel, M., Pileri, S., Tosi, P., Leoncini, L., Giordano, A. (2004). "CDK9/CYCLIN T1 expression during normal lymphoid differentiation and malignant transformation." *J. Pathol.* 203(4): 946-52.

[197] Wittmann, B. M., Wang, N., Montano, M.M. (2003). "Identification of a novel inhibitor of breast cell growth that is down-regulated by estrogens and decreased in breast tumors." *Cancer Res.* 63(16): 5151-8.

[198] Stone, B., Schummer, M., Paley, P.J., Thompson, L., Stewart, J., Ford, M., Crawford, M., Urban, N., O'Briant, K., Nelson, B.H. (2003). "Serologic analysis of ovarian tumor antigens reveals a bias toward antigens encoded on 17q." *Int. J. Cancer* 104(1): 73-84.

In: Cell Cycle Control: New Research
Editors: N. H. Leroy, N. T. Fournier
ISBN: 978-1-60456-796-0

Chapter VI

Trisomy of Human Chromosome 21 and Cell Cycle Control

***Mohammed Rachidi*[*1,2], *Charles Tetaria*[1] *and Lopes Carmela*[1,2,3]**
[1] Molecular Genetics of Human Diseases (GMMH), B
P 130228, F-98717 Punaauia, Tahiti, French Polynesia
[2] Laboratory of Genetic dysregulation models: Trisomy 21
and Hyperhomocysteinemia, EA 3508,
University Paris 7-Denis Diderot, Paris, France
[3] University of French Polynesia,
Department of Sciences, Tahiti, French Polynesia

Abstract

Down syndrome or Trisomy 21 is the most frequent genetic cause (1/700 new birth) of mental retardation. It is characterized by the triplication of human chromosome 21 in all cells and by overdosage of genes on this chromosome in all cells. Thus, trisomy 21 could provide a genetic model for the role of aneuploidy in leukemogenesis, tumorigenesis, cancerigenesis and neurogenesis abnormalities, and to decipher the molecular and cellular mechanisms involved in these different abnormalities. Genetic overdosage determines transcriptional alteration of most of the chromosome 21 genes. Thus, trisomic cells suffer a dosage imbalance of many proteins that can potentially impair all molecular pathways, including cell cycle control. Trisomy 21 is a unique example of the parallel amplification of multiple genes, which may act cooperatively in the same leukemogenic, tumorigenic, cancerigenic or neurogenic pathway. A better knowledge of the function of gene mapping to chromosome 21 may contribute to the understanding of the several mechanisms of diseases, acting in patients with DS, such as leukaemia, tumors, cancer, and neurogenesis alterations related to cell cycle control alterations. Studies of the three oncogenes AML1, ERG and ETS2, higly related to the

* Corresponding author: Mohammed RACHIDI; Molecular Genetics of Human Diseases (GMMH); BP 130228, F-98717 Punaauia, Tahiti; French Polynesia; mrachidi2@yahoo.fr

genesis of leukemias, are of the most interest to contribute to the general understanding of the oncogenic mechanisms of chromosomal aneuploidies, the most common abnormalities in cancer. In addition, the direct involvement of the human chromosome 21 Single minded 2 gene (SIM2) in the inhibition of gene expression, growth inhibition, inhibition of tumor growth as well as induction of apoptosis that could involve a block of cell cycle, establish SIM2 gene as a molecular target for cancer therapeutics and have additional important implications for the future diagnosis and treatment of specific solid tumors as well as for further understanding the cancer risk in Down syndrome patients. In addition, some members of the phosphorylation pathways that control the cell cycle are encoded by specific chromosome 21 genes, particularly the minibrain gene Dyrk1A, suggesting some alteration in cell cycle control in Down syndrome individuals. The involvement of Dyrk1A gene in the regulation of neural proliferation on the basis of alterations in the proliferation centres of brain of Drosophila minibrain (MNB) mutants and the altered levels of cell cycle regulators in transgenic mice overexpressing Dyrk1A as well as the phosphorylation of cell cycle regulators by DYRK1A support the evolutionary conservation of DYRK1A role as regulator of proliferation in neural progenitor and its role in cell cycle regulation. All these genes localized in the human chromosome 21 play important roles in cell cycle and cell growth and, with the recent developments in genomics and proteomics, could promise significant advancements in Down syndrome knowledge and considerable progresses to decode the molecular and cellular mechanisms involved in the DS associated diseases related to cell cycle alterations in view of new development of drugs and treatments.

Introduction

Trisomy 21 or Down syndrome is the most common autosomal trisomy, with an incidence of 1 in 700 live births, in comparison with other trisomies such as Trisomy 18 (1:3500) or Trisomy 13 (1:7000). The most common origin of the Down syndrome is an extra copy of chromosome 21 in all cells (LeJeune et al., 1959) generated by a chromosomal non-disjunction during meiosis which increases significantly with advancing maternal age (Connor and Ferguson-Smith, 1997).

This common genetic disorder is characterized by a complex phenotype, the main features of which are the morphological abnormalities of the head and limbs, short stature, joint hyperlaxity, hypotonia, skeletal abnormalities, frequent occurrence of visceral malformation, particularly of the heart, immunological defects, increased risk of leukaemia, haematological and endocrinal alterations, early occurrence of an Alzheimer-like neuropathology and mental retardation (Rahmani et al., 1989; Delabar et al., 1993; Korenberg et al., 1994; Epstein, 1995).

In addition to the specific mental retardation disease seen in Down syndrome patients, trisomy 21 is also a risk factor for a number of diseases. The incidence of congenital heart disease (Ferencz et al., 1989), the atherosclerosis (Murdoch et al., 1977; Yla-Herttuala et al., 1989), and the Hirschsprung disease are both significantly elevated in individuals with trisomy 21, and metabolic diseases are also observed including a higher frequency of diabetes and obesity (Puffenberger et al, 1994; Epstein, 1995; Cohen, 1999).

The most commonly accepted hypothesis for the molecular origin of Down syndrome is that the presence of three copies of all the genes on chromosome 21 that could be responsible for the complex phenotype seen in Down syndrome patients (Antonarakis, 1998; Epstein, 2001; Antonarakis et al., 2004)

The genetic overdosage determines alteration in transcriptional level of most of genes on chromosome 21 (for recent review see: Rachidi and Lopes, 2007). Most of the chromosome 21 genes are transcribed 1.5 fold, corresponding to the genetic overdosage. The other genes are both over-expressed or down-expressed at variable levels. Moreover, the chromosome 21 encodes several transcription factors and other proteins controlling the transcription. Their dosage alterations determine transcriptional variations of several genes located on other chromosomes (Rachidi and Lopes, 2007). Thus, trisomic cells suffer of dosage imbalance of a lot of proteins that can potentially impair all molecular pathways, including cell cycle control.

The cell cycle of eukaryotic cells can be divided into four successive phases (figure 1): M phase (mitosis), in which the nucleus and the cytoplasm divide; S phase (DNA synthesis), in which the DNA in the nucleus is replicated, and two gap phases, G1 and G2. The G1 phase is a critical stage, allowing responses to extracellular cues that induce either commitment to a further round of cell division or withdrawal from the cell cycle (G0) to embark on a differentiation pathway (Zetterberg et al., 1995). The G1 phase is also involved in the control of DNA integrity before the onset of DNA replication. Between S and M phases is the G2 phase during which the cell checks the completion of DNA replication and the genomic integrity before cell division starts. The transition from one phase of the cell cycle to the next is controlled by cyclin–CDK (cyclin-dependent kinase) complexes which ensure that all phases of the cell cycle are executed in the correct order.

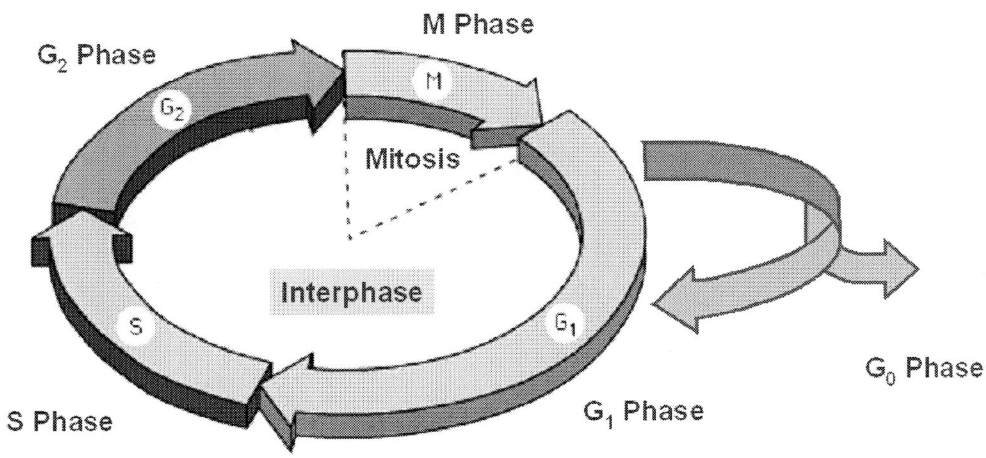

Figure 1. The cell cycle of eukaryotic cells can be divided into four successive phases: M phase (mitosis), in which the nucleus and the cytoplasm divide; S phase (DNA synthesis), in which the DNA in the nucleus is replicated, and two gap phases, G1 and G2. The G1 phase is a critical stage, allowing responses to extracellular cues that induce either commitment to a further round of cell division or withdrawal from the cell cycle (G0) to embark on a differentiation pathway. The G1 phase is also involved in the control of DNA integrity before the onset of DNA replication. Between S and M phases is the G2 phase during which the cell checks the completion of DNA replication and the genomic integrity before cell division starts.

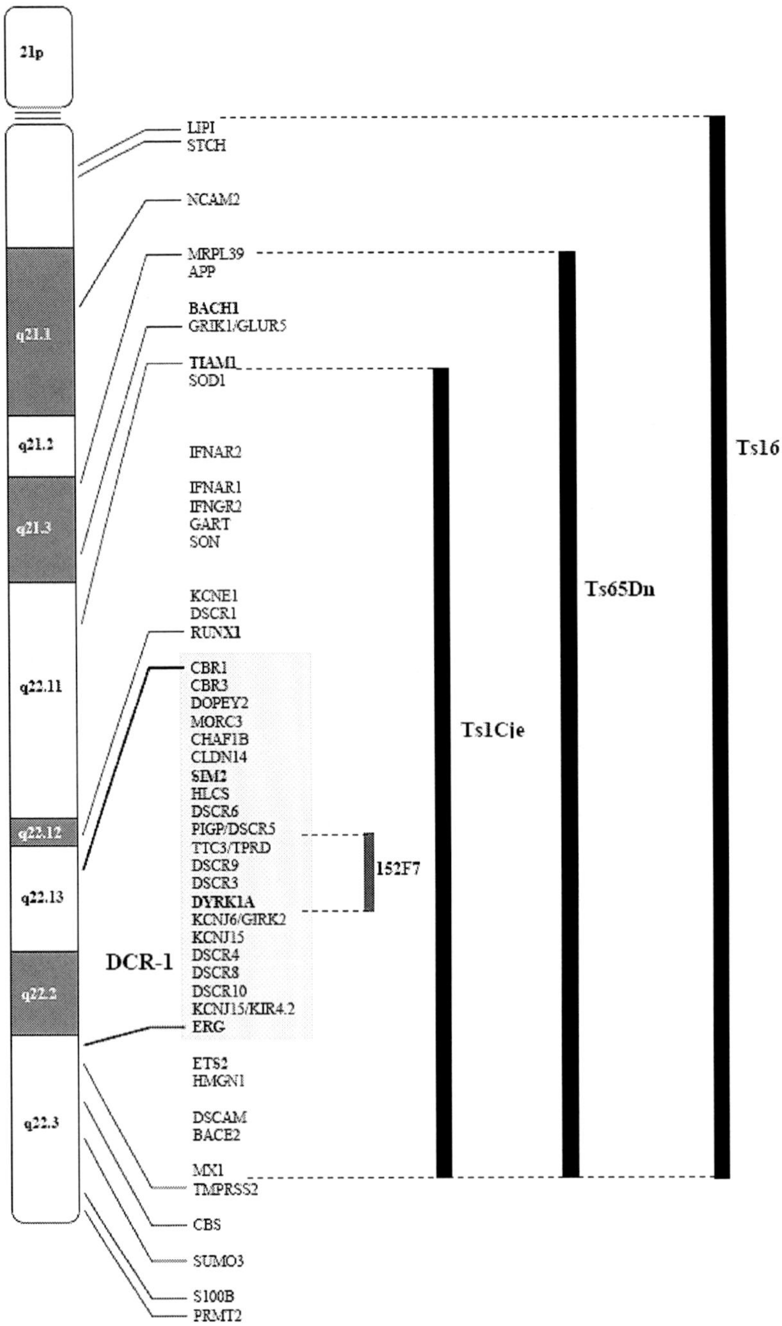

Figure 2. Chromosome 21 genes, the syntenic regions in segmental trisomic mouse models of Down syndrome and candidate genes for cell cycle control. Complete and partial trisomic mice (Black lines) and segmental transgenic mice carrying human YAC (grey line) are indicated on the right to the human chromosome 21. Their delimitations were indicated by gene localization. The Down syndrome Chromosomal Region-1 (DCR-1) was delimited between CBR and ERG genes. Genes in bold are involved in cell cycle control and may determine altered cell cycles in Down syndrome.

Interestingly, some members of the phosphorylation pathways that control the cell cycle are encoded by chromosome 21 genes, suggesting the possibility of alteration in cell cycle control in Down syndrome individuals (Groet et al., 2000; Yang et al., 2000; 2002, Hasle, 2001).

Several Down syndrome phenotypes are compatible with an alteration of the cell cycle control pathways. Here, we presented the interest of the trisomy 21, the chromosome 21 genes and their contribution to elucidate some mecanismes of diseases related to cell cycle pathways and we discussed the possible implication of the cell cycle control alteration in neurogenesis, tumorigenesis, cancerigenesis and leukemogenesis in Down syndrome patients.

Cell Cycle Control Alterations in Central Nervous System in Down Syndrome

Down syndrome is considered mainly the most frequent genetic cause of mental retardation that contributes to about 30% of all moderate-to-severe cases of mental retardation (Lejeune, 1990; Stoll et al., 1990; Pulsifer, 1996), characterized by low intelligence quotient (IQ), language defects, and cognitive impairment (Haxby, 1989; Brown et al., 1990; Hodapp and Zigler, 1990; Nadel, 2003; Chapman and Hesketh, 2000; Vicari, 2004; 2006; Brock and Jarrold, 2005; Nelson et al., 2005), short-term memory, long-term memory and learning deficits (Hodapp et al., 1999; Carlesimo et al., 1997; Hyde and Crnic, 2001; Brown et al., 2003; Clark and Wilson, 2003); developmental alterations in neurogenesis, neuronal differentiation, myelinization, and synaptogenesis (Marin-Padilla, 1976; Takashima et al., 1981; Dalton and Crapper-McLachlan, 1986; Coyle et al., 1986; Becker et al., 1986; Wisniewski and Schmidt-Sidor, 1989; Schmidt-Sidor et al., 1990; Wisniewski, 1990; Becker et al., 1991; Wisniewski and Bobinski, 1991; Takashima et al., 1994; Bahn et al., 2002), hypoplasia of the hippocampus and cortex, larger parahippocampal gyrus (Weis et al., 1991; Wang et al., 1992; Raz et al., 1995; Epstein, 1995; Schneider et al., 1997; Risser et al., 1997; Teipel et al., 2003; Teipel and Hampel, 2006), and smaller cerebellum (Jernigan et al., 1993; Raz et al., 1995; Aylward et al., 1997).

Reduction in neuron number and impairment of cortical and hippocampal structures can be detected in children and even fetuses with Down syndrome (Sylvester, 1983; Schmidt-Sidor et al., 1990; Golden and Hyman, 1994; Winter et al., 2000; Pinter et al., 2001; Contestabile et al., 2007). This suggests that the hypoplasia of the adult Down syndrome brain may be mainly related to developmental defects in the processes of neurogenesis rather than to late-occurring neurodegenerative processes.

Ts65Dn, the most widely studied animal model of Down syndrome (Davisson et al., 1990), is trisomic for major part of genes on human chromosome 21 that are highly conserved in the mouse (Antonarakis et al., 2004). Several phenotypes show direct parallels between Ts65Dn and Down syndrome, including reduced different brain areas (Baxter et al., 2000; Cooper et al., 2001) and cognitive impairments, in particular in tasks that require spatial and working memory that center on hippocampal function (Demas et al., 1996, 1998; Escorihuela et al., 1995, 1998; Granholm et al., 2002; Holtzman et al., 1996; Demas et al., 1998; Hunter et al., 2003; Hyde and Crnic, 2001; Hyde et al., 2001; Moran et al., 2002;

Reeves et al., 1995; Stasko and Costa, 2004), similar to those observed in DS patients (Pennington et al., 2003).

In neuronal progenitors, as in other somatic cells, proliferation and growth arrest are regulated by a balance of signals that control entry, progression and exit from the cell cycle (Oppenheim et al., 1989; Cunningham and Roussel, 2001). Alteration in these mechanisms can determine the hypoplasic phenotype observed in Down syndrome patients and mouse models.

In the postnatal Ts65Dn brain, a reduced mitotic index has been found in the external granule cells of the cerebellum (Roper et al., 2006), and proliferation of neural stem cells is reduced in the postnatal and adult dentate gyrus (Rueda et al., 2005; Clark et al., 2006; Contestabile et al., 2007). Thus, a neural cell proliferation phenotype is present in many cell types and central nervous system compartments. This proliferation defect persists after birth and likely continues to impact brain growth and function into adulthood in DS.

More recently, different marker for different cell cycle phases have been used to characterize the proliferating cells in hippocampus and dentate gyrus of Down syndrome fetuses and early and late post-natal Ts65Dn mice (Contestabile et al., 2007). Thus, they demonstrate altered proliferation of neuronal precursors in the hippocampus and dorsal neocortical wall in second trimester Down syndrome brain (Contestabile et al., 2007). Post-natally, the Down syndrome cerebral cortex contains less cyclin-dependent kinase (CDK) activity (Bernert et al., 1996) and both DNA synthesis and doubling times are slower in Down syndrome fibroblasts (Schneider and Epstein, 1972; Kimura et al., 2005).

Comparison of normogenic vs. trisomic subjects showed that Ts65Dn mice as well as DS fetuses had a larger percent number of cells in G2 compared with controls. The opposite was true for the number of cells in the M phase (Contestabile et al., 2007). A longer time spent by proliferating cells in G2 may result in a longer cell cycle and, hence, a reduced proliferation rate. Similarly, reduced corticogenesis with elongation of cell cycle of neocortical progenitors has been demonstrated in the Ts16 mouse model for Down syndrome (Haydar et al., 2000). These results strongly suggest that alterations in the dynamics of cell cycle causally contribute to reduce proliferation potency in Down syndrome.

The most common genetic origin of Down syndrome is the trisomy of the chromosome 21 in all cells of the Down syndrome patients. It is widely accepted that Down syndrome phenotypes are determined by the presence of three copies of genes located on the chromosome 21.

Several genes on chromosome 21 have been implicated in neurological phenotypes in Down syndrome (Rachidi and Lopes, 2007; 2008).

Among those genes, the most studied and known genes on chromosome 21 that have key roles in central nervous system development, in neurogenesis processes and neuronal differentiation, single minded, SIM2, that we described below, and the DYRK1A or minibrain gene.

DYRK1A Gene

DYRK1A is a member of the DYRK (dual-specificity tyrosine[Y]-regulated kinase) subfamily that constitutes an emerging evolutionarily conserved subfamily of proline- or arginine-directed protein kinases belonging to the CMGC family of cyclin-dependent kinases (CDKs), mitogen-activated protein kinases (MAPKs), GSK, and CDK-like kinases (CLKs) (Becker and Joost, 1999; Kannan and Neuwald, 2004; Lochhead et al., 2005). The DYRK protein kinases catalyze tyrosine-directed autophosphorylation, as well as phosphorylation of serine/threonine residues in exogenous substrates (Becker and Joost, 1999; Himpel et al., 2001).

DYRK1A is of the particular interest because it is known to function in brain processes relevant to Down syndrome (Hämmerle et al., 2003a) and it is overexpressed in the Down syndrome brain (Guimera et al., 1999; Dowjat et al., 2007). DYRK1A gene (Guimera et al., 1996) is the human orthologue of the Drosophila melanogaster minibrain (mnb) gene, which is involved in neurogenesis, learning and memory (Tejedor et al., 1995).

There is compelling evidence relating Dyrk1A to Down syndrome. Thus, the human DYRK1A gene is located within the Down syndrome critical region (Guimera et al., 1996), a genetically defined minimal region of chromosome 21 responsible for most of the relevant Down syndrome phenotypes, including mental retardation (Rahmani et al., 1989). Moreover, DYRK1A is overexpressed in Down syndrome brain (Guimera et al., 1999; Dowjat et al., 2007) and studies in mouse models have implicated DYRK1A in neurological, cognitive, and neurodegenerative phenotypes of Down syndrome (Altafaj et al., 2001; Branchi et al., 2004; Martinez de Lagran et al., 2004; Ferrer et al., 2005). Furthermore, the DYRK1A kinase has recently been proposed as a key element of a genetic regulatory circuit involved in several developmental Down syndrome phenotypes (Arron et al., 2006).

The involvement of Dyrk1A in the regulation of neural proliferation was initially suggested on the basis of alterations in the proliferation centres of the larval brain of Drosophila minibrain (Mnb) mutants (Tejedor et al., 1995). This supports the evolutionary conservation of this role as regulator of proliferation in neural progenitor (Hämmerle et al., 2008). This is also supported by the altered levels of cell cycle regulators found in transgenic mice overexpressing Dyrk1A (Branchi et al., 2004) and by the phosphorylation of cell cycle regulators by DYRK1A (de Graaf et al., 2004).

Dyrk was shown to be required for the proliferation of distinct neuronal cell types during postembryonic neurogenesis in Drosophila (Tejedor et al., 1995). Mutant flies with a reduced Dyrk1 expression have a reduced number of neurons in distinct areas of the adult brain and exhibit specific behavioural defects (Tejedor et al., 1995).

Homozygous null Dyrk1A mutant mouse embryos (Dyrk1A-/-) present delayed general growth with an overall reduction in organ growth, including neurogenesis decrease in the brain. Moreover, these DYRK1A-/- mice die during midgestation. Mice heterozygous for the mutation (DYRK1A+/- mice) show decreased neonatal viability and reduced body size from birth to adulthood and behavioral defects (Fotaki et al., 2004).

The reduction in brain size and behavioral defects observed provide evidence about the nonredundant vital role of DYRK1A and suggest a conserved mode of action that determines normal growth and brain size in both mice and flies (Fotaki et al., 2002).

Dyrk1A is specifically expressed in four sequential developmental phases: transient expression in preneurogenic progenitors, cell cycle-regulated expression in neurogenic progenitors, transient expression in recently born neurones, and persistent expression in late differentiating neurones (Hämmerle et al., 2008). These results also suggest that the subcellular localization of DYRK1A, including its translocation to the nucleus, is finely regulated (Hämmerle et al., 2008). Thus, the DYRK1A protein kinase could be a key element in the molecular machinery that couples sequential events in neuronal development. (from cell cycle progression to dendritogenesis).

Dyrk1A transcription seems to start during mitosis of neuroepithelial progenitor cells, maintains during G1, and stops before S-phase (Hämmerle et al., 2002). Dyrk1A proteins remain during S-and G2 phases, but seem to be downregulated to very low levels before the next mitosis (Hämmerle et al., 2002). The transient expression of Dyrk1A on a single cell cycle, which precedes the onset of neurogenesis, suggests that Dyrk1A may be restricted to a particular cell cycle within the lineage of neuroepithelial progenitor cells (Hämmerle et al., 2002).

Cell cycle progression is accompanied by transient DYRK1A expression, which begins in mitosis and stops before the S-phase (Hammerle et al., 2002). Thus, in 152F7 mice, DYRK1A overexpression prevents the transient down-regulation of DYRK1A expression, also maintaining its elevated levels in other phases of cell cycle, including the G2 phase (Branchi et al., 2004).

The fact that DYRK1A can phosphorylate cyclin L2 protein (de Graaf et al., 2004) indicates its role in cell cycle regulation. Moreover, DYRK1A phosphorylates a variety of substrates that play a key role in different biochemical pathways. In particular, the transcription factors of the Forkhead family (FKH-TFs), which are implicated in the control of mammalian cell cycle, and of the cyclic AMP response binding protein (CREB) family, which are important for neuronal function and survival, are known as DYRK1A substrates (Woods et al., 2001; Yang et al., 2001).

At a biochemical level, DYRK1A overexpression was associated with an increase in phosphorylation of the transcription factor FKHR and with high levels of cyclin B1, suggesting a correlation in vivo between DYRK1A overexpression and cell cycle protein alteration. In addition, altered phosphorylation of transcription factors of CREB family was observed, supporting a role of DYRK1A overexpression in the neuronal abnormalities seen in Down syndrome and suggesting that this pathology is linked to altered levels of proteins involved in the regulation of cell cycle (Branchi et al., 2004).

The high rate of FKHR phosphorylation and inactivation during the entire cell cycle averts the crucial action of FKHR in controlling the transient expression of cyclin B1 (Alvarez et al., 2001) required for progression of cell cycle through the G2-M phase (Sherwood et al., 1993; Murray, 1995). The final outcome of DYRK1A overexpression appears to be an altered cell cycle progression through the G2-M phase.

All these findings indicate an important role of DYRK1A in the pathways of cell cycle control and its overexpression in Down syndrome participate to neurogenesis alteration, and contribute to the neuronal phenotype seen in Down syndrome patients.

Human Chromosome 21 Genes and Tumorigenesis and Cancerogenesis

A genomic instability is commonly manifested by structural or numerical chromosomal aberrations (Roschke et al., 2003). Structural genomic aberrations leading to the activation of oncogenes or elimination of tumor suppression genes have been studied to understand the oncogenic role of numerical chromosomal aberrations, aneuploidy, which are the most common abnormalities in cancer.

The accumulation of multiple genetic abnormalities characterizes the evolution of the cancer cell resulting in a multi-step series of genetic changes that lead to essential alterations in cell physiology such as loss of growth controls and normal apoptotic response as well as sustained angiogenesis, invasion, and metastasis (Hanahan and Weinberg, 2000).

Thus, studies concerning mechanisms of trisomy 21 pathogenesis are likely to contribute to the general understanding of the oncogenic mechanisms of chromosomal aneuploidies, the most common abnormalities in cancer (Fults et al., 1990; Sakada et al., 1997; Yamamoto et al., 1999; von Deimling et al., 2000).

Increased transcription of the genes in the trisomic chromosome 21 may contribute to cancer in Down's patients (Satge et al., 1997; Chik et al., 1999). Oncogenes and tumor suppressor genes present in the chromosome 21 may further contribute to the cancers seen in these patients.

Interestingly, some genes located on human chromosome 21, such as SIM2 and TIAM1 genes, have been identified to have important roles in solid tumors and cancer that are related to an alteration of the cell cycle pathways.

SIM2 Gene

The human single minded gene SIM2 (Chen et al., 1995; Dahmane et al., 1995), located within the Down syndrome critical region (DSCR) of chromosome 21 associated with trisomy 21 which in triplication is associated with diverse phenotypic characteristics of Down syndrome (Rahmani et al, 1989; McCormick et al., 1989; Delabar et al., 1993; Korenberg et al., 1994).

SIM2 encodes transcription factor/helix-loop-helix protein (Crews et al., 1988), belong to a family of transcriptional repressors (Chrast et al., 1997; Moffett et al., 1997), involved in brain development and neuronal differentiation (Goshu et al., 2004; Rachidi et al., 2005).

SIM2 shows a high homology with the Drosophila single minded, sim, a master gene of the midline development in the central nervous system that functions as transcriptional regulator in cell fate determination (Crews et al., 1988; Thomas et al., 1988; Nambu et al., 1991).

The mammalian Sim2 is expressed in the embryonic brain in delimited regions of the neuroepithelium of D1 and D2 neuromeric regions and along the neural tube (Dahmane et al., 1995; Ema et al., 1996; Fan et al., 1996; Rachidi et al., 2005a). While, in later human fetal stages, it is found in discrete human brain regions, including the cortical layers, the hippocampus and the cerebellum (Rachidi et al., 2005a), that are key regions involved in learning and memory, and that are also altered in Down syndrome patients (Golden and Hyman, 1994; Raz et al., 1995; Milner et al., 1998; Ito, 2002; Miyashita, 2004). Sim2 is overexpressed about 1.5-fold in Ts1Cje mouse fetuses, a segmental trisomy 16 mouse model of Down syndrome (Vialard et al., 2000) and in trisomic tissues (Lyle et al., 2004)

Interestingly, Sim2 mutant mice are lethal in the early post-natal days and show skeletal alteration due probably to cell proliferation defects (Goshu et al., 2002). Functional studies indicated that SIM2 protein control the Shh expression in the brain (Epstein et al., 2000), involved in cell growth and differentiation in the brain. Moreover, SIM2 can inhibit cell cycle by inhibition of cyclin-E expression (Meng et al., 2006).

At date, in addition to a role in brain and neuronal development, SIM2 represent the strongest specific candidate gene involved in cancer being recently involved in the pathogenesis of solid tumors (Deyoung et al., 2002; Deyoung et al., 2003a) and overexpression of SIM2 was associated with tumors of the colon, pancreas, and prostate (Deyoung et al., 2003a; DeYoung et al., 2003b). Moreover, SIM2 gene was expressed in the colon, prostate and pancreatic carcinomas, but not in their corresponding normal tissues, and its expression was seen in a stage-specific manner in colon and prostate tumors (Deyoung et al., 2002).

Gene expression profiling showed that SIM2 is among the highly up-regulated genes in 29 prostate cancers (Halvorsen et al., 2005). SIM2 was significantly coexpressed and increased in prostate cancer. Tumor cell expression of SIM2 protein was associated with adverse clinicopathologic factors like increased preoperative serum prostate specific antigen, high histologic grade, invasive tumor growth with extra-prostatic extension, and increased tumor cell proliferation by Ki-67 expression. SIM2 protein expression was significantly associated with reduced patient survival and the increased expression of SIM2 protein is a novel marker of aggressive prostate cancer (Halvorsen, 2007).

In a colon tumor model, antisense inhibition of SIM2 expression in a colon cancer cell line resulted in inhibition of gene expression, growth inhibition and induction of apoptosis in vitro as well as inhibition of tumor growth in nude mouse tumoriginicity models (Deyoung et al., 2003a). The induction of apoptosis by the antisense SIM2 could involve a block of cell cycle, induction of differentiation or the activation of apoptotic cascades.

These results suggested that SIM2 might have both diagnostic and therapeutic utility. Moreover, the Inhibition by antisense technology of SIM2 expression in the CAPAN-1 pancreatic cancer cell line, causing a pronounced growth inhibition and induced cell death through apoptosis, provide a rationale for preclinical testing of the SIM2 antisense drug in pancreatic cancer models (DeYoung et al., 2003b).

The suggested stimulatory effect of SIM2 antisense on tumor cell apoptotic regulation (Aleman et al, 2005) and inhibition of cell cycle by SIM2 (Meng et al., 2006) indicate that inhibition of tumor growth by antisense blocking of the SIM2 in colon cancer may be due to an influence on cell cycle regulation.

TIAM1 Gene

TIAM1 gene (T-cell lymphoma invasion and metastasis 1 gene) is localized to chromosome 21 and has been identified to have important roles in the progression of epithelial cancers, especially carcinomas of the breast and colon.

Different insights into the potential role of TIAM1 in tumorigenesis and progression have been obtained by the generation of a strain of TIAM1(−/−) mice (Malliri et al., 2002). However, tumors that do occur in TIAM1(−/−) mice are more likely to progress, suggesting that in skin carcinogenesis, TIAM1 is an inhibitor of tumor development.

Several studies have examined TIAM1 role in breast carcinoma progression. The cellular adhesion molecule and receptor for hyaluronic acid (HA) binding receptor CD44 (Lesley et al., 1993; Underhill et al., 1997) binds to Tiam1 at the PHn–CC–Ex region in SP1 murine breast carcinoma cells. Furthermore, the binding of HA to CD44v3 causes an increase in Tiam1-mediated Rac activation and promotes cytoskeleton-mediated tumor cell migration (Bourguignon et al., 2000a). In addition, these finding provide more evidence for the importance of the PHn–CC–Ex domain in regulating TIAM1 localization to the plasma membrane. Similarly, in SP1 cells, Tiam1 binds to the cytoskeletal protein ankyrin, causing Tiam1-mediated Rac1 activation as well as an increase in breast tumor cell migration and invasion (Bourguignon et al., 2000b). TIAM1 is a target of heregulin (HRG) signaling. HRG-induced cellular ruffles, loss of intracellular adhesiveness, and increased cell migration all can be mimicked by overexpression of a fully functional Tiam1 construct, and overexpression of TIAM1 increases the migratory and invasive phenotype of breast tumor cell lines. Furthermore, in human breast carcinomas, a close correlation was observed between increased Tiam1 expression and increased tumor grade (Adam et al., 2001). These results suggest that increased TIAM1 expression and/or activity may promote progression of breast carcinoma.

Two colon carcinoma cell lines selected for increased metastatic potential in nude mice (Morikawa et al., 1988a; b; Minard et al., 2005; 2006) express more TIAM1 protein expression than their parental line. These data indicate that Tiam1 may have a role in the progression and metastasis of colon carcinomas and that TIAM1 regulates cell adhesion, migration and apoptosis in colon tumor cells.

To test the hypothesis that TIAM1 is a determinant of proliferation and metastasis in colorectal cancer, RNA interference (RNAi) study examined the effect of the inhibition of TIAM1 expression on proliferation and metastasis. It has been found that the silencing of TIAM1 resulted in the effective inhibition of in vitro cell growth and of the invasive ability of colorectal cancer cells. These results suggest that TIAM1 truly plays a causal role in the metastasis of colorectal cancer and that RNAi-mediated silencing of TIAM1 may provide an opportunity to develop a new treatment strategy for colorectal cancer (Liu et al., 2006).

Trisomy of the Human Chromosome 21 and Leukemogenesis

Down syndrome children have an approximately 20-fold increased risk of developing acute lymphoblastic leukemia and acute myeloid leukemia compared to normal children (Fong and Broder, 1987; Yang et al., 2000; Lange, 2000; Taub, 2001).

Acute myeloid leukemia in Down syndrome children displays unique characteristics including a predominance of the acute megakaryocytic leukemia phenotype (Ravindranath et al., 1992; 1996; Zipursky et al., 1994; 2001; Lange et al., 1998; Kojima et al., 2000; Gamis et al., 2003; Zeller et al., 2005; Creutzig et al., 2005) which is estimated to occur at a 600 fold greater frequency in DS children compared to non-DS children. (Zipursky et al., 1994; Hasle, 2001).

Moreover, three copies of chromosome 21 are common in childhood acute lymphoblastic leukaemia (Forestier et al, 2003; Harrison et al, 2001; Ross et al, 2003). This suggests that trisomy 21 plays an important role in leukemogenesis.

Cooperation of GATA1 with Chromosome 21 Genes

Association between DS and childhood megakaryoblastic disorders has led to an intensive search for leukemogenic genes on chromosome 21 that may cause the differentiation arrest and initiate the leukemia.

It has been observed that the transient myeloproliferative disorder and megakaryoblastic leukemia of Down syndrome are associated with mutations in the GATA1 gene, located on the X chromosome, in combination with trisomy 21.

Somatic mutations in exon 2 of GATA1, which encodes a zinc-finger transcription factor that is essential for normal erythroid and megakaryocytic differentiation, have been detected exclusively in almost all cases in Down syndrome–related acute megakaryocytic leukemia and transient myeloproliferative disorder, a "preleukemia" that may be present in as many as 10% of newborn infants with Down syndrome, but not in non- Down syndrome–related acute myeloid leukemia or non- acute megakaryocytic leukemia Down syndrome–related leukaemia cases (Wechsler et al., 2002; Mundschau et al., 2003; Hitzler et al., 2003; Groet et al., 2003; Rainis et al., 2003; Xu et al., 2003; Ahmed et al., 2004; Crispino et al 2005). The effect of the mutations is to introduce early stop codons that result in the synthesis of a shorter GATA1 (GATA1s, 40 kDa) protein translated from a downstream initiation site. GATA1 mutations in DS are believed to cause accumulation of poorly differentiated megakaryocytic precursors (Wechsler et al., 2002).

The mutations occur before birth and they were also found in fetal liver of aborted DS fetuses (Taub et al, 2004; Ahmed et al, 2004). It appears that mutagenesis of the GATA1 gene is a very early event in the development of Down syndrome–related acute megakaryocytic leukemia, during the process of multistep leukemogenesis.

The acquired mutations in GATA1 are probably responsible for the differentiation arrest and the initiation of clonal proliferation of immature megakaryoblasts.

A recent study Knock-in of the mutated GATA1s into the GATA1 locus surprisingly resulted in normal adult megakaryopoieis (Li et al., 2005). However, examination of the fetal liver revealed abundant proliferation of megakaryocytic progenitors. This study proposed the existence of a fetal hematopoietic progenitor that is sensitive to a dominant pro-proliferative effect of GATA1s. The presence of trisomy 21 enhances the survival and proliferation of these fetal cells resulting in a congenital pre-leukemia syndrome. Gene expression profiling of Down syndrome megakaryocytic leukemias (Bourquin et al., 2006) provides further support for a dominant pro-proliferative role for the GATA1s mutation. Recently a family with a germ line GATA1s mutation has been reported (Hollanda et al., 2006), in which the individuals do not develop leukemia. Thus, these mutations are necessary but insufficient for the development of the full blown acute megakaryocytic leukemia that affects some of these patients during early childhood and may not be the first event because the GATA1 gene is located on the X chromosome.

Some genes on chromosome 21 have been demonstrated to cooperate with the GATA1 mutant during leukemogenesis and seem to be involved in the mechanisms of a lineage-specific malignant conversion. These genes on chromosome 21 promotes proliferation and provides a survival advantage to cells that acquire mutations in GATA1 indicating the key role of trisomy 21 as a developmental genetic model.

The strongest candidate genes located on human chromosome 21 that are involved in leukemogenesis and regulating megakaryopoiesis, and that interact with the GATA1 mutation include the AML1, ERG, ETS2 and BACH1 genes.

AML1 or RUNX1 Gene

The oncogene AML1, also known as RUNX1 or CBFA2 (Levanon et al., 2001; Speck et al., 2002) a DNA-binding subunit of the transcription complex termed core-binding factor (Michaud et al., 2003; Peterson and Zhang, 2004), is an established regulator of hematopoiesis and megakaryopoiesis (Okuda et al., 1996; Song et al., 1999; Michaud et al., 2003).

AML1 haploinsufficiency in humans impairs some aspect of megakaryopoiesis, as is also seen in mouse models (Okuda et al., 1996; Song et al., 1999). AML1 is commonly mutated and involved in various translocations in both myeloid and lymphoid leukemias and most of these abnormalities cause loss of function.

In humans, hereditary loss-of-function mutations of AML1 cause the autosomal dominant familial platelet disorder with a predisposition to develop acute myeloid leukemia (Song et al., 1999). In addition, inherited mutations in AML1 causing haplo-insufficiency with a low level of expression in hematopoietic stem cells lead to a syndrome of familial thrombocytopenia and increased susceptibility to leukaemia (Escher et al., 2007).

Most recently, it was reported that AML1 transcripts, are downregulated in Down syndrome megakaryoblasts compared to non-Down syndrome megakaryoblasts, suggesting that AML1 is linked to the megakaryocytic lineage (Bourquin et al., 2006). In addition to its gene dosage effects, AML1 coexpresses and cooperates with GATA1 in megakaryocytic differentiation (Elagib et al., 2003).

The rare human syndromes determined by AML1 mutations and other functional studies (Elagib et al., 2003; Ichikawa et al., 2004) suggest that AML1 seem to play a major role in megakaryopoiesis and megakaryocytic differentiation in a dose-dependent manner.

ETS2 and ERG Genes

The Ets protein family consists of a large number of evolutionarily conserved transcription factors that control specific genes involved in diverse processes, including cell proliferation, apoptosis, differentiation, lymphoid cell development, angiogenesis, and invasiveness (Sementchenko and Watson, 2000).

ETS2 and ERG, both ETS transcription factor family members, are overexpressed in adult myeloid leukaemia cases with complex acquired karyotypes involving chromosome 21 (Baldus, 2004). This suggests that these two genes may also play important roles in leukemogenesis in individuals with Down syndrome, since both genes are located on chromosome 21q22 within the minimal Down syndrome region (Rahmani et al., 1989).

In hematopoietic cells, the proto-oncogene ETS2 is abundantly expressed in monocytes and macrophages of the myeloid lineage but not in granulocytes (Boulukos et al., 1990) and has been implicated in regulating megakaryocytic gene expression (Lemarchandel et al., 1993), macrophage differentiation (Aperlo et al., 1996), and T-cell development (Zaldumbide et al., 2002). Constitutive ETS2 expression transforms cultured cells and makes them tumorigenic (Seth et al., 1989). In prostate cancer, ETS2 is overexpressed (Liu et al., 1997) and its function is required to maintain the transformed state of human prostate cancer cells (Sementchenko et al., 1998).

Transgenic mice overexpressing ETS2 developed a smaller thymus and lymphocyte abnormalities, similar to features observed in Down syndrome. In all circumstances of ETS2 overexpression, the increased apoptosis correlated with increased p53 and alterations in downstream factors in the p53 pathway. In the human HeLa cancer cell line, transfection with functional p53 enables ETS2 overexpression to induce apoptosis. Furthermore, crossing the ETS2 transgenic mice with p53(-/-) mice genetically rescued the thymic apoptosis phenotype (Wolvetang et al., 2003). Therefore, we conclude that overexpression of human chromosome 21-encoded ETS2 induces apoptosis that is dependent on p53 (Wolvetang et al., 2003).

The proto-oncogene ERG, an ets transcription factor on chromosome 21q, involved in megakaryopoiesis and megakaryocytic leukemias is induced upon megakaryocytic differentiation of erythroleukemia cells (Rainis et al., 2005; Marcucci et al., 2005).

Overexpression of ERG in an erythroleukemia cell line causes a phenotypic shift from the erythroid into the megakaryocytic lineage (Rainis et al., 2005). This support that overexpressed ERG from trisomy 21 promotes fetal megakaryopoiesis for a differentiation arresting mutations. This promegakaryocytic pressure may, in turn, enhance the selection and proliferation of hematopoietic progenitor cells with GATA1 mutations, leading to an arrest in differentiation (Rainis et al., 2005).

BACH1 Gene

The human BACH1 gene (BTB and CNC homolog 1) transcription activator or repressor is localized to chromosome 21q22.1 (Blouin et al., 1998; Ohira et al., 1998) within the Down syndrome–associated gene region (Korenberg et al., 1994) and have been identified to be involved in leukemogenesis in trisomic patients.

Transgenic expression of BACH1 in megakaryocytes in vivo results in thrombocytopenia. This is most likely due to maturation arrest of megakaryocytic cells and resulting dysfunctions in platelet formation (Toki et al., 2005). Interestingly, the phenotype of BACH1 transgenic mice resembles that of the p45 (nuclear factor erythroid 2, NF-E2) null mice, with similarities in the megakaryocytic phenotype (Shivdasani et al., 1995; Lecine et al., 1998), suggesting the presence of common or overlapping mechanisms underlying the impaired megakaryopoiesis observed in these mutant mice.

The BACH1 transgenic mice show that terminal differentiation of the megakaryocytes, as well as platelet production, were markedly impaired by the overexpression of BACH1, indicating that the chromosome 21 gene BACH1 control the terminal megakaryocytic differentiation (Toki et al., 2005).

These results support that BACH1 is one of the candidates for genes that cooperate with the mutated GATA1 at a very early stage of leukemogenesis in Down syndrome. Moreover, it has been demonstrated that the mutation in GATA1, in addition to downregulate the expression of ALM1, also increases the expression of BACH1 (Bourquin et al., 2006) leading to a further block of megakaryocytic differentiation.

Conclusions

Trisomy 21 is characterized by the triplication of human chromosome 21 in all cells and by overdosage of genes on this small chromosome in all cells. Thus, trisomy 21 could provide a general model for the role of aneuploidy in leukemogenesis, tumorigenesis, cancerigenesis and neurogenesis abnormalities, and may be considered as an interesting developmental genetic model to study and decipher the molecular and cellular mechanisms involved in these different abnormalities.

Trisomy 21 is a unique example of collaborating pro-proliferation and differentiation arresting mutations in leukaemia, tumors and cancer. Also, it is a unique example of the parallel amplification of multiple genes, which may act cooperatively in the same leukemogenic, tumorigenic, cancerigenic or neurogenic pathway. Finally, it is a unique model for deciphering the targets genes and regulatory factors working in the cell cycle events in body and brain.

A better knowledge of the function of genes mapping to chromosome 21 may contribute to the understanding of the several mechanisms of diseases, acting in patients with DS, such as leukaemia, tumors, cancer, and neurogenesis alterations related to cell cycle control alterations.

Additional and complementary studies of the three oncogenes AML1, ERG and ETS2, higly related to the genesis of leukemias, could be of the most interest to contribute to the general understanding of the oncogenic mechanisms of chromosomal aneuploidies, the most common abnormalities in cancer.

In addition, the direct involvement of the human chromosome 21 Single minded 2 gene (SIM2) in the inhibition of gene expression, growth inhibition, inhibition of tumor growth as well as induction of apoptosis that could involve a block of cell cycle, establish SIM2 gene as a molecular target for cancer therapeutics and have additional important implications for the future diagnosis and treatment of specific solid tumors as well as for further understanding the cancer risk in Down syndrome patients.

In addition, some members of the phosphorylation pathways that control the cell cycle are encoded by specific chromosome 21 genes, particularly the minibrain gene Dyrk1A, suggesting some alteration in cell cycle control in Down syndrome individuals. The involvement of Dyrk1A gene in the regulation of neural proliferation on the basis of alterations in the proliferation centres of brain of Drosophila minibrain (MNB) mutants and the altered levels of cell cycle regulators in transgenic mice overexpressing Dyrk1A (MNB) as well as the phosphorylation of cell cycle regulators by DYRK1A support the evolutionary conservation of DYRK1A role as regulator of proliferation in neural progenitor and its role in cell cycle regulation.

All these genes localized in the human chromosome 21, AML1, ERG, ETS2, BACH-1, TIAM-1, SIM2 and DYRK1A play important roles in cell cycle and cell growth and could with the recent developments, especially in developmental genetics and neurosciences, in genomics and proteomics promise significant advancements in Down syndrome knowledge and considerable progresses to decode the molecular and cellular mechanisms involved in the DS associated diseases related to cell cycle alterations in view of new development of drugs and treatments.

Acknowledgments

We are grateful to our colleagues of Department of Molecular Biology-Jacques Monod at the Pasteur Insitute (Paris) for their advices and support. We also thank J.M. Delabar (University Paris 7) and L. Peltzer (University of French Polynesia) for their continuous support.

References

Adam, L., Vadlamudi, R. K., McCrea, P. & Kumar, R. (2001) Tiam1 overexpression potentiates heregulin-induced lymphoid enhancer factor-1/beta-catenin nuclear signaling in breast cancer cells by modulating the intercellular stability. *J. Biol. Chem., 276,* 28443–28450.

Ahmed, M., Sternberg, A., Hall, G., Thomas, A., Smith, O., O'Marcaigh, A., Wynn, R., Stevens, R., Addison, M., King, D., Stewart, B., Gibson, B., Roberts, I. & Vyas, P.

(2004) Natural history of GATA1 mutations in Down Syndrome. *Blood, 103,* 2480–2489.

Aleman, M. J., DeYoung, M. P., Tress, M., Keating, P., Perry, G. W. & Narayanan, R. (2005) Inhibition of Single Minded 2 gene expression mediates tumor-selective apoptosis and differentiation in human colon cancer cells. *Proc. Natl. Acad. Sci., 102,* 12765-12770.

Altafaj, X., Dierssen, M., Baamonde, C., Marti, E., Visa, J., Guimera, J., Oset, M., Gonzalez, J. R., Florez, J., Fillat, C. & Estivill, X. (2001) Neurodevelopmental delay, motor abnormalities and cognitive deficits in transgenic mice overexpressing Dyrk1A (minibrain), a murine model of Down's syndrome. *Hum. Mol. Genet., 10,* 1915–1923.

Alvarez, B., Martinez, A. C., Burgering, B. M. & Carrera, A. C. (2001) Forkhead transcription factors contribute to execution of the mitotic programme in mammals. *Nature, 413,* 744-747.

Antonarakis, S. E. (1998) 10 years of genomics, chromosome 21, and Down syndrome, *Genomics 51,* 1–16.

Antonarakis, S. E., Lyle, R., Dermitzakis, E. T., Reymond, A. & Deutsch, S. (2004) *Nature Rev. Genet., 5,* 725-738.

Aperlo, C., Pognonec, P., Stanley, E. R. & Boulukos, K. E. (1996) Constitutive c-ets2 expression in M1D+ myeloblast leukemic cells induces their differentiation to macrophages. *Mol. Cell. Biol., 16,* 6851–6858.

Arron, J. R., Winslow, M. M., Polleri, A., Chang, C. P., Wu, H., Gao, X., Neilson, J. R., Chen, L., Heit, J. J., Kim, S. K., Yamasaki, N., Miyakawa, T., Francke, U., Graef, I. A. & Crabtree, G. R. (2006) NFAT dysregulation by increased dosage of DSCR1 and DYRK1A on chromosome 21. *Nature, 441,* 595-600.

Aylward, E. H., Habbak, R., Warren, A. C., Pulsifer, M. B., Barta, P. E., Jerram, M. & Pearlson, G. D. (1997) Cerebellar volume in adults with Down syndrome. *Arch Neurol., 54,* 209-212.

Bahn, S., Mimmack, M., Ryan, M., Caldwell, M. A., Jauniaux, E., Starkey, M., Svendsen, C. N. & Emson, P. (2002) Neuronal target genes of the neuron-restrictive silencer factor in neurospheres derived from fetuses with Down's syndrome: a gene expression study. *Lancet, 359,* 310–315.

Baldus, C. D., Liyanarachchi, S., Mrozek, K., Auer, H., Tanner, S. M., Guimond, M., Ruppert, A. S., Mohamed, N., Davuluri, R. V., Caligiuri, M. A., Bloomfield, C. D. & de la Chapelle, A. (2004) Acute myeloid leukemia with complex karyotypes and abnormal chromosome 21: amplification discloses overexpression of APP, ETS2, and ERG genes. *Proc. Natl. Acad. Sci., 101,* 3915–3920.

Baxter, L. L., Moran, T. H., Richtsmeier, J. T., Troncoso, J. & Reeves, R. H. (2000) Discovery and genetic localization of Down syndrome cerebellar phenotypes using the Ts65Dn mouse. *Hum. Mol. Genet., 9,* 195–202.

Becker, L. E., Armstrong, D. L. & Chan, F. (1986) Dendritic atrophy in children with Down's syndrome. *Ann. Neurol., 20,* 520–532.

Becker, L., Mito, T., Takashima, S. & Onodera, K. (1991) Growth and development of the brain in Down syndrome. In C. H. Epstein (Ed.), *The Morphogenesis of Down syndrome* (pp. 133–152). New York, Wiley-Liss.

Becker, W. & Joost, H. G. (1999) Structural and functional characteristics of Dyrk, a novel subfamily of protein kinases with dual specificity. *Prog. Nucleic Acid Res. Mol. Biol., 62,* 1-17.

Bernert, G., Nemethova, M., Herrera-Marschitz, M., Cairns, N. & Lubec, G. (1996) Decreased cyclin dependent kinase in brain of patients with Down syndrome. *Neurosci. Lett., 216,* 68-70.

Blouin, J. L., Duriaux Sail, G., Guipponi, M., Rossier, C., Pappasavas, M. P. & Antonarakis, S. E. (1998) Isolation of the human BACH1 transcription regulator gene, which maps to chromosome 21q22.1. *Hum. Genet., 102,* 282-288.

Boulukos, K. E., Pognonec, P., Sariban, E., Bailly, M., Lagrou, C. & Ghysdael, J. (1990) Rapid and transient expression of Ets2 in mature macrophages following stimulation with cMGF, LPS, and PKC activators. *Genes Dev., 4,* 401–409.

Bourguignon, L. Y., Zhu, H., Shao, L. & Chen, Y. W. (2000a) CD44 interaction with tiam1 promotes Rac1 signaling and hyaluronic acid-mediated breast tumor cell migration. *J. Biol. Chem., 275,* 1829–1838.

Bourguignon, L. Y., Zhu, H., Shao, L. & Chen, Y. W. (2000b) Ankyrin–Tiam1 interaction promotes Rac1 signaling and metastatic breast tumor cell invasion and migration. *J. Cell. Biol., 150,* 177–191.

Bourquin, J. P., Subramanian, A., Langebrake, C., Reinhardt, D., Bernard, O., Ballerini, P., Baruchel, A., Cavé, H., Dastugue, N., Hasle, H., Kaspers, G. L., Lessard, M., Michaux, L., Vyas, P., van Wering, E., Zwaan, C. M., Golub, T. R. & Orkin, S. H. (2006) Identification of distinct molecular phenotypes in acute megakaryoblastic leukemia by gene expression profiling. *Proc. Natl. Acad. Sci., 103,* 3339–3344.

Branchi, I., Bichler, Z., Minghetti, L., Delabar, J. M., Malchiodi-Albedi, F., Gonzalez, M. C., Chettouh, Z., Nicolini, A., Chabert, C., Smith, D. J., Rubin, E. M., Migliore-Samour, D. & Alleva, E. (2004) Transgenic mouse in vivo library of human Down syndorme Critical Region 1: association between *DYRK1A* overexpession, brain development abnormalities, and cell cycle protein alteration. *J. Neuropathol. Exp. Neurol., 63,* 429-440.

Brock, J. & Jarrold, C. (2005) Serial order reconstruction in Down syndrome: evidence for a selective deficit in verbal short-term memory. *J. Child Psychol. Psychiatry, 46,* 304-316.

Brown, F. R., Greer, M. K., Aylward, E. H. & Hunt, H. H. (1990) Intellectual and adaptive functioning in individuals with Down syndrome in relation to age and environmental placement. *Pediatrics, 85,* 450–452.

Brown, J. H., Johnson, M. H., Patterson, S. J., Gilmore, R., Longhi, E. & Karmiloff-Smith, A. (2003) Spatial representation and attention in toddlers with Williams syndrome and Down syndrome. *Neuropsychologia, 41,* 1037-1046.

Carlesimo, G. A., Marotta, L. & Vicari, S. (1997) Long-term memory in mental retardation: evidence for a specific impairment in subjects with Down's syndrome. *Neuropsychologia, 35,* 71-79.

Chapman, R. S. & Hesketh, L. J. (2000) Behavioral phenotype of individuals with Down syndrome. *Ment. Retard. Dev. Disabil. Res. Rev., 6,* 84-95.

Chen, H., Chrast, R., Rossier, V., Gos, A., Antonarakis, S. E., Kudoh, J., Yamaki, A., Shindoh, N., Maeda, H., Minoshima, S., et al. (1995) Single-minded and Down syndrome? *Nat. Genet., 10,* 9-10.

Chik, K., Li, C., Shing, M. M., Leung, T. & Yuen, P. M. (1999) J. Pediatr. Hematol. Oncol. 21, 149–151.

Chrast, R., Scott, H. S., Chen, H., Kudoh, J., Rossier, C., Minoshima, S., Wang, Y., Shimizu, N. & Antonarakis, S. E. (1997) Cloning of two human homologs of the Drosophila single-minded gene SIM1on chromosome 6q and SIM2 on 21qwithin the Down syndrome chromosomal region. *Genome Res., 7,* 615-624.

Clark, D. & Wilson, G. N. (2003) Behavioral assessment of children with Down syndrome using the Reiss psychopathology scale. *Am. J. Med. Genet., 118,* 210-216.

Clark, S., Schwalbe, J., Stasko, M. R., Yarowsky, P. J. & Costa, A. C. (2006) Fluoxetine rescues deficient neurogenesis in hippocampus of the Ts65Dn mouse model for Down syndrome. *Exp. Neurol., 200,* 256–261.

Cohen, W. I. (1999) Health care guidelines for individuals with Down syndrome: 1999 revision. *Down Syndrome Quarterly, 4,* 1-16.

Connor, J. M., & Ferguson-Smith, M. A. (1997) *Essential medical genetics* (5th edn). Oxford, Blackwell Science.

Contestabile, A., Fila, T., Ceccarelli, C., Bonasoni, P., Bonapace, L., Santini, D., Bartesaghi, R. & Ciani, E. (2007) Cell cycle alteration and decreased cell proliferation in the hippocampal dentate gyrus and in the neocortical germinal matrix of fetuses with Down syndrome and in Ts65Dn mice. *Hippocampus, 17,* 665-678.

Cooper, J. D., Salehi, A., Delcroix, J. D., Howe, C. L., Belichenko, P. V., Chua-Couzens, J., Kilbridge, J. F., Carlson, E. J., Epstein, C. J., & Mobley, W. C. (2001) Failed retrograde transport of NGF in a mouse model of Down's syndrome: reversal of cholinergic neurodegenerative phenotypes following NGF infusion. *Proc. Natl. Acad. Sci., 98,* 10439-10444.

Coyle, J. T., Oster-Granite, M. L. & Gearhart, J. D. (1986) The neurobiologic consequences of Down syndrome. *Brain Res. Bull., 16,* 773–787.

Creutzig, U., Reinhardt, D., Diekamp, S., Dworzak, M., Stary, J. & Zimmermann, M. AML patients with Down syndrome have a high cure rate with AML-BFM therapy with reduced dose intensity. *Leukemia, 19,* 1355–1360.

Crews, S. T., Thomas, J. B. & Goodman, C. S. (1988) The Drosophila singleminded gene encodes a nuclear protein with sequence similarity to the per gene product, *Cell, 52,* 143–151.

Crispino, J. D. (2005) GATA1 in normal and malignant hematopoiesis. *Semin. Cell Dev. Biol., 16:* 137-147.

Cunningham, J. J. & Roussel, M. F. (2001) Cyclin-dependent kinase inhibitors in the development of the central nervous system. *Cell Growth Differ., 12,* 387–396.

Dahmane, N., Charron, G., Lopes, C., Yaspo, M. L., Maunoury, C., Decorte, L., Sinet, P. M., Bloch, B. & Delabar, J. M. (1995) Down syndrome critical region contains a gene homologous to Drosophila sim expressed during rat and human central nervous system development, *Proc. Natl. Acad. Sci. USA, 92,* 9191–9195.

Dalton, A. J. & Crapper-McLachlan, D. R. (1986) Clinical expression of Alzheimer's disease in Down syndrome. *Psychiatr. Clin. North. Am.*, *9,* 959–70.

Davisson, M. T., Schmidt, C. & Akeson, E. C. Segmental trisomy of murine chromosome 16: a new model system for studying Down syndrome. *Prog. Clin. Biol. Res., 360,* 263-280.

von Deimling, A., Fimmers, R., Schmidt, M. C., Bender, B., Fassbender, F., Nagel, J., Jahnke, R., Kaskel, P., Duerr, E. M., Koopmann, J., Maintz, D., Steinbeck, S., Wick, W., Platten, M., Müller, D. J., Przkora, R., Waha, A., Blümcke, B., Wellenreuther, R., Meyer-Puttlitz, B., Schmidt, O., Mollenhauer, J., Poustka, A., Stangl, A. P., Lenartz, D. & von Ammon, K. (2000) Comprehensive allelotype and genetic analysis of 466 human nervous system tumors. *J. Neuropathol. Exp. Neurol., 59,* 544–558.

Delabar, J. M., Theophile, D., Rahmani, Z., Chettouh, Z., Blouin, J. L., Prieur, M., Noel, B. & Sinet, P. M. (1993) Molecular mapping of twenty-four features of Down syndrome on chromosome 21. *Eur. J. Hum. Genet., 1,* 114–124.

Demas, G. E., Nelson, R. J., Krueger, B. K. & Yarowsky, P. J. (1996) Spatial memory deficits in segmental trisomic Ts65Dn mice. *Behav. Brain Res., 82,* 85-92.

Demas, G. E., Nelson, R. J., Krueger, B. K. & Yarowsky, P. J. (1998) Impaired spatial working and reference memory in segmental trisomy (Ts65Dn) mice. *Behav. Brain Res., 90,* 199–201.

DeYoung, M. P., Scheurle, D., Damania, H., Zylberberg, C. & Narayanan, R. (2002) Down's syndrome-associated single minded gene as a novel tumor marker. *Anticancer Res., 22,* 3149-3157.

DeYoung, M. P., Tress, M. & Narayanan, R. (2003a) Identification of Down's syndrome critical locus gene SIM2-s as a drug therapy target for solid tumors. *Proc. Natl. Acad. Sci., 100,* 4760-4765.

DeYoung, M. P., Tress, M. & Narayanan, R. (2003b) Down's syndrome- associated SingleMinded 2 gene as a pancreatic cancer drug therapy target. *Cancer Lett., 200,* 25-31.

Dowjat, W. K., Adayev, T., Kuchna, I., Nowicki, K., Palminiello, S., Hwang, Y. W. & Wegiel, J. (2007) Trisomy-driven overexpression of DYRK1A kinase in the brain of subjects with Down syndrome. *Neurosci. Letters, 413,* 77-81.

Elagib, K. E., Racke, F. K., Mogass, M., Khetawat, R., Delehanty, L. L. & Goldfarb, A. N. (2003) RUNX1 and GATA-1 coexpression and cooperation in megakaryocytic differentiation. *Blood, 101,* 4333–4341.

Ema, M., Suzuki, M., Morita, M., Hirose, K., Sogawa, K., Matsuda, Y., Gotoh, O., Saijoh, Y., Fujii, H., Hamada, H. & Fujii-Kuriyama, Y. (1996) cDNA cloning of a murine homologue of Drosophila single-minded, its mRNA expression in mouse development, and chromosome localization. *Biochem. Biophys. Res. Commun., 218,* 588-594.

Epstein, C. J. (1995) Down syndrome (trisomy 21). In C. R. Scriver, A. L., Beaudet, W. S., Sly, & D. Valle (Eds.), *The Metabolic and Molecular Bases of Inherited Disease* (pp.749–794) New York, McGraw-Hill.

Epstein, J. (2001) Down syndrome. In C. R. Scriver, A. L. Beaudet, W. S. Sly & D. Valle, (Eds), *The Metabolic and Molecular Bases of Inherited Disease,* 7th edn, Vol. I, (pp. 1223–1256). New York, McGraw-Hill, Inc.

Epstein, D. J., Matinu, L., Michaud, J. L., Losos, K. M., Fan, C. M. & Joyner, A. L. (2000) Members of the bHLH-PAS family regulate Shh transcription in forebrain regions of the mouse CNS. *Development, 127,* 1701-1709.

Escher, R., Wilson, P., Carmichael, C., Suppiah, R., Liu, M., Kavallaris, M., Cannon, P., Michaud, J. & Scott, H. S. (2007) *Blood Cells Mol. Dis., 39,* 107-114.

Escorihuela, R. M., Vallina, I. F., Martinez-Cue, C., Baamonde, C., Dierssen, M., Tobena, A., Florez, J. & Fernandez-Teruel, A. (1998) *Neurosci. Lett., 247,* 171–174.

Escorihuela, R. M., Fernandez-Teruel, A., Vallina, I. F., Baamonde, C., Dierssen, M., Tobena, A. & Florez, J. (1995) A behavioral assessment of Ts65Dn mice: a putative Down syndrome model. *Neurosci. Lett., 199,* 143–146.

Fan, C. M., Kuwana, E., Bulfone, A., Fletcher, C. F., Copeland, N. G., Jenkins, N. A., Crews, S., Martinez, S., Puelles, L., Rubenstein, J. L. & Tessier-Lavigne, M. (1996) Expression patterns of two murine homologs of Drosophila single-minded suggest possible roles in embryonic patterning and in the pathogenesis of Down syndrome. *Mol. Cell. Neurosci., 7,* 1-16.

Ferencz, C., Neill, C. A., Boughman, J. A., Rubin, J. D., Brenner, J. I. & Perry, L. W. (1989) Congenital cardiovascular malformations associated with chromosome abnormalities: An epidemiologic study. *J. Pediatr., 114,* 79-86.

Ferrer, I., Barrachina, M., Puig, B., Martinez de Lagran, M., Marti, E., Avila, J. & Dierssen, M. (2005) Constitutive Dyrk1A is abnormally expressed in Alzheimer disease, Down syndrome, Pick disease, and related transgenic models. *Neurobiol. Dis., 20,* 392-400.

Fong, C. T. & Brodeur, G. M. (1987) Down syndrome and leukemia: epidemiology, genetics, cytogenetics and mechanisms of leukemogenesis. *Cancer Genet. Cytogenet., 28,* 55-76.

Forestier, E., Heim, S., Blennow, E., Borgstrom, G., Holmgren, G., Heinonen, K., Johannsson, J., Kerndrup, G., Andersen, M. K., Lundin, C., Nordgren, A., Rosenquist, R., Swolin, B. & Johansson, B. (2003) Cytogenetic abnormalities in childhood acute myeloid leukaemia: a Nordic series comprising all children enrolled in the NOPHO-93-AML trial between 1993 and 2001. *Br. J. Haematol., 121,* 566–577.

Fotaki, V., Dierssen, M., Alcantara, S., Martinez, S., Marti, E., Casas, C., Visa, J., Soriano, E., Estivill, X. & Arbones, M. L. (2002) Dyrk1A haploinsufficiency affects viability and causes developmental delay and abnormal brain morphology in mice. *Mol. Cell. Biol., 22,* 6636-6647.

Fotaki, V., Martinez De Lagran, M., Estivill, X., Arbones, M. & Dierssen, M. (2004) Haploinsufficiency of Dyrk1A in mice leads to specific alterations in the development and regulation of motor activity. *Behav. Neurosci., 118,* 815–821.

Fults, D., Pedone, C. A., Thomas, G. A. & White, R. (1990) Allelotype of human malignant astrocytoma. *Cancer Res., 50,* 5784–5789.

Gamis, A. S., Woods, W. G., Alonzo, T. A., Buxton, A., Lange, B., Barnard, D. R., Gold, S., Smith, F. O., Children's Cancer Group Study 2891. (2003) Increased age at diagnosis has a significantly negative effect on outcome in children with Down syndrome and acute myeloid leukemia: a report from the Children's Cancer Group Study 2891. *J. Clin. Oncol., 21,* 3415–3422.

Golden, J. A. & Hyman, B. T. (1994) Development of the superior temporal neocortex is anomalous in trisomy 21. J. Neuropathol. *Exp. Neurol., 53,* 513–552.

Goshu, E., Jin, H., Fasnacht, R., Sepenski, M., Michaud, J. L. & Fan, C. M. (2002) Sim2 mutant have developmental defects not overlapping with those of the Sim1 mutants. *Mol. Cell Biol., 22,* 4147-4157.

Goshu, E., Jin, H., Lovejoy, J., Marion, J. F., Michaud, J. L. & Fan, C. M. (2004) Sim2 contributes to neuroendocrine hormone gene expression in the anterior hypothalamus. *Mol. Endocrinol., 18,* 1251-1262.

de Graaf, K., Hekerman, P., Spelten, O., Herrmann, A., Packman, L. C., Bussow, K., Muller-Newen, G. & Becker, W. (2004) Characterization of cyclin L2, a novel cyclin with an arginine/serine-rich domain: phosphorylation by DYRK1A and colocalization with splicing factors. *J. Biol. Chem., 279,* 4612–4624.

Granholm, A. C., Fork, K. A., Hyde, L. A., Bimonte, H. A., Hunter, C. L., Nelson, M., Alberck, D., Sanders, L. A., Mufson, E. J. & Crnic, L. S. (2002) Estrogen restores cognition and cholinergic phenotype in an animal model of Down syndrome. *Physiol. Behav., 77,* 371-385.

Groet, J., Ives, J. H., Jones, T. A., Danton, M., Flomen, R. H., Sheer, D., Hrascan, R., Pavelic, K. & Nizetic, D. (2000) Narrowing of the region of allelic loss in 21q11-21 in squamous non-small cell lung carcinoma and cloning of a novel ubiquitin-specific protease gene from the deleted segment. *Genes Chromosomes Cancer, 27,* 153-161.

Groet, J., McElwaine, S., Spinelli, M., Rinaldi, A., Burtscher, I., Mulligan, C., Mensah, A., Cavani, S., Dagna-Bricarelli, F., Basso, G., Cotter, F. E. & Nizetic, D. (2003) Acquired mutations in GATA1 in neonates with Down's syndrome with transient myeloid disorder. *Lancet, 361,* 1617-1620.

Guimera, J., Casas, C., Pucharcos, C., Solans, A., Domench, A., Planas, A. M., Ashley, J., Lovett, M., Estivill, X., & Pritchard, M. A. (1996) A human homologue of Drosophila minibrain (MNB) is expressed in the neuronal regions affected in Down syndrome and maps to the critical region. *Hum. Mol. Genet., 5,* 1305–1310.

Guimera, J., Casas, C., Estivill, X. & Pritchard, M. (1999) Human minibrain homologue (MNBH/DYRK1): characterization, alternative splicing, differential tissue expression, and overexpression in Down syndrome, Genomics, 57, 407–418.

Halvorsen, O. J., Oyan, A. M., Bo, T. H., Olsen, S., Rostad, K., Haukaas, S. A., Bakke, A. M., Marzolf, B., Dimitrov, K., Stordrange, L., Lin, B., Jonassen, I., Hood, L., Akslen, L. A. & Kalland, K. H. (2005) Gene expression profiles in prostate cancer: association with patient subgroups and tumour differentiation. *Int. J. Oncol., 26,* 329-336.

Halvorsen, O. J., Rostad, K., Oyan, A. M., Puntervoll, H., Bo, T. H., Stordrange, L., Olsen, S., Haukaas, S. A., Hood, L., Jonassen, I., Kalland, K. H. & Akslen, L. A. (2007) Increased expression of SIM2-s protein is a novel marker of aggressive prostate cancer. *Clin. Cancer Res., 13,* 892-897.

Hämmerle, B., Carnicero, A., Elizalde, C., Ceron, J., Martinez, S. & Tejedor, F.J. (2003) Expression patterns and subcellular localization of the Down syndrome candidate protein MNB/DYRK1A suggest a role in late neuronal differentiation. *Eur. J. Neurosci., 17,* 2277-2286.

Hämmerle, B., Vera-Samper, E., Speicher, S., Arencibia, R., Martinez, S. & Tejedor, F.J. (2002) Mnb/Dyrk1A is transiently expressed and asymmetrically segregated in neural progenito cells at the transition to neurogenic divisions. *Dev. Biol., 246,* 259-273.

Hämmerle, B., Elizalde, C. & Tejedor, F. J. (2008) The spatio-temporal and subcellular expression of the candidate Down syndrome gene Mnb/Dyrk1A in the developing mouse brain suggests distinct sequential roles in neuronal development. *Eur. J. Neurosci., 27,* 1061–1074.

Hanahan, D. & Weinberg, R. A. (2000) The hallmarks of cancer. *Cell, 100,* 57-70.

Harrison, C. J. (2001) The detection and significance of chromosomal abnormalities in childhood acute lymphoblastic leukaemia. *Blood Rev., 15,* 49–59.

Hasle, H. (2001) Pattern of malignant disorders in individuals with Down's syndrome. *Lancet Oncol., 2,* 429-436.

Haxby, J. V. (1989) Neuropsychological evaluation of adults with Down's syndrome: patterns of selective impairment in non-demented old adults. *J. Ment. Defic. Res., 33,* 193-210.

Haydar, T. F., Nowakowski, R. S., Yarowsky, P. J. & Krueger, B. K. (2000) Role of founder cell deficit and delayed neuronogenesis in microencephaly of the trisomy 16 mouse. *J. Neurosci., 20,* 4156-4164.

Himpel, S., Panzer, P., Eirmbter, K., Czajkowska, H., Sayed, M., Packman, L. C., Blundell, T., Kentrup, H., Grötzinger, J., Joost, H. G. & Becker, W. (2001) Identification of the autophosphorylation sites and characterization of their effects in the protein kinase DYRK1A. Biochem. J., 359, 497-505.

Hitzler, J. K., Cheung, J., Li, Y., Scherer, S. W. & Zipursky, A. (2003) GATA1 mutations in transient leukemia and acute megakaryoblastic leukemia of Down syndrome. *Blood, 101,* 4301-4304.

Hodapp, R. M., Ewans, D. E. & Gray, F. L. (1999) Intellectual development in children with Down syndrome. In J. Rondal, J. A. Perera, & L. Nadel (Eds) *Down syndrome: a review of current knowledge* (pp. 124-132). London, UK: Whurr publisher.

Hodapp, R. M. & Zigler, E. (1990) Applying the developmental perspective to individuals with Down syndrome. In D. Cicchetti, & M. Beeghly (Eds.), *Children with Down syndrome: a developmental perspective* (pp. 1-28). New York: Cambridge University Press.

Hollanda, L. M., Lima, C. S., Cunha, A. F., Albuquerque, D. M., Vassallo, J., Ozelo, M. C., Joazeiro, P. P., Saad, S. T. & Costa, F. F. (2006) An inherited mutation leading to production of only the short isoform of GATA-1 is associated with impaired erythropoiesis. *Nature Genet., 38,* 807-812.

Holtzman, D. M., Santucci, D., Kilbridge, J., Chua-Couzens, J., Fontana, D. J., Daniels, S. E., Johnson, R. M., Chen, K., Sun, Y., Carlson, E., Alleva, E., Epstein, C. J. & Mobley, W. C. (1996) Developmental abnormalities and age-related neurodegeneration in a mouse model of Down syndrome. *Proc. Natl. Acad. Sci. USA, 93,* 13333–13338.

Hunter, C. L., Isacson, O., Nelson, M., Bimonte-Nelson, H., Seo, H., Lin, L., Ford, K., Kindy, M. S. & Granholm, A. C. (2003) Regional alterations in amyloid precursor protein and nerve growth factor across age in a mouse model of Down's syndrome. *Neurosci, Res., 45,* 437-445.

Hyde, L. A. & Crnic, L. S. (2001) Age-related deficits in context discrimination learning in Ts65Dn mice that model Down syndrome and Alzheimer's disease. *Behav. Neurosci., 115,* 1239–1246.

Hyde, L. A., Frisone, D. F. & Crnic, L. S. (2001) Ts65Dn mice, a model for Down syndrome, have deficits in context discrimination learning suggesting impaired hippocampal function. Behav. *Brain Res., 118,* 53–60.

Ichikawa, M., Asai, T., Saito, T., Yamamoto, G., Seo, S., Yamazaki, I., Yamagata, T., Mitani, K., Chiba, S., Ogawa, S., Kurokawa, M. & Hirai, H. (2004) AML-1 is required for megakaryocytic maturation and lymphocytic differentiation, but not for maintenance of hematopoietic stem cells in adult hematopoiesis. *Nature Med., 10,* 299-304.

Ito, M. (2002) The molecular organization of cerebellar long-term depression. *Nature Rev. Neurosci., 3,* 896-902.

Jernigan, T. L., Bellugi, U., Sowell, E., Doherty, S. & Hesselink, J. R. (1993) Cerebral morphologic distinctions between Williams and Down syndromes. *Arch. Neurol., 50,* 186–191.

Kannan, N. & Neuwald, A. F. (2004) Evolutionary constraints associated with functional specificity of the CMGC protein kinases MAPK, CDK, GSK, SRPK, DYRK, and CK2alpha. *Protein Sci. 13,* 2059–2077.

Kimura, M., Cao, X., Skurnick, J., Cody, M., Soteropoulos, P. & Aviv, A. (2005) Proliferation dynamcs in cultured skin fibroblasts from Down syndrome subjects. *Free Radic. Biol. Med., 39,* 374-380.

Kojima, S., Sako, M., Kato, K., Hosoi, G., Sato, T., Ohara, A., Koike, K., Okimoto, Y., Nishimura, S., Akiyama, Y., Yoshikawa, T., Ishii, E., Okamura, J., Yazaki, M., Hayashi, Y., Egushi, M., Tsukimoto, I. & Ueda, K. (2000) An effective chemotherapeutic regimen for acute myeloid leukaemia and myelodysplastic syndrome in children with Down's syndrome. *Leukemia, 14,* 786–791.

Korenberg, J. R., Chen, X. N., Schipper, R., Sun, Z., Gonsky, R., Gerwehr, S., Carpenter, N., Daumer; C., Dignan, P., Disteche, C., Graham, Jr. J. M., Hugdins, L., McGillivray, B., Miyazaki, K., Ogasawara, N., Park, J. P., Pagon, R., Pueschel, S., Sack, G., Say, B., Schuffenhauer, S., Soukup, S. & Yamanaka, T. (1994) Down syndrome phenotypes: The consequences of chromosomal imbalance. *Proc. Natl. Acad. Sci. USA., 91,* 4997-5001.

Lange B. (2000) The management of neoplastic disorders of haematopoiesis in children with Down's syndrome. *Br. J. Haematol., 110:* 512-524.

Lange, B. J., Kobrinsky, N., Barnard, D. R., Arthur, D. C., Buckley, J. D., Howells, W. B., et al. (1998) Distinctive demography, biology, and outcome of acute myeloid leukemia and myelodysplastic syndrome in children with Down syndrome: Children's Cancer Group Studies 2861 and 2891. *Blood, 91:* 608–615.

Lecine, P., Villeval, J. L., Vyas, P., Swencki, B., Xu, Y. & Shivdasani, R. A. (1998) Mice lacking transcription factor NF-E2 provide in vivo validation of the proplatelet model of thrombocytopoiesis and show a platelet production defect that is intrinsic to megakaryocytes. *Blood, 92,* 1608-1616.

Lejeune, J. (1990) Pathogenesis of mental deficiency in trisomy 21. *Am. J. Med. Genet. Suppl., 7,* 20-30.

LeJeune, J., Gautier, M., & Turpin, R. (1959) Etude des chromosomes somatiques de neufs enfants Mongoliens, *C. R. Acad. Sci. Paris*, *248,* 1721-1722.

Lemarchandel, V., Ghysdael, J., Mignotte, V., Rahuel, C. & Romeo, P. H. GATA and Ets cis-acting sequences mediate megakaryocytespecific expression. *Mol. Cell. Biol., 13,* 668–676.

Lesley, J., Hyman, R. & Kincade, P. W. (1993) CD44 and its interaction with extracellular matrix. *Adv. Immunol., 54,* 271–335.

Levanon, D., Glusman, G., Bangsow, T., Ben-Asher, E., Male, D. A., Avidan, N., Bangsow, C., Hattori, M., Taylor, T. D., Taudien, S., Blechschmidt, K., Shimizu, N., Rosenthal, A., Sakaki, Y., Lancet, D. & Groner, Y. (2001) Architecture and anatomy of the genomic locus encoding the human leukemia-associated transcription factor RUNX1/AML1. *Gene, 262,* 23-33.

Li, Z., Godinho, F. J., Klusmann, J. H., Garriga-Canut, M., Yu, C. & Orkin, S. H. (2005) Developmental stage-selective effect of somatically mutated leukemogenic transcription factor GATA1. *Nature Genet., 37,* 613-619.

Liu, A. Y., Corey, E., Vessella, R. L., Lange, P. H., True, L. D., Huang, G. M., Nelson, G. M. & Hood, L. (1997) Identification of differentially expressed prostate genes: increased expression of transcription factor ETS-2 in prostate cancer. *Prostate, 30,* 145–153.

Liu, L., Zhang, Q., Zhang, Y., Wang, S. & Ding, Y. (2006) Lentivirus-mediated silencing of Tiam1 gene influences multiple functions of a human colorectal cancer cell line. *Neoplasia, 8,* 917-924.

Lochhead, P. A., Sibbet, G., Morrice, N. & Cleghon, V. (2005) Activation-loop autophosphorylation is mediated by a novel transitional intermediate form of DYRKs. *Cell, 121,* 925–936.

Lyle, R., Gehrig, C., Neergaard-Henrichsen, C., Deutsch, S. & Antonarakis, S. E. (2004) Gene expression from the aneuploid chromosome in a trisomy mouse model of down syndrome. *Genome Res., 14,* 1268-1274.

Malliri, A., van der Kammen, R. A., Clark, K., Van, D. V., Michiels, F. & Collard, J. G. (2002) Mice deficient in the Rac activator Tiam1 are resistant to Ras-induced skin tumours. *Nature, 417,* 867–871.

Marcucci, G., Baldus, C. D., Ruppert, A. S., Radmacher, M. D., Mrozek, K., Whitman, S. P., Kolitz, J. E., Edwards, C G., Vardiman, J. W., Powell, B. L., Baer, M. R., Moore, J. O., Perrotti, D., Caligiuri, M. A., Carroll, A. J., Larson, R. A., de la Chapelle, A. & Bloomfield, C. D. (2005) Overexpression of the ETS-related gene, ERG, predicts a worse outcome in acute myeloid leukaemia with normal karyotype: a Cancer and Leukemia Group B study. *J. Clin. Oncol., 23,* 9234-42.

Marin-Padilla, M. (1976) Pyramidal cell abnormalities in the motor cortex of a child with Down's syndrome. *J. Comp. Neurol., 167,* 63–75.

Martinez de Lagran, M., Altafaj, X., Gallego, X., Marti, E., Estivill, X., Sahun, I., Fillat, C. & Dierssen, M. (2004) Motor phenotypic alterations in TgDyrk1a transgenic mice implicate DYRK1A in Down syndrome motor dysfunction. *Neurobiol. Dis., 15,* 132-142.

McCormick, M. K., Schinzel, A., Petersen, M. B., Stetten, G., Driscoll, D. J., Cantu, E. S., Tranebjaerg, L., Mikkelsen, M., Watkins, P. C. & Antonarakis, S. E. (1989) Molecular genetic approach to the characterization of the "Down syndrome region" of chromosome 21. *Genomics, 5,* 325–331.

Meng, X., Shi, J., Peng, B., Zou, X. & Zhang, C. (2006) Effect of mouse Sim2 gene on the cell cycle of PC12 cells. *Cell Biol. Int., 30,* 349-353.

Michaud, J., Scott, H. S. & Escher, R. (2003) AML1 interconnected pathways of leukemogenesis. *Cancer Invest., 21,* 105–136.

Milner, B., Squire, L. R. & Kandel, E. R. (1998) Cognitive neuroscience and the study of memory, *Neuron, 20,* 445–468.

Minard, M. E., Herynk, M. H., Collard, J. G. & Gallick, G. E. (2005) The guanine nucleotide exchange factor Tiam1 increases colon carcinoma growth at metastatic sites in an orthotopic nude mouse model. *Oncogene, 24,* 2568-2573.

Minard, M. E., Ellis, L. M. & Gallick, G. E. (2006) Tiam1 regulates cell adhesion, migration and apoptosis in colon tumor cells. *Clin. Exp. Metastasis, 23,* 301-313.

Miyashita, Y. (2004) Cognitive memory: cellular and network machineries and their top-down control. *Science, 306,* 435-440.

Moffett, P., Reece, M. & Pelletier, J. (1997) The murine Sim-2 gene product inhibits transcription by active repression and functional interference. *Mol. Cell. Biol., 17,* 4933-4947.

Moran, T. H., Capone, G. T., Knipp, S., Davisson, M. T., Reeves, R. H. & Gearhart, J. D. (2002) The effects of piracetam on cognitive performance in a mouse model of Down's syndrome. *Physiol. Behav., 77,* 403-409.

Morikawa, K., Walker, S. M., Nakajima, M., Pathak, S., Jessup, J. M. & Fidler, I. J. (1988a) Influence of organ environment on the growth, selection, and metastasis of human colon carcinoma cells in nude mice. *Cancer Res., 48,* 6863–6871.

Morikawa, K., Walker, S. M., Jessup, J. M. & Fidler, I. J. (1988b) In vivo selection of highly metastatic cells from surgical specimens of different primary human colon carcinomas implanted into nude mice. *Cancer Res., 48,* 1943–1948.

Mundschau, G., Gurbuxani, S., Gamis, A. S., Greene, M. E., Arceci, R. J. & Crispino, J. D. (2003) Mutagenesis of GATA1 is an initiating event in Down syndrome leukemogenesis. *Blood, 101,* 4298-4300.

Murdoch, J. C., Rodger, J. C., Rao, S. S., Fletcher, C. D. & Dunnigan, M.G. (1977) Down's syndrome: An atheroma-free model. *Br. Med. J., 2,* 226-228.

Murray, A. (1995) Cyclin ubiquitination: the destructive end of mitosis. *Cell, 81,* 149-152.

Nadel, L. (2003) Down's syndrome: a genetic disorder in biobehavioral perspective. *Genes Brain Behav., 2,* 156-166.

Nambu, J. R., Lewis, J. O., Wharton, K. A. & Crews, S. T. (1991) The Drosophila single-minded gene encodes a helix-loop-helix protein that acts as a master regulator of CNS midline development. *Cell, 67,* 1157-1167.

Nelson, L., Johnson, J. K., Freedman, M., Lott, I., Groot, J., Chang, M., Milgram, N. W. & Head, E. (2005) Learning and memory as a function of age in Down syndrome: a study using animal-based tasks. *Prog. Neuropsychopharmacol. Biol. Psychiatry, 29,* 443-453.

Ohira, M., Seki, N., Nagase, T., Ishikawa, K., Nomura, N. & Ohara, O. (1998) Characterization of a human homolog (BACH1) of the mouse Bach1 gene encoding a BTB-basic leucine zipper transcription factor and its mapping to chromosome 21q22.1. *Genomics, 47,* 300-306.

Okuda T, van Deursen J, Hiebert SW, Grosveld G, Downing JR. AML1, the target of multiple chromosomal translocations in human leukemia, is essential for normal fetal liver hematopoiesis. Cell 1996; 84: 321–330.

Oppenheim, R. W., Cole, T. & Prevette, D. Early regional variations in motoneuron numbers arise by differential proliferation in the chick embryo spinal cord. *Dev. Biol. 133,* 468–474.

Pennington, P. F., Moon, J., Edgin, J., Stedron, J. & Nadel, L. (2003) The neuropsychology of Down syndrome: evidence for hippocampal dysfunction. *Child Dev., 74,* 75-93.

Pinter, J. D., Eliez, S., Schmitt, J. E., Capone, G. T. & Reiss, A. L. (2001) Neuroanatomy of Down's syndrome: a high-resolution MRI study. *Am. J. Psychiatry, 158,* 1659–1665.

Puffenberger, E. G., Kauffman, E. R., Bolk, S., Matise, T. C., Washington, S. S., Angrist, M., Weissenbach, J., Garver, K. L., Mascari, M. & Ladda, R. (1994) Identity-by-descent and association mapping of a recessive gene for Hirschsprung disease on human chromosome 13q22, *Hum. Mol. Genet., 3,* 1217–1225.

Pulsifer, M B. (1996) The neuropsychology of mental retardation. *J. Int. Neuropsychol. Soc., 2,* 159-176.

Rachidi, M., Lopes, C., Charron, G., Delezoide, A. L., Paly, E., Bloch, B. & Delabar, J. M. (2005) Spatial and temporal localization during embryonic and fetal human development of the transcription factor SIM2 in brain regions altered in Down syndrome. *Int. J. Dev. Neurosci., 23,* 475-484.

Rachidi, M. & Lopes, C. (2007) Molecular Mechanisms of Mental Retardation in Down Syndrome. In *Focus in Mental Retardation Research.* (pp. 77-134). Nova Science Publishers, Inc.

Rachidi, M. & Lopes, C. (2008) Mental retardation and associated meurological dysfunctions in Down syndrome: a consequence of dysregulation in critical chromosome 21 genes and associated molecular pathways. Eur. J. Paediatr. Neurol., 12, 168-182.

Rahmani, Z., Blouin, J. L., Creau-Goldberg, N., Watkins, P. C., Mattei, J. F., Poissonnier, M., Prieur, M., Chettouh, Z., Nicole, A., Aurias, A., Sinet, P. M. & Delabar, J. M. (1989) Critical role of the D21S55 region on chromosome 21 in the pathogenesis of Down syndrome. *Proc. Natl. Acad. Sci. USA, 86,* 5958–5962.

Rainis, L., Bercovich, D., Strehl, S., Teigler-Schlegel, A., Stark, B., Trka, J., Amariglio, N., Biondi, A., Muler, I., Rechavi, G., Kempski, H., Haas, O. A. & Izraeli, S. (2003) Mutations in exon 2 of GATA1 are early events in megakaryocytic malignancies associated with trisomy 21. *Blood, 102,* 981-986.

Rainis, L., Toki, T., Pimanda, J. E., Rosenthal, E., Machol, K., Strehl, S., Göttgens, B., Ito, E. & Izraeli, S. (2005) The proto-oncogene ERG in megakaryoblastic leukemias. *Cancer Res., 65,* 7596–7602.

Ravindranath, Y., Abella, E., Krischer, J. P., Wiley, J., Inoue, S., Harris, M., Chauvenet, A., Alvarado, C. S., Dubowy, R., Ritchey, A. K., Land, V., Steuber, C. P. & Weinstein, H. (1992) Acute myeloid leukemia (AML) in Down's syndrome is highly responsive to chemotherapy: experience on Pediatric Oncology Group AML study 8498. *Blood, 80,* 2210–2214.

Ravindranath, Y., Yeager, A. M., Chang, M. N., Steuber, C. P., Krischer, J., Graham-Pole, J., Carroll, A., Inoue, S., Camitta, B. & Weinstein, H. J. (1996) Acute myeloid leukemia in children: a randomized comparative study of purged autologous bone marrow transplantation versus intensive multiagent consolidation chemotherapy in first remission. Pediatric Oncology Group Study-POG 8821. *N. Engl. J. Med., 334,* 1428–1434.

Raz, N., Torres, I. J., Briggs, S. D., Spencer, W. D., Thornton, A. E., Loken, W. J., Gunning, F. M., McQuain, J. D., Driesen, N. R. & Acker, J.D. (1995) Selective neuroanatomic abnormalities in Down_s syndrome and their cognitive correlates: evidence from MRI morphometry. *Neurology, 45,* 356–366.

Reeves, R. H., Irving, N. G., Moran, T. H., Wohn, A., Kitt, C., Sisodia, S. S., Schmidt, C., Bronson, R. T. & Davisson, M. T. (1995) A mouse model for Down syndrome exhibits learning and behaviour deficits. *Nat. Genet., 11,* 177-184.

Risser, D., Lubec, G., Cairns, N. & Herrera-Marschitz, M. (1997) Excitatory amino acids and monoamines in parahippocampal gyrus and frontal cortical pole of adults with Down syndrome. *Life Sci., 60,* 1231–1237.

Roper, R. J. & Reeves, R.H. (2006) Understanding the basis for Down syndrome phenotypes. *PloS Genet., 2,* e50.

Roschke, A. V., Tonon, G., Gehlhaus, K. S., McTyre, N., Bussey, K. J., Lababidi, S., Scudiero, D. A., Weinstein, J. N. & Kirsch, I. R. (2003) Karyotypic complexity of the NCI-60 drug-screening panel. *Cancer Res., 63,* 8634–8647.

Ross, M. E., Zhou, X., Song, G., Shurtleff, S. A., Girtman, K., Williams, W. K., Liu, H. C., Mahfouz, R., Raimondi, S. C., Lenny, N., Patel, A. & Downing, J. R. (2003) Classification of pediatric acute lymphoblastic leukemia by gene expression profiling, *Blood, 102,* 2951–2959.

Rueda, N., Mostany, R., Pazos, A., Florez, J. & Martinez-Cué, C. (2005) Cell proliferation is reduced in the dentate gyrus of aged but not young Ts65Dn mice, a model of Down syndrome. *Neurosci. Lett., 380,* 197-201.

Sakata, K., Tamura, G., Nishizuka, S., Maesawa, C., Suzuki, Y., Iwaya, T., Terashima, M., Saito, K. & Satodate, R. (1997) Commonly deleted regions on the long arm of chromosome 21 in differentiated adenocarcinoma of the stomach. *Genes Chromosomes Cancer, 18,* 318–321.

Satge, D., Sasco, A. J., Cure, H., Leduc, B., Sommelet, D. & Vekemans, M. J. (1997) Cancer 80, 929–935.

Schmidt-Sidor, B., Wisniewski, K. E., Shepard, T. H. & Sersen, E. A. (1990) Brain growth in Down syndrome subjects 1–22 weeks of gestation age and birth to 60 months. *Clin. Neuropathol., 4,* 181–190.

Schneider, E. L. & Epstein, C. J. (1972) Replication rate and lifespan of cultured fibroblasts in Down's syndrome. *Proc. Soc. Exp. Biol. Med., 141,* 1092–94.

Schneider, C., Risser, D., Kirchner, L., Kitzmuller, E., Cairns, N., Prast, H., Singewald, N. & Lubec, G. (1997) Similar deficits of central histaminergic system in patients with Down syndrome and Alzheimer disease, *Neurosci. Lett., 222,* 183–186.

Sementchenko, V. I., Schweinfest, C. W., Papas, T. S. & Watson, D. K. (1998) ETS2 function is required to maintain the transformed state of human prostate cancer cells. *Oncogene, 17,* 2883–2888.

Sementchenko, V. I. & Watson, D. K. (2000) Ets target genes: past, present and future. *Oncogene, 19,* 6533-6548.

Seth, A., Watson, D. K., Blair, D. G., Papas, T. S. (1989) c-ets-2 protooncogene has mitogenic and oncogenic activity. *Proc. Natl. Acad. Sci., 86,* 7833–7837.

Sherwood, S. W., Kung, A. L., Roitelman, J., Simoni, R. D. & Schimke, R. T. (1993) In vivo inhibition of cyclin B degradation and induction of cell-cycle arrest in mammalian cells by the neutral cysteine protease inhibitor N-acetylleucylleucylnorleucinal. *Proc. Natl. Acad. Sci., 90,* 3353-3357.

Shivdasani, R. A., Rosenblatt, M. F., Zucker-Franklin, D., Jackson, C. W., Hunt, P., Saris, C. J. & Orkin, S. H. (1995) Transcription factor NF-E2 is required for platelet formation independent of the actions of thrombopoietin/MGDF in megakaryocyte development. *Cell, 81,* 695-704.

Song, W. J., Sullivan, M. G., Legare, R. D., Hutchings, S., Tan, X., Kufrin, D., Ratajczak, J., Resende, I. C., Haworth, C., Hock, R., Loh, M., Felix, C., Roy, D. C., Busque, L., Kurnit, D., Willman, C., Gewirtz, A. M., Speck, N. A., Bushweller, J. H., Li, F. P., Gardiner, K., Poncz, M., Maris, J. M. & Gilliland, D. G. (1999) Haploinsufficiency of CBFA2 causes familial thrombocytopenia with propensity to develop acute myelogenous leukaemia. *Nature Genet., 23,* 166–175.

Speck, N. A. & Gilliland, D. G. (2002) Core-binding factors in haematopoiesis and leukaemia. *Nature Rev. Cancer, 2,* 502-513.

Stasko, M. R. & Costa, A. C. (2004). Experimental parameters affecting the Morris water maze performance of a mouse model of Down syndrome. *Behav. Brain Res., 154,* 1–17.

Stoll, C., Alembik Y., Dott, B. & Roth, M. P. (1990) Epidemiology of Don syndrome in 118,265 consecutive births. *Am. J. Med. Genet., 7(Suppl.),* 79-83.

Sylvester, P. E. (1983) The hippocampus in Down's syndrome. *J. Ment. Defic. Res., 27,* 227-236.

Takashima, S., Becker, L. E., Armstrong, D. L. & Chan, F. (1981) Abnormal neuronal development in the visual cortex of the human fetus and infant with Down's syndrome. A quantitative and qualitative Golgi study. *Brain Res., 225,* 1–21.

Takashima, S., Iida, K., Mito, T. & Arima, M. (1994) Dendritic and histochemical development and aging in patients with Down's syndrome. *J. Intellect. Disabil. Res., 38(Pt3),* 265–273.

Taub, J. W. (2001) Relationship of chromosome 21 and acute leukemia in children with Down syndrome. *J. Pediatr. Hematol. Oncol., 23,* 175–178.

Taub, J. W., Mundschau, G., Ge, Y., Poulik, J. M., Qureshi, F., Jensen, T., James, S. J., Matherly, L. H., Wechsler, J. & Crispino, J. D. (2004) Prenatal origin of GATA1 mutations may be an initiating step in the development of megakaryocytic leukemia in Down syndrome. *Blood, 104,* 1588-1589.

Teipel, S. J. & Hampel, H. (2006) Neuroanaotmy of Down syndrome in vivo: a moldel of preclinical Alzheier's disease. *Behav. Genet., 36,* 405-415.

Teipel, S. J., Schapiro, M. B., Alexander, G. E., Krasuski, J. S., Horwitz, B., Hoehne, C., Moller, H. J., Rapoport, S. I. & Hampel, H. (2003) Relation of corpus callosum and hippocampal size to age in non-demented adults with Down's syndrome. *Am. J. Psychiatry, 160,* 1870-1878.

Tejedor, F., Zhu, X. M., Kaltenbach, E., Ackermann, A. Baumann, A., Canal, I., Heisenberg, M., Fischbach, K. F. & Pongs, O. (1995) Minibrain: a new protein kinase family involved in postembryonic neurogenesis in Drosophila. *Neuron, 14,* 287–301.

Thomas, J. B., Crews, S. T. & Goodman, C. S. (1988) Molecular genetics of the single-minded locus: a gene involved in the development of the Drosophila nervous system. *Cell, 52,* 133-141.

Toki, T, Katsuoka, F., Kanezaki, R., Xu, G., Kurotaki, H., Sun, J., Kamio, T., Watanabe, S., Tandai, S., Terui, K., Yagihashi, S., Komatsu, N., Igarashi, K., Yamamoto, M. & Ito, E. (2005) Transgenic expression of BACH1 transcription factor results in megakaryocytic impairment. *Blood, 105,* 3100-3108.

Underhill, C. B., Green, S. J., Comoglio, P. M. & Tarone, G. (1987) The hyaluronate receptor is identical to a glycoprotein of Mr 85,000 (gp85) as shown by a monoclonal antibody that interferes with binding activity. *J. Biol. Chem., 262,* 13142–13146.

Vialard, F., Toyama, K., Vernoux, S., Carlson, E. J., Epstein, C. J., Sinet, P. M. & Rahmani, Z. (2000) Over-expression of mSim2 gene in the zona limitans of the diencephalon of segmental trisomy 16 Ts1Cje fetuses, a mouse model for trisomy 21: a novel whole-mount RNA hybridization assay. *Dev. Brain Res., 121,* 73–78.

Vicari, S. (2004) Memory devlopment and intellecutal disabilities. *Acta Paediatr., Suppl. 93,* 60-63.

Vicari, S. (2006) Motor development and neuropsychological patterns in persons with trisomy 21. *Behav. Genet., 36,* 355-364.

Wang, P. P, Doherty, S., Hesselink, J. R. & Bellugi, U. (1992) Callosal morphology concurs with neurobehavioral and neuropathological findings in two neurodevelopmental disorders. *Arch. Neurol., 49,* 407–411.

Wechsler, J., Greene, M., McDevitt, M. A., Anastasi, J., Karp, J. E., Le Beau, M. M. & Crispino, J. D. (2002) Acquired mutations in GATA1 in the megakaryoblastic leukemia of Down syndrome. *Nature Genet., 32,* 148-152.

Weis, S., Weber, G., Neuhold, A. & Rett, A. (1991) Down syndrome: MR quantification of brain structures and comparison with normal control subjects. *Am. J. Neuroradiol., 12,* 1207–1211.

Winter, T. C., Ostrovsky, A. A., Komarniski, C. A. & Uhrich, S. B. (2000) Cerebellar and frontal lobe hypoplasia in fetuses with trisomy 21: usefulness as combined US markers. *Radiology, 214,* 533-538.

Wisniewski, K. E. (1990) Down syndrome children often have brain with maturation delay, retardation of growth, and cortical dysgenesis. *Am. J. Med. Genet. 7,* 274–281.

Wisniewski, K. E. & Bobinski, M. (1991) Hypothalamic abnormalities in Down syndrome, *Prog. Clin. Biol. Res., 373,* 153–167.

Wisniewski, K. E. & Schmidt-Sidor, B. (1989) Postnatal delay of myelin formation in brains from Down syndrome infants and children. *Clin. Neuropathol., 8*, 55–62.

Wolvetang, E. J., Wilson, T. J., Sanij, E., Busciglio, J., Hatzistavrou, T., Seth, A., Hertzog, P. J. & Kola, I. (2003) ETS2 overexpression in transgenic models and in Down syndrome predisposes to apoptosis via the p53 pathway. *Hum. Mol. Genet., 12,* 247-255.

Woods, Y. L., Rena, G., Morrice, N., Barthel, A., Becker, W., Guo, S., Unterman, T. G. & Cohen, P. (2001) The kinase DYRK1A phosphorylates the transcription factor FKHR at Ser329 in vitro, a novel in vivo phosphorylation site. *Biochem. J., 355,* 597–607.

Xu, G., Nagano, M., Kanezaki, R., Toki, T., Hayashi, Y., Taketani, T., Taki, T., Mitui, T., Koike, K., Kato, K., Imaizumi, M., Sekine, I., Ikeda, Y., Hanada, R., Sako, M., Kudo, K., Kojima, S., Ohneda, O., Yamamoto, M. & Ito, E. (2003) Frequent mutations in the GATA-1 gene in the transient myeloproliferative disorder of Down's syndrome. *Blood, 102,* 2960-2968.

Yamamoto, N., Uzawa, K., Miya, T., Watanabe, T., Yokoe, H., Shibahara, T., Noma, H. & Tanzawa, H. (1999) Frequent allelic loss/imbalance on the long arm of chromosome 21 in oral cancer: evidence for three discrete tumor suppressor gene loci. *Oncol. Rep., 6,* 1223–1227.

Yang, E. J., Ahn, Y. S. & Chung, K. C. Protein kinase Dyrk1 activates cAMP response element-binding protein during neuronal differentiation in hippocampal progenitor cells. *J. Biol. Chem., 276,* 39819–39824.

Yang, Q., Rasmussen, S. A. & Friedman, J. M. (2002) Mortality associated with Down's syndrome in the USA from 1983 to 1997: A population-bases study. *Lancet, 359,* 1019-1025.

Yla-Herttuala, S., Luoma, J., Nikkari, T. & Kivimäki, T. (1989) Down's syndrome and atherosclerosis. *Atherosclerosis, 76,* 269-272.

Zaldumbide, A., Carlotti, F., Pognonec, P. & Boulukos, K. E. (2002) The role of the Ets2 transcription factor in the proliferation, maturation, and survival of mouse thymocytes. *J. Immunol., 169,* 4873–4881.

Zeller, B., Gustafsson, G., Forestier, E., Abrahamsson, J., Clausen, N., Heldrup, J., Hovi, L., Jonmundsson, G., Lie, S. O., Glomstein, A. & Hasle, H.; Nordic Society of Paediatric Haematology and Oncology (NOPHO). (2005) Acute leukaemia in children with Down syndrome: a population-based Nordic study. *Br. J. Haematol., 128,* 797–804.

Zetterberg, A., Larsson, O. & Wiman, K. G. (1995) What is the restriction point? *Curr. Opin. Cell Biol., 7,* 835–842.

Zipursky, A., Poon, A. & Doyle, J. (2001) Leukemia in Down syndrome. A review. *Pediatr. Hematol. Oncol., 9,* 139-149.

Zipursky, A., Thorner, P., De Harven, E., Christensen, H. & Doyle, J. (1994) Myelodysplasia and acute megakaryoblastic leukemia in Down's syndrome. *Leuk. Res., 18,* 163–171.

In: Cell Cycle Control: New Research
Editors: N. H. Leroy, N. T. Fournier
ISBN: 978-1-60456-796-0

Chapter VII

Transforming Growth Factor Beta and Cell Cycle Regulation

Ota Fuchs*
Institute of Hematology and Blood Transfusion, U Nemocnice 1, 128 20 Prague 2, Czech Republic and Center of Experimental Hematology with Institute of Pathophysiology, First Medical Faculty, Charles University, Prague, Czech Republic.

Abstract

Progression through each phase of the cell cycle (G1, S, G2 and M) is precisely controlled by the activity of distinct cyclin-dependent kinases (CDKs) and their associated regulatory subunits known as cyclins. The composition and regulation of CDKs vary according to cell type and environment. The transition from the first growth G1 phase to the DNA synthesis S phase is regulated by the phosphorylation of retinoblastoma protein (pRb). The cyclin D1/CDK4/6 complex phosphorylates pRb, while $p16^{INK4a}$, a CDK inhibitor that interacts with CDK4, inhibits this process. Further CDK inhibitors $p15^{INK4b}$, $p21^{CIP1/WAF1}$, and $p27^{KIP1}$ affect pRb. The tumor suppressor activity of TGFβ depends on its ability to regulate the expression of a number of proteins that play key roles in the control of cell cycle progression from the G1 phase to the S phase. TGFβ is able to induce growth arrest in a variety of cell types including epithelial, endothelial, hematopoietic, neural, and certain types of mesenchymal cells. TGFβ induces the transcription of genes for $p15^{INK4b}$ and $p21^{CIP1/WAF1}$. The $p15^{INK4b}$ specifically blocks the activation of CDK4 and CDK6, whereas $p21^{CIP1/WAF1}$ and $p27^{KIP1}$ inhibit the activity of CDK2. Induction of $p15^{INK4b}$ and $p21^{CIP1/WAF1}$ is important for TGFβ mediated growth arrest in epithelial cells, but not in hematopoietic cells where upregulation of $p57^{KIP2}$ is essential for TGFβ-induced growth arrest. In epithelial cells in addition to regulating the activity of CDK complexes, TGFβ also inhibits the transcription of the protooncogene c-myc that plays a role in controlling cell cycle entry into S phase in higher eukaryotes. Downregulation of c-myc by TGFβ is

* Correspondence concerning this article should be addressed to: Ota Fuchs, PhD. Tel: +420 221977313; Fax: +420 221977370; E- mail: Ota.Fuchs@uhkt.cz.

frequently impaired in cancer cells. The identification and characterization of TGFβ signaling pathway increased our understanding of the dual role of TGFβ as a tumor suppressor and prooncogenic factor. In late stages of epithelial tumorigenesis tumor cells become resistant to growth inhibition by TGFβ due to inactivation of TGFβ signaling pathway or aberrant regulation of the cell cycle. Solid tumors frequently harbor mutations in members of the TGFβ signaling pathway, including in Smad4 and TβRII. On the other hand, hematologic cancers generally become TGFβ resistant through decreased expression of TGFβ signaling by oncoproteins including Evi1, Tax and fusion proteins Runx1-Evi1, Runx1-ETO and PML-RARα. Tumor cells often show increased expression or production of TGFβ. TGFβ can regulate extracellular matrix composition and degradation that play complex roles in tumor invasion and metastasis.

Keywords: TGFbeta, cell cycle, cyclin dependent kinases, cdk inhibitors, retinoblastoma protein

Introduction

The mammalian cell goes through four phases (G1, S, G2 and M) of a cell cycle to replicate itself and divide into two daughter cells. The progression through the cell cycle phases is tightly controlled. The next phase will not start until the previous one has been successfully completed. The cell proliferation rate is regulated by proteins involved in the control of cell cycle, mainly by kinase complexes consisting of a cyclin-dependent kinase (CDK) subunit and a cyclin subunit [1-4]. Transitions between different phases of the cell cycle are driven by sequential activation and inactivation of various CDKs in complees with cyclins.

Mitogenic factors (growth factors) stimulate the expression of protooncogenes, such as Ras or Myc, which result in transcriptional regulation leading to increased proliferation. The family of D-type cyclins is the first to be induced by these mitogenic signals, followed by the expression of E-type cyclins. The D-type cyclins stimulate the activity of CDK4/CDK6, whereas E-type cyclins enhance CDK2 activity. The progression from G1- to S-phase is a key regulation point during the cell cycle. Cells converge the mitogenic and antimitogenic signals at the restriction (R) point in the late G1-phase, and make the decision whether to continue the cell cycle or not [5,6]. Once the cells pass through the R point, they complete the cell cycle, independently of further growth factor signaling. The key event in the G1 phase is the phosphorylation of the pocket proteins (retinoblastoma protein (pRb), p107, p130) by CDK/cyclin D and E complexes. Each CDK complex phosphorylates presumably distinct residues of pRb and this specificity provides a mechanism for temporal control of pRb during G1 [7-10]. Phosphorylation of pRb leads to the release of E2F proteins, resulting in the transcription of genes essential for entry into S-phase of the cell cycle [11-13]. When mitogenic signals are not strong enough to enhance CDK activity and inactivate pRb, cells can not enter into S-phase. Eventually, cells will exit the cell cycle to enter the G0 quiescent, nonreplicative state.

The progression from the G1- to S-phase of the cell cycle can also be affected by inhibitory signals (different stresses, e.g. DNA damage, oxidative stress; cell contact inhibition; etc.). These inhibitory signals induce the expression of antiproliferative CDK inhibitors, which inhibit CDK activity. The INK4 ($p15^{INK4b}$, $p16^{INK4a}$, $p18^{INK4c}$ and $p19^{INK4d}$) and the CIP/KIP ($p21^{CIP1/WAF1}$, $p27^{KIP1}$ and $p57^{KIP2}$) families inhibit cyclin D and cyclin E complexes [14-17].

Transforming growth factor β (TGFβ) is the antimitogenic cytokine, which is able to induce growth arrest in a variety of cell types including epithelial, endothelial, hematopoietic, neural, and certain types of mesenchymal cells. On the other hand, TGFβ exerts growth stimulatory activity on other cells of mesenchymal origin [18]. The growth inhibitory effects of TGFβ take place at the G1 phase of the cell cycle, e.g. through negative regulation of pRb, c-myc or CDKs, such as CDK4, or stimulation and activation of CDK inhibitors, such as $p15^{INK4b}$, $p21^{CIP1/WAF1}$, or $p27^{KIP1}$ [19-22].

Transforming Growth Factor Beta Signaling

TGFβ mediates its effects on cells through a heteromeric receptor complex that consists of type I (TβRI) and type II (TβRII) components. TβRI and TβRII are serine-threonine kinases that phosphorylate downstream signaling proteins on activation [23-25]. The receptor complex is activated by TGFβ binding to the TβRII component of the receptor complex causing formation of the heteromeric TβRI/TβRII receptor complex. The activated TβRII component then phosphorylates the TβRI component in the GS box of TβRI, a glycine-serine-rich region of the receptor. TβRI then propagates the signal from the receptor through the activation of the Smad pathway (Figure 1) [26-29] or non-Smad signaling pathways [30,31].

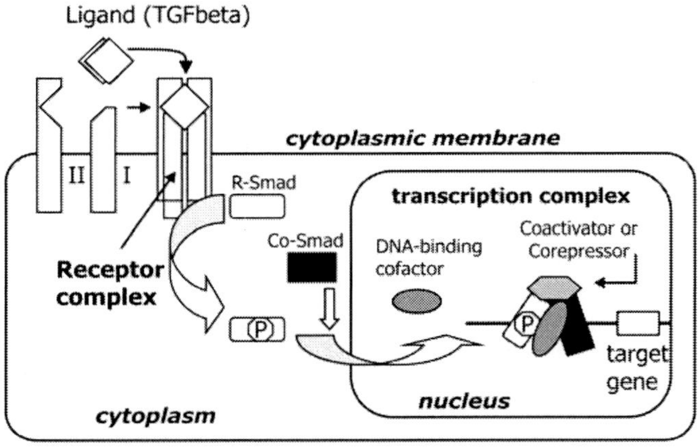

Figure 1. TGFβ- induced signaling pathway through Smad proteins. TGFβ binding to the type II receptor results in phosphorylation and activation of the type I receptor. The activated type I receptor then phosphorylates the receptor-regulated Smads (R-Smads, Smad2/3) that promote dimer or trimer formation with Co-Smad (Smad4) followed by translocation into the nucleus. Smad complex interacts with DNA-binding cofactor and corepressor or coactivator to modulate transcription of TGFβ target genes (for example genes for CDK inhibitors $p15^{INK4b}$ or $p21^{CIP1/WAF1}$).

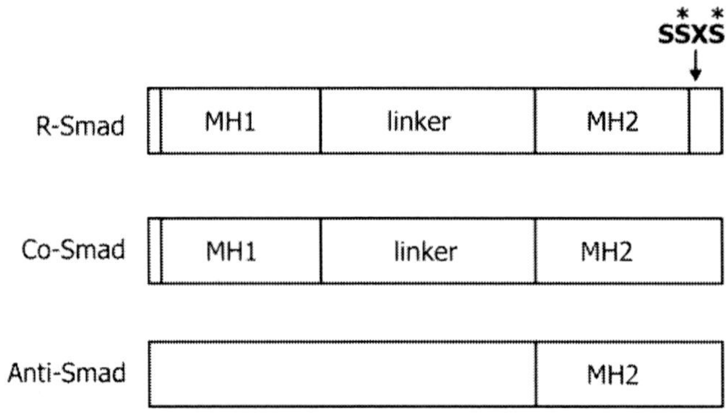

Figure 2. Classification of Smads and their functional domains. R-Smads are receptor-regulated Smads that contain the MH1 and MH2 domains and the linker region which contains phosphorylation sites for mitogen-activated kinases (MAPKs). The MH2 domain of R-Smads contains receptor type I phosphorylation sites (the SSXS motif). Co-Smads are common mediators for R-Smads to translocate into nucleus and to regulate transcription of target genes. Co-Smads have similar structure as R-Smads but lack the SSXS motif. Anti-Smads (inhibiting Smad6 and Smad7) inhibit TGFβ signaling and their structure is distinct from the structure of R-Smads and Co-Smads.

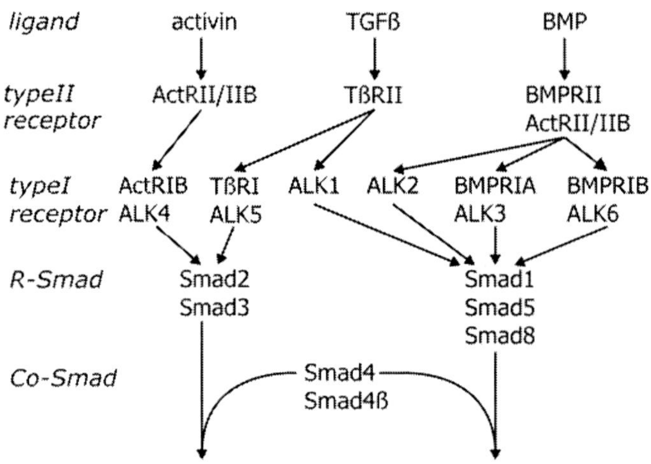

Figure 3. TGFβ, activins and BMP signal through various type I and type II receptors and Smad proteins. Type I receptors are also known as activin-receptor like kinases (ALKs). Smad1, Smad2, Smad3, Smad5 and Smad8 are receptor-regulated Smads (R-Smads). Smad4 and Smad4β (*Xenopus laevis*) are common partner Smads (Co-Smads). TGFβ can activate two distinct type I receptors/Smads signaling pathways (ALK5/Smad2, Smad3 or ALK1/Smad1,Smad5 and Smad8 pathways) in endothelial and hematopoietic cells. Abbreviations: ActR, activin receptor; BMPR, BMP receptor; TβR, TGFβ receptor.

The Smad pathway consists of members of the Smad family of proteins, including Smad1, Smad2, Smad3, Smad5 and Smad4 [26-29]. The Smad proteins compose an evolutionarily conserved signaling pathway that has been shown in *C.elegans*, *Drosophila melanogaster*, *Xenopus,* as well as humans [32]. These proteins contain two characteristic regions (Figure 2) that are homologous to the *Drosophila* orthologue, Mad, and that are located at the NH_2 and COOH termini of the protein. These regions are termed the Mad-homology domains MH1 and MH2, respectively, and are connected by a less well-conserved, proline-rich linker domain. Three major classes of Smad proteins have been identified: (a) the receptor-regulated Smads (R-Smads), which are direct targets of the TGFβ receptor family type I kinases (Figure 3) and include Smads1, 2, 3, 5 and 8; (b) the common Smads (Co-Smads; generally Smad4 or Smad4β in the *Xenopus*), which form heteromeric complexes with the R-Smads and propagate the TGFβ-mediated signal; and (c) the inhibitory Smads (I-Smads or Anti-Smads; Smad6 and Smad7), which antagonize TGFβ signaling through the Smad pathway [33,34]. Ligand (TGFβ) binding to the TGFβ receptor complex (Figure 1) results in TβRI-mediated phosphorylation of Smad2 and Smad3 on two serine residues in a conserved $SS^*X\ S^*$ motif located at the COOH terminus of the R-Smads (Figure 2). Activation of Smad2 and Smad3 induces their partnering with Smad4. This complex then translocates to the nucleus and regulates transcription of target genes that include cell cycle regulators. In a similar fashion, the bone morphogenetic protein (BMP) receptors activate Smad1, Smad5 and Smad8, which then partner with Smad4 and translocate to the nucleus to induce transcription of different genes than those induced by the Smad2/Smad3/Smad4 complexes (Figure 3).

Mechanism of Lengthening G1 Transit Time or Arrest in Late G1 Phase of the Cell Cycle by Transforming Growth Factor Beta

Inhibition of pRb Phosphorylation by TGFβ

Cells are sensitive to TGFβ during a limited period in early G1 phase of the cell cycle, until they reach a restriction point [35]. TGFβ is acting before the G1 to S phase transition and inhibits pRb kinase. TGFβ prevents or inhibits G1 cyclin-CDK activation through multiple mechanisms, leading to pRb dephosphorylation. TGFβ decreases the expression of the protooncogene c-myc, which functions as promoter of cell growth and proliferation [36-39]. TGFβ increases the expression of the CDK inhibitors $p15^{INK4b}$ and $p21^{CIP1/WAF1}$ [40-42]. Smads mediate these effects at the transcriptional level, and Smad3 plays a key role in these processes. The c-myc promoter contains a TGFβ inhibitory element (TIE), which consists of a Smad binding site and an E2F binding site. In the presence of TGFβ, a repressor complex containing Smad3, Smad4, E2F4/5, DP1 and p107, a member of the pRb family, binds to the TIE and inhibits c-myc transcription [38,39,43]. Therefore, repression of a major mitogenic transcription factor c-myc is a key event in the epithelial cytostatic program of TGFβ.

c-Myc suppression is required for the rapid induction by TGFβ of the expression of CDK inhibitors $p15^{INK4b}$ [44] and $p21^{CIP1/WAF1}$ [45]. The induction of $p15^{INK4b}$ in response to TGFβ is dependent on functional cooperativity of Sp1 with a complex of Smad2, Smad3 and Smad4 [46]. The induction of $p21^{CIP1/WAF1}$ by TGFβ requires not only transcription factor Sp1 but also mitogen-activated protein kinase (MAPK) and extracellular signal regulated kinase (ERK) pathways [47-49]. Enhanced expression of $p15^{INK4b}$ increases its binding to the cyclin D-dependent kinases, CDK4 and CDK6, inhibiting their activity (Figure 4). The association of p27[KIP1a] with the D-type cyclins is prevented and p27[KIP1a] binds to and inhibits cyclin E/CDK2 complexes [50]. Therefore, TGFβ inhibits both the cyclin D and cyclin E kinases, which are required for progression of cells through G /S transition.

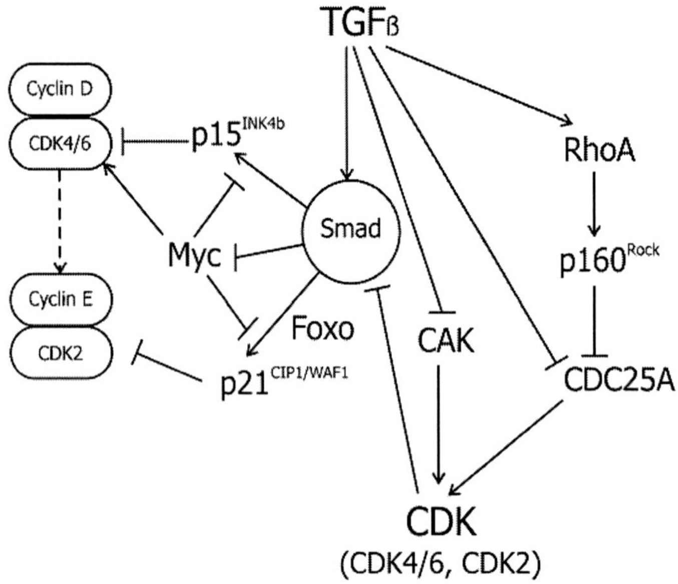

Figure 4. Schematic diagram of induction of cell cycle arrest by cell cycle arrest by TGFβ. TGFβ-induced cell cycle block mechanisms involve transcription of CDK inhibitors p15[INK4b] and p21[CIP1/WAF1]. p15[INK4b] specifically inhibits cyclinD-CDK4/6 complexes, whereas p21[CIP1/WAF1] inhibits cyclin E-CDK2 complexes. Activity of both these complexes is necessary for cell cycle progression. TGFβ decreases the expression of the protooncogene c-myc, which functions as promoter of cell growth and proliferation. The distinct mechanism of cell cycle arrest by TGFβ involves TGFβ-induced activation of the RhoA and its effector kinase p160[ROCK] which directly inhibits the activity of Cdc25A and thereby prevents activation of CDK2 and CDK4/6. TGFβ signaling can also lead to the inhibition of the CDK activating kinase (CAK).

c-Jun is one of the target proteins of TGFβ signaling and this means that the transcription of c-Jun is stimulated by TGFβ through Smad-dependent mechanism. c-Jun transactivates the promoter of the human p21[CIP1/WAF1] gene by activation of the transcription factor Sp1[51]. In addition, p53 and its relatives p63 and p73, major transactivators of p21[CIP1/WAF1] expression, bind to a distal enhancer element of p21[CIP1/WAF1] gene and via direct interaction with Smads a induce p21[CIP1/WAF1] gene expression [52]. Similar to the action of p53 and its relatives, the Smad3/Smad4 complex binds directly to a distal enhancer element of p21[CIP1/WAF1] gene. This binding is stabilized by transcription factors of the FOXO family [53]. TGFβ seems to additionally stabilize p21[CIP1/WAF1] protein levels posttranscriptionally [54]. The result is inhibition of CDK2 activity as a primary response to TGFβ (Figure 4) and the inhibition of phosphorylation of pRb by TGFβ as a secondary event [35-37].

A negative feed back control of Smad activity by CDK was also described (Figure 4) [55]. CDK phosphorylation of Smad3 inhibits its transcriptional activity and antiproliferative function [55].

The Role of the Family of Transcription Factors Id (inhibitors of differentiation) in the Effect of TGFβ on Cell Cycle Transition

Expression of Id family of transcription factors is repressed by TGFβ while induced by bone morphogenetic proteins (BMPs) family members in epithelial cells [56-58]. Id2 antagonizes the antiproliferative effects of the $p21^{CIP1/WAF1}$ protein. Therefore, the inhibition of Id2 expression by TGFβ is a prerequisite for induction of $p21^{CIP1/WAF1}$ by TGFβ and for the block of the G1 to S phase cell cycle transition [58,59].

Cooperation with other Tumor Suppressors or further Proteins in the Tissue-specific Induction of Cell Cycle Block by TGFβ

Cooperation with p53 family of transcription factors was desribed [52]. Another tumor suppressor which cooperates with Smads during $p21^{CIP1/WAF1}$ gene transcriptional induction in stomach epithelial cells is the transcription factor Runx3 [60]. Runx3 expression is lost during gastric carcinogenesis, leading to resistance to induction of cell cycle block by TGFβ [61].

Another tissue-specific mechanism for transcriptional regulation of $p21^{CIP1/WAF1}$ during prostate cell cycle block induced by TGFβ involves the tumor susceptibility protein ELAC2 [62]. *ELAC2* is a member of a gene family predicted to encode a metal-dependent hydrolase domain [63]. ELAC2 protein associates with activated Smad2. The C-terminal MH2 domain of Smad2 interacts with N-terminal region of ELAC2. Small interfering siRNA-mediated knockdown of ELAC2 in prostate cells suppressed TGFβ-induced growth arrest. Moreover, ELAC2 was shown to specifically associate with the nuclear Smad2 partner, Fast-1 (also known as FoxH1, a forkhead transcription factor) and to potentiate the interaction of activated Smad2 with transcription factor Sp1. Fast1 binds to Smad2/3 and mediates TGFβ signaling [64]. Furthermore, activation of the $p21^{CIP1/WAF1}$ gene promoter by TGFβ is potentiated by ELAC2.

The tumor suppressor protein tuberous sclerosis complex 2 (TSC2, tuberin) specifically interacts with the MH2 domain of Smad2 and Smad3 protein to regulate TGFβ-responsive genes such as $p21^{CIP1/WAF1}$. Inhibition of tuberin expression by TSC2 antisense greatly reduces the ability of TGFβ to transcriptionally regulate $p21^{CIP1/WAF1}$, $p27^{KIP1}$ and cyclin A leading to an abrogation of the antiproliferative effects of TGFβ in hematopoietic cells [65]. TSC2 is downstream effector of signaling by the serine-threonine tumor suppressor kinase LKB1 (also known as ATK11) [66-68]. TSC2 mediates G1 cell cycle arrest induced by LKB1 in epithelial cells.

Disruption of the adaptor protein ELF (embryonic liver fodrin) leads to inhibition of TGFβ signaling by Smad proteins in mice [69]. ELF was originally identified in endodermal stem/progenitor cells and possesses potent antioncogenic activity. ELF is frequently inactivated in human gastric cancer tissue samples. Inactivation of the ELF adaptor protein results in the loss of Smad3/Smad4 activation and therefore disrupts TGFβ signaling [70,71]. Expression of the ELF is lost in various human cancers, such as colon, liver and stomach [71,72].

The cytoplasmic form of promyelocytic leukemia (cPML) tumor supressor directly interacts with transforming growth factor β receptors regulated Smads (R-Smads) and promotes TGFβ receptors complex (TβRI and TβRII) endocytosis and R-Smads phosphorylation (Figure 1 and Figure 5). Cytoplasmic PML acts as a bridging factor between protein called Smad anchor of receptor activation (SARA) and Smad proteins (Figure 5). The loss of functional cPML induces not only acute promyelocytic leukemia, but it might influence behaviour of cancer cells and their resistance to TGFβ [73].

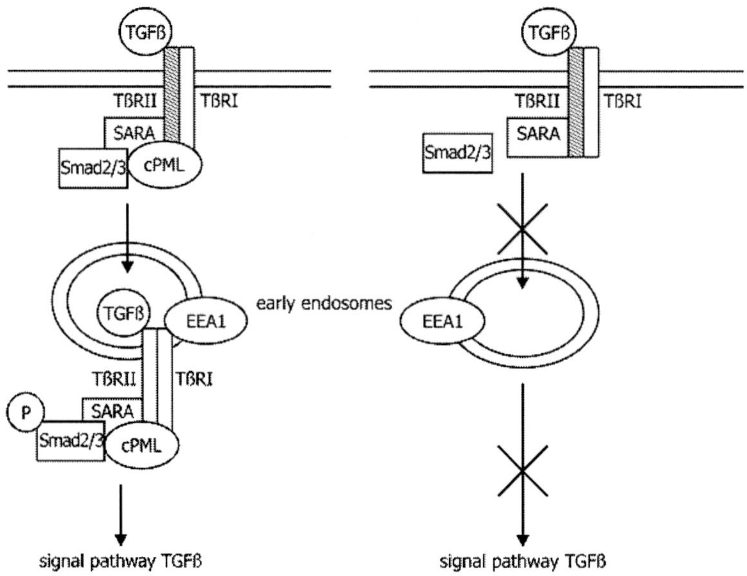

Figure 5. Model for the role of cytoplasmic promyelocytic leukemia PML (cPML) in TGFβ signaling. Protein SARA (Smad anchor of receptor activation) can bind to both R-Smad and TGFβ receptors. In wild type cels, cPML complexes with SARA and Smad2/3. This is essential for SARA and TβRI/II localization in the early endosomes and in turn allows Smad2/3 phosphorylation and translocation into the nucleus to regulate target genes. Smad2/3 phosphorylation and translocation into the nucleus are markedly reduced in cPML-null cells. In acute promyelocytic leukemia (APL) which is associated with reciprocal translocations involving retinoic acid receptor alpha (RARα), the fusion protein PML-RARα may interrupt the interaction of cPML with Smad complexes, leading to defect in TGFβ signal transduction.

3-Phosphoinositide-dependent kinase 1 (PDK1) negatively regulates TGFβ-induced signaling in a kinase-dependent manner through physical interaction with Smad proteins [74]. PDK1 physically interacts with STRAP (serine-threonine kinase receptor-associated protein) and enhances STRAP-induced inhibition of TGFβ signaling. STRAP, a WD domain protein, associates with both functional TβRI and TβRII *in vivo* [75,76]. PDK1 co-immunoprecipitates with Smad proteins and induces Smad2, Smad3, Smad4, and Smad7. The association between PDK1 and Smad proteins is increased by insulin treatment but decreased by TGFβ treatment. Interaction of Smad proteins with PDK1 inhibits TGFβ mediated cell growth arrest.

The basic helix-loop-helix transcription factor TFE3 functions as a potent inhibitor of anti-proliferative signals emanating from the $p16^{INK4A}$, pRb and TGFβ pathways [77].

Because a TGFβ growth arrest depends on the presence of pRb, it is tempting to speculate that TFE3 confers resistance to TGFβ signalling by interfering with the repression of of pRb target genes. However, TFE3 was also shown to synergize with TGFβ in activating certain Smad target genes, notably the negative feedback protein Smad7. Interestingly, this suggests that TFE3 can be both an enhancer of TGFβ signalling and an inhibitor of TGFβ-induced cell cycle arrest. TFE3 binds specifically to E2F3, a transcription factor, which appears to be particularly important for cell proliferation (Figure 6). Immunodepletion or mutation of E2F3 dramatically impairs the mitogen-induced transcriptional activation of numerous E2F-responsive genes (B-myb, cyclin A, cyclin E, CDC2, and CDC6) [78,79]. The cooperative binding of TFE3/E2F3 dimers to target promoters result in transcriptional activation of cyclin E which can control S-phase entry independently of pRb phosphorylation status [77]. More likely, TFE3 acts on additional transcriptional targets to render cells insensitive to pRb-dependent inhibitory signals. TFE3 functions as co-factor for a number of other transcription factors, for example LEF1 and ETS-1. Both these transcription factors play a role in proliferation (Figure 6). TFE3 is the transcription factor that can uncouple cell cycle progression from pRb phosphorylation [77].

Increased expression of Erbin (ErbB2/Her2-interacting protein), which interacts specifically with Smad3 and, to a lesser extent, with Smad2, physically sequesters Smad2/Smad3 from thein association with Smad4 and hence negatively modulates TGFβ-dependent transcriptional responses and cell growth inhibition [80].

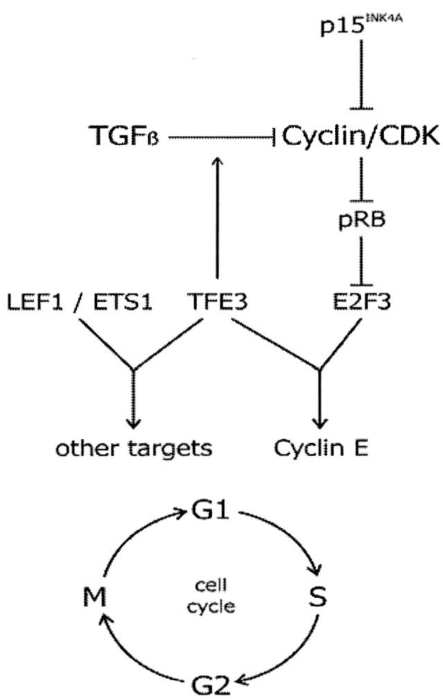

Figure 6. Model for the uncoupling of pRb function from cell cycle progression by the helix-loop-helix transcription factor TFE3. TFE3 acts as a cofactor for E2F3 to regulate genes, such as for example gene for cyclin E. TFE3 also cooperates with other transcription factors (LEF-1 and ETS1), playing a role in proliferation. Overexpression of TFE3 desensitizes the cell for pRb-dependent growth inhibitory signals. As TFE3 can also cooperate with Smads to enhance TGFβ signal transduction, TFE3 may act both upstream and downstream of pRb.

Smad-independent Pathways in TGFβ Mediated cell Growth Arrest

Many lines of evidence indicate that other signaling pathways, in addition to the prominent TβRI/ TβRII, Smad2/Smad3/Smad4 signaling pathway, are involved in the growth inhibitory response to TGFβ. It has been suggested that the Ras/ERK signaling pathway and protein phosphatase 2A (PP2A) are involved in mediating TGFβ-induced G1 cell cycle arrest in certasin epithelial cell systems [30,31,81-83].

ERK phosphorylation sites in Smad3 have been mapped to serines 203 and 207 and to threonine 178 [84]. CDK and ERK may synergize to phosphorylate and inhibit Smad3 in certain biological systems and in particular during oncogenesis. Cyclin D1 synthesis is induced by the Ras/ERK pathway. The oncogenic effect of Ras is dependent on cyclin D1. For example, mice lacking cyclin D1 are resistant to Ras [85]. Smad2 is phosphorylated to a greater extent by ERK than Smad3.

TGFβ1 at low doses (picogram levels) stimulated the colony formation from $CD34^+$ cells isolated from human bone marrow [86,87]. Low TGFβ1 induced activation of signal transducers and activators of transcription (STAT) proteins. On the other hand, the high inhibitory concentrations of TGFβ1 induced a strong phosphorylation of Smad3 and also activated stress kinase/related transcription factors, namely c-Jun and ATF2.

TGFβ induces association of its receptor with PP2A and modulates the posttranscriptional control of cell cycle progression. PP2A associates with p70 ribosomal S6 kinase, a serine/threonine kinase, which induces translation of mRNAs and is essential for G1/S phase progression [83]. This association is leading to TβR dephosphorylation and inactivation that is sufficient to induce G1 cell cycle arrest [83].

TGFβ signaling can also lead to the inhibition of the CDK activating kinase (CAK) [88] and down-regulation of the CDK-activating tyrosine phosphatase Cdc25A [89] (Figure 4). Cdc25A is a member of the Cdc25 family of dual-specificity protein phosphatases that function as positive regulators of cell cycle progression. Cdc25 phosphatases activate CDKs by dephosphorylating their inhibitory threonine and tyrosine phospho residues. Repression of Cdc25A and induction of $p15^{INK4b}$ are independent effects mediating the inhibition of CDK4/6 by TGFβ. Bhowmick et al. [90-92] described a new TGFβ-regulated pathway to cell cycle arrest. TGFβ rapidly activates RhoA, a small GTPase involved in actin rearrangement, in several epithelial cells but not in NIH 3T3 fibroblasts. RhoA and its effector kinase $p160^{ROCK}$ directly inhibit the activity of Cdc25A (Figure 4) to induce cell cycle arrest on TGFβ treatment in epithelial cells but not in fibroblasts. The inhibition of $p160^{ROCK}$ by the use of either a specific inhibitor or by siRNA abrogates the effect of TGFβ.

Five TβRs (I-V) co-express in all normal cell types [93,94]. Many human carcinoma cells express little or noTβRV. Growth of these cells is not inhibited by TGFβ. This suggests that TβRV may be involved in mediating the growth inhibitory response and that its loss may contribute to the malignant phenotype. Recently, TβRV has been identified as the IGFBP3 receptor, which mediates IGF-independent growth inhibition in response to TGFβ in epithelial cells and other responsive cells (Figure 7) [95,96]. Structural and functional analyses of TβRV revealed that TβRV is identical to the low density lipoprotein receptor-related protein 1 (LRP1) and that TβRV/LRP1 is required for growth inhibition by either

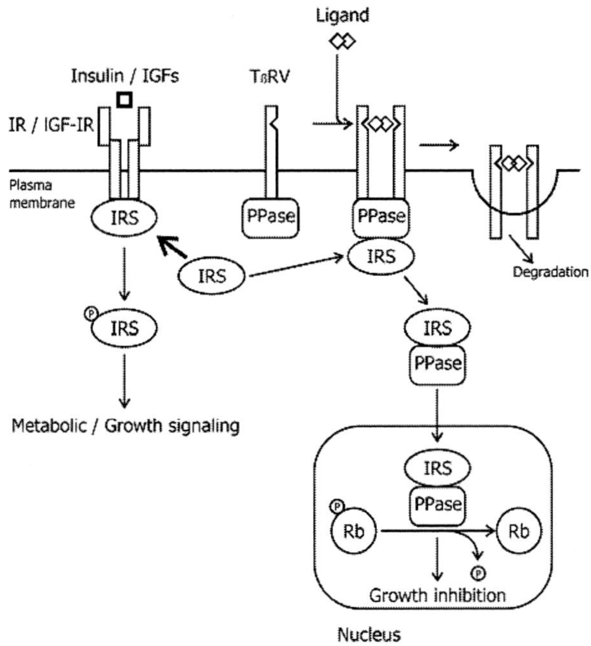

Figure 7. Model for mechanism by which insulin growth factor binding protein 3 (IGFB3) and TGFβ induce growth inhibition mediated by transforming growth factor receptor V (TβRV) and insulin receptor substrate (IRS) proteins. Dimeric ligands (IGFB3 and TGFβ) stimulate the Ser/Thr-sprecific protein phosphatase (PPase) associated with the cytoplasmic tail of TβRV. After stimulation PPase forms complexes with IRS proteins and translocates with them into the nucleus and dephosphorylates pRb and other cell cycle regulators., resulting in growth arrest. Insulin and insulin growth factor 1 (IGF1) analogs reverse this growth arrest by stimullating tyrosine phosphorylation of IRS proteins via interaction with their cognate receptors. This phosphorylation prevents the formation of the IRS-stimulated PPase complex and subsequent serine-specific dephosphorylation of IRS proteins by stimulated PPase. Under nonstimulating conditions, IRS proteins are mainly phosphorylated at serine residues.

IGFBP3 or TGFβ [97]. The insulin receptor substrate (IRS) proteins, the signaling molecules with an cytoplasmic localization, shared by both the insulin receptor and IGF-I receptor signaling cascades, are directly involved in IGFBP3-induced growth inhibition and its reversal by insulin and IGFI analogs [98]. IGFB3 inhibits cell growth by stimulating an okaidic acid-sensitive Ser/Thr-specific protein phosphatase (PPase) which dephosphorylates IRS proteins. Insulin does not affect the TGFβ-induced growth inhibitory response. However, insulin partially reverses TGFβ growth inhibition in the presence of α5β1 integrin antagonists [98]. Growth inhibition by TGFβ involves IRS proteins. TGFβ is capable of stimulating a Ser/Thr-specific PPase associated with TβRV (Figure 7). TβRV requires the TβRI/TβRII/Smad2/3/4 signaling cascade for mediating TGFβ-induced growth inhibition. The weak phosphatase activity stimulated by TGFβ alone may not lead to growth arrest. Following dimeric ligand (IGFB3 or TGFβ) binding to TβRV, the ligand –induced dimerization or oligomerization of TβRV results in activation of the PPase and the subsequent association of the activated PPpase with IRS proteins. The PPase-IRS complexes translocate to the nuclei and then dephosphorylate pRb and other cell cycle regulators, resulting in growth inhibition (Figure 7). The integrin signaling cascade initiated by extracellular matrix (ECM) proteins (e.g., fibronectin and collagen) induced by TGFβ impairs the ability of insulin to block TGFβ-stimulated TβRV-mediated signaling by down-regulating insulin-stimulated tyrosine phosphorylation of IRS proteins (Figure 8). The c-Met signaling cascade induced by the c-Met ligand hepatocyte growth factor (HGF) reverses growth inhibition by IGFBP3 and TGFβ at the cell cycle regulation in the nucleus (Figure 8). c-Met protooncogene product is receptor tyrosine kinase.

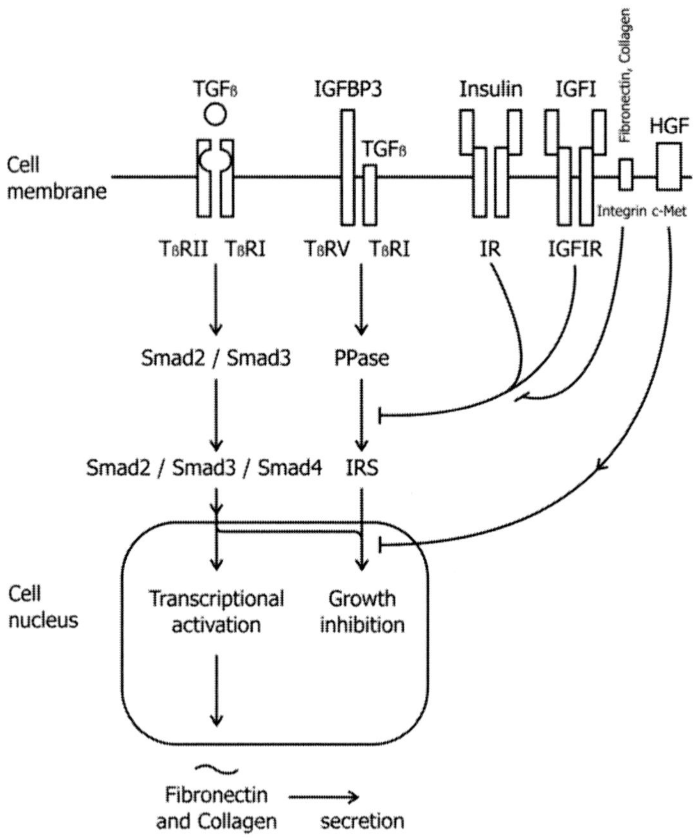

Figure 8. Schematic diagram of cross-talk of the TGFβ receptor (TβRII/TβRI and TβRV), insulin receptor (IR), IGF1 receptor (IGF1R, integrin and receptor tyrosine kinase c-Met signaling cascades. TGFβ signaling through TβRII/TβRI and Smad2/3/4 proteins induces transcription of extracellular matrix (ECM)-related genes (e.g. collagen, fibronectin, and PAI-1) and growth inhibition in responsive cells. In addition to the TβRII/TβRI signaling cascade, the TβRV/PPase/IRS signaling cascade is required for TGFβ-induced growth inhibition. The TβRV/PPase/IRS signaling cascade is negatively regulated by the IR and IGF1R signaling cascades. Insulin or TGFs (IGFI and IGFII) attenuate the TβRV signaling cascade and TGFβ- induced growth inhibition by stimulating tyrosine phosphorylation of IRS proteins. The integrin signaling cascade initiated by ECM proteins induced by TGFβ, impairs the ability of insulin to attenuate the TβRV signaling cascade. Increased ECM synthesis mediated by the TβRII/TβRI signaling cascade further attenuates insulin signaling and enhances TGFβ- induced growth inhibition. The c-Met receptor tyrosine kinase signaling cascade induced by the c-Met ligand hepatocyte growth factor (HGF) reverses growth inhibition by IGFBP3 and TGFβ at the cell cycle regulation in the nucleus.

Corepressors as Negative Regulators of Transforming Growth Factor- Beta Signaling

Ski and Sno Protooncoproteins

Ski (Sloan-Kettering Institute protooncogene) and SnoN (*Ski*-related novel gene) are corepressors for Smad-mediated transcription [99,100]. The oncogene v-ski was originally identified in avian Sloan-Kettering viruses and was found to transform chicken embryo fibroblasts [101]. The *ski* gene family consists of two members, *ski* and *sno*, which share significant homology in their N- and C- terminal regions [102,103]. Ski and Sno form a complex via their C-terminal regions [103]. c-Ski and Sno bind directly to the corepressors N-CoR/SMRT and mSin3A and to Smad proteins, and are involved in the histone deacetylase (HDAC) complex (Figure 9) [104,105]. Both c-Ski and Sno inhibit Smad-induced transcriptional activation by recruiting the HDAC complex and act as oncoproteins. On the other hand, c-Ski and Sno are required for transcriptional repression also by pRb, which also interacts with HDAC [106,107].

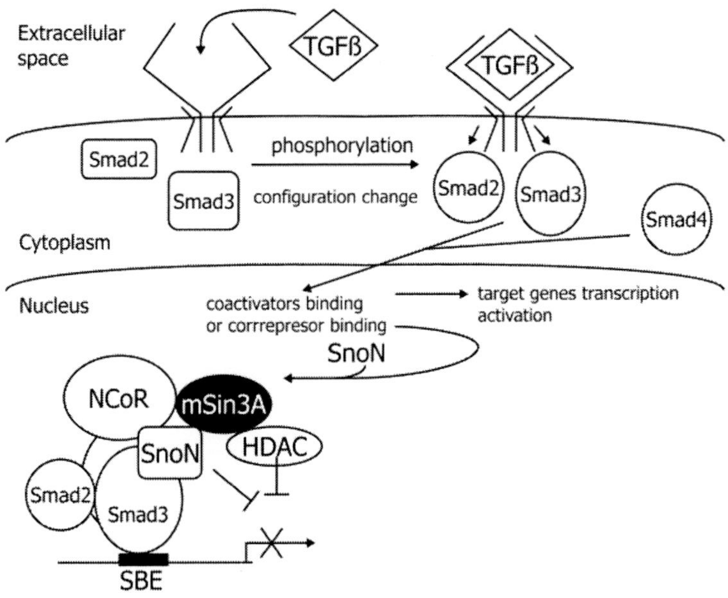

Figure 9. The repression of transcription of TGFβ target genes by correpressor SnoN. SnoN interacts directly with Smad2, Smad3, and Smad4 proteins and represses their ability to activate TGFβ target genes transcription and to induce growth arrest. SnoN is incorporated into the histone deacetylase 1 complex through binding to corepressors N-CoR and mSin3A.

Unexpectedly, Sarker et al. [108] found that SnoN knock-down specifically inhibited TGFβ-induced transcription in the lung epithelial cell line Mv1Lu but not in HeLa or HaCaT cells. SnoN knock-down was also found to block TGFβ -dependent cell cycle arrest in Mv1Lu cells. Collectively, these data indicate that rather than suppressing TGFβ-induced responses, endogenous SnoN acts as a positive mediator of TGFβ-induced transcription and cell cycle arrest in lung epithelial cells. This study also shows that SnoN couples the TGFβ signal to gene expression in a cell-specific manner [108].

Activation of TGFβ signaling leads to rapid degradation of SnoN and, to a lesser extent, of the related Ski protein by proteasomes [109-111].

Repression of TGFβ Signaling by Oncoproteins Evi1, Tax and Leukemia-associated Fusion Proteins

The oncoprotein Evi1 (ecotropic virus integration site 1), which is highly expressed in myelodysplastic syndromes and acute myeloid leukemias with worse prognosis, represses TGFβ signaling through inhibition of Smad3 DNA binding and by recruiting the transcription repressor, CtBP (C-terminal binding protein), HDAC and histone lysine methyltransferases [112-115]. The attenuation of TGFβ signaling is a physiologically relevant function of Evi1 and understanding of its relevance to leukemogenesis requires further studies. Evi1 is also overexpressed in some ovarian cancers and human colon cancer cell lines and overexpression is associated with amplification of the *Evi1* gene. Evi1 did not inhibit TGFβ-mediated downregulation of cyclin D1 or block TGFβ-mediated growth inhibition in non-transformed intestinal epithelial cells [116].

Human T-cell leukemia virus I (HTLV-I) is the etiologic agent of adult T-cell leukemia (ATL), an aggressive T-cell malignancy. The viral oncoprotein Tax is a transcriptional regulator of critical genes for T-cell homeostasis. Tax significantly decreases transcriptional activity and growth inhibitory response to TGFβ [117-119]. Tax inhibits TGFβ signaling, in part, by disrupting the interaction of the Smads with the transcriptional coactivator p300. Tax directly interacts with Smad2, Smad3, and Smad4. Furthermore, Tax inhibits Smad3-Smad4 complex formation and its DNA binding [118]. Tax also induces c-Jun N-terminal kinase (JNK) activity and c-Jun phosphorylation, leading to the formation of Smad3-c-Jun complexes which abrogate Smad3 DNA binding [119].

Chimeric transcription factors that result from the (3;21) translocation Runx1/Evi1 and Runx1/MDS1/Evi1 interfere with TGFβ signaling and antagonize the growth-inhibitory effects of TGFβ similarly as Evi1 [120-122]. The other hand, fusion protein MDS1/Evi1 enhances TGFβ signaling and strengthens its growth-inhibitory effect [123].

Runx1-ETO (known also as AML1-ETO or AML1-MTG8) is a chimeric transcription factor generated as a consequence of the t(8;21) chromosomal translocation in acute myeloid leukemia. This fusion protein binds Smad3 and the complex of corepressors (Sin3A and N-CoR) with HDCA and interferes with TGFβ signaling and antagonizes the growth-inhibitory effects of TGFβ [124].

However, the disruption of TGFβ signaling alone is not sufficient for the development of leukemia. This fact is supported by experiments with TβRI- or Smad3- null mice, which did not exhibit any sign of leukemia throughout their lifespan. Therefore, these studies suggest that disruption of TGFβ signaling may not play an important role in leukemia initiation, but instead may be involved in the progression of leukemia [125].

Transforming Growth Factor-Beta Induces Epithelial-to-mesenchymal Transition and Apoptosis VIA Cell Cycle-dependent Mechanism

Epithelial-to-mesenchymal transition (EMT) is a morphological switch of cells from epithelial polarized phenotype to the mesenchymal fibroblastoid phenotype. EMT plays pivotal roles in various physiological and pathological events, ranging from embryonic development to tumor metastasis [126-129]. The roles of TGFβ in the regulation of concurrent apoptosis and EMT remains poorly understood. TGFβ1 induces EMT in cells non-synchronized or those synchronized at G1/S phase of the cell cycle, whereas it did not induce EMT in cells synchronized at G2/M phase [130]. Synchronization of cells at G1/S phase markedly reduces apoptotic response. Dramatic increase of apoptosis is induced in cells synchronized at G2/M phase as measured by fluorescence-activated cell sorter and DNA fragmentation. Consistent with these results, TGFβ1-induced caspase activity is significantly higher in cells synchronized at G2/M phase [130]. These data show that TGFβ1/induced apoptosis and EMT are related with the cell cycle state. The apoptosis is induced mostly in cells at G2/M phase, whereas EMT is only induced in cells at G1/S phase [130].

Protein kinase D (PKD) plays an important role in TGFβ1-induced EMT. TGFβ1-induced increase of PKD phosphorylation can only be detected at G1/S phase but not at G2/M phase. Transfection of cells with small interference RNA (siRNA) specific for PKD, which reduced the expression level of PKD, suppressed the induction of EMT [130].

Conclusion and Perspectives

The primary signaling pathway downsteam of TGFβ is the Smad pathway. This pathway is subject to complex regulation, much of which is at the level of posttranslational modifications of pathway components, in particular, the Smads. The enzymes responsible may be constitutively active, may be cell type-specific or may be regulated by other signaling pathways or by the cell cycle. In this way, signals from TGFβ are integrated with signals from other growth factors and cytokines, are regulated by the cell cycle and are dependent on cell type. The size of the Smad pool in cells is tightly controlled also by ubiquitylation. Ubiquitin ligases mediate Lys 48 polyubiqutylation of Smads targeting them for subsequent proteasomal degradation. Different E3 ubiquitin ligases trigger ubiquitylation and degradation of inactive and C-end phosphorylated, active receptor regulated Smad. Co-Smad, Smad4, is also ubiquitylated with different E3 ubiquitin ligases for wild type Smad4 and for cancer-associated point mutated Smad4. TGFβ induced G1 cell cycle arrest requires the activity of proteosomal pathway. Proteasomal inhibitors modulate TGFβ signaling in a gene-specific manner.

TGFβ is known to demonstrate tumor suppressor activity. In many human cancers, there are inactivating mutations in components of the TGFβ pathway resulting in uncontrolled proliferation. In addition, through its actions on both the tumor cells and surrounding stromal cells, TGFβ can enhance invasion, motility and metastasis. TGFβ can thus play a dual role in the initiation and malignant progression of human cancer. TGFβ is critical to control cell proliferation by inducing G1-growth arrest through multiple pathways leading to inhibition of E2F transcription activity. The retinoblastoma protein pRb is a key controller of E2F activity and G1/S transition. TGFβ downregulates also the expression of several G2 checkpoint kinases but wihout causing cell acumulation in G2/M phases. TGFβ downregulates their expression at mRNA levels.

Acknowledgements

This work was supported by the Internal Grant Agency of the Ministry of Health of the Czech Republic NR/9045-3, VZ 00023736 from the Ministry of Health of the Czech Republic, and grant LC 06044 from Ministry of Education, Youth and Sport of the Czech Republic.

References

[1] Grana, X.; Reddy, EP. Cell cycle control in mammalian cells: role of cyclins,cyclin dependent kinases (CDKs), growth suppressor genes and cyclin-dependent kinase inhibitors (CKIs). *Oncogene* 1995, 11, 211-21

[2] Morgan, DO. Cyclin-dependent kinases: engines, clocks and microprocessors. *Annu Rev Cell Dev Biol* 1997, 13, 261-291.

[3] Nurse, P. A long twentieth century of the cell cycle and beyond. Cell 2000, 7, 71-78.

[4] Berthet, C.; Kaldis, P. Cell-specific responses to loss of cyclin-dependent kinases. *Oncogene* 2007, 26, 4469-4447.

[5] Peeper, DS.; van der Eb, AJ.; Zantema, A. The G1/ S cell-cycle checkpoint in eukaryotic cells. *Biochim Biophys Acta* 1994, 1198, 215-230.

[6] Blagosklonny, MV.; Pardee AB. The restriction point of the cell cycle. *Cell Cycle* 2002, 1, 103-110.

[7] Bartek, J.; Bartkova, J; Lukas, J. The retinoblastoma protein pathway and the restriction point. *Curr Opin Cell Biol* 1996, 8, 805-814.

[8] Bartek, J.; Bartkova, J; Lukas, J. The retinoblastoma protein pathway in cell cycle control and cancer. *Exp Cell Res* 1997, 237, 1-6.

[9] Mittnacht, S. Control of pRB phosphorylation. *Curr Opin Genet Dev* 1998, 8, 21-27.

[10] Knudsen, ES.; Knudsen, KE. Retinoblastoma tumor suppressor: where cancer meets the cell cycle. *Exp Biol Med* 2006, 231, 1271-1281.

[11] Dyson, N. The regulation of E2F by pRb-family proteins. *Genes Dev* 1998, 12, 2245-2262.

[12] Ohtani, K.; DeGregori, J.; Nevins, JR. Regulation of the cyclin E gene by the transcription factor E2F1. *Proc Natl Acad Sci USA* 1995, 92, 12146-12150.

[13] Stevens, C.; La Thangue NB. E2F and cell cycle control: a double-edged sword. *Arch Biochem Biophys* 2003, 412, 157-169.

[14] Sherr, CJ.; Roberts, JM. Inhibitors of mammalian G1 cyclin-dependent kinases. *Genes Dev* 1995, 9, 1149-1163.
[15] Sherr, CJ.; Roberts, JM. CDK inhibitors: Positive and negative regulators of G1- phase progression. *Genes Dev* 1999, 13, 1501-1512.
[16] Denicourt, C.; Dowdy, SF. Cip/Kip proteins: more than just CDKs inhibitors. *Genes Dev* 2004, 18, 851-855.
[17] Cheng, T. Cell cycle inhibitors in normal and tumor stem cells. *Oncogene* 2004, 23, 7256-7266.
[18] Roberts, AB.; Sporn, MB. Transforming growth factor β. *Adv Cancer Res* 1988, 51, 107-145.
[19] Hocevar, BA.; Howe, PH. Mechanisms of TGF-beta-induced cell cycle arrest. *Miner Electrolyte Metab* 1998, 24, 131-135.
[20] Hu, X.; Zhang, X.; Zhong, Q.; Fisher, AB.; Bryington, M.; Zuckerman, KS. Differential effects of transforming growth factor on cell cycle regulatory molecules in human myeloid leukemia cells. *Oncogene* 2001, 20, 6840-6850.
[21] Massague, J.; Gomis RR. The logic of TGFβ signaling. *FEBS Lett* 2006, 580, 2811-2820.
[22] Pardali, K.; Moustakas, A. Actions of TGF-β as tumor suppressor and pro-metastatic factor in human cancer. *Biochim Biophys Acta* 2007, 1775, 21-62.
[23] Wrana, JL. TGF-beta receptors and signalling mechanisms. *Miner Electrolyte Metab* 1998, 24, 120-130.
[24] Massague, J. TGFbeta signaling: receptors, transducers and Mad proteins. *Miner Electrolyte Metab* 1998, 24, 131-135.
[25] Shi, Y.; Massague, J. Mechanisms of TGF-β signaling from cell membrane to the nucleus. *Cell* 2003, 113, 685-700.
[26] Kretzschmar, M.; Massague, J. SMADs: mediators and regulators of TGF-beta signaling. *Curr Opin Genet Dev* 1998, 8, 103-111.
[27] Feng, X-H.; Derynck, R. Specificity and versatility in TGF-β signaling through Smads. *Annu Rev Cell Dev Biol* 2005, 21, 659-693.
[28] Massague, J.; Seoane, J.; Wotton, D. Smad transcription factors. *Genes Dev* 2005, 19, 2783-2810.
[29] Brown, KA.; Pietenpol, JA.; Moses, HL. A tale of two proteins: differential roles and regulation of Smad2 and Smad3 in TGF-β signaling. *J Cell Biochem* 2007, 101, 9-33.
[30] Derynck, R.; Zhang, YE. Smad-dependent and Smad-independent pathways in TGF-β family signalling. *Nature* 2003, 425, 577-584.
[31] Moustakas, A.; Heldin, C-H. Non-Smad TGF-β signals. *J Cell Sci* 2005, 118, 3573-3584.
[32] Massague, J. TGF-beta signal transduction. *Annu Rev Biochem* 1998, 67, 753-791.
[33] Bai, S.; Cao, X. A nuclear antagonistic mechanism of inhibitory Smads in transforming growth factor-beta signaling. *J Biol Chem* 2002, 277, 4176-4182.
[34] Zhang, S.; Fei, T; Zhang, L.; Zhang, R.; Chen, T.; Ning, Y. et al. Smad 7 antagonizes transforming growth factor beta signaling in the nucleus by interfering with functional Smad-DNA complex formation. *Mol Cell Biol* 2007, 27, 4488-4499.

[35] Laiho, M.; DeCaprio, JA.; Ludlow, JW., Livingston, DM.; Massague, J. Growth inhibition by TGFβ 1 linked to suppression of retinoblastoma protein phosphorylation. *Cell* 1990, 62, 175-185.

[36] Pietenpol, JA.; Stein, RW.; Moran, E.; Yaciuk, R.; Schlegel, R.; Lyons, RM. et al. TGF-β1 inhibition of c-myc transcription and growth in keratinocytes is abrogated by viral transforming proteins with pRb binding domains. *Cell* 1990, 61, 777-785.

[37] Műnger, K; Pietenpol, JA.; Pittelkow, MR.; Holt, JT.; Moses, HL. Transforming growth factor factor β1 regulation of c-myc expression, pRB phosphorylation, and cell cycle progression in keratinocytes. *Cell Growth Differ* 1992, 3, 291-298.

[38] Yagi, K.; Furuhashi, M.; Aoki, H.; Goto, D.; Kuwano, H., Sugamura, K. et al. c-Myc is a downstream target of the Smad pathway. *J Biol Chem* 2002, 277, 854-861.

[39] Frederick, JP.; Liberati, NT.; Waddell, DS.; Shi, Y.; Wang, X-F. Transforming growth factor β-mediated transcriptional repression of c-myc is dependent on direct binding of Smad3 to a novel repressive Smad binding element. *Mol Cell Biol* 2004, 24, 2546-2559.

[40] Hannon, GJ.; Beach, D. $p15^{INK4B}$ is a potential effector of TGF-β-induced cell cycle arrest. *Nature* 1994, 371, 257-261.

[41] Li, JM.; Nichols, MA.; Chandrasekharan, S.; Xiong, Y.; Wang, X-F. Transforming growth factor β activates the promoter of cyclin-dependent kinase inhibitor $p15^{INK4B}$ through an Sp1 consensus site. *J Biol Chem* 1995, 270, 26750-26753.

[42] Datto, MB.; Li, Y.; Panus, JF.; Howe, DJ.; Xiong, Y.; Wang, XF. Transforming growth factor β induces the cyclin-dependent kinase inhibitor p21 through a p53-independent mechanism. *Proc Natl Acad Sci USA* 1995, 92, 5545-5549.

[43] Chen, C-R.; Kang, Y.; Siegel, PM.; Massague, J. E2F4/5 and p107 as Smad cofactors linking the TGF-β receptor to c-myc repression. *Cell* 2002, 110, 19-32.

[44] Warner, BJ.; Blain, SW.; Seoane, J.; Massague, J. Myc downregulation by transforming growth factor beta required for activation of the p15(Ink4b) G(1) arrest pathway. *Mol Cell Biol* 1999, 19, 5913-5922.

[45] Claassen, GF.; Hann, SR. A role for transcriptional repression of p21CIP1 by c-Myc in overcoming transforming growth factor beta-induced cell-cycle arrest. *Proc Natl Acad Sci USA* 2000, 97, 9498-9503.

[46] Feng, XH.; Lin, X.; Derynck, R. Smad2, Smad3 and Smad4 cooperate with Sp1 to induce p15(Ink4B) transcription in response to TGF-beta. *EMBO J* 2000, 19, 5178-5193.

[47] Hu, PP.; Shen, X.; Huang, D.; Liu, Y.; Counter, C.; Wang, XF. The MEK pathway is required for stimulation of p21(WAF1/CIP1) by transforming growth factor- beta. *J Biol Chem* 1999, 274, 35381-35387.

[48] Moustakas, A., Kardassis, D. Regulation of the human p21/WAF1/Cip1 promoter in hepatic cells by functional interaction between Sp1 and Smad family members. *Proc Natl Acad Sci USA* 1998, 95, 6733-6788.

[49] Pardali, K.; Kurisaki, A.; Moren, A.; Dijke, P.; Kardassis, D.; Moustakas, A. Role of Smad proteins and transcription factor Sp1 in p21(WAF1/Cip1) regulation by transforming growth factor- beta. *J Biol Chem* 2000, 275, 29244-29256.

[50] Massague, J.; Blain, SW.; Lo, R.S. TGF beta signaling in growt control, cancer, and heritable disorders. *Cell* 2000, 103, 295-309.

[51] Kardassis, D.; Papakosta, P.; Pardali, K.; Moustakas, A. c-Jun transactivates the promoter of the human p21(WAF1/Cip1) gene by acting as a superactivator of the ubiquitous transcription factor Sp1. *J Biol Chem* 1999, 274, 29572-29581.

[52] Cordenonsi, M.; Dupont, S; Maretto, S.; Insinga, A.; Imbriano, C.; Piccolo, S. Links between tumor suppressors: p53 is required for TGF-β gene responses by cooperating with Smads. *Cell* 2003, 113, 301-314.

[53] Seoane, J.; Le, HV.; Shen, L.; Anderson, SA.; Massague, J. Integration of Smad and forkhead pathways in the control of neuroepithelial and glioblastoma cell proliferation. *Cell* 2004, 117, 211-223.

[54] Gong, J.; Ammanamanchi, S.; Ko, TC.; Brattain, MG. Transforming growth factor β1 increases the stability of $p21^{WAF1/Cip1}$ protein and inhibits CDK2 kinase activity in human colon carcinomas FET cells. *Cancer Res* 2003, 63, 3340-3346.

[55] Matsuura, I.; Denissova, NG.; Wang, G.; He, D.; Long, J.; Liu, F. Cyclin-dependent kinases regulate the antiproliferative function of Smads. *Nature* 2004, 430, 226-231.

[56] Kang, Y.; Chen, CR.; Massague, J. A self-enabling TGFβ response coupled to stresss signaling. Smad engages stress response factor ATF3 for Id1 repression in epithelial cells. *Mol Cell* 2003, 11, 915-926.

[57] Siegel, PM.; Shu, W.; Massague, J. Mad upregulation and Id2 repression accompany transforming growth factor (TGF)-β-mediated epithelial cell growth suppression. *J Biol Chem* 2003, 278, 35444-35450.

[58] Kowanetz, M.; Valcourt, U.; Bergström, R.; Heldin, C-H.; Moustakas, A. Id2 and Id3 define the potency of cell proliferation and differentiation responses to transforming growth factor β and bone morphogenetic protein. *Mol Cell Biol* 2004, 4241-4254.

[59] Pardali, K.; Kowanetz, M.; Heldin, C-H.; Moustakas, A. Smad pathway-specific transcriptional regulation of the cell cycle inhibitor $p21^{WAF1/Cip1}$. *J Cell Physiol* 2005, 204, 260-272.

[60] Chi, XZ.; Yang, JO.; Lee, KY.; Ito, K.; Sakakura, C.; Li, HR. et al. RUNX3 suppresses gastric epithelial cell growth by inducing $p21^{WAF1/Cip1}$ expression in cooperation with transforming growth factor β-activated SMAD. *Mol Cell Biol* 2005, 25, 8097-8107.

[61] Li, QL.; Ito, C.; Sakakura, C.; Fukamachi, H.; Inoue, K.; Chi, XZ. et al. Causal relationship between the loss of RUNX3 expression and gastric cancer. *Cell* 2002, 109, 113-124.

[62] Noda, D.; Itoh, S.; Watanabe, Y.; Inamitsu, M.; Dennler, S.; Itoh, F. et al. ELAC2, a putative prostate cancer susceptibility gene product, potentiates TGF-β/Smad-induced growth arrest of prostate cells. *Oncogene* 2006, 25, 5591-5600.

[63] Tavtigian, SV.; Simard, J.; Teng, DHF.; Abtin, V.; Baumgard, M.; Beck, A. et al. A candidate of prostate cancer susceptibility gene at chromosome 17p. *Nat Genet* 2001, 27, 172-180.

[64] Attisano, L.; Silvestri, C.; Izzi, L.; Labbe, E. The transcriptional role of Smads and FAST (FOXH1) in TGFbeta and activin signalling. *Mol Cell Endocrinol* 2001, 180, 3-11.

[65] Birchenall-Roberts MC.; Fu, T.; Bang, OS.; Dambach, M.; Resau, JH.; Sadowski, CL. et al. Tuberous sclerosis complex 2 gene product interacts with human SMAD proteins. A molecular link of two tumor suppressor pathways. *J Biol Chem* 2004, 279, 25605-25613.

[66] Tiainen, M.; Ylikorkala, A.; Mäkelä, TP. Growth suppression by Lkb1 is mediated by a G1 cell cycle arrest. *Proc Natl Acad Sci USA* 1999, 96, 9248-9251.

[67] Tiainen, M.; Vaahtomeri, K.; Ylikorkala, A.; Mäkelä, TP. Growth arrest by the LKB1 tumor suppressor induction of $p21^{WAF1/CIP1}$. *Hum Mol Genet* 2002, 11, 1497-1504.

[68] Alessi, DR.; Sakamoto, K.; Bayascas, JR. LKB1-dependent signaling pathways. *Annu Rev Biochem* 2006, 75, 137-163.

[69] Tang, Y.; Katuri, V.; Dillner, A.; Mishra, B., Deng, CX., Mishra, L. Disruption of transforming growth factor-beta signaling in ELF beta-spectrin-deficient mice. *Science* 2003, 299, 574-577.

[70] Katuri, V.; Tang, Y.; Marshall, B.; Rashid, A.; Jogunoori, W.; Volpe, EA. et al. Inactivation of ELT/TGF-beta signaling in human gastrointestinal cancer. *Oncogene* 2005, 24, 8012-8024.

[71] Katuri, V.; Tang, Y.; Li, C.; Jogunoori, W.; Deng, CX.; Rashid, A. et al. Critical interactions between TGF-beta signaling/ELF, and E-cadherin/beta-catenin mediated tumor suppression. *Oncogene* 2006, 25, 1871-1886.

[72] Tang, Y.; Katuri, V.; Srinivasan, R.; Fogt, F.; Redman, R.; Anand, G. et al. Transforming growth factor-β suppresses nonmetastatic colon cancer through Smad4 and adaptor protein ELF at an early stage of tumorigenesis. *Cancer Res* 2005, 65, 4228-4237.

[73] Lin, KH.; Bergmann, S.; Pandolfi, PP. Cytoplasmic PML function in TGF-Phosphoinositide-dependent kinase 1 (PDK1) negatively regulates TGFβ-induced signaling in a kinase-dependent manner through physical interaction with Smad proteins signalling. *Nature* 2004, 431, 205-211.

[74] Seong, HA.; Jung, H.; Kim, KT.; Ha, H. 3-Phosphoinositide-dependent PDK1 negatively regulates transforming growth factor-beta-induced signaling in a kinase-dependent manner through physical interaction with Smad proteins. *J Biol Chem* 2007, 282, 12272-12289.

[75] Datta, PK.; Chytil, A.; Gorska, AE.; Moses, HL. Identification of STRAP, a novel WD domain protein in transforming growth factor-β signaling. *J Biol Chem* 1998, 273, 34671-34674.

[76] Seong, HA.; Jung, H.; Choi, HS.; Kim, KT.; Ha, H. Regulation of transforming growth factor-β signaling and PDK1 kinase activity by physical interaction between PDK1 kinase activity by physical interaction between PDK1 and serine-threonine kinase receptor-associated protein. *J Biol Chem* 2005, 280, 42897-42908.

[77] Nijman, SMB.; Hijmans, EM.; El Messaoudi, S., van Dongen, MMW.; Sardet, C.; Bernards, R. A functional genetic screen identifies *TFE3* as a gene that confers resistance to the anti-proliferative effects of the retinoblastoma protein and transforming growth factor-β. *J Biol Chem* 2006, 281, 21582-21587.

[78] Leone, G.; DeGregori, J.; Yan, Z.; Jakoi, L.; Ishida, S.; Williams, RS.; Nevins, JR. E2F3 activity is regulated during the cell cycle and is required for the induction of S phase. *Genes Dev* 1998, 12, 2120-2130.

[79] Humbert, PO.; Verono, R.; Trimarchi, JM., Rogers, C., Dandapani, S.; Lees, JA. E2f3 is critical for normal cellular proliferation. *Genes Dev* 2000, 14, 690-703.

[80] Dai, F.; Chang, C.; Lin, X.; Dai, P.; Mei, L.; Feng, X-H. Erbin inhibits transforming growth factor β signaling through a novel Smad-interacting domain. *Mol Cell Biol* 2007, 27, 6183-6194.

[81] Mulder, KM. Role of Ras and Mapks in TGFβ signaling. *Cytokine Growth Factor Rev* 2000, 11, 23-35.

[82] Yue, J.; Mulder, KM. Activation of the mitogen-activated protein kinase pathway by transforming growth factor-β. *Methods Mol Biol* 2000, 142, 125-131.

[83] Petritsch, C.; Beug, H.; Balmain, A.; Oft, M. TGF-beta inhibits p70 S6 kinase via protein phosphatase 2A to induce G(1) arrest. *Genes Dev* 2000, 14, 3093-3101.

[84] Matsuura, I.; Wang, G.; He, D.; Liu, F. Identification and characterization of ERK MAP kinase phosphorylation sites in Smad3. *Biochemistry* 2005, 44, 12546-12553.

[85] Yu, Q.; Geng, Y.; Sicinski, P. Specific protection against breast cancers by cyclin D1 ablation. *Nature* 2001, 411, 1017-1021.

[86] Kale, VP. Differential activation of MAPK signaling pathways by TGF-β1 forms the molecular mechanism behind its dose-dependent bidirectional effects on hematopoiesis. *Stem Cells Dev* 2004, 13, 27-38.

[87] Kale, VP.; Vaidya, AA. Molecular mechanisms behind the dose-dependent differential activation of MAPK pathways induced by transforming growth factor-β1 in hematopoietic cells. *Stem Cells Dev* 2004, 13, 536-547.

[88] Nagahara, H.; Ezhevsky, SA.; Vocero-Akbani, AM.; Kaldis, P.; Solomon, MJ.; Dowdy, SF. Transforming growth factor β targeted inactivation of cyclin E: cyclin-dependent kinase 2 (Cdk2) complexes by inhibition of Cdk2 activating kinase activity. *Proc Natl Acad Sci USA* 1999, 96, 14961-14966.

[89] Lavarone, A.; Massague, J. Repression of the CDK activator Cdc25A and cell-cycle arrest by cytokine TGF-β in cells lacking the CDK inhibitor p15. *Nature* 1997, 387, 417-422.

[90] Bhowmick, NA.; Ghiassi, M,; Bakin, A.; Aakre, M.; Lundquist CA.; Engel, ME. et al. Transforming growth factor-β1 mediates epithelial to mesenchymal transdifferentiation through a RhoA-dependent mechanism. *Mol Biol Cell* 2001, 12, 27-26.

[91] Bhowmick, NA.; Ghiassi, M,; Aakre, M.; Brown, K.; Singh, V.; Moses, HL. TGF-β-induced RhoA and p160^{ROCK} activation is involved in the inhibition of Cdc25A with resultant cell-cycle arrest. *Proc Natl Acad Sci USA* 2003, 100, 15548-15553.

[92] Brown, K.; Bhowmick, NA. Linking TGF-β-mediated Cdc25A inhibition and cytoskeletal regulation through RhoA/ p160^{ROCK} signalling. *Cell Cycle* 2004, 3, 408-410.

[93] Leal, SM.; Liu, Q.; Huang, SS.; Huang, JS. The type V transforming growth factor β receptor is the putative insulin-like growth factor-binding protein 3 receptor. *J Biol Chem* 1997, 272, 20572-20576.

[94] Liu, Q.; Huang, SS.; Huang, JS. Function of the type V transforming growth factor β receptor in transforming growth factor β-induced growth inhibition of mink lung epithelial cells. *J Biol Chem* 1997, 272, 18891-18895.

[95] Wu, HB.; Kumar, A.; Tsai, WC.; Mascarenhas, D.; Healey, J.; Rechler, MM. Characterization of the inhibition of DNA synthesis in proliferating mink lung epithelial cells by insulin-like growth factor binding protein-3. *J Cell Biochem* 2000, 77, 288-297.

[96] Huang, SS.; Huang, JS. TGF-β control of cell proliferation. *J Cell Biochem* 2005, 96, 447-462.

[97] Huang, SS.; Ling, TY.; Tseng, WF.; Huang, YH.; Tang, FM.; Leal, SM.; Huang, JS. Cellular growth inhibition by IGFBP3 and TGF-β1 requires LRP1. *FASEB J* 2003, 17, 2068-2081.

[98] Huang, SS.; Leal, SM.; Liu, I-H.; Huang, JS. Identification of insulin receptor substrate as key molecule for the TβR-V/LRP-1 mediates growth inhibition signaling cascade. *FASEB J* 2004, 18, 1719-1721.

[99] Liu, X.; Sun, Y.; Weinberg, RA.; Lodish, HF. Ski/Sno and TGF-beta signalling. *Growth Factor Rev* 2001, 12, 1-8.

[100] He, J.; Tegen, SB.; Krawitz, AR.; Martin, GS.; Luo, K. The transforming activity of Ski and SnoN is dependent on their ability to repress the activity of Smad proteins. *J Biol Chem* 2003, 278, 30540-30547.

[101] Li, Y.; Turck, CM.; Teumer, JK.; Stavnezer, E. Unique sequence, ski, in Sloan-Kettering avian retroviruses with properties of a new cell-derived oncogene. *J Virol* 1986, 57, 1065-1072.

[102] Nomura, N.; Sasamoto, S.; Ishii, S.; Date, T.; Matsui, M.; Ishizaki, R. Isolation of human cDNA clones of *ski* and *ski*-related gene *sno. Nucleic Acids Res* 1989, 17, 5489-5500.

[103] Nagase, T.; Nomura, N.; Ishii, S. Complex formation between proteins encoded by the *ski* gene family. *J Biol Chem* 1993, 268, 13710-13716.

[104] Nomura, N.; Khan, MM.; Kaul, SC.; Dong, H-D.; Wadhwa, R; Colmanares, C. et al. Ski is a component of the histone deacetylase complex required for transcriptional repression by Mad and thyroid hormone receptor. *Genes Dev* 1999, 13, 412-423.

[105] Shinagawa, T.; Dong, H-D.; Xu, M.; Maekawa, T.; Ishii, S. The *sno* gene, which encodes a component of the histone deacetylase complex, acts as a tumor suppressor in mice. *EMBO J* 2000, 19, 2280-2291.

[106] Brehm, A.; Miska, EA.; McCance, DJ.; Reid, JL.; Bannister, AJ.; Kouzarides, T. Retinoblastoma protein recruits histone deacetylase to repress transcription. *Nature* 1998, 391, 597-601.

[107] Magnaghi-Jaulin, L.; Groisman, R.; Naguibneva, I.; Robin, P.; Lorain, S.; Le Villain, JP. et al. Retinoblastoma protein represses transcription by recruiting a histone deacetylase. *Nature* 1998, 391, 601-605.

[108] Sarker, KP.; Wilson, SM.; Bonni, S. SnoN is a cell type-specific mediator of transforming growth factor-beta responses. *J Biol Chem* 2005, 280, 13037-13046.

[109] Sun, Y.; Liu, X.; Ng-Eaton, E.; Lodish, HF.; Weinberg, RA. SnoN and Ski protooncoproteins are rapidly degraded in response to transforming growth factor-β signaling. *Proc Natl Acad Sci USA* 1999, 96, 12442-12447.

[110] Fukasawa, H.; Yamamoto, T.; Togawa, A.; Ohashi, N.; Fujigaki, Y.; Oda, T. et al. Ubiquitin-dependent degradation of SnoN and Ski is increased in renal fibrosis induced by obstructive injury. *Kidney Int* 2006, 69, 1733-1740.

[111] Tan, R.; Zhang, J.; Tan, X.; Zhang, X.; Yang, J.; Liu, Y. Downregulation of SnoN expression in obstructive nepropathy is mediated by an enhanced ubiquitin-dependent degradation. *J Am Soc Nephrol* 2006, 17; 2781-2791.

[112] Kurokawa, M.; Mitani, K.; Irie, K.; Matsuyama, T.; Takahashi, T.; Chiba, S. et al. The oncoprotein Evi-1 represses TGF-β signalling by inhibiting Smad3. *Nature* 1998, 394, 92-96.

[113] Izutsu, K.; Kurokawa, M.; Imai, Y.; Maki, K.; Mitani, K.; Hirai, H. The corepressor CtBP interacts with Evi-1 to repress transforming growth factor β signaling. *Blood* 2001, 97, 2815-2822.

[114] Vinatzer, U.; Taplick, J.; Seiser, C.; Fonatsch, C.; Wieser, R. The leukemia-associated transcription factors EVI-1 and MDS1/EVI1 repress transcription and interact with histone deacetylase. *Br J Haematol* 2001, 114, 566-573.

[115] Nitta, E.; Izutsu, K.; Yamaguchi, Y.; Imai, Y.; Ogawa, S.; Chiba, S. et al. Oligomerization of Evi-1 regulated by the PR domain contributes to recruitment of corepressor CtBP. *Oncogene* 2005, 24, 6165-6173.

[116] Liu, Y.; Chen, L.; Ko, TC.; Fields, AP.; Thompson, EA. Evi1 is a survival factor which conveys resistance to both TGFβ- and taxol-mediated cell death via PI3K/AKT. *Oncogene* 2000, 25, 3565-3675.

[117] Mori, N.; Morishita, M.; Tsukazaki, T.; Giam, C-Z.; Kumatori, A.; Tanaka, Y.; Yamamoto, N. Human T-cell leukaemia virus type I oncoprotein Tax represses Smad-dependent transforming growth factor β signalling through interaction with CREB-binding protein/p300. *Blood* 2001, 97, 2137-2144.

[118] Lee, DK.; Kim, B-C.; Brady, JN.; Jeang, K-T.; Kim, S-J. Human T-cell lymphotropic virus type 1 Tax inhibits transforming growth factor-β signaling by blocking the association of Smad proteins with Smad-binding element. *J Biol Chem* 2002, 277, 33766-33775.

[119] Arnulf, B.; Villemain, A.; Nicot, C.; Mordelet, E.; Charneau, P.; Kersual, J. et al. Human T-cell lymhotropic virus oncoprotein Tax represses TGF-β1 signaling in human T cells via c-Jun activation: a potential mechanism of HTLV-I leukemogenesis. *Blood* 2002, 100, 4129-4138.

[120] Izutsu, K.; Kurokawa, M.; Imai, Y.; Ichikawa, M.; Asai, T.; Maki, K. et al. The t(3;21) fusion product, AML1/Evi-1 blocks AML1-induced transactivation by recruiting CtBP. *Oncogene* 2002, 21, 2695-2703.

[121] Senyuk, V.; Chakraborty, S.; Mikhail, FM.; Zhao, R.; Chi, Y.; Nucifora, G. The leukemia-associated transcription repressor AML1/MDS1/Evi-1 requires CtBP to induce abnormal growth and differentiation of murine hematopoietic cells. *Oncogene* 2002, 21, 3232-3240.

[122] Mitani, K. Molecular mechanisms of leukemogenesis by AML1/EVI-1. *Oncogene* 2004, 23, 4263-4269.

[123] Sood, R.; Talwar-Trikha, A.; Chakrabarti, SR.; Nucifora, G. MDS1/EVI1 enhances TGF-β1 signaling and strengthens its growth-inhibitory effect, but the leukemia associated fusion protein AML1/MDS1/Evi-1, product of the t(3;21), abrogates growth-inhibition in response to TGF-β1. *Leukemia* 1999, 13, 348-357.

[124] Jakubowiak, A.; Pouponnot, C.; Berguido, F.; Frank, R.; Mao, S.;Massague, J.; Nimer, SD. Inhibition of the transforming growth factor β1 signaling pathway by the AML1/ETO leukemia-associated fusion protein. *J Biol Chem* 2000, 275, 40282-40287.

[125] Lin, H-K.; Bergmann, S.; Pandolfi, PP. Deregulated TGF-β signaling in leukemogenesis. *Oncogene* 2005, 24, 5693-5700.

[126] Thiery, JP. Epithelial-mesenchymal transitions in development and pathologies. *Curr Opin Cell Biol* 2003, 15, 740-746.

[127] Huber, MA.; Kraut, N.; Beug, H. Molecular requirements for epithelial-mesenchymal transition during tumor progression. *Curr Opin Cell Biol* 2005, 17, 548-558.

[128] Zavadil, J.; Böttinger, EP. TGF-β and epithelial-to-mesenchymal transitions. *Oncogene* 2005, 24, 5764-5774.

[129] Rahimi, RA.; Leof, EB. TGF-β signaling: a tale of two responses. *J Cell Biochem* 2007, 102, 593-608.

[130] Yang, Y.; Pan, X.; Lei, W.; Wang, J.; Song, J. Transforming growth factor-β1 induces epithelial-to-mesenchymal transition and apoptosis via a cell cycle-dependent mechanism. *Oncogene* 2006, 25, 7235-7244.

In: Cell Cycle Control: New Research
Editors: N. H. Leroy, N. T. Fournier
ISBN: 978-1-60456-796-0

Chapter VIII

Transcription Factor C/EBPα and its Effect on Cell Cycle Regulation

Ota Fuchs*

Institute of Hematology and Blood Transfusion, U Nemocnice 1, 128 20 Prague 2, Czech Republic and Center of Experimental Hematology with Institute of Pathophysiology, First Medical Faculty, Charles University, Prague, Czech Republic.

Abstract

The CCAAT/enhancer binding protein alpha (C/EBPα) is the founding member of a family of related leucine zipper transcription factors that play important roles in differentiation of liver, lung, adipocyte and myeloid cells. C/EBPα controls differentiation-dependent gene expression and inhibits proliferation in these types of cells. The capacity of C/EBPα to promote growth arrest has been studied *in vitro*, by analysis of knockout mice and by examination of leukemic cells. Several models of C/EBPα-induced growth arrest have been described. These include C/EBPα-mediated (1) stabilization of the cyclin-dependent kinase 2 (CDK2) inhibitor, $p21^{WAF1/CIP1}$, (2) regulation of growth-inhibiting Rb-E2F complexes, (3) interaction with free E2F, inhibition of E2F activity and down-regulation of the E2F target gene c-Myc, (4) interaction with Max, a member of the basic region –helix-loop-helix-leucine zipper proteins, that belongs to a network of transcription factors, which includes the Myc and Mad families of protein, (5) inhibition of CDK2 and CDK4 activity, and (6) interaction with the SWI/SNF chromatin remodelling complexes. C/EBPα may promote growth arrest by cell specific mechanisms. This means that different models of C/EBPα-mediated cell cycle arrest operate in different cell types. Different regions of C/EBPα are involved in growth inhibition. According to the third model, C/EBPα interacts with free transcription factor E2F through the non-DNA binding surface of its basic region and N-terminal region of the C/EBPα also plays role in growth inhibition. The basic DNA-binding region of the C/EBPα is also involved in the C/EBPα and Max interaction. In the

* Correspondence concerning this article should be addressed to: Ota Fuchs, PhD. Tel: +420 221977313; Fax: +420 221977370; E- mail: Ota.Fuchs@uhkt.cz.

CDK2/CDK4 inhibition model, C/EBPα interacts with and inhibits the activity of CDK2/CDK4 through proline- and histidine-rich region located in the central part of C/EBPα. According to the SWI/SNF recruitment model, C/EBPα interacts with SWI/SNF components through a centrally located 75-amino acid region overlapping with the CDK2/CDK4 binding region. Acute myeloid leukemias (AMLs) are clonal disorders that are characterized by a block in differentiation along one or more haematopoietic lineage. Molecular abnormalities are frequently detected in AML. The genetic alterations in AML often affect transcription factors that also have an important role in normal hematopoiesis. C/EBPα mutations have been observed in AML patients with the frequency approximately 5-14% and are often associated with the induction of proliferation.

Keywords: C/EBPα, cell cycle, retinoblastoma protein, E2F, cyclin-dependent kinase, SWI/SNF chromatin remodeling complex, differentiation, leukemia.

Introduction

The CCAAT/enhancer binding proteins (C/EBPs) are a family of six basic region leucine zipper (bZIP) transcription factors playing an important role in controlling cell proliferation and differentiation. C/EBPα is the most well studied founding member of this family, which plays a critical role in the regulation of mitotic growth arrest and differentiation in numerous cell types, including preadipocytes, myeloid cells, hepatocytes, keratinocytes, and pneumocytes [1-7]. Knockout experiments producing C/EBPα null mice display cell proliferative defects in the liver and lung and die at birth because of energy imbalance presumably from impaired glucose metabolism and altered fat metabolism with a failure of adipocytes to accumulate lipids [8-13]. Several human tumour types display reduction in the levels of C/EBPα suggesting that C/EBPα is a tumour suppressor [14-17]. However, genetic evidence supporting the tumor suppressor function of C/EBPα has been only obtained for myeloid leukemias [18-21]. Downregulation or functional inactivation of C/EBPα might be a required step in the development in several tumor types, indicating a general role for C/EBPα as a tumor suppressor. C/EBPα is abundantly expressed in mouse epidermal keratinocytes. On the other hand, C/EBPα mRNA and protein levels were greatly diminished in squamous cell carcinoma cell lines and also in the skin squamous cell carcinoma itself. Oncogenic Ras, which is present in squamous cell carcinoma cells, negatively regulates C/EBPα expression and the loss of C/EBPα expression may contribute to the development of skin squamous cell carcinoma [22]. Mice lacking C/EBPα in the epidermis, although showing normal differentiation and keratinocyte proliferation, were highly susceptible to skin tumorigenesis [23]. The role of impaired C/EBPα function in tumor progression is supported by experiments with reintroduction of C/EBPα. Reintroduced C/EBPα caused the growth arrest blocked *in vivo* tumorigenicity of AML and skin carcinomas [24-26]. On the other hand, overexpression of apparently normal C/EBPα RNA or protein was observed in 6 patients with B-cell precursor acute lymphoblastic leukemia (BCP-ALL) harboring the translocation t(14; 19)(q32; q13). *C/EBPα* is activated in these BCP-ALL cells by juxtaposition to the immunoglobulin gene enhancer upon this rearrangement and exhibits oncogenic properties [27,28].

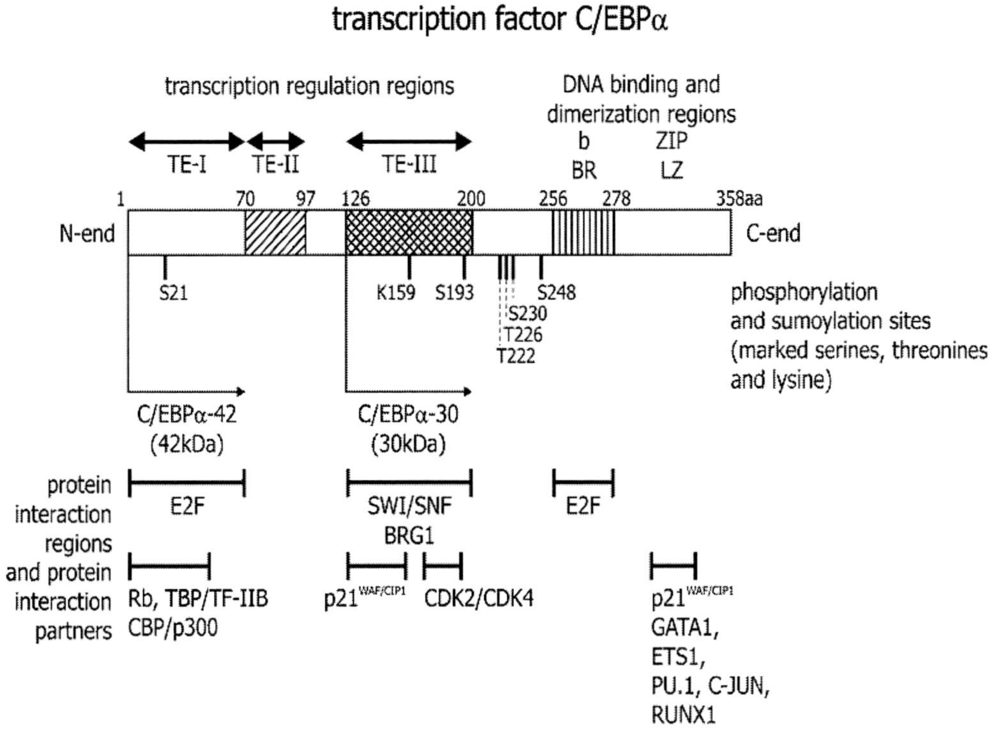

Figure 1. Schematic presentation of the domain structure of C/EBPα protein, phosphorylation and sumoylation sites, protein interaction regions and protein interaction partners. Numbers directly above the schema of protein indicate the amino acids of the rat C/EBPα protein. The full-length, 42 kDa form of C/EBPα protein and the shorter 30 kDa form of the protein are also shown.

Domain Organization of C/EBPα

C/EBPα gene and mRNA

C/EBPα is encoded by an intronless gene that is 2783 bp long and maps to human chromosome 19q13.1 [29]. Expression patterns of C/EBPα mRNA are similar in the mouse and human with measurable levels in liver, adipose tissue, intestine, lung, adrenal gland, skeletal muscle, pancreas, placenta, prostate gland, mammary glands and peripheral blood mononuclear cells. However, the expression was undetectable or very low in brain, kidney, thymus, testis and ovary [30]. In liver and adipose tissue, highest levels of C/EBPα mRNA are detected only in differentiated tissue.

C/EBPα mRNA is translated into two major proteins, C/EBPα p42 (42 kDa) and C/EBPα p30 (30 kDa) by a ribosome-scanning mechanism in which a fraction of ribosomes ignore the first two AUG codons and initiate translation at the third AUG codon located 351 nucleotides downstream of the first one (Figure 1). Translation start site multiplicity of the C/EBPα mRNA is dictated by a small 5' open reading frame [31,32]. The translation product C/EBPα -30 initiated at the third AUG codon is devoid of the potent transcription-regulation domain contained in C/EBPα p42 and stimulates transcription of the target gene (e.g. albumin gene) much less efficiently than the C/EBPα p42 [33].

Transcription factor C/EBPα

The full-length, 42 kDa form of C/EBPα contains three transactivation domains (TE-I, TE-II and TE-III) as well as the basic region/leucine zipper (bZIP) [34]. This bZIP domain contains a basic region (BR), which mediates DNA binding, and a leucine zipper region (LZ) for homodimerization and heterodimerization of C/EBPα with other C/EBP and different proteins (Figure 1). TE-I and TE-II domains (sometimes marked as TAD1, transription activation domain 1) mediate cooperative binding of C/EBPα to TBP (TATA box-binding protein) and another basal transcription initiation factor TFIIB [35]. Both, TBP and TFIIB are essential components of the RNA polymerase II basal transcription apparatus. The TE-III domain (also known as TAD2) contains a negative regulatory subdomain [34,36]. This negative regulatory region or transcriptional attenuator domain also inhibits transcriptional synergy of multiple DNA-binding regulators and was named synergy control (SC) motif [37].

The truncated C/EBPα p30 isoform acts as an inhibitor of C/EBPα p42-mediated transactivation of transcription of target genes. This inhibition occurs by formation of heterodimers of both C/EBPα isoforms. These heterodimers have impaired DNA-binding ability and transcription transactivation capacity compared with C/EBPα p42 homodimers [33,38]. There are examples where the C/EBPα p30 isoform has some transcriptional activity [39]. In these cases the truncated C/EBPα p30 isoform might stabilize binding of other transcription factors and activate transcription. This 30 kDa product also lacks the antimiotic activity exhibited by the full-length C/EBPα p42 [40].

The basic region (BR) of C/EBP which mediates DNA binding, preferentially recognizes the palindromic DNA sequence 5'-ATTGCGCAAT-3' [41,42]. Although the C/EBPs possess similar DNA binding specificities and dimerization properties, each protein exhibits unique functional properties *in vivo.* The appearance of specific phenotypes in each C/EBP-deficient mouse shows that these proteins do not have fully redundant functions. Direct evidence for specific functions has come from experiments where the coding sequence for C/EBPα was replaced with the coding sequence C/EBPβ. This gene replacement strategy to generate a viable and fertile C/EBPα-null mouse line rescued hepatic-specific function to maintain normal blood glucose levels, but could not rescue function in white adipose tissue to regulate fat storage [43].

The Post-Translational Modifications of C/EBPα

Phosphorylation and sumoylation of C/EBPα are important regulatory mechanisms which cause changes in function of this protein. Phosphorylation and sumoylation sites in the C/EBPα protein are shown in Figure 1.

Phosphorylation of C/EBPα

The McDonough strain of feline sarcoma virus contains an oncogene called v-*fms* with tyrosine kinase activity. Fms-like tyrosine kinase 3 (Flt3) encodes a receptor tyrosine kinase for which activating mutations have been identified in a proportion of acute myelogenous leukemia (AML) patients. These mutations activate the Flt3 kinase activity constitutively, and result in increased cellular proliferation and viability. Activation of Flt3 inhibits C/EBPα function by extracellular signal receptor kinase (ERK)1/2-mediated phosphorylation on serine 21, which affects the ability of C/EBPα to induce granulocytic differentiation and may explain the differentiation block of leukaemic blasts [44].

Glycogen synthase kinase-3 (GSK3), an insulin-inhibited protein kinase phosphorylates C/EBPα on two threonine residues (T222 and T226, refer to rodent sequence) and on serine (S230). The functional importance of these phosphorylation events in the regulation of C/EBPα activity is not yet clear but plays some role in preadipocyte diferentiation [45].

Ras signalling phosphorylates C/EBPα on serine 248 of the transactivation domain, resulting in an enhancement of the ability of C/EBPα to transactivate the granulocyte-colony stimulating factor (G-CSF) receptor promoter, which contributes to the induction of granulocyte differentiation [46].

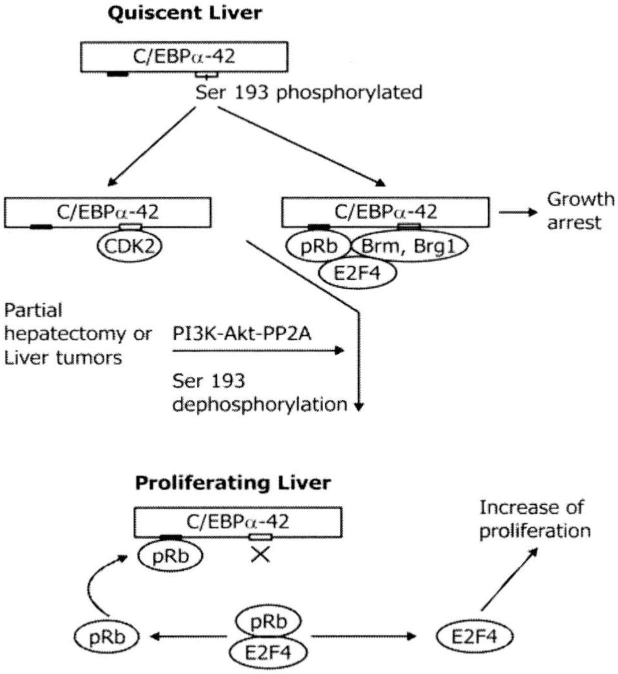

Figure 2. A proposed pathway by which Ser 193-dephosphorylated C/EBPα accelerates liver proliferation. In quiescent livers, C/EBPα arrests proliferation through inhibition of CDK2 and E2F transcription. Partial hepatectomy and liver tumors activate the PI3K/Akt/protein phosphatase PP2A pathway, which dephosphorylates Ser 193 and blocks the interactions of C/EBPα with CDK2 and catalytic subunit Brm (Brahma) of chromatin remodeling complex SWI/SNF. The Ser 193-dephosphorylated C/EBPα interacts with pRb and sequesters pRb from the E2F-pRb complexes, leading to accelerated proliferation.

The biological function of C/EBPα in liver cells also depends on phosphorylation-dephosphorylation of a single serine 193 (S193) residue within the C/EBPα growth-inhibitory region [47]. S193-phosphorylated C/EBPα binds to cyclin-dependent kinase 2 (CDK2) and to brahma (Brm, named according to the ATPase of the Drosophila SWI/SNF complex involved in chromatin remodelling during transcription) and inhibits proliferation. S193-dephosphorylated C/EBPα accelerates proliferation by neutralization growth-inhibitory activity of retinoblastoma protein (Rb) through sequestering Rb from E2F-Rb complex repressors (Figure 2). The E2F transcription factor plays a crucial role in the control of cell cycle progression and regulates the expression of genes required for G1/S transition (Figure 3). E2F activity is modulated by multiple mechanisms including negative regulation by interaction with the product of the *Rb* tumour suppressor gene expression. The binding of Rb to E2F results in active transcriptional repression of E2F-regulated genes and in growth suppression.

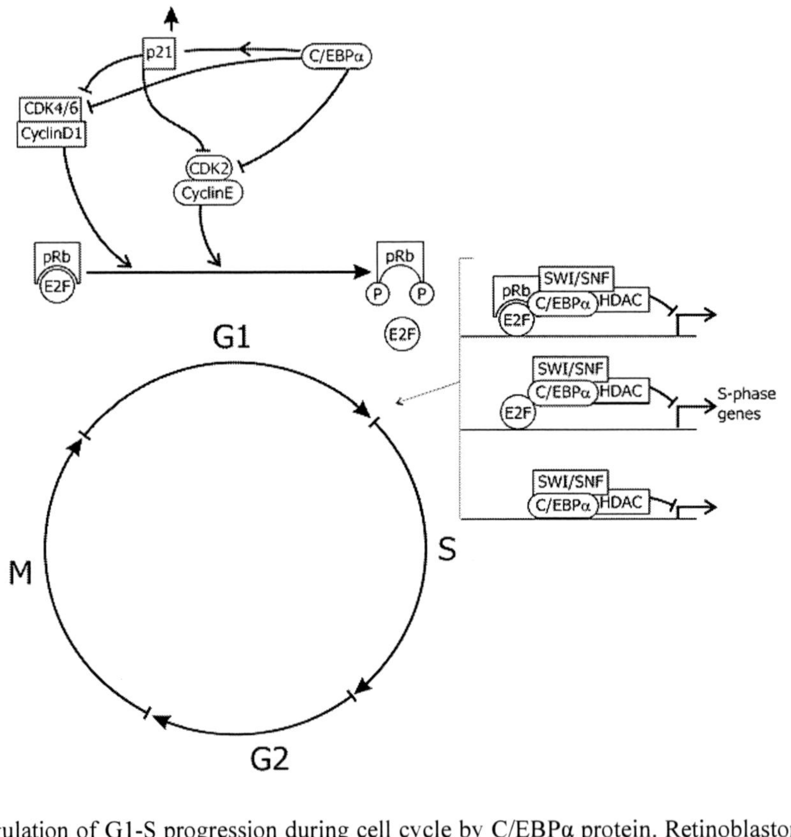

Figure 3. Regulation of G1-S progression during cell cycle by C/EBPα protein. Retinoblastoma protein (pRb) is phosphorylated by Cdks that are inhibited by C/EBPα protein and p21$^{WAF1/CIP1}$ Cdk inhibitor induced by C/EBPα protein. Phosphorylation of pRb and its release from transcription factor E2F derepresses S-phase genes, which are otherwise inhibited by the pRb-E2F complex through recruitment of the SWI/SNF chromatin-remodelling complex and histone deacetylases (HDACs). C/EBPα protein may be bound to promoters of target S-phase genes indirectly by E2F or it could bind directly to target S-phase gene promoters.

Mass spectrometry-based proteomic approach was applied to identify putative coactivator proteins interacting with the DNA-binding domain (DBD) of C/EBP transcription factors [48]. c-Jun N-terminal kinase 1 (JNK1) among others was detected as protein interacting with DBD of C/EBPs from nuclear extract of myelomonocytic U937 cells. JNK1 phosphorylates C/EBPα at unidentified amino-acid residue and inhibits its ubiquitination, probably owing to altered conformation of C/EBPα [48]. This results in increased C/EBPα stability and availability which is reflected by enhanced DNA binding and transcriptional activity of C/EBPα. In some patients with acute myeloid leukemia (AML), the JNK1 mRNA expression and its kinase activity are decreased which suggests a possible reason for C/EBPα inactivation in AML.

The activation of phospatidylinositol 3-kinase (PI3K)/Akt (protein kinase B) in liver tumours leads to accumulation of protein phosphatase 2A (PP2A) in the nuclei, where PP2A dephosphorylates C/EBPα on S193 and blocks its growth-inhibitory activity [49]. This PI3K/Akt-mediated block of C/EBPα inhibition leads to the lack of negative control of proliferation in the liver and to development of tumours.

Sumoylation of C/EBPα

C/EBPα can be sumoylated at the lysine residue (K159, see Figure 1) of SC motif within the transcriptional attenuator domain [50,51]. Sumoylation of this motif can affect the inhibitory function by influencing protein-protein interactions, a mechanism by which sumoylation probably regulates the activity of the transcription factor. Sato et al. [51] investigated the level and functional roles of sumoylated C/EBPα during the differentiation of hepatocytes. SUMO-1 (small ubiquitin-related modifier-1) masks BRG1 (product of expression of brahma-related gene 1)-binding site of C/EBPα. BRG-1 is the core subunit of an ATP-dependent chromatin remodelling complex. Sumoylation of C/EBPα dramatically decreases the stimulation of C/EBPα-mediated transactivation of the liver-specific albumin gene by BRG1. Sumoylated C/EBPα failed to induce proliferation arrest because its interaction with with BRG1 was inhibited.

Mutations in C/EBPα occur in approximately 5-14% of AML patients and are often associated with the induction of proliferation. N-terminal mutations cause truncation of the 42 kDa C/EBPα and overproduction of 30 kDa C/EBPα isoform that lacks TE-I and TE-II transactivation domains but retains TE-III and bZIP domains. 30 kDa C/EBPα isoform predominates and functions in a dominant negative manner resulting in a loss of C/EBPα function. One of target proteins of 30 kDa C/EBPα isoform is Ubc9, an essential E2 enzyme required for SUMO conjugation, or sumoylation [52]. Ubc9-mediated enhanced sumoylation of C/EBPα p42 decreases the transactivation capacity and thus the function of C/EBPα p42 (Figure 4).

Regulation of Cell Cycle Progression by C/EBPα

Growth-Inhibiting Activity of C/EBPα

C/EBPα is a strong inhibitor of cell proliferation when overexpressed in cultured cells [20,53]. C/EBPα mediates differentiation in several organ systems, including liver, adipose, lung and the haematopoietic tissue. C/EBPα promotes differentiation by the up-regulation of lineage-specific gene products and by the exit from cell cycle that means proliferation arrest. The capacity of C/EBPα to promote growth arrest has been studied *in vitro*, by analysis of knockout mice and by examination of leukaemic cells [20,54]. Several models of C/EBPα-induced growth arrest have been described [20,53]. These include C/EBPα-mediated (1) stabilization of the cyclin-dependent kinase 2 (CDK2) inhibitor, $p21^{WAF1/CIP1}$ [55], (2) regulation of growth-inhibiting Rb-E2F complexes [56], (3) interaction with free E2F, inhibition of E2F activity and down-regulation of the E2F target gene c-*myc* [57,58], (4) interaction with Max, a member of the basic region-helix-loop-helix-leucine zipper proteins, that belongs to a network of transcription factors including the Myc and Mad families of protein [59,60]. (5) inhibition of CDK2 and CDK4 activity [61], and (6) interaction with the SWI/SNF chromatin remodelling complexes [62,63].

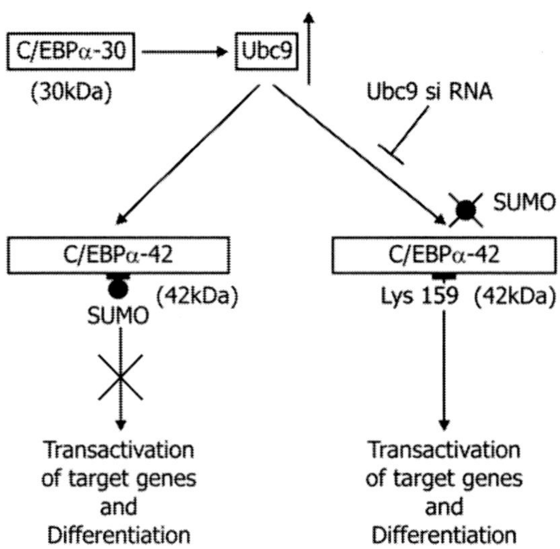

Figure 4. Model for the role of the E-2 conjugating enzyme Ubc9,which is essential for sumoylation, in AML with *C/EBPα* N-terminal mutation, where expression of C/EBPα p30 dominant negative isoform is increased. C/EBPα p30 enhances C/EBPα p42 sumoylation via upregulation of Ubc9 and blocks the transcription and differentiation potential of C/EBPα p42. Small interference RNA for Ubc9 (Ubc9 si RNA) overcomes the C/EBPα p30-mediated block of transcription and differentiation potential of C/EBPα p42.

Models (1) and (2) have been questioned by experiments performed in $p21^{WAF1/CIP1}$ [64] and Rb [65] null cell lines. Although the $p21^{WAF1/CIP1}$ protein is a key element of growth regulation in cells in culture, $p21^{WAF1/CIP1}$ knockout mice are morphologically normal and do not exhibit increased tumor formation. Overexpression of $p21^{WAF1/CIP1}$ in liver resulted in a dramatic inhibition of liver proliferation during development as well as in response to the proliferative stimulus of partial hepatectomy. C/EBPα controls hepatocyte proliferation *in vivo* in the developing liver. In the absence of C/EBPα, $p21^{WAF1/CIP1}$ protein levels are reduced and changes in $p21^{WAF1/CIP1}$ protein levels do not correlate with $p21^{WAF1/CIP1}$ mRNA levels. In the regenerating rat liver, C/EBPα and $p21^{WAF1/CIP1}$ protein levels are coordinated.

Pocket domain proteins (pRb, p107 and p130) plays a key role in controlling cell proliferation. They function mainly through their interaction with E2F transcription factors, forming complexes that repress transcription of S-phase genes (Figure 2 and Figure 3). Proteins p107 and p130 preferentially associate with E2F4 and E2F5, whereas pRb can bind to all the E2Fs [66]. C/EBPα regulates the formation of growth-inhibiting E2F-pRb complexes or enhances their activity. C/EBPα disrupts E2F-p107 complexes associated with proliferating liver cells [56]. However, C/EBPα increases the abundance of E2F-p130 complexes in 3T3-L1 preadipocytes [67]. This response involves induction or/and stabilization of $p21^{WAF1/CIP1}$, which inhibits CDK activity.

According to the model (3) C/EBPα suppresses cell proliferation through interaction with free E2F, as opposed to regulating pRb-E2F complexes. Different regions of C/EBPα are involved in growth inhibition (Figure 1). C/EBPα was found in complex with E2F and this complex was bound to E2F sites in genes such as dihydrofolate reductase (*DHFR*) in murine fibroblasts where C/EBPα inhibits proliferation [68]. E2F-1 was upregulated during the G1-S transition and C/EBPα also repressed transcription from reporter constructs containing the *DHFR* or *E2F-1* promoters [68]. C/EBPα can repress transcription from the *Myc* promoter, which also contains an E2F-binding element [57]. C/EBPα in complex with E2F are thus present in repressors which inhibit S-phase gene transcription. Further support for the E2F corepression model was described by Porse et al. [58]. They identified transactivation element TE-I at the N-terminus of C/EBPα as a critical inhibitory domain. They also studied basic region of the C/EBPα which plays an important role in protein-protein interactions of C/EBPα. C/EBPα interacts with free E2F through the non-DNA binding surface of its basic region. These interactions are necessary for C/EBPα activity. Mutations in this basic region cause decreased repression of E2F-driven transcription compared with wild-type C/EBPα. These mutations also inhibit differentiation of adipocytes and neutrophils, whereas other hematopoietic lineages develop normally. No definitive evidence for C/EBPα-E2F complexes bound to E2F sites via the E2F has been reported till now [20,54,69].

The basic DNA-binding region of the C/EBPα is also involved in the C/EBPα and Max interaction, likely through C/EBPα R297, K298 and/or K302 in the fourth model [60]. C/EBPα R297 is known to participate in the interaction between C/EBPα and E2F [58].

Model (4) based on interaction of C/EBPα with the basic region-helix-loop-helix-leucine zipper protein Max is conected with model (3) through the protein c-Myc, the E2F target gene. c-Myc is a basic helix-loop-helix leucine zipper protein that dimerizes with its partner Max to activate gene transcription through consensus E-box elements located on the promoters of certain genes [70,71]. Myc was discovered to be an oncogene causing leukemia in birds and inducing *in vitro* transformation of avian myeloid cells [72]. Dysregulated c-Myc expression has been implicated in the development of lymphoid malignancies and other tumors [73,74], as well as in the induction of genomic instability [75]. This demonstates an important role of c-Myc in the regulation of the cell cycle [76]. Interaction of C/EBPα with

Max involves the basic region of C/EBPα. Endogenous C/EBPα and Max but not Myc and Max, colocalize in intranuclear structures during granulocyte differentiation of myeloid U937 cells. Max enhanced the transactivation capacity of C/EBPα on a minimal promoter. A chromatin immunoprecipitation assay revealed occupancy of the human C/EBPα promoter *in vivo* by Max and Myc under cellular settings and by C/EBPα and Max under retinoic acid induced granulocytic differentiation [60]. Enforced expression of Max and C/EBPα results in granulocytic differentiation of the human hematopoietic CD34$^+$ cells. Silencing of Max by short hairpin RNA in CD34$^+$ and U937 cells strongly reduced the differentiation-inducing potential of C/EBPα, indicating the importance of C/EBPα-Max in myeloid progenitor differentiation [60].

In the CDK2/CDK4 inhibition model (5), C/EBPα interacts with and inhibits the activity of CDK2/CDK4 through the 15-amino acid proline- and histidine-rich region (PHR) located in the central part of C/EBPα [61]. CDK2 and CDK4 are key regulators of cell cycle progression. They phosphorylate pocket domain proteins thereby releasing their inhibition of E2F family members. C/EBPα binds to, and inhibits the activities of, both CDK2-cyclin E and CDK4-cyclin D complexes. Model (5) has also been questioned by control experiments with mice homozygous for the deletion of PHR [77]. C/EBPα$'^{PHR}$ is responsible for interaction of C/EBPα with Cdk2/4. Mice homozygous for the ΔPHR allele did not display any phenotype that could be related to the role of C/EBPα as a growth repressor [77]. C/EBPα may promote growth arrest by cell-specific mechanisms. This means that different models of C/EBPα-mediated cell-cycle arrest (Figure 2) operate in different cell types.

Finally, according to the SWI/SNF recruitment model (6), C/EBPα interacts with SWI/SNF components through a centrally located 75-amino acid region overlapping with the CDK2/CDK4 binding region [62,63]. C/EBPα interacts with the SWI/SNF chromatin remodeling complex during the regulation of differentiation-specific genes. C/EBPα fails to suppress proliferation in SWI/SNF defective cell lines after knock-down of SWI/SNF core components or after deletion of the SWI/SNF interaction domain in C/EBPα respectively. Reconstitution of SWI/SNF function restores C/EBPα-dependent proliferation arrest. Antiproliferative activity of C/EBPα critically depends on components of the SWI/SNF core complex and suggest that the functional interaction between SWI/SNF and C/EBPα is a prerequisite for proliferation arrest.

Although overexpression experiments and analysis of cells lacking known cell-cycle regulators are useful for identifying the pathways in which C/EBPα functions, these approaches do not necessarily reveal the primary target. A unifying mechanism of how C/EBPα suppresses proliferation has yet to be found.

The Role of C/EBPα in Myeloid Differentiation and Leukemia

A ctritical role for the function of C/EBPα in granulopoiesis was demonstrated in mice harbouring a disruption of the *C/EBPα* gene [8]. These mice show a selective early block in granulopoiesis, with the appearance of many myeloid blasts in fetal liver and peripheral blood [78]. C/EBPα cooperates with additional factors to direct monocytic commitment of primary myeloid progenitors [79]. C/EBPα binds and activates the endogenous *PU.1* gene in myeloid cells [80,81]. Acute leukemias are clonal disorders that are characterized by a block of differentiation along one or more hematopoietic lineages. C/EBPα mutations have been observed in AML patients with the approximate frequency 5-14% [18-21,82]. In the absence of specific *C/EBPα* mutations, decreased expression of C/EBPα is an alternative mechanism that disrupts C/EBPα function [18-21]. C/EBPα is not only a critical tumor suppressor in the hematopoietic system but acts as oncogen in human precursor-B lymphoblastic acute leukaemia (BCP-ALL) cells by juxtaposition to the immunoglobulin gene enhancer upon the t(14;19)(q32;q13) chromosomal rearrangement. Translocations involving the immunoglobulin heavy chain locus (*IGH*) at chromosomal band 14q32 in BCP-ALL cells is a rare but recurrent event. These translocations can be connected with the translocation of the *C/EBPα* gene on chromosome 19q13.1. This t(14;19)(q32;q13) chromosomal rearrangement leads to overexpression of C/EBPα protein, usually of normal sequence, which exhibits oncogenic properties [27,28]. Thus it appears that either loss of function of C/EBPα or gain of function of C/EBPα has leukemogenic potential (Figure 5).

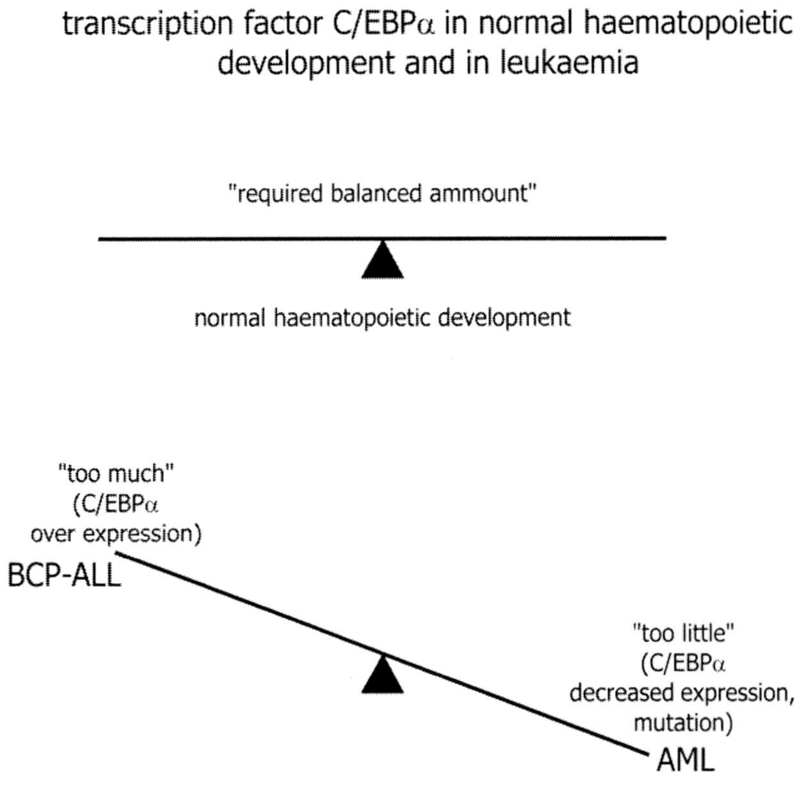

Figure 5. The effect of the amount of transcription factor C/EBPα in leukemogenesis. "Just the right" amount of C/EBPα is needed for maintenance of normal hematopoiesis. "Too much or too little" of C/EBPα can contribute to leukemogenesis. Apparently normal C/EBPα is over-expressed in BCP-ALL harbouring the translocation t(14; 19)(q32; q13). Mutations or decreased expression of C/EBPα cause AML.

Conclusion and Perspectives

Transcription factor C/EBPα is in addition to its transcriptional role also a strong inhibitor of cell proliferation. It inhibits proliferation of cultured cells when overexpressed, and it inhibits proliferation of newborn liver and in liver of adult animals. It is apparent now that C/EBPα is able to regulate cellular proliferation and differentiation through a variety of mechanisms. Although overexpression experiments and analysis of cells lacking known cell-cycle regulators are useful for identifying the pathways in which C/EBPα functions, these approaches do not necessarily reveal the primary target. A unifying mechanism of how C/EBPα suppresses proliferation has yet to be found. C/EBPα coordinates terminal differentiation by the upregulation of differentiation specific gene products and by mediating exit from the cell cycle. Both these functions of C/EBPα are conceivable targets in tumorigenic processes. Mutations of C/EBPα play an important role in the development of acute myeloid leukemia but not in other carcinomas such as lung, liver, skin and mammary gland. Downregulation or functional inactivation of C/EBPα is a prerequisite in some types of these carcinomas. A novel mechanism of C/EBPα p42 transcriptional repression in leukemia by Ubc9-mediated sumoylation has been described. Interestingly, either by mutating the SUMO site or by disrupting Ubc9 we could overcome the C/EBPα p30-mediated block of transcription and the differentiation potential of C/EBPα p42. Therefore, understanding such mechanisms leads to the development of new therapeutics that stimulate C/EBPα activity to overcome the differentiation block observed in AML. On the other hand, C/EBPα can also stimulate proliferation and behave as oncogen in human precursor-B lymphoblastic acute leukaemia (BCP-ALL). The next challenge is to find a more comprehensive list of C/EBPα targets in different cell types and cellular states as well as networks of protein interactions and regulatory patways that control their activities.

Acknowledgements

This work was supported by the Internal Grant Agency of the Ministry of Health of the Czech Republic NR/9045-3, VZ 00023736 from the Ministry of Health of the Czech Republic, and grant LC 06044 from Ministry of Education, Youth and Sport of the Czech Republic.

References

[1] Cao, Z; Umek, RM; McKnight, SL. Regulated expression of three C/EBP isoforms during adipose conversion of 3T-L1 cells. *Genes Dev* 1991, 5, 1538-1552.

[2] Scott, LM; Civin, CI; Rorth, P; Friedman AD. A novel temporal expression pattern of three C/EBP family members in differentiating myelomonocytic cells. *Blood* 1992, 80, 1725-1735.

[3] Freytag, SO; Paielli, D; Gilbert, JD. Ectopic expression of the CCAAT/enhancer binding protein α promotes the adipogenic program in a variety of mouse fibroblastic cells. *Genes Dev* 1994, 8, 1654-1663.

[4] Flodby, P; Barlow, C; Kylefjord, H; Ährlund-Richter, L; Xanthopoulos, KG. Increased hepatic cell proliferation and lung abnormalities in mice deficient in CCAAT/enhancer binding protein α. *J Biol Chem* 1996; 271, 24753-24760.

[5] Oh, HS; Smart, RC. Expression of CCAAT/enhancer binding proteins (C/EBP) is associated with squamous differentiation in epidermis and isolated primary keratinocytes and is altered in skin neoplasms. *J Invest Dermatol* 1998, 110, 939-945.

[6] Radomska, HS; Huettner, CS; Zhang, P; Cheng, T; Scadden, DT; Tenen, DG. CCAAT/enhancer binding protein α is a regulatory switch sufficient for induction of granulocyte development from bipotential myeloid progenitors. *Mol Cell Biol* 1998, 18, 4301-4304.

[7] Nerlov, C. The C/EBP family of transcription factors: a paradigm for interaction between gene expression and proliferation control. *Trends Cell Biol* 2007, 17, 318-324

[8] Wang, ND; Finegold, MJ; Bradley, A; Ou, CN; Abdelsayed, SV; Wilde, M et al. Impaired energy homeostasis in C/EBP alpha knockout mice. *Science* 1995, 269, 1108-1112.

[9] Zhang, DE; Zhang, P; Wang, ND; Hetherington, CJ; Darlington, GJ; Tenen, DG. Absence of granulocyte colony-stimulating factor signaling and neutrophil and development in CCAAT enhancer binding protein alpha-deficient mice. *Proc Natl Acad Sci USA* 1997, 94, 569-574.

[10]Kimura, T; Christoffels, VM; Chowdhury, S; Iwase, K; Matsuzaki, H; Mori, M et al. Impaired energy homeostasis in C/EBP alpha knockout mice. *Science* 1995, 269, 1108-1112.

[11]Inoue, Y; Inoue, J; Lambert, G; Yim, SH; Gonzalez, FJ. Disruption of hepatic C/EBPα results in impaired glucose tolerance and age-dependent hepatosteatosis. *J Biol Chem* 2004, 279, 44740-44748.

[12]Yang, J; Croniger, CM; Lekstrom-Himes, J; Zhang, P; Fenyus, M; Tenen, DG et al. Metabolic response of mice to a postnatal ablation of CCAAT/enhancer binding protein α. *J Biol Chem* 2005, 280, 38689-38699.

[13]Bassères, DS; Levantini, E; Ji, H; Monti, S; Elf, S; Dayaram, T et al. Respiratory failure due to differentiation arrest and expansion of alveoral cells Following lung-specific loss of the transcription factor C/EBPα in mice. *Mol Cell Biol* 2006, 26, 1109-1123.

[14]Halmos, B; Huettner, CS, Kocher, O; Ferenczi, K; Karp, DD; Tenen, DG. Down-regulation and antiproliferative role of C/EBPα in lung cancer. *Cancer Res* 2002, 62, 528-534.

[15]Gery, S; Tanosaki, S; Bose, S; Vadgama, J; Koeffler, HP. Down-regulation and growth inhibitory role of C/EBPα in breast cancer. *Clin Cancer Res* 2005, 11, 3184-3190.

[16]Takai, N; Kawamata, N; Walsh, CS; Gery, S; Desmond, JC; Whittaker, S et al. Discovery of epigenetically masked tumor suppressor genes in endometrial cancer. *Mol. Cancer Res* 2005, 3, 261-269.

[17]Tada, Y; Brena, RM; Hackanson, B; Morrison, C; Otterson, GA; Plass, C et al. Epigenetic modulation of tumor suppressor CCAAT/enhancer binding protein α activity in lung cancer. *J Nat Cancer Inst* 2006, 98, 396-406.

[18]Nerlov, C. C/EBPα mutations in acute myeloid leukaemias. *Nat Rev Cancer* 2004, 4, 394-400.

[19]Mueller, BU; Pabst, T. C/EBPα and the pathophysiology of acute myeloid leukemia. *Curr Opin Hematol* 2006, 13, 7-14.

[20]Schuster, MB; Porse, BT. C/EBPα: a tumour suppressor in multiple tissues? *Biochim Biophys Acta* 2006, 1766, 88-103.

[21]Fuchs, O. Growth-inhibiting activity of transcription factor C/EBPα, its role in haematopoiesis and its tumour suppressor or oncogenic properties in leukaemias, *Folia Biol (Praha)* 2007, 53, 97-108.

[22]Shim, M; Powers, KL; Ewing, SJ; Zhu, S; Smart, RC. Diminished expression of C/EBPα in skin carcinomas is linked to oncogenic Ras and reexpression of C/EBPα in carcinoma cells inhibits proliferation. *Cancer Res* 2005, 65, 861-867.

[23]Loomis, KD; Zhu, S; Yoon, K; Johnson, PF ; Smart, RC. Genetic ablation of CCAAT/enhancer binding protein α in epidermis reveals its role in suppression of epithelial tumorigenesis. *Cancer Res* 2007, 67, 6768-6776.

[24]Truong, BT; Lee, YJ; Lodie, TJ; Park, DJ; Perrotti, D; Watanabe, N et al. CCAAT/Enhancer binding proteins repress the leukemic phenotype of acute myeloid leukemia. *Blood* 2003, 101, 1141-1148.

[25]Schepers, H; Wierenga, ATJ; van Gosliga, D; Eggen, BJL; Vellenga, E; Schuringa, JJ. Reintroduction of C/EBPα in leukemic $CD34^+$ stem/progenitor cells impairs self-renewal and partially restores myelopoiesis. *Blood* 2007, 110, 1317-1326.

[26]Tan, EH; Hooi, SC; Laban, M; Wong, E; Ponniah, S; Wee, A; Wang ND. CCAAT/Enhancer-binnding protein alpha knock-in mice exhibit early liver glycogen storage and reduced susceptibility to hepatocellular carcinoma. *Cancer Res* 2005, 65, 10330-10337.

[27]Chapiro, E; Russell, L; Radford-Weiss, I; Bastard, C; Lessard, M; Struski, S. et al. Overexpression of *CEBPA* resulting from the translocation t(14; 19)(q32; q13) of human precursor-B-cell acute lymphoblastic leukaemia. *Blood* 2006, 108, 3560-3563.

[28]Akasaka, T; Balasas, T; Russell, LJ; Sugimoto, KJ.; Majid, A; Walewska, et al. Five members of the *CEBP* transcription factor family are targeted by recurrent *IGH* translocations in B-cell precursor acute lymphoblastic leukemia (BCP-ALL). *Blood* 2007, 109, 3451-3461.

[29]Hendricks-Taylor, LR; Bachinski, LL; Sicilliano, MJ; Fertitta, A; Trask, B; de Jong, PJ; Ledbetter, DH, Darlington, GJ. The CCAAT/enhancer binding protein (C/EBPα) gene (CEBPA) maps to human chromosome 19q13.1 and the related nuclear factor NF-IL6 (C/EBP ß) gene (CEBPB) maps to human chromosome 20q13.1. *Genomics 14*, 12-17.

[30]Birkenmeier, EH; Gwynn, B; Howard, S; Jerry, J;. Gordon, JI; Landschulz, WH; McKnight, SL. Tissue-specific expression, developmental regulation and genetic mapping of the gene encoding CCAAT/enhancer binding protein. *Genes Dev. 3*, 1146-1156.

[31]Calkhoven, CF; Bouwman, PRJ; Snippe, L; Geert, AB. Translation start site multiplicity of the CCAAT/enhancer binding protein α mRNA is dictated by a small 5′ open reading frame. *Nucleic Acids Res* 1994, 22, 5540-5547.

[32]Calkhoven, C. F., Muller, C., Leutz, A. Translational control of C/EBPα and C/EBPβ isoform expression. *Genes Dev* 2000, 14, 1920-1932.

[33]Ossipow, V; Decombes, P; Schibler, U. CCAAT/enhancer-binding protein mRNA is translated into multiple proteins with different transcription activation potentials. *Proc Natl Acad Sci USA* 1993, 90, 8219-8223.

[34]Nerlov, C; Ziff, EB. Three levels of functional interaction determine the activity of CCAAT/enhancer binding protein-α on the serum albumin promoter. *Genes Dev* 1994, 8, 350-362.

[35]Nerlov, C; Ziff, EB. CCAAT/enhancer binding protein-α amino acid motifs with dual TBP and TFIIB binding ability co-operate to activate transcription in both yeast and mammalian cells. *EMBO J* 1995, 14, 4318-4328.

[36]Pei, DQ; Shih, CH. An "attenuator domain" is sandwiched by two distinct trans-activation domains in the transcription factor C/EBP. *Mol Cell Biol* 1991, 11, 1480-1487.

[37]Iniguez-Lluhí, J; Pearce, D. A common motif within the negative regulatory regions of multiple factors inhibit their transcriptional synergy. *Mol Cell Biol* 2000, 20, 6040-6050.

[38]Cleaves, R; Wang, QF; Friedman AD. C/EBPα p30, a myeloid leukemia oncoprotein limits G-CSF receptor expression but not terminal granulopoiesis via site-selective inhibition of C/EBP DNA binding. *Oncogene* 2004, 23, 716-772.

[39]Calkhoven, CF; Snippe, L; Ab, G. Differential stimulation by CCAAT/enhancer-binding protein α isoforms of the estrogen-activated promoter of the very-low-density apolipoprotein II gene. *Eur J Biochem* 1997, 249, 113-120.

[40]Lin, FT; McDougold, CA; Diehl, AM; Lane, MD. A 30-kDa alternative translation product of the CCAAT/enhancer binding protein α message: transcriptional activator lacking antimiotic. activity. *Proc Natl Acad Sci USA* 1993, 90, 9606-9610.

[41]Johnson, PF. Identification of C/EBP basic region residues involved in DNA sequence recognition and half-site spacing preference. *Mol Cell Biol* 1993, 13, 6916-6930.

[42]Koldin, B; Suckow; M, Seydel, A; von Wilcken-Bergmann, B; Müller-Hill, B. A comparison of the different DNA binding specificities of the bZIP proteins C/EBP and GCN4. *Nucleic Acids Res* 1995, 23, 4162-4169.

[43]Chen, S-S; Chen, J-F; Johnson, PF; Muppala, V; Lee, Y-H. C/EBPβ, when expressed from the *C/ebpα* gene locus, can functionally replace C/EBPα in liver but not in adipose tissue. *Mol Cell Biol* 2000, 20, 7292-7299.

[44]Radomska, HS; Basseres, DS; Zheng, R; Zhang, P; Dayaram, T; Yamamoto, Y. et al. Block of C/EBPα function by phosphorylation in acute myeloid leukemia with FLT3 activating mutations. *J Exp Med 2006*, 203, 371-381.

[45]Liu, H-K; Perrier, S; Lipina, C; Finlay, D; McLauchlan, H; Hastie, CJ; Hundal, H. S., Sutherland, C. Functional characterisation of the regulation of CAAT enhancer binding protein alpha by GSK-3 phosphorylation of threonines 222/226. *BMC Mol Biol* 2006, 7, 1-12.

[46]Behre, G; Singh, SM; Liu, H; Bortolin, LT; Christopei, M;Radomska, HS et al. Ras signaling enhances the activity of C/EBPα to induce granulocytic differentiation by phosphorylation of serine 248. *J Biol Chem* 2002, 277, 26293-26299.

[47]Wang, G-L; Timchenko, NA. Dephosphorylated C/EBPα accelerates cell proliferation through sequestering retinoblastoma protein. *Mol Cell Biol* 2005, 25, 1325-1338.

[48]Trivedi, AK; Bararia, D; Christopeit, M; Peerzada, AA; Singh, SM; Kieser, A et al. Proteomic identification of C/EBP-DBD multiprotein complex: JNK1 activates stem cell regulator C/EBPalpha by inhibiting its ubiquitination. *Oncogene* 2007, 26**,** 1789-1801.

[49]Wang, G-L; Iakova, P.; Wilde, M; Awad, S; Timchenko, NA. Liver tumors escape negative control of proliferation via PI3K/Akt-mediated block of C/EBPα growth inhibitory activity. *Genes Dev* 2004, 18, 912-925.

[50]Subramanian, L; Benson, MD; Iniguez-Lluhí, JA. A synergy control motif within the the attenuator domain of CCAAT/enhancer-binding protein α inhibits transcriptional synergy through its PIASy-enhanced modification by SUMO-1 or SUMO-3. *J Biol Chem* 2003, 278, 9134-9141.

[51]Sato, Y; Miyake, K; Kaneoka, H; Iijima, S. Sumoylation of CCAAT/enhancer-binding protein α and its functional roles in hepatocyte differentiation. *J Biol Chem* 2006, 281, 21629-21639.

[52]Geletu, M; Balkhi, MY; Zada, AAP; Christopeit, M; Pulikkan, JA; Trivedi, AK et al. Target proteins of C/EBPα p30 in AML: C/EBPα p30 enhances sumoylation of C/EBPα p42 via up-regulation of Ubc9. *Blood* 2007, 110, 3301-3309.

[53]Hendricks-Taylor, LR; Darlington, GJ. Inhibition of cell proliferation by C/EBPα occurs in many cell types, does not require the presence of p53 or Rb, and is not affected by large T-antigen. *Nucleic Acids Res* 1995, 23, 426-433.

[54]Johnson, PF. Molecular stop signs: regulation of cell cycle arrest by C/EBP transcription factors. *J Cell Sci* 2005, 118, 2545-2555.

[55]Timchenko, NA; Harris, TE; Wilde, M; Bilyeu, TA; Burgess-Beusse, BL; Finegold, MJ; Darlington, GJ. CCAAT/enhancer-binding protein α regulates p21 protein and hepatocyte proliferation in newborn mice. *Mol Cell Biol* 1997, 17, 7353-7361.

[56]Timchenko, NA; Wilde, M; Darlington, GJ. C/EBPα regulates formation of S-phase-specific E2F-p107 complexes in livers of newborn mice. *Mol Cell Biol* 1999, 19, 2936-2945.

[57]Johansen, LM; Iwama, A; Lodie, TA; Sasaki, K; Felsher, DW; Golub, TR; Tenen DG. c-Myc is a critical target for C/EBPα in granulopoiesis. *Mol Cell Biol* 2001, 21, 3789-3806.

[58]Porse, BT; Pedersen, TA; Xu, X; Lindberg, B; Wewer, UM; Friis-Hansen, L; Nerlov, C. E2F repression by C/EBPα is required for adipogenesis and granulopoiesis in vivo. *Cell* 2001, 107, 247-258.

[59]Grandori, C; Cowley, SM; James, LP; Eisenman, RN. The Myc/Max/Mad network and the transcriptional control of cell behavior. *Annu Rev Cell Dev Biol* 2000, 16, 653-699

[60]Zada, AA; Pulikkan, JA; Bararia, D; Geletu, M; Trivedi, AK; Balkhi, MY et al. Proteomic discovery of Max as a novel interacting partner of C/EBPα: a Myc/Max/Mad link. *Leukemia* 2006, 20, 2137-2146.

[61]Wang, H; Iakova, P; Wilde, M; Welm, A; Goode, T; Roesler, W.J; Timchenko, NA. C/EBPα arrests cell proliferation through direct inhibition of Cdk2 and Cdk4. *Mol Cell* 2001, 8, 817-828.

[62]Pedersen, TA; Kowenz-Leutz, E; Leutz, A; Nerlov, C. Cooperation between C/EBPα, TBP/TFIIB and SWI/SNF recruiting domains is required for adipocyte differentiation. *Genes Dev* 2001, 15, 3208-3216.

[63]Műller, C; Calkhoven, C. F; Sha, X; Leutz, A. The CCAAT enhancer-binding protein α (C/EBPα) requires a SWI/SNF complex for proliferation arrest. *J Biol Chem* 2004, 279, 7353-7358.

[64]Műller, C; Alunni-Fabbroni, M; Kowenz-Leutz, E; Mo, X; Tommasino, M; Leutz, A. Separation of C/EBPα-mediated proliferation arrest and differentiation pathways. *Proc Natl Acad Sci USA* 1999, 96, 7276-7281.

[65]Hendricks-Taylor, LR; Darlington, GJ. Inhibition of cell proliferation by C/EBPα occurs in many cell types, does not require the presence of p53 or Rb, and is not affected by large T-antigen. *Nucleic Acids Res* 1995, 23, 426-433.

[66]Classon, M; Harlow, E. The retinoblastoma tumour suppressor in development and cancer. *Nat Rev Cancer* 2002, 2, 910-917.

[67]Timchenko, NA; Wilde, M; Iakova, P; Albrecht, JH; Darlington, GJ. E2F/p107 and E2F/p130 complexes are regulated by C/EBPα in 3T3-L1 adipocytes. *Nucleic Acids Res* 1999, 27, 3621-3630.

[68]Slomiany, BA; D'Arigo, KL; Kelly, MM; Kurtz, DT. C/EBPα inhibits cell growth via direct repression of E2F-DP-mediated transcription. *Mol Cell Biol* 2000, 20, 5986-5997.

[69]Porse, BT; Bryder, D; Theilgaard-Monch, K; Hasemann, MS; Anderson, K; Damgaard, I et al. Loss of C/EBPα cell cycle control increases myeloid progenitor Proliferation and transforms the neutrophil granulocyte lineage. *J Exp Med* 2005, 202, 85-96.

[70]Blackwell, TK; Kretzner; L; Blackwood, EM; Eisenman, RN; Weintraub, H. Sequence-specific DNA binding by the c-Myc protein. *Science* 1990, 250, 1149- 1151.

[71]Blackwood, EM; Eisenman, RN. Max: a helix-loop-helix zipper protein that forms a sequence-specific DNA-binding complex with c-Myc. *Science* 1991, 251, 1211-1217.

[72]Sheiness, D; Fanshier, L; Bishop, JM. Identification of nucleotide sequences which may encode the oncogenic capacity of avian retrovirus MC29. *J Virol* 1978, 28, 600-610.

[73]Dalla-Favera, R; Martinotti, S; Gallo, RC; Erikson, J; Croce, CM. Translocation and rearrangements of the c-myc oncogene locus in human undifferentiated B-cell lymphomas. *Science* 1983, 219, 963-967.

[74]Leder, A; Pattengale, PK; Kuo, A; Stewart, TA; Leder, P. Consequences of widespread deregulation of the c-myc gene in transgenic mice: multiple neoplasms and normal development. *Cell* 1986, 45, 485-495.

[75]Felsher, DW; Bishop, JM. Transient excess of MYC activity can elicit genomic instability and tumorigenesis. *Proc Natl Acad Sci USA* 1999, 96, 3940-3944.

[76]Obaya, AJ; Mateyak, MK; Sedivy JM. Mysterious liaisons: the relationship between c-Myc and the cell cycle. *Oncogene* 1999, 18, 2934-2941.

[77]Porse, BT; Pedersen, TA; Hasemann, MS; Schuster, MB; Kirstetter, P., Luedde, T et al. The proline-histidine-rich CDK2/CDK4 interaction region of C/EBPα is dispensable for C/EBPα-mediated growth regulation in vivo. *Mol Cell Biol* 2006, 26, 1028-1037.

[78]Zhang, DE; Yhang, P; Wang, ND; Hetherington, CJ; Darlington, GJ; Tenen, DG. Absence of granulocyte colony-stimulating factor signaling and neutrophil development in CCAAT enhancer binding protein α-deficient mice. *Proc Natl Acad Sci USA* 1997, 94, 569-574.

[79]Wang, D; D'Costa, J; Civin, CI; Friedman, A. D. C/EBPα directs monocytic commitment of primary myeloid progenitors. *Blood* 2006, 108, 1223-1229.

[80]Kummalue, T; Friedman, AD. Cross-talk between regulators of myeloid development: C/EBPα binds and activates the promoter of the PU.1 gene. *J Leukoc Biol* 2003, 74, 464-470.

[81]Friedman, AD. C/EBPα induces PU.1 and interacts with AP-1 and NF-κB to regulate myeloid development. *Blood Cells Mol Dis* 2007, 39, 340-343.

[82]Pabst, T; Mueller, BU; Zhang, P; Radomska, HS; Narravula, S; Schnittger, S et al. Dominant negative mutations of CEBPA, encoding CCAAT/enhancer binding protein-α (C/EBPα) in acute myeloid leukemia. *Nat Genet* 2001, 27, 263-270.

In: Cell Cycle Control: New Research
Editors: N. H. Leroy, N. T. Fournier
ISBN: 978-1-60456-796-0

Chapter IX

The Non-Hemodynamic Effect of Angiotensin II in Renal Fibrogenesis

Bi-Cheng Liu[*] and Jian-Dong Zhang
Institute of Nephrology, Southeast University, Nanjing, 210009, China.

Abstract

Renal fibrosis is a common pathway of progressive renal disease and parallels the reduction in organ function that necessitates dialysis or transplantation. In the last decades, accumulating experimental and clinical evidence convincingly suggests that the renin-angiotensin system (RAS) is a key player in the progression of chronic kidney disease. Angiotensin II (ANG II), the main peptide of the RAS, was classically viewed as a vasoactive factor participating in the local and systemic hemodynamic regulation. With the progression of researches, it has emerged as a multifunctional cytokine exhibiting many non-hemodynamic effects such as growth, profibrotic, matrix metabolism and even proinflammation. These new understandings of ANG II have important clinical implications. It predicted and explained why blockade of the RAS would significantly slow the progression of fibrotic disease. However, it further suggested that higher doses and/or a combination of ANG II blockade with another agent or agents might truly halt the progressive fibrosis.

This review will focus on the non-hemodynamic aspect of ANG II and its contribution in renal fibrogenesis.

Introduction

The critical role of the circulating RAS in the regulation of arterial pressure and sodium homeostasis has been recognized for many years [1]. ANG II is the most powerful

[*] Correspondence concerning this article should be addressed to: Professor Bi-Cheng Liu, MD, PhD, Email: liubc64@yahoo.com.cn.

biologically active product of the RAS, although there are other bioactive angiotensin peptides, including ANG III, ANG IV, and ANG 1-7. ANG II directly constricts vascular smooth muscle cells, enhances myocardial contractility, stimulates aldosterone production, stimulates release of catecholamines from the adrenal medulla and sympathetic nerve endings, increases sympathetic nervous system activity, and stimulates thirst and salt appetite. ANG II also regulates sodium transport by epithelial cells in intestine and kidney [2]. There has also been a growing appreciation of the organ-specific roles exerted by ANG II acting as a paracrine factor. In addition to its physiological roles, locally produced ANG II induces inflammation, cell growth, mitogenesis, apoptosis, migration, and differentiation, regulates the gene expression of bioactive substances, and activates multiple intracellular signaling pathways, all of which might contribute to tissue injury [3]. Clinical and preclinical studies on the pharmacological effects of ACEIs and ARBs support the notion that ANG II exerts a central role in the pathogenesis of renal injury via activation of AT1 receptors when inappropriately activated [4].

Biological Characters of Angiotensin II in the Kidney

From the classical textbook, we can clearly find that ANG II is generated from angiotensinogen--the substrate of RAS system by two steps. One is from angiotensinogen to angiotensin I with the help of renin, another is from angiotensin I to angiotensin II catalyzed by Angiotensin Converting Enzyme (ACE). Besides this concept, other ANG II-generating enzymes include the serine protease chymase, which is supposed to mediate over 80% of ANG II formation in the heart and less than 60% in the vessels [5]. ACE inhibitors do not reduce chymase activity. Upregulation of chymase, mainly in the tubules, is observed in renal biopsies of patients with diabetic nephropathy [6].

The origin of ANG II in the kidney had not been made clear until the definition of local RAS was introduced. In the lumen of the proximal tubule, ANG II levels are in the nanomolar range, or approximately 1,000-fold greater than levels in plasma [7]. A local RAAS including all its components could have been shown in the proximal tubular cells of the kidney. Proximal tubular cells actively produce ANG II and also secrete angiotensinogen into the urine. Intraluminal angiotensinogen may be converted in the distal tubules to AngII, and recent observations suggest that it leads to induction of sodium channels independent of aldosterone [8,9]. It appears that the glomerulus also contains a local RAS. Components of the RAS have been detected in cultured rat and human mesangial cells, and glomerular endothelial cells express ACE [10]. De novo synthesis of ANG II in isolated rat glomeruli has been shown by Atiyeh et al [11]. In addition, ANG II concentrations in glomerular ultrafiltrate are significantly greater than those in plasma, suggesting that glomerular ANG II synthesis occurs in vivo [12]. Renal injury can activate the local RAAS in both directly and indirectly manners [13].

The multiple effects of ANG II are mediated by different receptors. The two major ANG II receptors, AT1 and AT2, are differentially expressed within the kidney. Both are characterized by the configuration of a seven-transmembrane receptor but share only approximately 30% homology on the protein level [14].

ANG II is metabolized by peptidases, such as aminopeptidase A, into ANG III and further into ANG IV [15]. ANG III interacts with AT1 and AT2 receptors, albeit with a lower affinity, and exhibits principally similar effects as ANG II [16]. ANG IV binds to a specific receptor called AT4, which is widely expressed in the kidney, including endothelial cells and proximal as well as convoluted tubules. The AT4 receptor, by protein purification and peptide sequencing, has been identified to be insulin-regulated aminopeptidase [17].

Angiotensin II and its direct Effect on Kidney Cells

Resident kidney cells play a diverse role in the whole process of fibrosis. It was defined as a "victim", "accessary" and potential "saviour". Therefore, the direct influence of ANG II on resident kidney cells emerged into our sights as the first part of current discussion.

Research in the past several decades has provided ample evidence that ANG II modulates cell growth [18]. Depending on the cell type, subtype of ANG II receptors, and the presence of other growth factors and cytokines, ANG II may either stimulate growth (proliferation, hypertrophy) or act as a growth suppressor (apoptosis, phenotype change). These different cellular responses play a vital role in renal fibrosis.

In vitro and in vivo studies have provided clear evidence that ANG II exerts growth stimulatory actions. Depending on the target cell and presence of other growth factors, ANG II may induce either proliferation or hypertrophy. That various cyclin-CDK complexes must be activated and CDKIs should be suppressed in proliferation by ANG II is not surprising. In the kidney, ANG II stimulates proliferation of cultured mesangial cells, tubular epithelial cells and interstitial fibroblasts [19-21]. Mervaala et al investigated the growth effects in double-transgenic rats harboring human renin and human angiotensinogen genes. These animals have high local tissue ANG II levels, and organ injury is partly independent of hypertension. There was a significant increase in Ki-67–positive cells in the heart and kidney in double-transgenic rats; however, the exact cell type remains unclear [22].

Hypertrophy is defined as an increase in cell size, RNA content, and protein synthesis, without concomitant changes in cellular DNA content. Geisterfer et al reported in 1988 that ANG II induces hypertrophy, but not proliferation in confluently cultured rat aortic smooth muscle cells [23]. In the kidney, it was well defined that mesangial cell and proximal tubular epithelial cell could be induced to hypertrophy by ANG II too [24,25]. The hypertrophic action of ANG II is mediated through high-affinity AT_1 receptor and some growth factor such as TGF-beta1 and CTGF [26,27]. Cells undergoing ANG II-induced hypertrophy re-enter the cell cycle from G_0 but do not progress into the S phase. ANG II-mediated CDKI inductions such as $p27^{Kip1}$ play a pivotal role in this cell cycle arrest [28].

In some systems, ANG II mediates apoptosis. Although accumulating data suggest that apoptosis requires active cell cycle-dependent processes, few data exist concerning the association of ANG II-induced apoptosis and regulation of cell cycle events. Shankland's group showed a marked increase in glomerular cell apoptosis in nephritic $p27^{Kip1-/-}$ mice compared with nephritic p27$^{Kip1+/+}$ controls. Apoptosis was also increased in tubulointerstitial cells after obstruction in p27$^{Kip1-/-}$ mice compared with control p27$^{Kip1+/+}$ mice. Interestingly, inhibiting cyclin A-CDK2 activation reduced apoptosis in p27$^{Kip1-/-}$ cells. In apoptotic p27$^{Kip1-/-}$ mesangial cells, cyclin A-CDK2 was also activated without a proceeding increase in cyclin E-CDK2 activity, suggesting that uncoupling of CDK2 activity from the scheduled sequence of cell cycle protein expression may lead to an inappropriate and premature initiation of G_1/S phase transition, causing the cell to respond by apoptosis rather than proliferation [29].

ANG II induces apoptosis under certain conditions. Renal tubular apoptosis is a prominent feature of diabetic Ren-2 rats that overexpressed renin and ANG II [30]. It has been suggested that activation of AT_2 receptors is responsible for proapoptotic effect of ANG II, but this issue is controversial because AT_1 receptor antagonists also attenuated apoptosis in certain models [31,32]. Overexpression of AT_2-receptors in fibroblasts had antiproliferative effects without causing apoptosis. This effect was associated with an inhibition of CDK4 and cyclin E expression. CKI expression was not influenced in the AT_2-receptors expressing fibroblasts [33].

Finally, ANG II could exhibit a key cellular event that calls phenotypical change. These phenotypical changes could initiate in various cellular types [34-36]. One important issue is to obtain alpha-SMA expression in mesangial cells, tubular epithelial cells and interstitial fibroblasts.

In some experimental models and human membranous nephropathy, the expression of mesangial α-SMA (also called mesangioblast) is associated with the mesangial expansion and glomerular sclerosis, and blocking of ANG II could also ameliorate these changes [37]. In some aspects, the appearance of the mesangioblasts within glomeruli undergoing repair appears to reenact the embryologic steps of glomerulogenesis. The mesangioblast phenotype is also associated with changes in the ECM synthetic profile of mesangial cells. Changes in mesangial phenotype and the subsequent changes in glomerular ECM profile may also affect mesangial cell survival. It is likely to note that ANG II can also induce α-SMA expression in tubular epithelial cells and interstitial fibroblasts via direct and indirect pathways [38]. These activated tubular epithelial cells (this phenomenon is also called epithelial to mesenchymal transition, EMT) and interstitial fibroblasts will contribute as an alternative resource of the myofibroblast pool which is considered as a determinant in renal fibrosis [39].

Another important issue is the effect of ANG II on podocyte. As reported, ANG II could induce the apoptosis and detachment of podocyte [40]. It should be noticed that the interpretation of results from *in vitro* studies is largely limited, due to the properties of podocytes are largely influenced by the cell culture conditions used. By employing an isolated glomeruli, it was indicated that ANG II could depolarize podocytes, most probably by activating a Ca^{2+}-dependent Cl^- conductance, via an ANG II type 1 (AT1) receptor. Besides these direct evidences, blocking ANG II is proved to be a promising therapy to protect function of podocyte [41].

These differences in ANG II-mediated effects could be exclusively explained by activation of different ANG II receptors or different signal transduction systems. For example, depending on the cell type, hypertrophy and apoptosis could be mediated by either

AT_1- or AT_2-receptor activation [42]. Improved understanding of ANG II-mediated cell cycle regulation is necessary to develop improved therapeutic strategies.

Angiotensin II and Extracellular Matrix

Excessive deposition of extracellar matrix (ECM) has been accepted as a significant feature of renal fibrosis. The amount of ECM deposited in the kidney depends on the balance between synthesis and degradation of ECM [43]. ANG II can directly influence the final deposition via mentioned two pathways.

Angiotensin II and the Synthesis of ECM Components

Nearly one decade ago, studies suggested that ANG II up-regulates collagen expression in cultured mesangial cells [44]. After that, other molecules involved in matrix expansion including fibronectin, proteoglycans and osteopontin were identified to be induced by ANG II, as well as other kidney cells such as fibroblast and tubular epithelia were documented to be potential target of ANG II [45]. Besides these molecules, it is important to note that novel proteins include splice variants of fibronectin and novel chains of laminin and type IV collagen were synthesized under the stimulation of ANG II. In addition to providing structural support, these matrix proteins may also interact with cellular receptors especially of the integrin family. As a consequence, the nature of the interstitial matrix may significantly influence the behavior of the neighboring cells. The interstitial scar can be viewed as a dynamic scaffold that changes as the renal disease progresses. Initially the scaffold is relatively unstable and susceptible to degradation. With time, cross-linking bridges add stability and confer relative resistance to protease activity. Changes in matrix composition such as the ratio of collagen I to collagen III may contribute to the stabilizing process [46].

Angiotensin II and the Degradation of ECM

Two major ECM protease systems, matrix metalloproteinases (MMP)/tissue inhibitors of matrix metalloproteinases (TIMP) system and plasminogen activator (PA)/plasmin/PA inhibitors (PAI) system, are interrelated and involved in matrix degradation. The physiology of each system is complex, and the activities are tightly regulated at many levels, including gene expression, activation, and inhibition by specific inhibitors.

MMPs are classified into six groups based on substrate and sequence homology in which collagenases, gelatinases are the most important two. Collagenases [MMP-1, -8 (neutrophil collagenase), -13, and -18 (*Xenopus laevis* collagenase)] cleave collagens I, II, and III at a specific site, generating two fragments. Gelatinases [MMP-2 (gelatinase A) and -9 (gelatinase B)] cleave the denatured collagens (gelatins) and laminin, as well as some chemokines. Tissue inhibitors of metalloproteinases (TIMPs) are endogenous, specific inhibitors that bind and inhibit MMPs [47]. ANG II is reported to be a mediator of high glucose induced alteration of MMPs/TIMPs which results in a decreased degradation ability [48]. In STZ treated rats, we found that blockade of ANG II could restore the ratio of MMP2/TIMP2 and subsequently attenuate the increasing thickness of glomerular basement [49]. Recently, Ozawa et al employed a chronic infusion system to present a direct *in vivo* evidence of influence of ANG II on degradation system. They found that after 12 days of ANG II infusion, mRNA levels of MMP-9 and tPA were significantly increased by 117 and 42%, respectively, compared with the sham-operated rats. Following cessation of ANG II infusion, mRNA levels of MMP-9 and tPA also returned to the normal range within 3 days [50].

Plasmin is generated from plasminogen by the enzymatic activity of tissue-type PA (tPA) and urokinase-type PA (uPA). PA activity is inhibited by PAI [51]. A strong positive correlation between plasmin activity and ECM degradation in cultured mesangial cells and the ability of plasmin inhibitors to inhibit ECM degradation suggest the importance of plasmin in ECM degradation by mesangial cells [52]. Plasmin not only degrades several matrix proteins such as fibronectin, laminin, proteoglycan, and type IV collagen but also activates pro-MMP. PAI-1 seems to play an important role in ECM degradation in mesangial cells, because a larger molar excess of PAI-1 over tPA and uPA is detected in mesangial cell culture supernatant and because anti-PAI-1 antibody increases ECM degradation [53,54]. Tremendous data from both *in vitro* and *in vivo* indicate that ANG II can induce PAI-1 expression in mRNA and protein levels [55-57]. Molecular studies indicate that angiotensin-dependent, TGF-beta1 independent induction of PAI-1 gene expression involves the AT1 receptor and a glucocorticoid response element in the PAI-1 promoter [58]. AT1 receptor blockade but not nonspecific antihypertensive treatment reduces renal PAI-1 expression that is induced by ANG II infusion [59,60].

Interaction between ANG II and other Growth Factors

Numerous studies have demonstrated that renal fibrosis is orchestrated by many growth factors including ANG II. Thereby, an important issue is the relationship between ANG II and other growth factors.

Angiotensin II and TGF-beta1

TGF-beta1 mediates tissue fibrosis and is overexpressed in fibrotic renal disease, and anti-TGF-beta1 strategies have been shown to prevent fibrosis in experimental renal fibrotic disease [61].

Chronic renal disease is almost always associated with activation of the RAS system and elevated ANG II levels, which in turn will induce expression of TGF-beta1. In cultured renal mesangial and tubular epithelial cells, ANG II induces TGF-beta1 mRNA and protein production and increases the conversion of inactive TGF-beta1 to the biologically active form, and infusion of ANG II *in vivo* induces glomerular TGF-beta1 mRNA expression [62,63]. By employing an elegant transgenic model, Fern et al showed that TGF-beta1 mRNA levels linearly increased with number of angiotensinogen genes [64]. These data show that ANG II is the main force that drives TGF-beta1 production in tissue injury.

A number of experimental studies strongly suggested that the benefit of reducing ANG II is due to reduce the TGF-beta1 levels [65,66]. It was evidenced by part of CSGCT(Collaborative Study Group Captopril Trial) that lower TGF-beta1 levels after ANG II inhibition is beneficial to humans [67] Similar results were obtained by Laviades et al, who analyzed TGF-beta1 levels and biochemical markers of collagen metabolism in hypertensive patients treated by losartan [68]. These data indicated that TGF-beta1 as a mediator of profibrotic actions of ANG II. This deduction was strengthened by the evidence obtained by Wolf G et al. that anti-TGF-beta1 antibody abolished the release of collagen induced by ANG II [69]. Besides these, ANG II also was demonstrated to facilitate the action of TGF-β1 by increasing the expression of its receptor and activation of Smad without directly stimulation of TGF-β1 [70].

In contrast, TGF-β1 can also regulate the ANG II production. The secretion of renin, which shows elevated tissue levels in some forms of fibrotic renal disease, is enhanced by TGF-β1 *in vitro* and *in vivo* [71]. Colocalized overexpression of TGF-β1 and angiotensinogen has been shown in a model of chronic renal disease induced by renal ablation [72]. Expression of ACE is increased in a model of chronic kidney disease induced by combination of ablation and hyperaldosteronism and again colocalizes with TGF-β1 expression [73]. Further evidence comes from an *in vitro* study showing that TGF-β1 directly induces expression of ACE protein and activity in cultured cardiac fibroblast [74]. If confirmed in vivo, these data suggest that TGF-β1 could enhance ANG II production by local tissue RAS and thus form a positive feedback loop with ANG II.

Angiotensin II and CTGF

Connective tissue growth factor (CTGF) is a novel profibrotic factor that is upregulated in different human kidney diseases and contributes to renal fibrosis. We observed that Angiotensin II could increase CTGF expression in mRNA and protein in tubular epithelial cell line, which was also confirmed by other groups from different cell lines [75-77]. It was reported that ANG II-mediated CTGF expression requires Erk 1,2, and Rho signaling and a calcineurin-associated pathway [77]. We have also demonstrated that the blockade of CTGF, by an antisense CTGF oligonucleotide and anti-CTGF antibody diminished AngII-induced collagen release and hypertrophy [78]. By infusing Ang II, Ruiz-Ortega et al found that CTGF expression was elevated in aorta and kidney accompanying with excessive ECM synthesis [79]. In models of renal injury, ACE inhibitors and AT_1 antagonists diminished CTGF upregulation and fibrosis [80]. In diabetic patients, authors also analyzed that urinary CTGF could be reduced by the administration of ARB or ACEI [81]. All of these data clearly indicate that CTGF acts as a downstream mediator of ANG II induced fibrosis.

Angiotensin II and other Factors

In vivo observations have shown that ANG II infusion increases VEGF expression in the renal tissue by regulating its transcription [82]. In podocytes, ANG II stimulates the expression of the VEGF [83]. Vascular endothelial growth factor, in turn, leads to an upregulation of the TGF-β1 receptor type II and increased Smad2 phosphorylation. The increase of VEGF induced by ANG II was also detected in renal epithelial cells.

Signal transduction pathways of PDGF and epidermal growth factor are activated by ANG II without increasing the binding of respective ligands to the receptors, and this activation should account for the mechanism of ANG II induced mitosis [84].

Angiotensin II and Inflammation

Progressive renal disease is histologically characterized by an interstitial infiltration of mononuclear cells [85]. As a source of several profibrotic molecules, monocytes and macrophages in particular are thought to directly contribute to the fibrogenic process. While this interstitial infiltrate may originate in small part from in situ proliferation of resident interstitial macrophages, most of these cells have migrated from the circulation through peritubular capillary endothelium into the interstitial space. The stimulus for this efflux is thought to be chemoattractant molecules that are either secreted across the basolateral membrane of tubules or that pass between tubular cells moving from the tubular lumina directly into the interstitium. Chemoattractant cytokines (chemokines) such as monocyte chemoattractant protein-1 (MCP-1) are thought to play an important role.

ANG II contributes to the recruitment of infiltrating cells into the kidney [86]. ANG II causes the adhesion of circulating cells to endothelial and mesangial cells, and the migration of inflammatory cells into the kidney. This process is mediated by upregulation of adhesion molecules, cytokines and chemokines. Emerging evidences suggest that transcription factors like NF-kB and novel defined toll like receptor (TLR) are also involved in ANG II induced inflammation.

Systemic ANG II infusion into normal rats caused mononuclear cell infiltration in glomeruli and interstitium, while pharmacological blockade of RAS reduced the inflammatory cells infiltration in several models of renal injury. Double-transgenic rats of renin and angiotensinogen genes presented elevated ICAM-1 and VCAM-1, as well as their ligands, such as $\alpha5\beta2$ and $\alpha4\beta1$ integrin in leukocytes [87]. In humans, the blockade of the RAS diminished the expression of adhesion molecules. In hypertensive transgenic rats only AT_1 blockade, but not the vasodilator hydralazine, diminished the number of activated monocytes adhered to the endothelium, as well as endothelial cell injury [88]. ANG II also upregulated the expression of chemokines that are small proteins with chemoattractant properties. ANG II increased monocyte chemoattractant protein type 1 (MCP-1), the main chemokine for monocyte/macrophage, and interleukin-8 and interferon gamma inducible protein-10, which are potent chemoattractants and activators of neutrophils.

In vivo, ANG II activates the renal NF-κB pathway that was partially diminished by AT_1 or AT_2 antagonists alone, and was abolished by combination of both receptor antagonists and

ACE inhibition [89]. In mesangial cells, NF-κB activation was mediated by both AT_1 and AT_2 receptors; in tubuloepithelial cells, this was mainly by AT_1, while in endothelial cells it was via AT_2. In the UUO model, investigators have found that blockade of renal NF-κB activation by treatment with two different NF-κB inhibitors, pyrrolidine dithiocarbamate (PDTC) and parthenolide, diminished the inflammatory cell infiltration and downregulated gene expression of several proinflammatory factors [90]. In spontaneously hypertensive rats, NF-κB inhibition attenuated renal interstitial inflammation and hypertension [91]. These data suggest that in some experimental renal diseases, the blockade of ANG II generation by an ACE inhibitor or by combined blockade of both AT_1 and AT_2 receptors, as well as by the inhibition of the NF-κB pathway, is necessary to completely stop the inflammatory process.

TLRs belong to an expanding group of innate pattern recognition receptors that can recognize microbes in all body and cellular compartments [92]. Innate immune activation via TLRs serves as an attractive research hypothesis in order to address the molecular mechanisms of infection-associated renal pathology and noninfectious inflammatory responses in the kidney. TLRs recognize molecular structures of pathogens that infect the kidney, and TLR signaling initiates local inflammation. Alternatively, TLRs trigger renal inflammation upon binding circulating components of pathogens (e.g. those embedded in immune complexes). This mechanism may contribute to the onset and progression of glomerulonephritis and possibly to other types of chronic kidney disease [93]. A new discovery by Wolf et al indicated that ANG II upregulates on mesangial cells Toll-like 4 receptors that bind LPS. This ANG II-mediated Toll-like 4 receptor upregulation resulted in enhanced NF-κB activation [94].

On the other hand, inflammatory cells express all the components of the RAS, and can produce ANG II. One of the early responses to inflammation is an upregulation of angiotensinogen levels. During the differentiation process of monocytes to macrophages, there is an activation of the RAS and re-expression of the ANG II type 2 receptor (AT_2). Moreover, monocytes/macrophages present at inflammatory sites have high ACE activity. These data show that ANG II can activate inflammatory cells, which in turn can activate the RAS and increase local ANG II generation, and therefore contribute to the progression and perpetuation of inflammation.

Recently, an elegant designed experiment showed deletion of AT1a receptor in macrophages resulted in increasing renal fibrosis [95]. Other studies, however, have suggested an opposing view because AT1a receptor-deficient mice with UUO have less fibrosis than wild-type mice [96]. Further studies obviously need to be performed to clarify this important issue of ANG II and imflammation.

Angiotensin II II and Proteinuria

Proteinuria is a marker of renal injury, reflecting loss of normal permeaselectivity. Further, proteinuria itself has been proposed to contribute to progressive renal injury [97]. ANG II can not only augment proteinuria but also enhance the profibrotic signal of proteinuria.

Numerous studies in experimental animals and humans have documented that inhibition of the renin-angiotensin system by ACEIs or ARBs reduces urinary protein excretion rate and retards the development of renal injury [98]. By using isolated perfused kidney (IPK) technique, Clavant and his colleagues found that ANG II could induce clinical levels of albuminuria [99]. One possible mechanism is that ANG II preferentially raises efferent glomerular arteriole resistance resulting in enhancement of capillary filtration pressure [100]. During the past ten years, podocyte has been characterized as an important controller of secretion of proteinuria. ANG II causes podocyte apoptosis, actin rearrangement, increased VEGF synthesis, increased TGF-β1 levels, reduction of ZO-1 and nephrin levels leading to proteinuria [101,102].

Several lines of evidence suggested a role of glomerular ultrafiltrate of plasma proteins or protein-associated factors in chronic tubulointerstitial damage. In vitro studies using proximal tubular cells as a model to assess effects of apical exposure to plasma proteins proved highly valuable to approaching direct causal relationships [103]. In the proximal tubule, albumin and other ultrafiltered proteins are reabsorbed by endocytosis involving megalin and cubulin [104]. ANG II stimulates albumin endocytosis in proximal tubule cells via AT_2 receptor–mediated protein kinase B activation [105]. Tojo A et al found ANG II infusion increased megalin expression and albumin reabsorption in the proximal tubule, while ANG II blockade restored albumin reabsorption via amelioration of megalin expression in the proximal tubules of early stage diabetic rats [106,107].

Another important issue is albumin could be also responsible for the activation of local RAS. Wolf G et al indicated that an increase in tubular albumin reabsorption increase ANG II and in turn upregulate TGF-β1 receptor expression [108]. Accordingly, a group in Chile found that components of the intrarenal RAS are selectively activated in the human diabetic kidney with different levels of proteinuria [109]. Putting these data together, it is easily to believe there lies a loop between ANG II and proteinuria. Further studies need to be performed on better understanding this loop.

Angiotensin II and Oxidative Stress

ROS are composed of a series of oxygen intermediates, including the free radical superoxide anion ($\cdot O_2^-$), the nonradical hydrogen peroxide (H_2O_2), the highly reactive hydroxyl free radical ($\cdot OH$), peroxynitrite ($ONOO^-$), and singlet oxygen ($^1O_2$), in which one of the electrons is raised to an orbital of higher energy with an inversion of spin [110]. ROS modify the activity of tyrosine kinases and phosphatases, activate MAPK, and stimulate transcription factors [111]. Moreover, adhesion molecule expression is induced and matrix metalloproteinase activity is influenced. ROS can also directly change the activity of ion channels and alter the generation of NO by uncoupling of endothelial NOS (eNOS). All of these effects contribute to renal injury by inflammation and foster the development of renal fibrosis.

More than a decade ago, it was concluded from indirect evidence that ANG II may induce ROS formation [112]. Acute ANG II infusion experiments into naive rats in the presence or absence of various free radical scavengers were performed, and these scavengers partly inhibited vascular hyperpermeability and cellular damage [113,114]. Subsequent studies demonstrated that treatment of cultured vascular smooth muscle cells with ANG II for 4 to 6 hours increased $\cdot O_2^-$ [115]. These effects were mediated by the AT_1 receptor and were induced by an activation of membrane-bound NAD(P)H oxidase because the flavoprotein inhibitor diphenylene iodinium and p22phox antisense oligonucleotides attenuated this response [116,117]. AT_1 receptor–transduced ROS formation, depending on NAD(P)H oxidase, has also been observed in several renal cells in culture [118]. Similar observations have been made in the kidney when the endogenous renin-angiotensin system was stimulated using the two kidney–one clip (2K-1C) model [119]. Using Losartan, Manucha W et al reported a significant downregulation of ROS with recovery of the SOD activity in obstructed kidneys [120]. Interestingly, Welch et al showed that Tempol, a SOD mimetic, reverses effects of prolonged ANG II treatment such as renal vasoconstriction, renal cortical hypoxia, activation of NADPH oxidase, increased expression of p22(phox), and reduced expression of EC-SOD [121].

ROS is considered as an important mediator of ANG II's actions. ANG II stimulates phosphorylation and activation of extracellular signal–regulated kinase 1,2 (ERK1,2) in renal tubular cells depending on NAD(P)H oxidase–mediated $\cdot O_2^-$ formation [122]. ANG II-mediated phosphorylation of the EGF receptor is sensitive to inhibition by antioxidants, presumably by interfering with the proteolytic cleavage of heparin-binding EGF that transactivates the EGF receptor [123]. Src tyrosine kinases are also activated by ANG II-stimulated ROS formation [124]. Various tyrosine phosphatases are susceptible to oxidation and inactivation by ROS because these enzymes contain a conserved cysteine that is oxidized to sulfenic acid in the presence of ROS. In addition, transcription factors such as NF-κB and AP-1 are induced by oxidative stress [125]. Substantial experimental evidence indicates that ANG II-induced ROS are potential regulators of adhesion molecule expression. Induction of intercellular adhesion molecule-1 and vascular cellular adhesion molecule-1 by ANG II involves NAD(P)H oxidase activation. ANG II-mediated ROS formation is important for renal growth processes that are part of an adaptive process of surviving nephrons during chronic renal injury [126].

Several studies have described the oxidative stress in the development of various human renal diseases. Women with preeclampsia have an agonistic antibody in the plasma that binds to AT1 receptors that activate NAD(P)H oxidase by inducing p22phox, p47phox, and p67phox [127]. Wolf G and others have found signs of oxidative stress associated with the apoptosis of renal cells in biopsies from patients with different types of renal allograft rejection [128]. Despite there was few studies supported that anti-oxidant therapy would benefit human kidney diseases, several studies showed that angiotensin-converting enzyme inhibitor or AT1 receptor antagonist treatment reduces oxidative stress in patients with renal diseases [129,130].

In summary, the concept that ANG II mediates oxidative stress in the kidney has been validated in experimental models, however, the exact role is still incompletely understood in human renal diseases.

Angiotensin II Receptors and Signal Transduction Pathways Involved in Renal Fibrosis

ANG II exerts its effects by binding to membrane AT_1 or AT_2 receptors [131]. AT_1 regulates blood pressure, cell proliferation, and the production of cytokines and extracellular matrix proteins [132]. AT_2 also regulates blood pressure control, renal natriuresis, cell growth inhibition, and renal inflammatory cell infiltration [133]. Although the rodent contains subtypes of the AT_1 receptor (AT_{1A}, AT_{1B}) encoded by separate genes, a single gene encodes one class of AT_1 receptors in humans. Within the kidney, AT_1 receptor distribution is widespread [134]. AT_1 receptors are present in the vasculature, on plasma membranes of smooth muscle and endothelial cells, including afferent and efferent arterioles. In the glomerulus, AT_1 receptors are present in abundance on mesangial cells, endothelial cells, and podocytes. Tubular epithelial cells along the entire nephron also express AT_1 receptors, with a predominance in the proximal tubule along apical and basolateral membranes [135].

The AT_2 receptor is expressed in relative abundance in fetal tissues, with a rapid decline in mRNA levels after birth [136]. Until recently, expression of AT_2 receptors in the adult was found to confine in certain organs, such as the heart, brain, or adrenal gland [137,138]. However, both functional and immunohistochemical studies indicate the presence of AT_2 receptors in adult kidney, accounting for up to 10% of total ANG II receptors. The cellular localization of AT_2 receptors in the adult kidney remains incompletely understood. Although Ozono et al described glomerular, tubular, and interstitial cell localization [139]. Immunohistochemical studies by Miyata et al showed AT_2 -receptor expression in all nephron segments except for the medullary thick ascending limb and glomerulus [140]. The reason for this discrepancy is unclear but may relate to variable antibody sensitivity or specificity. Within glomeruli, AT_2 -receptor expression has been found in mesangial cells and podocytes. In rat glomerular endothelial cells in culture, AT_2 receptors mediate production of the chemokine, RANTES, suggesting a role in the influx of monocytes and macrophages in glomerular inflammation [141]. In rabbit afferent arterioles, activation of AT_2 receptors causes vasodilatation, which counteracts the vasoconstricting effect of ANG II mediated by AT_1 receptors [142].

ANG II, via AT_1, activates various nuclear transcription factors, including the activator protein-1 (AP-1), STAT family of transcription factors, and cAMP-response element-binding protein (CREB) [125,143-145]. ANG II also increases calcium release and activates protein kinase C (PKC), protein tyrosine kinases (PTK), mitogen-activating protein kinases (MAPK), extracellular signal-regulated kinase (ERK), c-Jun amino terminal kinase (JNK), p38 MAP kinase (p38 MAPK), and, as mentioned before, their downstream effector AP-1 [146-149]. MAPK are involved in proliferation, differentiation, fibrosis, and inflammation. In this sense, Mezzano et al have observed an elevation of renal AP-1 activity in ANG II-infused rats, which caused the tubular damage and attenuated by the AT_1 antagonist losartan [150], suggesting that ANG II may regulate the fibrotic process via AT_1/AP-1 pathway. Different intracellular mechanisms regulate the AT_2-mediated cellular responses. AT_2 is linked to inhibition of MAPK, activation of protein-phosphotyrosine phosphatase (PPase), and ceramide production [151,152]. The temporal and spatial patterns of signaling pathway activation are the most likely determinants of a particular functional response. Multiple studies show that the activation of different pathways by ANG II is time dependent. For example, activation of the G protein-dependent pathway and generation of IP_3 occurs in seconds, while MAP kinase and JAK/STAT activation occurs in minutes to hours after initial activation of AT_1R [153,154].

ANG II also cross-talks with several tyrosine kinases via AT_1Rs, including receptor tyrosine kinases [155]. Among these, transactivation of EGFR is a major mechanism by which ANG II influences growth-related signaling pathways. ANG II-mediated EGFR activation occurs in the cholesterol-rich domains of caveolae in the Src-dependent and redox-sensitive manner [156,157]. Studies show that this activation is also dependent on intact microtubules and occurs via calcium-dependent and -independent pathways [158]. These pathways lead to activation of a disintegrin and metalloproteinase (ADAMs), causing release of heparin-binding EGF (HB-EGF) [159,160]. Recently, in COS-7 cells, ANG II-activated ADAM-17 has been identified as the metalloproteinase that promotes HB-EGF shedding [161,162]. HB-EGF induces conformational changes in EGFRs, allowing them to dimerize and autophosphorylate on tyrosine. Once activated, EGFRs serve as a docking site for Grb2/Shc/Sos complexes, inducing two major transduction pathways: the PI3K/PDK1/Akt cascade, which leads to cellular metabolism, growth, survival, and remodeling, and the Ras/Raf/ERK pathway which leads to cell growth, hypertrophy, and inflammation [163]. Kagiyama et al. reported that EGFR activation is required for ANG II-mediated hypertension and left ventricular hypertrophy, both of which are attenuated when rats are treated with an intravenous infusion of antisense oligodeoxynucleotide to EGFR [164]. These studies and many others point to EGFR as an important factor in growth and hypertrophy caused by ANG II.

Other Important Roles Related to Angiotensin II

Besides of the above mentioned effects of ANG II, we would like to briefly introduce the other important issues related to ANG II, that is the novel effect of intracellular ANG II and the active fragments of ANG II.

Evidence is emerging that binding of ANG II to its cell surface AT1-receptors also activates endocytotic (or internalisation) processes that promote trafficking of both the effector and the receptor into intracellular compartments [165]. The potential role of intracellular ANG II could be traced back to an early study in which radiolabelled ANG II was found in the nuclei of vascular smooth muscle and cardiac cells following systemic administration [166]. Re defined the concept of intracrine ANG II as the effector peptide that is either synthesized within a cell via actions of renin, angiotensinogen and ACE or internalised from extracellular ANG II [167]. De Mello and colleagues have consistently shown that intracellular dialysis of renin, angiotensinogen or ANG II into hamster cardiomyocytes significantly altered cell-to-cell communications, supporting an important role for intracellular ANG II in regulating cardiac function [168]. As for the kidney, there are also quite a few lines of evidence. Li et al recently reported that extracellular ANG II accumulates in proximal tubule cells via AT_1 receptor-mediated endocytosis and that microinjection of ANG II into single proximal tubule cells stimulates intracellular calcium mobilization by acting on intracellular AT_1 receptors [169]. Interestingly, Chen et al demonstrated that ANG II induced internalization of an AT1a-receptor-green fluorescent protein complex, which is then translocated to the nucleus in transfected CHO cells [170]. It is likely that internalized Ang II activates nuclear transcription factor NF-κB by binding to cytoplasmic and nuclear AT1-receptors, thereby increasing expression of the Na^+/H^+ exchanger NHE-3, which promotes sodium transport in proximal tubules, and pro-inflammatory cytokines, chemokines and growth factors including angiotensinogen, MCP-1, IL-1β and TGF-β1 [171]. NF-κB activation by intracellular ANG II may be particularly relevant because activation of NF-κB by ANG II is one of the major mechanisms by which ANG II acts as an inflammatory cytokine to induce tubulointerstitial fibrosis. In a newly published article, Zhou et al showed that ANG II induced growth of proximal tubule cells and activated NF-κB, and these responses were blocked by inhibition of AT1-receptor-mediated endocytosis of extracellular ANG II with losartan or colchicine [172]. Taken together, these studies suggest that internalised or intracellular ANG II may play an important role in renal cellular growth and fibrotic responses by activating NF-κB signaling. This information may aid future development of drugs to prevent and treat ANG II-induced target organ injury in cardiovascular and renal diseases by blocking intracellular and/or nuclear actions of ANG II.

Although ANG II has long been considered to represent the end product of the RAS, there is accumulating evidence that this system encompasses additional effector peptides with diverse functions [173]. Shorter ANG II fragments such as Ang-(1-7) (deletion of the C terminalamino acid), angiotensin III (ANG III) (deletion of the N terminal amino acid) and angiotensin IV (ANG IV) (deletion of the two N terminal amino acids) are formed via the activity of ACE and several other peptidases and have been found to accomplish central, cardiovascular and renal actions.

ANG III plays an important role in brain and cardiovascular physiologic processes [174]. Some actions originally attributed to ANGII, such as vasopressin release, are attributable to ANG III [175]. Data suggested that ANG III could be involved in renal pathophysiologic processes. In renal cells, ANG III increases angiotensinogen, transforming growth factor-beta, fibronectin, and monocyte chemoattractant protein-1 (MCP-1) gene expression; therefore, ANG III could contribute to inflammation and fibrosis in the kidney [176,177]. ANG II and ANG III elicit cellular responses through their binding to specific receptors, namely AT_1 and AT_2 receptors. ANG III binds mainly to AT_2 receptors and exhibits lower affinity for AT_1 receptors, compared with that of ANG II [178]. By using selective nonpeptide antagonists for AT_1 and AT_2 receptors and cells from AT_1 receptor-knockout mice, Lorenzo et al demonstrated that in mesangial cells ANG III activates NF-kB mainly via AT_2 receptors.

Angiotensin-(1-7) [Ang-(1-7)] is a bioactive component of the renin-angiotensin system that is endogenously formed in the circulation and various tissues by several enzymatic pathways from either ANG I or ANG II. ANG II and Ang-(1-7) exhibit counter-regulatory effects in the systemic circulation, as well as in tissues important in cardiovascular regulation [179]. It is well documented that ANG II promotes cell proliferation and vasoconstriction, whereas Ang-(1-7) has antiproliferative actions on cardiac myocytes and vascular smooth muscle, and is a vasodilator [180]. The kidneys, which are closely linked to cardiovascular diseases, are also a main target for Ang-(1–7) actions. Ang-(1–7) is present in the kidney in concentrations comparable to ANG II and produces complex renal effects. In water-loaded rats and mice, Ang-(1–7) produces antidiuresis, probably through a mechanism involving cross-talk between Mas and AT_1, AT_2 and AVP V_2 receptors [181]. In contrast, in isolated preparation and anesthetized animals Ang-(1–7) is reported to produce natriuresis – an action which is in keeping with its *in vitro* effects on proximal tubular cells [182]. Collectively, these studies show that Ang-(1-7) exerts effects opposite to those of ANG II on the kidney, and that an attenuation of Ang-(1-7) in the kidney may be associated with human hypertension and injury.

ANG IV has sparked great interest because of its wide range of physiological effects such as memory acquisition and natriuretic effect [183]. Beyond these effects, Gesualdo et al found that ANG IV stimulates plasminogen activator inhibitor-1 expression in proximal tubular epithelial cells which may be implicated in the pathogenesis of renal interstitial fibrosis in several forms of chronic glomerulonephritides like ANG II [184]. At the molecular level, ANG IV interacts with AT1- and AT2-receptors, but only with low affinity (EC50 within the micromolar range) [185]. Yet, most of its physiological effects are already observed at nanomolar concentrations and, most importantly, they cann't be blocked by AT1- and AT2-receptor antagonists, losartan and PD 123177. This has led to the concept of a novel angiotensin receptor subtype: the 'AT4-receptor'. By using a ^{125}I-labelled ANG IV, Handa R identified the expression and properties of the angiotensin AT_4 receptor in epithelial HK-2 cells (an immortalized cell line derived from adult human proximal tubules) [186], and this receptor was shown to have similar pharmacologic and coupling properties as that described in more distal segments of the nephron and to be linked to the activation of several intracellular signaling and ion regulating systems [187,188].

Summary

There is now ample evidence to support the central role of RAS in the progression of chronic kidney diseases. Although hemodynamic effects of ANG II certainly contribute to renal damage, researches from the last decades clearly suggest this vasoactive agent is a multifarious factor engaged in many non-hemodynamic processes such as cell growth, apoptosis, metabolism of matrix, inflammation, proteinuria, oxidative stress, and inflammation. It should be emphasized that the non-hemodynaic effect is one of the important pathologic effect of ANG II, intervention focusing on this by larger dosage use of RAS blockers or combining use ACEIs and ARBs might be a key strategies in targeting the reverse of renal fibrosis. Detailed experiments have suggested that there are a complicated signal transduction networks in modulating the former processes including receptors, various kinases and transcription factors. Some new concepts such as intracellular ANG II and active fragments of ANG II are emerging to broaden the former field. Clarification of the mechanisms underlying the non-hemodynamic actions of ANG II will open up exciting new research avenues, and will offer the prospect of new diagnostic and therapeutic opportunities.

Acknowledgement

The authors would like to appreciate for the financial support from National Natural Scientific Fund (N0. 30471732), the key project of Jiangsu Provincial Natural Scientific Fund (BK2007709) and Jiangsu Provincial Metical Talent Program to Professor Bi-Cheng Liu.

References

[1] Ibarra-Rubio ME, Pedraza-Chaverri J: The current concept of the renin-angiotensin system. *Rev.Invest Clin. 45*:165-177, 1993.

[2] Kobori H, Nangaku M, Navar LG, Nishiyama A: The intrarenal renin-angiotensin system: from physiology to the pathobiology of hypertension and kidney disease. *Pharmacol. Rev. 59*:251-287, 2007.

[3] Kim S, Iwao H: Molecular and cellular mechanisms of angiotensin II-mediated cardiovascular and renal diseases. *Pharmacol.Rev. 52*:11-34, 2000.

[4] Komers R, Anderson S: Optimal strategies for preventing progression of renal disease: should angiotensin converting enzyme inhibitors and angiotensin receptor blockers be used together? *Curr.Hypertens.Rep. 2*:465-472, 2000.

[5] Petrie MC, Padmanabhan N, McDonald JE, Hillier C, Connell JM, McMurray JJ: Angiotensin converting enzyme (ACE) and non-ACE dependent angiotensin II generation in resistance arteries from patients with heart failure and coronary heart disease. *J.Am.Coll.Cardiol. 37*:1056-1061, 2001.

[6] Lansang MC, Stevanovic R, Price DA, Laffel LM, Hollenberg NK: ACE and non-ACE pathways in the renal vascular response to RAS interruption in type 1 diabetes mellitus. *Kidney Int. 67*:1033-1037, 2005.

[7] van Kats JP, Schalekamp MA, Verdouw PD, Duncker DJ, Danser AH: Intrarenal angiotensin II: interstitial and cellular levels and site of production. *Kidney Int. 60*:2311-2317, 2001.

[8] Beutler KT, Masilamani S, Turban S, Nielsen J, Brooks HL, Ageloff S, Fenton RA, Packer RK, Knepper MA: Long-term regulation of ENaC expression in kidney by angiotensin II. *Hypertension 41*:1143-1150, 2003.

[9] Brooks HL, Allred AJ, Beutler KT, Coffman TM, Knepper MA: Targeted proteomic profiling of renal Na(+) transporter and channel abundances in angiotensin II type 1a receptor knockout mice. *Hypertension 39*:470-473, 2002.

[10] Douglas JG, Hopfer U: Novel aspect of angiotensin receptors and signal transduction in the kidney. *Annu.Rev.Physiol 56*:649-669, 1994.

[11] Atiyeh BA, Arant BS, Jr., Henrich WL, Seikaly MG: In vitro production of angiotensin II by isolated glomeruli. *Am.J.Physiol 268*:F266-F272, 1995.

[12] Ingelfinger JR, Dzau VJ: Molecular biology of renal injury: emphasis on the role of the renin-angiotensin system. *J.Am.Soc.Nephrol. 2*:S9-20, 1991.

[13] Navar LG, Harrison-Bernard LM, Imig JD, Wang CT, Cervenka L, Mitchell KD: Intrarenal angiotensin II generation and renal effects of AT1 receptor blockade. *J.Am.Soc.Nephrol.* 10 Suppl 12:S266-S272, 1999.

[14] Ardaillou R, Chansel D, Chatziantoniou C, Dussaule JC: Biology and functions of renal receptors for angiotensin II and its active fragments. *Adv.Nephrol.Necker Hosp.* *28*:225-257, 1998.

[15] Wolf G, Thaiss F, Scherberich JE, Schoeppe W, Stahl RA: Glomerular angiotensinase A in the rat: increase of enzyme activity following renal ablation. *Kidney Int.* *38*:862-868, 1990.

[16] Reaux-Le GA, Iturrioz X, Fassot C, Claperon C, Roques BP, Llorens-Cortes C: Role of angiotensin III in hypertension. *Curr.Hypertens.Rep.* *7*:128-134, 2005.

[17] Wright JW, Krebs LT, Stobb JW, Harding JW: The angiotensin IV system: functional implications. *Front Neuroendocrinol.* *16*:23-52, 1995.

[18] Wolf G, Neilson EG: Angiotensin II as a renal growth factor. *J.Am.Soc.Nephrol.* *3*:1531-1540, 1993.

[19] Wolf G, Haberstroh U, Neilson EG: Angiotensin II stimulates the proliferation and biosynthesis of type I collagen in cultured murine mesangial cells. *Am.J.Pathol.* *140*:95-107, 1992.

[20] Wolf G, Ziyadeh FN, Helmchen U, Zahner G, Schroeder R, Stahl RA: ANG II is a mitogen for a murine cell line isolated from medullary thick ascending limb of Henle's loop. *Am.J.Physiol 268*:F940-F947, 1995.

[21] Schuttert JB, Liu MH, Gliem N, Fiedler GM, Zopf S, Mayer C, Muller GA, Grunewald RW: Human renal fibroblasts derived from normal and fibrotic kidneys show differences in increase of extracellular matrix synthesis and cell proliferation upon angiotensin II exposure. *Pflugers Arch.* *446*:387-393, 2003.

[22] Mervaala EM, Muller DN, Park JK, Schmidt F, Lohn M, Breu V, Dragun D, Ganten D, Haller H, Luft FC: Monocyte infiltration and adhesion molecules in a rat model of high human renin hypertension. *Hypertension 33*:389-395, 1999.

[23] Geisterfer AA, Peach MJ, Owens GK: Angiotensin II induces hypertrophy, not hyperplasia, of cultured rat aortic smooth muscle cells. *Circ.Res.* *62*:749-756, 1988.

[24] Wolf G, Neilson EG: Angiotensin II induces cellular hypertrophy in cultured murine proximal tubular cells. *Am.J.Physiol 259*:F768-F777, 1990.

[25] Orth SR, Weinreich T, Bonisch S, Weih M, Ritz E: Angiotensin II induces hypertrophy and hyperplasia in adult human mesangial cells. *Exp.Nephrol.* *3*:23-33, 1995.

[26] Liu BC, Sun J, Liu H, Chen L, Chen Q, Fan LM: [Role of connective tissue growth factor in angiotensin II induced human proximal tubular cell hypertrophy]. *Zhonghua Yi.Xue.Za Zhi.* *84*:1470-1474, 2004.

[27] Wolf G: Link between angiotensin II and TGF-beta in the kidney. *Miner.Electrolyte Metab 24*:174-180, 1998.

[28] Wolf G, Jablonski K, Schroeder R, Reinking R, Shankland SJ, Stahl RA: Angiotensin II-induced hypertrophy of proximal tubular cells requires p27Kip1. *Kidney Int.* *64*:71-81, 2003.

[29] Ophascharoensuk V, Fero ML, Hughes J, Roberts JM, Shankland SJ: The cyclin-dependent kinase inhibitor p27Kip1 safeguards against inflammatory injury. *Nat.Med.* *4*:575-580, 1998.

[30] Kelly DJ, Cox AJ, Tolcos M, Cooper ME, Wilkinson-Berka JL, Gilbert RE: Attenuation of tubular apoptosis by blockade of the renin-angiotensin system in diabetic Ren-2 rats. *Kidney Int. 61*:31-39, 2002.

[31] Yamada T, Horiuchi M, Dzau VJ: Angiotensin II type 2 receptor mediates programmed cell death. *Proc.Natl.Acad.Sci.U.S.A 93*:156-160, 1996.

[32] Dimmeler S, Rippmann V, Weiland U, Haendeler J, Zeiher AM: Angiotensin II induces apoptosis of human endothelial cells. Protective effect of nitric oxide. *Circ.Res. 81*:970-976, 1997.

[33] Gingras B, Rodier G, Giasson E, Coulombe P, Chassagne C, Meloche S: Expression of angiotensin type II receptor downregulates Cdk4 synthesis and inhibits cell-cycle progression. *Oncogene 22*:2633-2642, 2003.

[34] Long X, Xiong SD, Xiong WN, Xu YJ, Zhang ZX, Cao Y, Chen J, Xu XH: Losartan inhibited angiotensin induced human lung fibroblast transformation and collagen synthesis. *Zhonghua Jie.He.He.Hu Xi.Za Zhi. 30*:273-278, 2007.

[35] Kurikawa N, Suga M, Kuroda S, Yamada K, Ishikawa H: An angiotensin II type 1 receptor antagonist, olmesartan medoxomil, improves experimental liver fibrosis by suppression of proliferation and collagen synthesis in activated hepatic stellate cells. *Br.J.Pharmacol. 139*:1085-1094, 2003.

[36] Kobayashi S, Moriya H, Nakabayashi I, Nishiyama J, Fukuda T: Angiotensin II and IGF-I may interact to regulate tubulointerstitial cell kinetics and phenotypic changes in hypertensive rats. *Hypertens.Res. 25*:257-269, 2002.

[37] Mezzano SA, Aros CA, Droguett A, Burgos ME, Ardiles LG, Flores CA, Carpio D, Vio CP, Ruiz-Ortega M, Egido J: Renal angiotensin II up-regulation and myofibroblast activation in human membranous nephropathy. *Kidney Int.*Suppl S39-S45, 2003.

[38] Chen L, Liu BC, Zhang XL, Zhang JD, Liu H, Li MX: Influence of connective tissue growth factor antisense oligonucleotide on angiotensin II-induced epithelial mesenchymal transition in HK2 cells. *Acta Pharmacol.Sin. 27*:1029-1036, 2006.

[39] Wu K, Zhou T, Sun G, Wang W, Zhang Y, Zhang Y, Hao L, Chen N: Valsartan inhibited the accumulation of dendritic cells in rat fibrotic renal tissue. *Cell Mol.Immunol. 3*:213-220, 2006.

[40] Gloy J, Henger A, Fischer KG, Nitschke R, Bleich M, Mundel P, Schollmeyer P, Greger R, Pavenstadt H: Angiotensin II modulates cellular functions of podocytes. *Kidney Int.*Suppl 67:S168-S170, 1998.

[41] Nitschke R, Henger A, Ricken S, Gloy J, Muller V, Greger R, Pavenstadt H: Angiotensin II increases the intracellular calcium activity in podocytes of the intact glomerulus. *Kidney Int. 57*:41-49, 2000.

[42] Sequeira Lopez ML, Gomez RA: The role of angiotensin II in kidney embryogenesis and kidney abnormalities. *Curr.Opin.Nephrol.Hypertens. 13*:117-122, 2004

[43] Ha H, Lee HB: Reactive oxygen species and matrix remodeling in diabetic kidney. *J.Am.Soc.Nephrol. 14*:S246-S249, 2003.

[44] Wolf G, Haberstroh U, Neilson EG: Angiotensin II stimulates the proliferation and biosynthesis of type I collagen in cultured murine mesangial cells. *Am.J.Pathol. 140*:95-107, 1992.

[45] Wolf G, Killen PD, Neilson EG: Intracellular signaling of transcription and secretion of type IV collagen after angiotensin II-induced cellular hypertrophy in cultured proximal tubular cells. *Cell Regul. 2*:219-227, 1991.

[46] Eddy AA: Molecular basis of renal fibrosis. *Pediatr.Nephrol. 15*:290-301, 2000

[47] Bode W, Maskos K: Structural basis of the matrix metalloproteinases and their physiological inhibitors, the tissue inhibitors of metalloproteinases. *Biol.Chem. 384*:863-872, 2003.

[48] Zaoui P, Cantin JF, imardani-Bessette M, Monier F, Halimi S, Morel F, Cordonnier D: Role of metalloproteases and inhibitors in the occurrence and progression of diabetic renal lesions. *Diabetes Metab* 26 Suppl 4:25-29, 2000.

[49] Liu BC, Chen Q, Luo DD, Sun J, Phillips AO, Ruan XZ, Liu NF: Mechanisms of irbesartan in prevention of renal lesion in streptozotocin-induced diabetic rats. *Acta Pharmacol.Sin. 24*:67-73, 2003.

[50] Zhang Y, Wang C, Mizukami H, Itoh H, Kusama M, Ozawa K, Jinbu Y: Increased expression and activation of matrix metalloproteinase-2 (MMP-2) in O-1N: hamster oral squamous cell carcinoma with high potential lymph node metastasis. *J.Exp.Clin.Cancer Res. 25*:417-423, 2006.

[51] Eddy AA, Fogo AB: Plasminogen activator inhibitor-1 in chronic kidney disease: evidence and mechanisms of action. *J.Am.Soc.Nephrol. 17*:2999-3012, 2006.

[52] Rerolle JP, Hertig A, Nguyen G, Sraer JD, Rondeau EP: Plasminogen activator inhibitor type 1 is a potential target in renal fibrogenesis. *Kidney Int. 58*:1841-1850, 2000.

[53] Huang Y, Border WA, Lawrence DA, Noble NA: Noninhibitory PAI-1 enhances plasmin-mediated matrix degradation both in vitro and in experimental nephritis. *Kidney Int. 70*:515-522, 2006.

[54] Huang Y, Noble NA: PAI-1 as a target in kidney disease. *Curr.Drug Targets. 8*:1007-1015, 2007.

[55] Haraguchi M, Border WA, Huang Y, Noble NA: t-PA promotes glomerular plasmin generation and matrix degradation in experimental glomerulonephritis. *Kidney Int. 59*:2146-2155, 2001.

[56] Rerolle JP, Hertig A, Nguyen G, Sraer JD, Rondeau EP: Plasminogen activator inhibitor type 1 is a potential target in renal fibrogenesis. *Kidney Int. 58*:1841-1850, 2000.

[57] Oikawa T, Freeman M, Lo W, Vaughan DE, Fogo A: Modulation of plasminogen activator inhibitor-1 in vivo: a new mechanism for the anti-fibrotic effect of renin-angiotensin inhibition. *Kidney Int. 51*:164-172, 1997.

[58] Fintha A, Sebe A, Masszi A, Terebessy T, Huszar T, Rosivall L, Mucsi I: Angiotensin II activates plasminogen activator inhibitor-I promoter in renal tubular epithelial cells via the AT1 receptor. *Acta Physiol Hung. 94*:19-30, 2007.

[59] Haraguchi M, Border WA, Huang Y, Noble NA: t-PA promotes glomerular plasmin generation and matrix degradation in experimental glomerulonephritis. *Kidney Int. 59*:2146-2155, 2001.

[60] Abrahamsen CT, Pullen MA, Schnackenberg CG, Grygielko ET, Edwards RM, Laping NJ, Brooks DP: Effects of angiotensins II and IV on blood pressure, renal function, and PAI-1 expression in the heart and kidney of the rat. *Pharmacology 66*:26-30, 2002.

[61] Border WA, Noble NA: Targeting TGF-beta for treatment of disease. *Nat.Med. 1*:1000-1001, 1995.

[62] Kagami S, Border WA, Miller DE, Noble NA: Angiotensin II stimulates extracellular matrix protein synthesis through induction of transforming growth factor-beta expression in rat glomerular mesangial cells. *J.Clin.Invest 93*:2431-2437, 1994.

[63] Wolf G, Ziyadeh FN, Stahl RA: Angiotensin II stimulates expression of transforming growth factor beta receptor type II in cultured mouse proximal tubular cells. *J.Mol.Med. 77*:556-564, 1999.

[64] Fern RJ, Yesko CM, Thornhill BA, Kim HS, Smithies O, Chevalier RL: Reduced angiotensinogen expression attenuates renal interstitial fibrosis in obstructive nephropathy in mice. *J.Clin.Invest 103*:39-46, 1999.

[65] Peters H, Noble NA, Border WA: Transforming growth factor-beta in human glomerular injury. *Curr.Opin.Nephrol.Hypertens. 6*:389-393, 1997.

[66] Border WA, Noble NA: TGF-beta in kidney fibrosis: a target for gene therapy. *Kidney Int. 51*:1388-1396, 1997.

[67] Strippoli GF, Craig M, Deeks JJ, Schena FP, Craig JC: Effects of angiotensin converting enzyme inhibitors and angiotensin II receptor antagonists on mortality and renal outcomes in diabetic nephropathy: systematic review. *BMJ 329*:828, 2004.

[68] Laviades C, Varo N, Diez J: Transforming growth factor beta in hypertensives with cardiorenal damage. *Hypertension 36*:517-522, 2000.

[69] Wolf G, Zahner G, Schroeder R, Stahl RA: Transforming growth factor beta mediates the angiotensin-II-induced stimulation of collagen type IV synthesis in cultured murine proximal tubular cells. *Nephrol.Dial.Transplant. 11*:263-269, 1996.

[70] Rodriguez-Vita J, Sanchez-Lopez E, Esteban V, Ruperez M, Egido J, Ruiz-Ortega M: Angiotensin II activates the Smad pathway in vascular smooth muscle cells by a transforming growth factor-beta-independent mechanism. *Circulation 111*:2509-2517, 2005.

[71] Krag S, Nielsen AH, Wogensen L: High plasma concentrations of prorenin in a transgenic animal model of nephropathy with overexpression of transforming growth factor-beta1 in the kidneys. *Clin.Exp.Pharmacol.Physiol 27*:724-726, 2000.

[72] Miyazaki Y, Tsuchida S, Fogo A, Ichikawa I: The renal lesions that develop in neonatal mice during angiotensin inhibition mimic obstructive nephropathy. *Kidney Int. 55*:1683-1695, 1999.

[73] Gaedeke J, Peters H, Noble NA, Border WA: Angiotensin II, TGF-beta and renal fibrosis. *Contrib.Nephrol.153*-160, 2001.

[74] Petrov VV, Fagard RH, Lijnen PJ: Transforming growth factor-beta(1) induces angiotensin-converting enzyme synthesis in rat cardiac fibroblasts during their differentiation to myofibroblasts. *J.Renin.Angiotensin.Aldosterone.Syst. 1*:342-352, 2000.

[75] Liu BC, Sun J, Chen Q, Ma KL, Ruan XZ, Phillips AO: Role of connective tissue growth factor in mediating hypertrophy of human proximal tubular cells induced by angiotensin II. *Am.J.Nephrol.* *23*:429-437, 2003.

[76] Ruperez M, Ruiz-Ortega M, Esteban V, Lorenzo O, Mezzano S, Plaza JJ, Egido J: Angiotensin II increases connective tissue growth factor in the kidney. *Am.J.Pathol.* *163*:1937-1947, 2003.

[77] Iwanciw D, Rehm M, Porst M, Goppelt-Struebe M: Induction of connective tissue growth factor by angiotensin II: integration of signaling pathways. *Arterioscler.Thromb.Vasc.Biol.* *23*:1782-1787, 2003.

[78] Liu BC, Chen L, Sun J, Huang HQ, Ma KL, Liu H, Liu DG, Zhang XL: Connective tissue growth factor-mediated angiotensin II-induced hypertrophy of proximal tubular cells. *Nephron Exp.Nephrol.* *103*:e16-e26, 2006.

[79] Ruperez M, Lorenzo O, Blanco-Colio LM, Esteban V, Egido J, Ruiz-Ortega M: Connective tissue growth factor is a mediator of angiotensin II-induced fibrosis. *Circulation* *108*:1499-1505, 2003.

[80] Liu N, Shimizu S, Ito-Ihara T, Takagi K, Kita T, Ono T: Angiotensin II receptor blockade ameliorates mesangioproliferative glomerulonephritis in rats through suppression of CTGF and PAI-1, independently of the coagulation system. *Nephron Exp.Nephrol.* *105*:e65-e74, 2007.

[81] Ruiz-Ortega M, Ruperez M, Esteban V, Rodriguez-Vita J, Sanchez-Lopez E, Carvajal G, Egido J: Angiotensin II: a key factor in the inflammatory and fibrotic response in kidney diseases. *Nephrol.Dial.Transplant.* *21*:16-20, 2006.

[82] Rizkalla B, Forbes JM, Cooper ME, Cao Z: Increased renal vascular endothelial growth factor and angiopoietins by angiotensin II infusion is mediated by both AT1 and AT2 receptors. *J.Am.Soc.Nephrol.* *14*:3061-3071, 2003.

[83] Chen S, Lee JS, Iglesias-de la Cruz MC, Wang A, Izquierdo-Lahuerta A, Gandhi NK, Danesh FR, Wolf G, Ziyadeh FN: Angiotensin II stimulates alpha3(IV) collagen production in mouse podocytes via TGF-beta and VEGF signalling: implications for diabetic glomerulopathy. *Nephrol.Dial.Transplant.* *20*:1320-1328, 2005.

[84] Clark MA, Gonzalez N: Angiotensin II stimulates rat astrocyte mitogen-activated protein kinase activity and growth through EGF and PDGF receptor transactivation. *Regul.Pept.* *144*:115-122, 2007.

[85] Noronha IL, Fujihara CK, Zatz R: The inflammatory component in progressive renal disease--are interventions possible? *Nephrol.Dial.Transplant.* *17*:363-368, 2002.

[86] Ozawa Y, Kobori H, Suzaki Y, Navar LG: Sustained renal interstitial macrophage infiltration following chronic angiotensin II infusions. *Am.J.Physiol Renal Physiol* *292*:F330-F339, 2007.

[87] Theuer J, Dechend R, Muller DN, Park JK, Fiebeler A, Barta P, Ganten D, Haller H, Dietz R, Luft FC: Angiotensin II induced inflammation in the kidney and in the heart of double transgenic rats. *BMC.Cardiovasc.Disord.* *2*:3, 2002.

[88] Strawn WB, Gallagher PE, Tallant EA, Ganten D, Ferrario CM: Angiotensin II AT1-receptor blockade inhibits monocyte activation and adherence in transgenic (mRen2)27 rats. *J.Cardiovasc.Pharmacol.* *33*:341-351, 1999.

[89] Ruiz-Ortega M, Lorenzo O, Ruperez M, Blanco J, Egido J: Systemic infusion of angiotensin II into normal rats activates nuclear factor-kappaB and AP-1 in the kidney: role of AT(1) and AT(2) receptors. *Am.J.Pathol. 158*:1743-1756, 2001.

[90] Esteban V, Lorenzo O, Ruperez M, Suzuki Y, Mezzano S, Blanco J, Kretzler M, Sugaya T, Egido J, Ruiz-Ortega M: Angiotensin II, via AT1 and AT2 receptors and NF-kappaB pathway, regulates the inflammatory response in unilateral ureteral obstruction. *J.Am.Soc.Nephrol. 15*:1514-1529, 2004.

[91] Sanz-Rosa D, Cediel E, de las HN, Miana M, Balfagon G, Lahera V, Cachofeiro V: Participation of aldosterone in the vascular inflammatory response of spontaneously hypertensive rats: role of the NFkappaB/IkappaB system. *J.Hypertens. 23*:1167-1172, 2005.

[92] Tse K, Horner AA: Update on toll-like receptor-directed therapies for human disease. *Ann.Rheum.Dis*. 66 Suppl 3:iii77-iii80, 2007.

[93] Anders HJ, Schlondorff D: Toll-like receptors: emerging concepts in kidney disease. *Curr.Opin.Nephrol.Hypertens. 16:*177-183, 2007.

[94] Wolf G, Bohlender J, Bondeva T, Roger T, Thaiss F, Wenzel UO: Angiotensin II upregulates toll-like receptor 4 on mesangial cells. *J.Am.Soc.Nephrol. 17*:1585-1593, 2006.

[95] Nishida M, Fujinaka H, Matsusaka T, Price J, Kon V, Fogo AB, Davidson JM, Linton MF, Fazio S, Homma T, Yoshida H, Ichikawa I: Absence of angiotensin II type 1 receptor in bone marrow-derived cells is detrimental in the evolution of renal fibrosis. *J.Clin.Invest 110*:1859-1868, 2002.

[96] Satoh M, Kashihara N, Yamasaki Y, Maruyama K, Okamoto K, Maeshima Y, Sugiyama H, Sugaya T, Murakami K, Makino H: Renal interstitial fibrosis is reduced in angiotensin II type 1a receptor-deficient mice. *J.Am.Soc.Nephrol. 12*:317-325, 2001

[97] Wolf G, Ziyadeh FN: Cellular and molecular mechanisms of proteinuria in diabetic nephropathy. *Nephron Physiol 106*:26-31, 2007.

[98] Wolf G, Ritz E: Combination therapy with ACE inhibitors and angiotensin II receptor blockers to halt progression of chronic renal disease: pathophysiology and indications. *Kidney Int. 67*:799-812, 2005.

[99] Clavant SP, Forbes JM, Thallas V, Osicka TM, Jerums G, Comper WD: Reversible angiotensin II-mediated albuminuria in rat kidneys is dynamically associated with cytoskeletal organization. *Nephron Physiol 93*:51-60, 2003.

[100] Denton KM, Anderson WP, Sinniah R: Effects of angiotensin II on regional afferent and efferent arteriole dimensions and the glomerular pole. *Am.J.Physiol Regul.Integr.Comp Physiol 279*:R629-R638, 2000.

[101] Chen S, Lee JS, Iglesias-de la Cruz MC, Wang A, Izquierdo-Lahuerta A, Gandhi NK, Danesh FR, Wolf G, Ziyadeh FN: Angiotensin II stimulates alpha3(IV) collagen production in mouse podocytes via TGF-beta and VEGF signalling: implications for diabetic glomerulopathy. *Nephrol.Dial.Transplant. 20*:1320-1328, 2005.

[102] Langham RG, Kelly DJ, Cox AJ, Gow RM, Holthofer H, Gilbert RE: Angiotensin II-induced proteinuria and expression of the podocyte slit pore membrane protein, nephrin. *Nephrol.Dial.Transplant. 19*:262-263, 2004.

[103] Russo LM, Bakris GL, Comper WD: Renal handling of albumin: a critical review of basic concepts and perspective. *Am.J.Kidney Dis.* *39*:899-919, 2002.

[104] Christensen EI, Birn H: Megalin and cubilin: multifunctional endocytic receptors. *Nat.Rev.Mol.Cell Biol.* *3*:256-266, 2002.

[105] Caruso-Neves C, Kwon SH, Guggino WB: Albumin endocytosis in proximal tubule cells is modulated by angiotensin II through an AT2 receptor-mediated protein kinase B activation. *Proc.Natl.Acad.Sci.U.S.A* *102*:17513-17518, 2005.

[106] Tojo A, Onozato ML, Kurihara H, Sakai T, Goto A, Fujita T: Angiotensin II blockade restores albumin reabsorption in the proximal tubules of diabetic rats. *Hypertens.Res.* *26*:413-419, 2003.

[107] Tojo A, Onozato ML, Ha H, Kurihara H, Sakai T, Goto A, Fujita T, Endou H: Reduced albumin reabsorption in the proximal tubule of early-stage diabetic rats. *Histochem.Cell Biol.* *116*:269-276, 2001.

[108] Wolf G, Schroeder R, Ziyadeh FN, Stahl RA: Albumin up-regulates the type II transforming growth factor-beta receptor in cultured proximal tubular cells. *Kidney Int.* *66*:1849-1858, 2004.

[109] Mezzano S, Droguett A, Burgos ME, Ardiles LG, Flores CA, Aros CA, Caorsi I, Vio CP, Ruiz-Ortega M, Egido J: Renin-angiotensin system activation and interstitial inflammation in human diabetic nephropathy. *Kidney Int.*SupplS64-S70, 2003.

[110] Fridovich I: Superoxide anion radical (O2-.), superoxide dismutases, and related matters. *J.Biol.Chem.* *272*:18515-18517, 1997.

[111] Shah SV, Baliga R, Rajapurkar M, Fonseca VA: Oxidants in chronic kidney disease. *J.Am.Soc.Nephrol.* *18*:16-28, 2007.

[112] Wilson SK: Role of oxygen-derived free radicals in acute angiotensin II--induced hypertensive vascular disease in the rat. *Circ.Res.* *66*:722-734, 1990.

[113] Wang D, Chen Y, Chabrashvili T, Aslam S, Borrego Conde LJ, Umans JG, Wilcox CS: Role of oxidative stress in endothelial dysfunction and enhanced responses to angiotensin II of afferent arterioles from rabbits infused with angiotensin II. *J.Am.Soc.Nephrol.* *14*:2783-2789, 2003.

[114] Haugen EN, Croatt AJ, Nath KA: Angiotensin II induces renal oxidant stress in vivo and heme oxygenase-1 in vivo and in vitro. *Kidney Int.* *58*:144-152, 2000.

[115] Griendling KK, Minieri CA, Ollerenshaw JD, Alexander RW: Angiotensin II stimulates NADH and NADPH oxidase activity in cultured vascular smooth muscle cells. *Circ.Res.* *74*:1141-1148, 1994.

[116] Lassegue B, Sorescu D, Szocs K, Yin Q, Akers M, Zhang Y, Grant SL, Lambeth JD, Griendling KK: Novel gp91(phox) homologues in vascular smooth muscle cells : nox1 mediates angiotensin II-induced superoxide formation and redox-sensitive signaling pathways. *Circ.Res.* *88*:888-894, 2001.

[117] Griendling KK, Sorescu D, Ushio-Fukai M: NAD(P)H oxidase: role in cardiovascular biology and disease. *Circ.Res.* *86*:494-501, 2000.

[118] Ushio-Fukai M, Alexander RW, Akers M, Yin Q, Fujio Y, Walsh K, Griendling KK: Reactive oxygen species mediate the activation of Akt/protein kinase B by angiotensin II in vascular smooth muscle cells. *J.Biol.Chem.* *274*:22699-22704, 1999.

[119] Heitzer T, Wenzel U, Hink U, Krollner D, Skatchkov M, Stahl RA, MacHarzina R, Brasen JH, Meinertz T, Munzel T: Increased NAD(P)H oxidase-mediated superoxide production in renovascular hypertension: evidence for an involvement of protein kinase C. *Kidney Int. 55*:252-260, 1999.

[120] Manucha W, Oliveros L, Carrizo L, Seltzer A, Valles P: Losartan modulation on NOS isoforms and COX-2 expression in early renal fibrogenesis in unilateral obstruction. *Kidney Int. 65*:2091-2107, 2004.

[121] Welch WJ, Blau J, Xie H, Chabrashvili T, Wilcox CS: Angiotensin-induced defects in renal oxygenation: role of oxidative stress. *Am.J.Physiol Heart Circ.Physiol 288*:H22-H28, 2005.

[122] Hannken T, Schroeder R, Zahner G, Stahl RA, Wolf G: Reactive oxygen species stimulate p44/42 mitogen-activated protein kinase and induce p27(Kip1): role in angiotensin II-mediated hypertrophy of proximal tubular cells. *J.Am.Soc.Nephrol. 11*:1387-1397, 2000.

[123] Thannickal VJ, Hassoun PM, White AC, Fanburg BL: Enhanced rate of H2O2 release from bovine pulmonary artery endothelial cells induced by TGF-beta 1. *Am.J.Physiol 265*:L622-L626, 1993.

[124] Denu JM, Tanner KG: Specific and reversible inactivation of protein tyrosine phosphatases by hydrogen peroxide: evidence for a sulfenic acid intermediate and implications for redox regulation. *Biochemistry 37*:5633-5642, 1998.

[125] Bondeva T, Roger T, Wolf G: Differential regulation of Toll-like receptor 4 gene expression in renal cells by angiotensin II: dependency on AP1 and PU.1 transcriptional sites. *Am.J.Nephrol. 27*:308-314, 2007.

[126] Wolf G: Role of reactive oxygen species in angiotensin II-mediated renal growth, differentiation, and apoptosis. *Antioxid.Redox.Signal. 7*:1337-1345, 2005.

[127] Dechend R, Viedt C, Muller DN, Ugele B, Brandes RP, Wallukat G, Park JK, Janke J, Barta P, Theuer J, Fiebeler A, Homuth V, Dietz R, Haller H, Kreuzer J, Luft FC: AT1 receptor agonistic antibodies from preeclamptic patients stimulate NADPH oxidase. *Circulation 107*:1632-1639, 2003.

[128] Minz M, Heer M, Arora S, Sharma A, Khullar M: Oxidative status in stable renal transplantation. *Transplant.Proc. 38*:2020-2021, 2006.

[129] Fliser D, Wagner KK, Loos A, Tsikas D, Haller H: Chronic angiotensin II receptor blockade reduces (intra)renal vascular resistance in patients with type 2 diabetes. *J.Am.Soc.Nephrol. 16*:1135-1140, 2005.

[130] Aslam S, Santha T, Leone A, Wilcox C: Effects of amlodipine and valsartan on oxidative stress and plasma methylarginines in end-stage renal disease patients on hemodialysis. *Kidney Int. 70*:2109-2115, 2006.

[131] Inagami T: Recent progress in molecular and cell biological studies of angiotensin receptors. *Curr.Opin.Nephrol.Hypertens. 4*:47-54, 1995.

[132] Griendling KK, Alexander RW: The angiotensin (AT1) receptor. *Semin.Nephrol. 13*:558-566, 1993.

[133] Carey RM, Wang ZQ, Siragy HM: Role of the angiotensin type 2 receptor in the regulation of blood pressure and renal function. *Hypertension 35*:155-163, 2000.

[134] Gasc JM, Shanmugam S, Sibony M, Corvol P: Tissue-specific expression of type 1 angiotensin II receptor subtypes. An in situ hybridization study. *Hypertension 24*:531-537, 1994.
[135] Burns KD: Angiotensin II and its receptors in the diabetic kidney. *Am.J.Kidney Dis. 36*:449-467, 2000.
[136] Shanmugam S, Monnot C, Corvol P, Gasc JM: Distribution of type 1 angiotensin II receptor subtype messenger RNAs in the rat fetus. *Hypertension 23*:137-141, 1994.
[137] Shanmugam S, Corvol P, Gasc JM: Angiotensin II type 2 receptor mRNA expression in the developing cardiopulmonary system of the rat. *Hypertension 28*:91-97, 1996.
[138] Wang ZQ, Moore AF, Ozono R, Siragy HM, Carey RM: Immunolocalization of subtype 2 angiotensin II (AT2) receptor protein in rat heart. *Hypertension 32*:78-83, 1998.
[139] Ozono R, Wang ZQ, Moore AF, Inagami T, Siragy HM, Carey RM: Expression of the subtype 2 angiotensin (AT2) receptor protein in rat kidney. *Hypertension 30*:1238-1246, 1997.
[140] Miyata N, Park F, Li XF, Cowley AW, Jr.: Distribution of angiotensin AT1 and AT2 receptor subtypes in the rat kidney. *Am.J.Physiol 277*:F437-F446, 1999.
[141] Wolf G, Schroeder R, Stahl RA: Angiotensin II induces hypoxia-inducible factor-1 alpha in PC 12 cells through a posttranscriptional mechanism: role of AT2 receptors. *Am.J.Nephrol. 24*:415-421, 2004.
[142] Endo Y, Arima S, Yaoita H, Tsunoda K, Omata K, Ito S: Vasodilation mediated by angiotensin II type 2 receptor is impaired in afferent arterioles of young spontaneously hypertensive rats. *J.Vasc.Res. 35*:421-427, 1998.
[143] Chen X, Wang J, Zhou F, Wang X, Feng Z: STAT proteins mediate angiotensin II-induced production of TIMP-1 in human proximal tubular epithelial cells. *Kidney Int. 64*:459-467, 2003.
[144] Wolf G, Wenzel U, Burns KD, Harris RC, Stahl RA, Thaiss F: Angiotensin II activates nuclear transcription factor-kappaB through AT1 and AT2 receptors. *Kidney Int. 61*:1986-1995, 2002.
[145] Nahman NS, Jr., Rothe KL, Falkenhain ME, Frazer KM, Dacio LE, Madia JD, Leonhart KL, Kronenberger JC, Stauch DA: Angiotensin II induction of fibronectin biosynthesis in cultured human mesangial cells: association with CREB transcription factor activation. *J.Lab Clin.Med. 127*:599-611, 1996.
[146] Feng Z, Wei C, Chen X, Wang J, Cheng H, Zhang X, Hong Q, Shi S, Fu B, Wei R: Essential role of Ca2+ release channels in angiotensin II-induced Ca2+ oscillations and mesangial cell contraction. *Kidney Int. 70*:130-138, 2006.
[147] Venema RC, Venema VJ, Eaton DC, Marrero MB: Angiotensin II-induced tyrosine phosphorylation of signal transducers and activators of transcription 1 is regulated by Janus-activated kinase 2 and Fyn kinases and mitogen-activated protein kinase phosphatase 1. *J.Biol.Chem. 273*:30795-30800, 1998.
[148] Venema RC, Ju H, Venema VJ, Schieffer B, Harp JB, Ling BN, Eaton DC, Marrero MB: Angiotensin II-induced association of phospholipase Cgamma1 with the G-protein-coupled AT1 receptor. *J.Biol.Chem. 273*:7703-7708, 1998.

[149] Marrero MB, Schieffer B, Li B, Sun J, Harp JB, Ling BN: Role of Janus kinase/signal transducer and activator of transcription and mitogen-activated protein kinase cascades in angiotensin II- and platelet-derived growth factor-induced vascular smooth muscle cell proliferation. *J.Biol.Chem. 272*:24684-24690, 1997.

[150] Ruiz-Ortega M, Ruperez M, Lorenzo O, Esteban V, Blanco J, Mezzano S, Egido J: Angiotensin II regulates the synthesis of proinflammatory cytokines and chemokines in the kidney. *Kidney Int.*Suppl12-22, 2002.

[151] Esteban V, Ruperez M, Vita JR, Lopez ES, Mezzano S, Plaza JJ, Egido J, Ruiz-Ortega M: Effect of simultaneous blockade of AT1 and AT2 receptors on the NFkappaB pathway and renal inflammatory response. *Kidney Int.*SupplS33-S38, 2003.

[152] Ruiz-Ortega M, Esteban V, Suzuki Y, Ruperez M, Mezzano S, Ardiles L, Justo P, Ortiz A, Egido J: Renal expression of angiotensin type 2 (AT2) receptors during kidney damage. *Kidney Int.*SupplS21-S26, 2003.

[153] Marrero MB, Fulton D, Stepp D, Stern DM: Angiotensin II-induced insulin resistance and protein tyrosine phosphatases. *Arterioscler.Thromb.Vasc.Biol. 24*:2009-2013, 2004.

[154] Schmitz U, Ishida T, Ishida M, Surapisitchat J, Hasham MI, Pelech S, Berk BC: Angiotensin II stimulates p21-activated kinase in vascular smooth muscle cells: role in activation of JNK. *Circ.Res. 82*:1272-1278, 1998.

[155] Yin G, Yan C, Berk BC: Angiotensin II signaling pathways mediated by tyrosine kinases. *Int.J.Biochem.Cell Biol. 35*:780-783, 2003.

[156] Ushio-Fukai M, Hilenski L, Santanam N, Becker PL, Ma Y, Griendling KK, Alexander RW: Cholesterol depletion inhibits epidermal growth factor receptor transactivation by angiotensin II in vascular smooth muscle cells: role of cholesterol-rich microdomains and focal adhesions in angiotensin II signaling. *J.Biol.Chem. 276*:48269-48275, 2001.

[157] Ushio-Fukai M, Griendling KK, Becker PL, Hilenski L, Halleran S, Alexander RW: Epidermal growth factor receptor transactivation by angiotensin II requires reactive oxygen species in vascular smooth muscle cells. *Arterioscler.Thromb.Vasc.Biol. 21*:489-495, 2001.

[158] Eguchi S, Numaguchi K, Iwasaki H, Matsumoto T, Yamakawa T, Utsunomiya H, Motley ED, Kawakatsu H, Owada KM, Hirata Y, Marumo F, Inagami T: Calcium-dependent epidermal growth factor receptor transactivation mediates the angiotensin II-induced mitogen-activated protein kinase activation in vascular smooth muscle cells. *J.Biol.Chem. 273*:8890-8896, 1998.

[159] Andreev J, Galisteo ML, Kranenburg O, Logan SK, Chiu ES, Okigaki M, Cary LA, Moolenaar WH, Schlessinger J: Src and Pyk2 mediate G-protein-coupled receptor activation of epidermal growth factor receptor (EGFR) but are not required for coupling to the mitogen-activated protein (MAP) kinase signaling cascade. *J.Biol.Chem. 276*:20130-20135, 2001.

[160] Blobel CP: ADAMs: key components in EGFR signalling and development. *Nat.Rev.Mol.Cell Biol. 6*:32-43, 2005.

[161] Mifune M, Ohtsu H, Suzuki H, Nakashima H, Brailoiu E, Dun NJ, Frank GD, Inagami T, Higashiyama S, Thomas WG, Eckhart AD, Dempsey PJ, Eguchi S: G protein coupling and second messenger generation are indispensable for metalloprotease-

dependent, heparin-binding epidermal growth factor shedding through angiotensin II type-1 receptor. *J.Biol.Chem. 280*:26592-26599, 2005.

[162] Ohtsu H, Dempsey PJ, Frank GD, Brailoiu E, Higuchi S, Suzuki H, Nakashima H, Eguchi K, Eguchi S: ADAM17 mediates epidermal growth factor receptor transactivation and vascular smooth muscle cell hypertrophy induced by angiotensin II. *Arterioscler.Thromb.Vasc.Biol. 26*:e133-e137, 2006.

[163] Prenzel N, Zwick E, Daub H, Leserer M, Abraham R, Wallasch C, Ullrich A: EGF receptor transactivation by G-protein-coupled receptors requires metalloproteinase cleavage of proHB-EGF. *Nature 402*:884-888, 1999.

[164] Kagiyama S, Eguchi S, Frank GD, Inagami T, Zhang YC, Phillips MI: Angiotensin II-induced cardiac hypertrophy and hypertension are attenuated by epidermal growth factor receptor antisense. *Circulation 106*:909-912, 2002.

[165] Zhuo JL, Li XC: Novel roles of intracrine angiotensin II and signalling mechanisms in kidney cells. *J.Renin.Angiotensin.Aldosterone.Syst. 8*:23-33, 2007

[166] Robertson AL, Jr., Khairallah PA: Angiotensin II: rapid localization in nuclei of smooth and cardiac muscle. *Science 172*:1138-1139, 1971.

[167] Re RN: On the biological actions of intracellular angiotensin. *Hypertension 35*:1189-1190, 2000.

[168] De Mello WC: Intracellular angiotensin II regulates the inward calcium current in cardiac myocytes. *Hypertension 32*:976-982, 1998.

[169] Li XC, Zhuo JL: In vivo regulation of AT1a receptor-mediated uptake of [125I]-Val5-angiotensin II in the kidney and adrenals of AT1a receptor-deficient mice. *Am.J.Physiol Renal Physiol* 2007 Nov 28; [Epub ahead of print].

[170] Chen R, Mukhin YV, Garnovskaya MN, Thielen TE, Iijima Y, Huang C, Raymond JR, Ullian ME, Paul RV: A functional angiotensin II receptor-GFP fusion protein: evidence for agonist-dependent nuclear translocation. *Am.J.Physiol Renal Physiol 279*:F440-F448, 2000.

[171] Mehta PK, Griendling KK: Angiotensin II cell signaling: physiological and pathological effects in the cardiovascular system. *Am.J.Physiol Cell Physiol 292*:C82-C97, 2007.

[172] Zhuo JL, Li XC: Novel roles of intracrine angiotensin II and signalling mechanisms in kidney cells. *J.Renin.Angiotensin.Aldosterone.Syst. 8*:23-33, 2007.

[173] Ardaillou R, Chansel D: Synthesis and effects of active fragments of angiotensin II. *Kidney Int. 52*:1458-1468, 1997.

[174] Reaux-Le GA, Iturrioz X, Fassot C, Claperon C, Roques BP, Llorens-Cortes C: Role of angiotensin III in hypertension. *Curr.Hypertens.Rep. 7*:128-134, 2005.

[175] Reaux A, Fournie-Zaluski MC, Llorens-Cortes C: Angiotensin III: a central regulator of vasopressin release and blood pressure. *Trends Endocrinol.Metab 12*:157-162, 2001.

[176] Ruiz-Ortega M, Lorenzo O, Egido J: Angiotensin III increases MCP-1 and activates NF-kappaB and AP-1 in cultured mesangial and mononuclear cells. *Kidney Int. 57*:2285-2298, 2000.

[177] Ruiz-Ortega M, Lorenzo O, Egido J: Angiotensin III up-regulates genes involved in kidney damage in mesangial cells and renal interstitial fibroblasts. *Kidney Int.Suppl 68*:S41-S45, 1998.

[178] Lorenzo O, Ruiz-Ortega M, Suzuki Y, Ruperez M, Esteban V, Sugaya T, Egido J: Angiotensin III activates nuclear transcription factor-kappaB in cultured mesangial cells mainly via AT(2) receptors: studies with AT(1) receptor-knockout mice. *J.Am.Soc.Nephrol. 13*:1162-1171, 2002.

[179] Santos RA, Ferreira AJ: Angiotensin-(1-7) and the renin-angiotensin system. *Curr.Opin.Nephrol.Hypertens. 16*:122-128, 2007.

[180] Trask AJ, Ferrario CM: Angiotensin-(1-7): pharmacology and new perspectives in cardiovascular treatments. *Cardiovasc.Drug Rev. 25*:162-174, 2007.

[181] Baracho NC, Simoes-e-Silva AC, Khosla MC, Santos RA: Effect of selective angiotensin antagonists on the antidiuresis produced by angiotensin-(1-7) in water-loaded rats. *Braz.J.Med.Biol.Res. 31*:1221-1227, 1998.

[182] Caruso-Neves C, Lara LS, Rangel LB, Grossi AL, Lopes AG: Angiotensin-(1-7) modulates the ouabain-insensitive Na+-ATPase activity from basolateral membrane of the proximal tubule. *Biochim.Biophys.Acta 1467*:189-197, 2000.

[183] Hamilton TA, Handa RK, Harding JW, Wright JW: A role for the angiotensin IV/AT4 system in mediating natriuresis in the rat. *Peptides 22*:935-944, 2001.

[184] Gesualdo L, Grandaliano G, Schena FP: Angiotensin IV and renal diseases. *Contrib.Nephrol.* 63-71, 2001.

[185] Dulin N, Madhun ZT, Chang CH, Berti-Mattera L, Dickens D, Douglas JG: Angiotensin IV receptors and signaling in opossum kidney cells. *Am.J.Physiol 269*:F644-F652, 1995.

[186] Handa RK, Krebs LT, Harding JW, Handa SE: Angiotensin IV AT4-receptor system in the rat kidney. *Am.J.Physiol 274*:F290-F299, 1998.

[187] Chai SY, Fernando R, Peck G, Ye SY, Mendelsohn FA, Jenkins TA, Albiston AL: The angiotensin IV/AT4 receptor. *Cell Mol.Life Sci. 61*:2728-2737, 2004.

[188] Krishnan R, Hanesworth JM, Wright JW, Harding JW: Structure-binding studies of the adrenal AT4 receptor: analysis of position two- and three-modified angiotensin IV analogs. *Peptides 20*:915-920, 1999.

In: Cell Cycle Control: New Research
Editors: N. H. Leroy, N. T. Fournier
ISBN: 978-1-60456-796-0

Chapter X

Cyclin-Dependent Kinase Inhibitors and Aging

Xiangmei Chen and Jianzhong Wang
Departments of Nephrology, Kidney Institute and Key Lab of PLA, General Hospital of PLA, 28 Fuxing Road, Beijing 100853, P.R.China.

Abstract

Aging is a complex set of phenotypes characterized by reduced repair and/or regeneration of lost or damaged cells. Mammalian ageing is associated with reduced regenerative capacity in tissues that contain stem cells. It has been proposed that this is at least partially caused by the senescence of progenitors with age. Mitogen-dependent progression through the first gap phase (G_1) and initiation of DNA synthesis (S phase) during the mammalian cell division cycle are cooperatively regulated by several classes of cyclin-dependent kinases (CDKs) whose activities are in turn constrained by CDK inhibitors (CKIs). CKIs are important negative regulators of the cell cycle. The Ink4a/Arf locus encodes 2 tumor suppressor molecules, p16INK4a and Arf, which are principal mediators of cellular senescence. Expression of the INK4a/ARF locus, is not only a major suppressor of cancer, but also an effector of mammalian aging. Ink4a/Arf expression is a biomarker of aging. The cell cycle inhibitor p16 (INK4A) is expressed in vivo in many tissues with age. The exposure of certain chronic stresses can trigger p16 (INK4A) expression and a senescence-like phenotype. Recent data suggest that expression of p16 (INK4a) induces an age-dependent decrease in the proliferative capacity of certain tissue-specific stem cells and unipotent progenitors. Telomere shortening limits the proliferative lifespan of human cells by activation of DNA damage pathways, including upregulation of the cell cycle inhibitor p21. Cellular senescence induced by different stresses and telomere shortening appears to play an important role in the aging process.

Introduction

Aging, which takes place in all tissues of systems of the body, can be defined as a progressive deterioration of biological functions after the organism has attained its maximal reproductive competence, being displayed as post-maturational processes that lead to diminished homeostasis and increased organismic vulnerability in vivo. The in vitro counterpart of aging is senescence. Normal somatic cells have a limited proliferative capacity in vitro: after a finite number of cell divisions they eventually enter a non-proliferative state referred to as senescence. Senescence is thought to be a major tumor suppressor mechanism, and many cancers contain cells that have escaped from senescence and become immortalized [1]. Aging process involves the cell cycle machinery functioning in all systems through out the body with aging, which will be elucidated system by system in the present chapter.

Cell Cycle

Organismal life is orchestrated by proliferation, diferentiation, senescence, and apoptosis as is executed by the cell cycle including accurate duplication of the genome during the DNA synthesis phase (*S phase*), and segregation of complete sets of chromosomes to each of the daughter cells in *M phase*. The somatic cell cycle also contains "Gap" phases, known as G_1, which connects the completion of M phase to initiation of S phase in the next cycle, and G_2, which separates the S and M phases. Dependent on environmental and developmental signals, cells in G_1 may temporarily or permanently leave the cell cycle and enter a quiescent or arrested phase known as *G0* [2]. This cell cycle is coordinated by actions of several kinases (CDKs) whose activities are regulated positively by cyclins [3] and negatively by cyclin-dependent kinase inhibitors (CKIs) [4].

There are at least 11 different mammalian cyclins termed A, B1,B2, C, D1, D2, D3, E, F, G, and H. The different cyclins attain peak activity during different phases of the cell cycle—for example, cyclins C, D1–3, and E reach their maximum activity during the G1 phase, and apparently regulate transition from G1 to S, whereas cyclins A and B1–2 are most active during the S and G2 phases when they regulate transition to the mitotic phase of the cell cycle. the fluctuations in cyclin D1 are more subtle than those characteristic of the other cyclins, and may represent redistribution during the S phase rather than enhanced activity during the G1 phase [5].

There are 11 members of the CDK family known till now. Among these, CDK1, 2, 3, 4, and 6 are known to play important roles in the cell cycle. CDK5 activity is restricted to the nervous system. CDK7 plays an indirect role in the cell cycle by activating the CDKs involved in the cell cycle. CDKs 7, 8, and 9 are regulators of transcription. Not much is known about the function of CDK10, but it has been shown to interact with transcription factor Ets2. As suggested by their name, the CDKs associate with cyclins for their activation, except for CDK5, which associates with p35 and p39. Cyclins A and E activate with CDK2; cyclins D1, D2, and D3 associate with CDK4 and CDK6; cyclins A and B interact with CDK1; and cyclin H, with CDK7. The association of the CDKs with their requisite cyclin partner results in the CDKs adopting a substrate-specific catalytic subunit. The timing of expression of specific cyclins controls the activation of their interacting CDKs. The CDKs are additionally controlled by phosphorylation, rigidly binding inhibitors, ubiquitin-mediated proteolysis of cyclins, and CKIs [1,6].

CKIs have been assigned to one of two families based on their structures and CDK targets. The first class includes the INK4 proteins (inhibitors of CDK4), so named for their ability to specifically inhibit the catalytic subunits of CDK4 and CDK6. Four such proteins p16INK4a, p15INK4b, p18INK4c, and p19INK4d bind only to CDK4 and CDK6 but not to other CDKs or to D-type cyclins. The INK4 proteins can be contrasted with more broadly acting inhibitors of the Cip/Kip family whose actions affect the activities of cyclin D-, E-, and A-dependent kinases. The latter class includes p21Cip1, p27Kip1, and p57Kip2, all of which contain characteristic motifs within their amino-terminal moieties that enable them to bind both to cyclin and CDK subunits [7,8].

Members of the INK4 protein family specifically inhibit CDK4 and CDK6-mediated phosphorylation of the retinoblastoma susceptibility gene product (Rb). Inactivation of p16INK4A in tumors expressing wild-type Rb is thought to be required in order for many malignant cell types to enter S phase efficiently or to escape senescence [9]. p16INK4A is strongly expressed in adult tissues including proliferative endometrium, breast ductal epithelium, squamous and tubal metaplastic epithelium of the uterine cervix, esophageal squamous epithelium, salivary glands, and antral gastric glands. p16 is also widely expressed in endocrine glands, including Langerhans cells in the pancreas and anterior pituicytes and Leydig and Sertoli cells in testis. In infants, p16 staining was limited to thymic Hassall's corpuscles, occasional thymic lymphocytes, and only rare pancreatic epithelial cells [10]. Whereas p16Ink4a is an inhibitor of cyclin D-dependent kinases, p19Arf (p14ARF in humans) antagonizes the E3 ubiquitin protein ligase activity of Mdm2 to activate p53. p19Arf induces p53-independent sumoylation of a variety of cellular target proteins with which it interacts, including both Mdm2 and NPM [11]. Genome integrity and cell proliferation and survival are regulated by an intricate network of pathways that includes cell cycle checkpoints, DNA repair and recombination, and programmed cell death. p19INK4d plays a crucial role in regulating genomic stability and overall cell viability under conditions of genotoxic stress [12]. Expression of the gene encoding p19(INK4D) is induced by the hormonal form of vitamin D(3) and by retinoids, both of which signal through related nuclear receptor transcription factors. Induction of p19(INK4D) expression contributed to cell cycle arrest by both ligands. Knockdown of p19(INK4D) rendered cells sensitive to autophagic cell death, a remarkable phenotype given the hyperproliferative responses to loss of other INK4 proteins [13].

p21WAF1 is up-regulated in response to different DNA damaging agents [14]. The occurrence of DNA double strand breaks induces cell cycle arrest in mortal and immortal human cells. In normal, mortal fibroblasts this block to proliferation is permanent. Following DNA damage in mortal fibroblasts, the induction of p21 is to a large degree shortlived. A transient increase in p21 is followed by a delayed induction of p16INK4a [15]. Cell cycle arrest in response to DNA damage involves protein stabilization and consequent upregulation of p53, which induces transcription of CDKI p21 (CDKN1A). p21 can also act as a negative regulator of the cellular levels and stability of p53 in different cell types [16]. It has been demonstrated that reactive oxygen species (ROS) can induce p21(Cip1) degradation through an NES-, Skp2-, and ubiquitin-dependent pathway [17]. p21 was originally described as functioning as a cell cycle regulator via inhibition of both cyclin-dependent kinases and processive DNA replication. Nowadays it is recognized to play other fundamental roles including transcriptional regulation and the modulation of apoptosis [18]. The role of the CKI p21 as a mediator of p53-induced growth arrest is well established. In addition, recent data provide strong evidence for new emerging functions of p21, including a role as a modulator of apoptosis [19]. Emerging evidence now suggests additional functions for p21 in diverse cellular processes including a role as a modulator of apoptosis. p21 protects cells from irradiation-induced apoptosis by suppression of CDK activity that appears to be required downstream of the mitochondria for an efficient activation of the caspase cascade [20]. p21 cell cycle inhibitor has a dual role in stem cell aging: On one hand, p21 protects adult stem cells from acute genotoxic stress by preventing inappropriate cycling of acutely damaged stem cells. On the other hand, p21 activation impairs stem cell function and survival of aging telomere dysfunctional mice indicating that p21 checkpoint function is disadvantageous in the context of chronic and persistent damage, which accumulates during aging [21]. Among the CKIs, p27Kip1 represents a protein of central activity for the control of several phenotypes, including proliferation, differentiation and malignant transformation [22]. The p57(Kip2) gene has been suggested to be a tumor suppressor gene, being inactivated in various types of human cancers. Selective p57(Kip2) expression sensitizes cancer cells to apoptotic agents such as cisplatin, etoposide and staurosporine (STS) via a mechanism, which does not require p57(Kip2)-mediated inhibition of CDK. p57(Kip2) influences the mitochondrial apoptotic cell death pathway in cancer cells [23].

Physiological nonproliferation states encompass reversible quiescence and permanent postmitotic conditions such as terminal differentiation and replicative senescence, which might share a core proliferation-restricting mechanism. It has been found that all sorts of nonproliferating cells can be mitotically reactivated by the sole suppression of histotype-specific CKIs in the absence of exogenous mitogens [24]. Non-proliferating cells, even those destined never to reenter the cell cycle, continue to express functionally significant levels of pre-assembled cyclin-CDK complexes, making cell cycle-arrest a state that must be constantly maintained by active expression of CKIs [25].

CKIs and Aging

Respiratory System

Age-associated changes in lung structure and function affect both the lung itself and the "respiratory pump" [26]. Alveolar cells of the lung are injured and killed with aging. Damaged tissue architecture and pulmonary function is restored during recovery as endothelial and type II epithelial cells proliferate [27].

It has been reported that the ambient air pollution particles (particulate matter; PM) can induce cell cycle arrest in alveolar epithelial cells (AEC). Exposure of PM to AEC induced cells cycle arrest in G1 phase, inhibited DNA synthesis, blocked cell proliferation and caused decrease in cyclin E, A, D1 and Cyclin E- cyclin-dependent kinase (CDK)-2 kinase activity after 4h. PM induced upregulation of CDK inhibitor, p21 protein and p21 activity in AEC. SiRNA-p21 blocked PM-induced downregulation of cyclins and AEC G1 arrest [28]. Severe hyperoxia can inhibit proliferation, and kills cells, resulting in morbidity and mortality. Oxidants are involved in a large variety of pulmonary diseases. Among the various cell types that compose the respiratory system, the epithelial cells appear to be a major target for oxidative stress. When cells are exposed to DNA-damaging agents such as oxidants, a feedback control is activated that acts as a brake on the cell cycle to inhibit entry into the S phase until DNA repair is completed. Progression through the G1 phase and the G1-S transition involves sequential assembly and activation of key regulators of the cell cycle machinery, CDKs (9). The CKI p21(CIP1) appears to play an important role in the response of pulmonary epithelial cells to stress factors like oxidants. p21Cip1/Waf1/Sdi1 protects the lung against Hyperoxia. Data showed that p21-mediated cytoprotection against hyperoxia involves regulation of Bcl-XL and is uncoupled from its ability to inhibit proliferation [29]. Lung epithelium in cystic fibrosis (CF) patients is characterized by structural damage and altered repair due to oxidative stress. Modulating p21WAF1/CIP1 degradation may have the therapeutic potential of reducing lung epithelial damage related to oxidative stress in CF patients [30]. Studies suggest that p21 protects the lung from oxidative stress, in part, by inhibiting DNA replication and thereby allowing additional time to repair damaged DNA (10). In another study, the mRNA level of p21(Cip1/Waf1) decreased in lung of mice during maturation and aging (11) Prolonged exposure to hyperoxia inhibits cell proliferation in G1 via increased expression of p21. While p21 inhibits proliferating cell nuclear antigen (PCNA)-dependent DNA synthesis, it can also directly lower PCNA abundance [31].

Hepatocyte growth factor (HGF) stimulates proliferation of human bronchial epithelial cells, which can be completely abrogated by interferon-gamma (IFN-gamma). HGF by itself is capable of inducing a G1 cyclin, cyclin D1, but fails to downregulate p27(kip1) CKI. IFN-gamma does not interfere with the effect of HGF on cyclin D1 induction; however, it prevents the downregulation of p27(kip1) CKI [27]. Heparin which inhibits pulmonary artery smooth muscle cell (PASMC) proliferation increases the levels of two CKIs, p21 and p27, although only p27 is important in inhibition of PASMC growth in vitro and in vivo. The expression profile was investigated of of all the cell cycle regulating genes, including all seven CDKIs (p21, p27, p57, p15, p16, p18, and p19), in the lungs of mice with hypoxia-induced pulmonary hypertension. Hypoxic exposure significantly inhibited p27 expression and stimulated p18 activity, while other CDIs, p21, p57, p15, p16, and p19 were not affected significantly in response to hypoxia. Heparin treatment restored p27 expression, but did not influence p18 in which ERK1/2 and p38 were mediators in heparin upregulation of p27 [32]. Neutrophil elastase (NE), a serine protease present in high concentrations in the airways of cystic fibrosis patients, injures the airway epithelium. It was reported that NE treatment of airway epithelial cells causes a marked decrease in epithelial DNA synthesis and proliferation. Results demonstrate that p27 is regulated by NE and is critical for NE-induced cell cycle arrest [33]. Heparin has growth inhibitory effects on pulmonary artery smooth muscle cell (PASMC) in vitro and in vivo. p27, but not p21, is required for the inhibition of hypoxic pulmonary vascular remodeling by heparin [34].

Circulating System

The altered cardiac structure that evolves with aging in rodents includes an increase in left ventricle mass, due to myocyte enlargement, and proliferation of the matrix where the myocytes reside, which may be linked to alteration of cardiac fibroblast number or function. The number of cardiac myocytes becomes reduced, owing to necrosis predominately, or apoptosis [35]. It has been demondtrated myocyte mitotic division can occur in the adult mammalian heart. A significant fraction of myocytes divides repeatedly from birth to senescence, counteracting the continuous death of cells in the aging mammalian rat heart [36]. Atherosclerosis is thought to be a characteristic feature of aging in humans. Aging is usually associated with a decrease in proliferative ability in most cell types, while in atherosclerosis and restenosis, aging has been shown to be associated with a higher level of vascular smooth muscle cell (VSMC) proliferation and neointimal lesion formation after angioplasty. Age-associated increase in c-fos activity contributes to augmented cyclin A expression and VSMC proliferation in old animals, contributing to the higher prevalence and severity of atherosclerosis in the elderly [37]. Vascular SMC proliferation is a crucial event in occlusive cardiovascular diseases. PPARalpha is a nuclear receptor controlling lipid metabolism and inflammation. PPARalpha controls SMC cell-cycle progression at the G1/S transition by activating p16(INK4a), resulting in an inhibition of Rb phosphorylation. PPARalpha activation inhibits SMC growth in vivo, which requires p16 expression [38].

Endothelial cells (EC) express constitutively two major isoforms (Nox2 and Nox4) of the catalytic subunit of NADPH oxidase, which is a major source of endothelial reactive oxygen species. Nox2-derived O2.-, through the modulation of p21cip1 and p53 expression, participates in endothelial cell cycle regulation and apoptosis [39]. The p21(WAF1) regulates cell-cycle progression, senescence, and differentiation in injured blood vessels. Histone deacetylase (HDAC) inhibitors have shown utility in controlling proliferation in a wide range of tumor cell lines. Trichostatin A (TSA), a specific and potent HDAC inhibitor, inhibits VSMC proliferation via the induction of p21(WAF1) expression and subsequent cell-cycle arrest with reduction of the phosphorylation of Rb protein at the G1-S phase [40]. The expression of TP53 (p53) and CDKN1A (CIP1; p21) were investigated in the arterial wall in giant cell arteritis (GCA). All p53-positive giant cells also contained p21-positive nuclei. The presence of p53 and p21 in the same types of cell in GCA indicates that the former protein is functional; p21 expression is induced by wild-type, functional p53 but not by its mutant form [41]. Overexpression of p21Cip1 significantly inhibited VSCM proliferation. p21Cip1 silencing also inhibited proliferation and increased apoptotic cell death. This data demonstrates that p21Cip1 regulates VSMC proliferation in a biphasic manner. Low levels of p21Cip1 are also essential to counter apoptosis during cell-cycle progression [42]. Mitomycin C (MMc) is an antibiotic that exerts a potent antiproliferative effect in tumor cells. The antiproliferative action of MMc was associated with a selective increase in the expression of the p21, and with a decrease in cyclin B1-CDK-1 complex activity. Finally, the local perivascular delivery of MMc immediately after balloon injury of rat carotid arteries induced p21 expression and markedly attenuated neointima formation [43].

VSMCs can proliferate in response to extrinsic mitogens and accumulate in neointima after arterial injuries. Cell proliferation is not only positively regulated by cyclin/CDK complex, but also negatively controlled by CKIs such as p27(kip1) and p57(kip2). Both CKIs co-expressed in quiescent VSMCs and rapidly diminished by stimulation. Protein level of p27(kip1) was regulated by both transcription and post-transcription, but that of p57(kip2) was mainly by post-transcription. Supplemental overexpression of p57(kip2) inhibited the activations of G1 cyclin/CDKs and subsequent hyperphosphorylations of all three retinoblastoma pocket proteins as well as G1/S transition of cell cycle. The downregulations of p27(kip1) and p57(kip2) responding to mitogenic stimulation, play key roles in the cell cycle progression of VSMCs [44]. Inhibition of proliferative neointima formed by vascular smooth muscle cells is a potential target in preventing angioplasty-induced restenosis. A potent antiproliferative by fusing the active regions of the p27 and p16 cell cycle inhibitors. Intravascular delivery of a replication-deficient adenoviral vector (AV) encoding this p27-p16 fusion protein, named W9, inhibited balloon injury-induced neointimal hyperplasia in rabbit carotid arteries. The AV-W9 treatment was safe and well tolerated. AV-W9 gene therapy may be useful in preventing angioplasty-induced intimal hyperplasia in the coronary artery [45,46]. Postnatal cardiomyocytes normally grow by hypertrophy but show a limited proliferate response to certain stimuli. Although the proliferative capacity declines shortly after birth, neonatal cardiomyocytes can grow both by hypertrophy and by proliferation. The hypertrophic and proliferative responses are both activated by G1 cell cycle molecules. The difference between the two responses appears to be that high amounts of p27 are present during hypertrophic growth, whereas proliferation involves downregulation of p27 [47].

Digestive System

In the gastrointestinal tract there is a general modest decline with aging in the function of the esophagus, stomach, colon, pancreas and liver. In the small intestine, there may be subtle alterations in the intestinal morphology, as well as a decline in the uptake of fatty acids and sugars. The age-associated malabsorption of nutrients that occurs with aging may be one of the several factors which contribute to the malnutrition that occurs with aging [48].

Retinoic acid (RA) plays an important role in the regulation of cell growth and differentiation. RA induces the in vitro life-span extension of oral keratinocytes, which is linked to a decreased cellular level of p16(INK4A) and the maintenance of telomerase activity [49]. Proliferating cell nuclear antigen (PCNA) is a nuclear protein associated with the cell cycle. PCNA expression in inflamed gingiva is higher in older subjects. There was a significant correlation between aging and PCNA expression in inflamed gingiva [50].

In the gastric mucosa, aging enhances transition of G(1) to S phase as well as progression through the S phase of the cell cycle. The age-related decline in p21(Waf1/Cip1) in the gastric mucosa appears to be independent of p53 status [51]. The tumor suppressor gene p16 is over-expressed in gastric epithelial cells of H. pylori infected patients and this is associated with an increase in apoptosis [52]. In aging animals the small intestine is also in a hyperproliferative state. There was a significant positive correlation between age in years and percent of positive PCNA enterocytes both at the level of crypts and villi. Moreover, the percentage of PCNA+ enterocytes was significantly higher in elderly versus adult subjects, both at the level of villi and of crypts. PCNA reactivity increases with advancing age both in crypts and villi. This abnormality of the proliferation pattern may explain the coexistence of normal morphology and impaired absorptive function in the elderly [53]. In Fischer-344 rats, aging has been shown to be associated with increased crypt cell production in the colonic mucosa. Levels of mucosal p21(Wafl/Cip1) were found to be lower in aged than in young rats. This was accompanied by a parallel decrease in mucosal p53, a tumor suppressor protein that is known to regulate p21(Wafl/Cip1). Levels of phosphorylated Rb protein, a form which is involved in regulating progression of cells through the S phase, are increased in the colonic mucosa of 24-month-old rats, suggesting that, G(1) to S phase transition, as well as progression through the S phase of the cell cycle are accelerated in the colonic mucosa of aged rats [54].

Liver volume and blood flow decrease with age. Aging is also associated with a decline in the intrinsic metabolic activity of the hepatic parenchyma, and in the gene expression of proteins involved in intermediary metabolism, mitochondrial respiration and drug metabolism. Aged hepatocytes accumulate oxidative DNA damage, responsible for the increase in mutations, particularly in the mitochondrial genome. Histologically, aged hepatocytes are characterized by accumulation of ageing pigments into the cytoplasm and by pseudocapillarization of the sinusoid [55]. The proportion of PCNA-strongly positive hepatocytes and PCNA-positive hepatocytes declined with age. Dietary restriction diminished PCNA-strongly positive hepatocytes significantly but not PCNA-positive hepatocytes in young rats, but the proportion of PCNA-strongly positive hepatocytes was significantly higher following dietary restriction than that in rats fed ad libitum in advanced age. Aging was associated with a progressive increase in the proportion of TUNEL-positive hepatocytes. Age and dietary restriction induce proliferative inhibition. The inhibition depends on PCNA expression, suggesting that suppression of cell proliferation and cell death are enhanced in hepatocytes of senile rats [56]. Interferon gamma (IFNgamma) is an important mediator of inflammatory liver damage as part of a complex cytokine network. In vitro, IFNgamma induces hepatocyte apoptosis. Hepatocyte response to IFNgamma is modulated by external survival factors and that this survival signalling requires p21(waf-1) [57]. IL-6 directly inhibits hepatocyte proliferation via a p21cip1-dependent mechanism and indirectly enhances hepatocyte proliferation via stimulating nonparenchymal cells to produce HGF. SOCS3 negatively regulates liver regeneration [58].

Urological System

The function of the kidney, as well as its morphology, changes markedly with age. Concomitant with a reduction in the number of viable glomeruli, there is an increase in the number and percentage of glomeruli showing sclerosis with age. Subcapsular cortical glomeruli are particularly susceptible to sclerosis in comparison to juxtamedullary glomeruli. In addition to alterations in overall numbers of glomeruli, there may also be an increase in the glomerular basement membrane and an increase in the mesangial matrix volume/material. These are likely to result from alterations in the balance between formation and breakdown of the extracellular matrix in the glomerulus [59].

The cell cycle inhibitor p16(INK4A) (also known as CKI2A) is expressed in vivo in many tissues with age. The exposure of certain chronic stresses can trigger p16(INK4A) expression and a senescence-like phenotype. In glomerular disease (GD), p16(INK4A) expression was strikingly increased in glomerular and interstitial cell nuclei compared to normals and TIN, and could not be attributed to age ($P<0.05$), suggesting a role for somatic cell senescence mechanisms in progression of GD [60]. Renal ischemia/reperfusion is associated with over-expression of CKIs p16(INK4a), p21(WAF1/CIP1) and p27(Kip1), indicating an substantial DNA damage and/or accelerated tissue senescence. [61]. Some features of kidney transplants with dysfunction overlap the lesions of aging, such as tubular atrophy and interstitial fibrosis (TA/IF) without major glomerular abnormalities. Somatic cell limitations could contribute to deterioration in aging and disease states. The finding of increased p16(INK4a) expression in renal transplants and diseased kidneys with TA/IF and impaired function supports the concept that some cell senescence changes that accompany aging are also induced by injury and disease stresses [62]. It has been shown that human kidney cortex displays telomere shortening with age. Some changes observed in cellular senescence in vitro do occur in human kidney with age, particularly in the renal cortex, in some cases correlating with histologic features. P16INK4a emerged with the most consistent correlations with age and histologic changes and inversely correlated with cell replication [63]. Expression of p16(INK4a) and p27(Kip1) CKI genes is increased in cortical cells of the aging human kidney and in chronic allograft rejection, supporting the senescence theory of chronic allograft nephropathy [64]. CKI inhibitor Seliciclib, given at an early phase of disease, prolonged survival, delayed the onset of proteinuria and renal function impairment, and protected the kidney against glomerular hypercellularity, tubulointerstitial damage, and inflammation, suggesting that CDK activity may be a useful target in the treatment of systemic lupus erythematosus. A direct immunomodulatory action of seliciclib on T cells and B cells may be one of the mechanisms underlying the beneficial effects [65].

It is well documented that females live longer than males and more renal damage occurs in males. Cellular senescence is known to play a major role in nephropathology, and as such, a range of senescence markers were compared in male and female renal tissue. A reduction in oxidative damage protection may be responsible for accelerated telomere shortening over time, resulting in increased cellular senescence, loss of renal function, and death in male rats [66]. Connexin 43 (Cx43) plays an important role in cell differentiation and growth control. High glucose not only downregulated Cx43 mRNA and protein expression in glomerular mesangial cells (GMC) but also increased expression of p21cip1 and p27kip1. Knocking down Cx43 gene expression significantly increased p27kip1 and p21cip1 expression, whereas overexpression of Cx43 significantly decreased the percentage of SA-beta-gal stained cells. Downregulation of Cx43 expression by high glucose promotes the senescence of GMC, which may be involved in the pathogenesis of diabetic nephropathy [67]. Transforming growth factor-beta (TGF-beta), Smads, and p21(WAF1) are important in the pathogenesis of diabetic tubular hypertrophy. Phosphoinositide 3 kinase (PI3K)/Akt kinase activity is increased in diabetic glomerular hypertrophy. PI3K is required for high glucose-induced cell cycle-dependent hypertrophy, p21(WAF1) transcription and expression, p21(WAF1) binding to the cyclin E/cdk2 complex, TGF-beta bioactivity, and Smad2/3 activity in LLC-PK1 cells [68]. NANOG is essential for mouse and human embryonic stem cell (ESC) pluripotency and selfrenewal. The kidney has its own cells expressing NANOG, and loss of NANOG expression occurs in an age-dependent manner in the kidney, either due to developmental factors or aging, particularly in renal papillary tissue [69]. Recent data indicated that aging accelerated glomerular fibrin deposition induced by lipopolysaccharide (LPS) in mice. Fibrin deposition contributed to glomerular inflammatory responses, which could be exacerbated by aging [70]. There was an aging-related increase of tubular cell apoptosis in the renal I/R model, which may be, at least partly, due to an enhanced mitochondrial pathway resulting possibly from increased oxidative stress [71]. Sodium-dependent dicarboxylate cotransporter (NaDC), which is responsible for the transportation of intermediates of the Krebs cycle, has been implicated in extending the life span of drosophila. WI-38 cells transfected with NaDC cDNA showed significant suppression of growth rate (by 40%), increase of positive rate of SA-beta-gal staining, decrease of mitochondrial membrane potential, shortening of telomere length, and increase of P16 and P21 expression. High expression of NaDC may be able to promote the cellular senescence of human diploid fibroblasts [74]. NaDC-3 expression was increased in aged Wistar rats and aged people, indicating that NaDC-3 may have a role in the process of kidney aging [75]. Integrin-linked kinase (ILK) is an integrin-binding cytoplasmic protein that has been implicated in regulating numerous cellular processes. ILK expression was positively correlated with senescence-associated beta-gal positive staining and negatively correlated with cellular proliferation. ILK may be involved in the fibrotic or senescent process in the aging kidney [76]. Signal transducers and activators of transcription (STAT) participate in inflammation, apoptosis, and tumorigenesis. STAT3 is the major mediator of the IL-6-induced signals regulating growth and differentiation. Apoptosis has been implicated in renal ischemic injury. STAT3 mediates resistance to ATP depletion-induced apoptosis of human proximal tubular epithelial cell (HKC), which may be a potential target in treatment of renal ischemic injury [77]. p19INK4D mRNA and protein were rapidly induced by IL-6 treatment without requiring de novo protein synthesis and the induction was fully suppressed by dominant-negative forms of STAT3. STAT3-regulated induction of p19INK4D, is likely to be involved in IL-6-induced growth arrest in cells [78]. Tissue inhibitor of metalloproteinase-1 (TIMP-1) is an important inhibitor of extracellular matrix degradation. Recently, it was reported that TIMP-1 also could inhibit apoptosis in B type lymphocyte. TIMP-1 inhibits the serum deprivation-induced apoptosis in RMC, in which Bax might be involved [79].

Enalapril treatment for 3 weeks, started after the onset of diabetes, reduced the glomerular expression of p16INK4 and p27Kip1 but not of p21Cip1. Enalapril attenuates the glomerular expression of cyclin-dependent kinase inhibitors in diabetes and suggests a molecular mechanism of how angiotensin-converting enzyme inhibitors prevent renal hypertrophy in diabetes [80]. Endothelin (ET)-1 induces proliferation of mesangial cells. ET-1 also induced cyclin D1 expression and stimulated CDK4 activity and cell cycle progression via the A-type receptor in rat mesangial cells. These effects were regulated by the expression of cyclin D1, p16INK4, p21cip1, and phosphorylatable form of pRb [81]. Autosomal dominant polycystic kidney disease (ADPKD) is a common genetic disease with few treatment options other than renal replacement therapy. Levels of p21 are inversely correlated with renal tubular epithelial cell proliferation. Enhanced epithelial proliferation and apoptosis mediated by p21 supports further investigation of p21 as a target for future therapy [82].

Oxidative stress, the result of cellular production of reactive oxygen species (ROS), has been implicated in causing many renal diseases. ATP, in part, blocked H(2)O(2)-induced increase of p21(WAF1/Cip1) and p27(Kip1) expression through PI3K and Akt signal pathway in renal PTCs [83]. Renal ischemia reperfusion injury is known to shorten telomeres and upregulate stress-induced genes, such as p21. p21 was constantly upregulated in kidneys after long ischemic intervals. Mainly, tubular cells contributed to the observed increase in p21 expression. Targeting p21 via the selective p53 inhibitor pifithrin-alpha was able to prevent acute renal failure when administered immediately before ischemia [84]. Systemic lupus erythematosus (SLE) is a progressive autoimmune disease. A peptidyl mimic of p21WAF1/CIP1 that inhibits the interaction between CDK 4 and type D cyclins abrogated the in vitro proliferative response of T cells and B cells. p21WAF1/CIP1 mimic were sufficient to inhibit the progression of the lupus-like syndrome in (NZB x NZW)F1 mice. Those observations open new avenues for the treatment of SLE and prompt us to evaluate the potential interest of peptidic therapy in human SLE [85]). CDK-inhibitor p21 regulates the MC proliferative response to immune-mediated injury [86].

p27Kip1 and p21CIP1 may play an important role in hypertrophy of renal tubule cells by reducing pRb phosphorylation. On the other hand, p16INK4 was not found to cause hypertrophic changes in EGF-treated LLC-PK1 cells [87]. Deletion of p27(Kip1) attenuates the functional and morphologic features of diabetic nephropathy. p27(Kip1) plays an important role in diabetic nephropathy [88].

Neurological System

Neurodegeneration and cell death in Alzheimer's disease might be associated with aberrant proliferative mechanisms and activation of cell-cycle related events. p16INK4a, p15INK4b, p18INK4c and p19INK4d are all elevated in Alzheimer's disease. Inhibitors of the INK4-family were strongly expressed in tangle-bearing neurones and neuritic components of plaques. A dysfunction of cell cycle regulation is of critical importance in the pathomechanism of Alzheimer's disease [89].

CDK7 plays a crucial role as a CDK-activating kinase and activates all of the major CDK-cyclin substrates. CDK7 levels in Alzheimer disease patients are higher than in controls [90]. Ageing p16INK4a-deficient mice showed a significantly smaller decline in subventricular zone proliferation, olfactory bulb neurogenesis, and the frequency and self-renewal potential of multipotent progenitors [91]. Inhibition of Schwann cell proliferation is the overriding crucial function of p21 and p16 in peripheral nerves [92]. Overexpression of p15 has a similar ability to inhibit cell proliferation, to cause replicative senescence, and to inhibit telomerase activity as p16 in glioma cells with an intact RB protein pathway [93].

Corticosterone-mediated upregulation of p21(Wafl/Cipl) and neuroprotection are completely abolished by glucocorticoid and mineralocorticoid receptor antagonists as well as inhibitors of PI3- and Akt-kinase. Overexpression of p21(Wafl/Cipl) suffices to protect neurons from apoptosis. p21(W afl/Cipl) is a novel antiapoptotic factor for postmitotic neurons as the molecular target of neuroprotection by high-dose glucocorticoids [94]. Reduction of endogenous p27 expression induces cell death through cell cycle progression in cultured cortical neurons. Ablation of p27 enhance kainate-induced seizure and hippocampal cell death in vivo [95]. Transfection of p27 siRNA reduced neuronal viability in a time-dependent manner. p27 siRNA induced phosphorylation of Rb. A decrease in endogenous p27 induces neuronal cell death [96].

Endocrinology System

It has been found that cyclin D2, and to a minor degree cyclin D1, are required for adult beta-cell growth. Although cyclin D2/CDK4 activity is critically important for beta-cell growth, it was unclear how much ongoing replication is required to maintain beta-cell mass. Recent long-term beta-cell labeling studies reveal that adult beta-cells could conceivably live for the life of the organism [97]. Overexpression of TIMP-1 in ß-cells enhances the replication of pancreatic islets ß-cells and counteracts type 1 diabetes, indicating that the TIMP-1 gene may be a potential target to prevent, or even reverse, type 1 diabetes [98].

p16INK4a is an effector of senescence and a potent inhibitor of the proliferative kinase CDK4, which is essential for pancreatic beta-cell proliferation in adult mammals. It has been shown that p16INK4a constrains islet proliferation and regeneration in an age-dependent manner. Expression of the p16INK4a transcript is enriched in purified islets compared with the exocrine pancreas, and islet-specific expression of p16INK4a, but not other cyclin-dependent kinase inhibitors, increases markedly with ageing. Transgenic mice that overexpress p16INK4a to a degree seen with ageing demonstrated decreased islet proliferation. Similarly, islet proliferation was unaffected by p16INK4a deficiency in young mice, but was relatively increased in p16(INK4a)-deficient old mice. Survival after toxin-mediated ablation of beta-cells, which requires islet proliferation, declined with advancing age; however, mice lacking p16INK4a demonstrated enhanced islet proliferation and survival after beta-cell ablation. An age-induced increase of p16INK4a expression limits the regenerative capacity of beta-cells with ageing [99]. p21(cip1), a regulatory molecule upstream of the G(1/0) checkpoint, is increased in beta-cells in response to mitogenic stimulation. Isolated p21(cip1) loss, as for pRb, p53, p18, and p27 and other inhibitors, results in normal beta-cell development and function, either because it is not essential or because its function is subserved or complimented by another protein [100]. p21(cip) was present in, and upregulated in, proliferating human islets, localizing specifically in beta-cells and translocating to the nucleus on mitogenic stimulation. Homozygous p21(cip) loss releases islets from growth inhibition, markedly enhancing proliferation in response to HGF and placental lactogen [101]. Diabetes results from an inadequate mass of functional beta cells, due to either beta cell loss caused by immune assault or the lack of compensation to overcome insulin resistance. Skp2-mediated degradation pathway regulating the cellular degradation of p27 is essential for establishing beta cell mass and to respond to increased metabolic demand associated with insulin resistance [102]. Beta-cell cycle progression and proliferation are critical to maintain beta-cell mass in adult mice. Of the cell cycle inhibitors, p27Kip1 is thought to be the primary modulator of the proliferative status in most cell types. p27 plays a role in beta-cell adaptation in genetic models of insulin resistance. p27 is a critical molecule for beta-cell proliferation during beta-cell development and early postnatal life but not for maintenance of adult mass [103].

An important role for the cyclin-dependent kinase inhibitors (CDKIs), p27Kip1 and p57Kip2, in the proliferation and differentiation of adrenal cells has been suggested by their knockout mice. Adrenal development and function are primarily regulated by ACTH. The primary site for mitogenic action of ACTH in rat adrenocortex is the zona glomerulosa, and that ACTH modulates proliferation of adrenocortical cells by regulating p27Kip1 and p57Kip2 expression in a time- and site-specific manner [104].

Genital System

Both aging and diabetes mellitus (DM) have depressing effects on the expression of PCNA in testicular tissue. The decline in PCNA expression is parallel with advancement of age and progress of DM. However, DM has a more suppressing effect than aging on that expression. This suppression could explain the decrease in male fertility potential that may be associated with both conditions [105].The expression of pseudogene cyclin D2 mRNA in the human ovary increases with age, as a novel marker for decreased ovarian function associated with the aging process [106]. All biomarkers of urogenital ageing studied were significantly increased in old compared to young-adult sham rats. Ovariectomy significantly increased these changes further in old versus young-adult rats with either smaller or larger differential effect than ageing compared to young-adult sham animals. Ovariectomy significantly exacerbates normative urogenital ageing changes in rats [107]. The initiation of oocyte growth, which seems to lead follicle growth, is associated with diminished p16 expression in the mouse ovary [108].

A reduction in Ink4 family function by the loss of p19(Ink4d) is sufficient to induce mild reproductive defects in male mice with a mixed genetic background [109]. The loss of p18(Ink4c) and p19(Ink4d) leads to male reproductive defects. Possible cell autonomous and paracrine mechanisms for the spermatogenic defects in mice lacking p18(Ink4c) or p19(Ink4d) are supported by expression of these CKIs in spermatogenic cells and in somatic cells of the testis and pituitary [110]. Male mice lacking both the Ink4c and Ink4d genes are infertile, whereas female fecundity is unaffected. These two CKIs collaborate in regulating spermatogenesis, helping to ensure mitotic exit and the normal meiotic maturation of spermatocytes [111]. INK4c and INK4d are expressed during mouse embryogenesis in stereotypic tissue-specific patterns and are also detected, together with INK4b, in tissues of young mice. INK4a is expressed neither before birth nor at readily appreciable levels in young animals. The absence of tumors in INK4d-deficient animals demonstrates that, unlike INK4a, INK4d is not a tumor suppressor but is instead involved in spermatogenesis [112].

Levels of p21CIP1 diminished over the proliferative life span of senescent human prostatic epithelial cells (HPECs). p27KIP1 levels fell as cells approached senescence. Progression to senescence was accompanied by a progressive increase in the level of p16(INK4A) [113]. p57kip2, a KIP family cyclin-dependent kinase (Cdk) inhibitor, blocks the cell cycle by acting on multiple cyclin-Cdk complexes. In the seminiferous tubule, strong immunoreactivity of p57kip2 was found in nuclei of early spermatocytes, but not in the spermatogonia. In Leydig cells, moderate immunoreactivity of p57kip2 was largely found in the cytoplasm, suggesting the noble function of p57kip2 in the differentiation of adult Leydig cells [114].

Locomotor System

p16(INK4a) is instrumental in the senescence of human articular chondrocytes or osteoarthritis (OA). The reduction of p16(INK4a) by RNA interference (RNAi) contributing to the recovery of osteoarthritic chondrocytes, suggested that p16(INK4a) may be a viable future therapeutic candidate [115]. Synovial tissue affected by rheumatoid arthritis is characterized by proliferation, which leads to irreversible cartilage and bone destruction. Synovial cells derived from hypertrophic synovial tissue readily expressed p16INK4a when they were growth-inhibited. In vivo adenoviral gene therapy with the p16INK4a gene efficiently inhibited the pathology in an animal model of rheumatoid arthritis. The induction of p16INK4a may provide a new approach to the effective treatment of rheumatoid arthritis [116].

Histone deacetylase inhibitors (HDIs), a new class of anti-cancer agents, have been reported to suppress formation of osteoclast precursors and their fusion into multinucleated cells. Treatment of osteoclasts with the HDI trichostatin A (TSA) caused apoptosis. TSA causes osteoclast apoptosis, which involves, in part, TSA-induced upregulation of p21(WAF1), and strongly supports HDIs as potential therapeutic agents for excessive bone resorption [117]. Rheumatoid arthritis (RA) is characterized by synovial hyperplasia and destruction of cartilage and bone. The fibroblast-like synoviocyte (FLS) population is central to the development of pannus by migrating into cartilage and bone. p21 plays a novel role in normal FLS, namely to repress migration. Loss of p21 expression that occurs in RA FLS may contribute to excessive invasion and subsequent joint destruction [118]. Senescence of human articular chondrocytes leads to the decrease in the proliferative capacity and phenotypic stability. In addition, PCNA and p21 could be molecular markers that represent the status of these age-related properties of human articular chondrocytes in vitro [119].

p27/Kip1 acts normally to inhibit periosteal bone growth. p27 has modest inhibitory effects on growth plate chondrocyte proliferation but is not required for the spatial or temporal regulation of proliferation or the conditional regulation by glucocorticoid [120]. p57(Kip2) and p21(Cip1/Waf1) are members of CKIs which play critical roles in the terminal differentiation of skeletal muscle. The mRNA levels of p57(Kip2) and p21(Cip1/Waf1) decreased in skeletal muscle of mice during maturation and aging except that the level of p21(Cip1/Waf1) mRNA in skeletal muscle of mice showed an increase only during maturation. The downregulation of p57(Kip2) and p21(Cip1/Waf1) during aging is contrary to the upregulation of CKIs during cellular replicative senescence, indicating that aging in an organismal level is mediated by mechanisms different from replicative senescence of cultured cells [121].

Sense Organs

A study has been conducted to determine whether small interfering (si)RNA downregulation of p27kip1 overcomes G(1)-phase arrest and promotes cell-cycle progression in human corneal endothelial cells (HCECs) from young (<30 years old) and older (>60 years old) donors. Transfection of p27kip1 siRNA was sufficient to promote proliferation in confluent cultures of HCECs from younger, but not older donors, suggesting that inhibition of proliferation in older donors is regulated by other mechanisms in addition to p27kip1 [122]. HCECs are considered to be nonreplicative in vivo; however, isolated HCECs can be cultured and grown successfully, indicating that they retain proliferative capacity. This capacity to replicate tends to decrease with donor age. p16(INK4a), p21WAF1/Cip1, p27Kip1, and Rb were all expressed in HCECs, regardless of donor age. Age-related differences in the relative expression of p16INK4a and p21WAF1/Cip1 and in the kinetics of Rb hyperphosphorylation led to the conclusion that, in addition to the normal inhibitory activity of p27Kip1, there is an age-dependent increase in negative regulation of the cell cycle by p16INK4a and p21WAF1/Cip1. This additional molecular mechanism may be responsible, at least in part, for the reduced proliferative response observed in HCECs from older donors [123].

The absence of a functional p27(Kip1) gene results in enhanced retinal pigment epithelium (RPE) nuclear division, without a commensurate increase in cell division. Although the mutant epithelium as a whole appears structurally normal, individual cells exhibit cytoskeletal changes and their interaction with the neural retina is compromised [124]. Homozygous P27(Kip1) knock-out mice had hypertrophic, hyperplastic, and dystrophic optic nerve (ON) [125]. p63 and p16 have been described as stem-cell markers of squamous epithelium. Overexpression of p63 and p16 reinforces likelihood of involvement of these genes in the pathogenesis of pterygium, perhaps related to the intense cellular turnover with substitution of superficial epithelial cells by less differentiated forms. This loss of normal cellular differentiation of the epithelial layers could explain the high rates of recurrence overall in the recurrent pterygia [126]. The pterygium is a growth onto the cornea of fibrovascular tissue that is continuous with the conjunctiva. Several pterygium epithelial cells were p27(KIP1) positive, whereas nuclear immunoreactivity for cyclin D1 and Ki-67 was detected in many epithelial cells. By contrast, immunoreactivity for p27(KIP1), cyclin D1 and Ki-67 was hardly detected in the pterygium stroma. It is suggested that pterygium growth and development are associated with the proliferation of epithelium, which is possibly involved in the expression of cell cycle-related molecules [127].

Sensorineural hearing loss, which is often caused by degeneration of hair cells in the auditory epithelium, is permanent because lost hair cells are not replaced. Several conceptual approaches can be used to place new hair cells in the auditory epithelium. One possibility is to enhance proliferation of non-sensory cells that remain in the deaf ear and induce transdifferentiation of some of these cells into the hair cell phenotype. Several genes, including p27(Kip1), have been shown to regulate proliferation and differentiation in the developing auditory epithelium [128]. When lost as a result of environmental insult or genetic abnormality, hair cells do not regenerate, and this loss is a common cause of deafness in humans. In Ink4d-/- animals, hair cells are observed to aberrantly re-enter the cell cycle and subsequently undergo apoptosis, resulting in progressive hearing loss [129]. Once hair cells in the cochlea are formed, p27 is expressed in non-sensory cells of the organ of Corti and prevents their re-entry into the cell cycle. Manipulations to regulate the time and place of p27 inhibition will be necessary for inducing functionally useful hair cell regeneration [130]. The down-regulation of p27 is a key player in cell cycle control and plays an undefined role in the pathogenesis of all types of cholesteatoma [131]. p16INK4A expression directly correlates with chronological aging of human skin in vivo. p16INK4A therefore is a biomarker for human aging in vivo [132].

Conclusion

CKIs have their distinctive differential expression in all the systems throughout the body exerting cell type-specific elaborate control of not only the development, tumorigenesis but also apoptosis, differentiation, homeostasis, and senescence/aging, which should be highlighted for the critical need of further exploration and elucidation.

References

[1] Arendt T, Holzer M, Gärtner U. Neuronal expression of cycline dependent kinase inhibitors of the INK4 family in Alzheimer's disease. *J Neural Transm*. 1998;105(8-9):949-60.

[2] van den Heuvel, S. Cell-cycle regulation. *WormBook*. 2005; 1-16.

[3] Murray, A.W. 2004. Recycling the cell cycle: cyclins revisited. *Cell*. 116:221–234.

[4] Harper, J.W. 1997. Cyclin dependent kinase inhibitors. *Cancer Surv*. 29:91–107.

[5] R Donnellan and R Chetty. Cyclin D1 and human neoplasia. *Molecular Pathology*, 1998; 51(1):1-7.

[6] Jayalakshmi Sridhar, Nagaraju Akula, and Nagarajan Pattabiraman. Selectivity and Potency of Cyclin-dependent Kinase Inhibitors. *The AAPS Journal* 2006; 8 (1): E204-E221.

[7] Charles J. Sherr, and James M. Roberts. CDK inhibitors: positive and negative regulators of G1-phase progression. 1999; 13(12): 1501-1512.

[8] Pavletich NP. Mechanisms of cyclin-dependent kinase regulation: structures of Cdks, their cyclin activators, and Cip and INK4 inhibitors. *J Mol Biol*. 1999 Apr 16;287(5):821-8.

[9] Park JY, Schutzer WE, Lindsley JN, Bagby SP, Oyama TT, Anderson S, Weiss RH. p21 is decreased in polycystic kidney disease and leads to increased epithelial cell cycle progression: roscovitine augments p21 levels. *BMC Nephrol*. 2007 Aug 22;8:12.

[10] Charles J. Sherr, and James M. Roberts. CDK inhibitors: positive and negative regulators of G1-phase progression. 1999; 13 (12): 1501-1512.

[11] Chen P, Zindy F, Abdala C, Liu F, Li X, Roussel MF, Segil N. Progressive hearing loss in mice lacking the cyclin-dependent kinase inhibitor Ink4d. *Nat Cell Biol*. 2003 May;5(5):422-6.

[12] Zhang J, Ghio AJ, Gao M, Wei K, Rosen GD, Upadhyay D. Ambient particulate matter induces alveolar epithelial cell cycle arrest: role of G1 cyclins. *FEBS Lett.* 2007; 581(27):5315-20.

[13] Takami K, Takuwa N, Okazaki H, Kobayashi M, Ohtoshi T, Kawasaki S, Dohi M, Yamamoto K, Nakamura T, Tanaka M, Nakahara K, Takuwa Y, Takizawa H. Interferon-gamma inhibits hepatocyte growth factor-stimulated cell proliferation of human bronchial epithelial cells: upregulation of p27(kip1) cyclin-dependent kinase inhibitor. *Am J Respir Cell Mol Biol.* 2002;26(2): 231-8.

[14] Clement A, Henrion-Caude A, Besnard V, Corroyer S1. Role of cyclins in epithelial response to oxidants. *Am J Respir Crit Care Med.* 2001;164(10 Pt 2): S81-4.

[15] O'Reilly MA, Staversky RJ, Watkins RH, Reed CK, de Mesy Jensen KL, Finkelstein JN, Keng PC. The cyclin-dependent kinase inhibitor p21 protects the lung from oxidative stress. *Am J Respir Cell Mol Biol.* 2001; 24(6): 703-10.

[16] Park CW, Chung JH. Age-dependent changes of p57(Kip2) and p21(Cip1/Waf1) expression in skeletal muscle and lung of mice. *Biochim Biophys Acta.* 2001;1520(2):163-8.

[17] Rivard A, Principe N, Andrés V. Age-dependent increase in c-fos activity and cyclin A expression in vascular smooth muscle cells. A potential link between aging, smooth muscle cell proliferation and atherosclerosis. *Cardiovasc Res.* 2000;45(4):1026-34.

[18] Nakano N, Urasawa K, Takagi Y, Saito T, Kaneta S, Ishikawa S, Higashi H, Tsutsui H, Hatakeyama M, Kitabatake A. Downregulation of cyclin-dependent kinase inhibitor; p57(kip2), is involved in the cell cycle progression of vascular smooth muscle cells. *Biochem Biophys Res Commun.* 2005; 338(3):1661-7.

[19] Anversa P, Fitzpatrick D, Argani S, Capasso JM. Myocyte mitotic division in the aging mammalian rat heart. *Circ Res.* 1991;69(4):1159-64.

[20] Kajstura J, Pertoldi B, Leri A, Beltrami CA, Deptala A, Darzynkiewicz Z, Anversa P. Telomere shortening is an in vivo marker of myocyte replication and aging. *Am J Pathol.* 2000;156(3):813-9.

[21] Celenligil-Nazliel H, Ayhan A, Uzun H, Ruacan S. The effect of age on proliferating cell nuclear antigen expression in oral gingival epithelium of healthy and inflamed human gingiva. *J Periodontol.* 2000;71(10):1567-74.

[22] Xiao ZQ, Yu Y, Khan A, Jaszewski R, Ehrinpreis MN, Majumdar AP. Induction of G(1) checkpoint in the gastric mucosa of aged rats. *Am J Physiol.* 1999; 277(5 Pt 1):G929-34.

[23] Corazza GR, Ginaldi L, Quaglione G, Ponzielli F, Vecchio L, Biagi F, Quaglino D. Proliferating cell nuclear antigen expression is increased in small bowel epithelium in the elderly. *Mech Ageing Dev.* 1998;104(1):1-9.

[24] Xiao ZQ, Jaszewski R, Majumdar AP. Aging enhances G(1) phase in the colonic mucosa of rats. *Mech Ageing Dev.* 2000;116(1):1-14.

[25] Higami Y, Shimokawa I, Okimoto T, Tomita M, Yuo T, Ikeda T. Effect of aging and dietary restriction on hepatocyte proliferation and death in male F344 rats. *Mech Ageing Dev.* 2000;116(1):1-14.

[26] Sis B, Tasanarong A, Khoshjou F, Dadras F, Solez K, Halloran PF. Accelerated expression of senescence associated cell cycle inhibitor p16INK4A in kidneys with glomerular disease. *Kidney Int*. 2007;71(3):218-26.

[27] Melk A, Schmidt BM, Vongwiwatana A, Rayner DC, Halloran PF. Increased expression of senescence-associated cell cycle inhibitor p16INK4a in deteriorating renal transplants and diseased native kidney. *Am J Transplant*. 2005;5(6):1375-82.

[28] Chkhotua AB, Gabusi E, Altimari A, D'Errico A, Yakubovich M, Vienken J, Stefoni S, Chieco P, Yussim A, Grigioni WF. Increased expression of p16(INK4a) and p27(Kip1) cyclin-dependent kinase inhibitor genes in aging human kidney and chronic allograft nephropathy. *Am J Kidney Dis*. 2003;41(6):1303-13.

[29] Zoja C, Casiraghi F, Conti S, Corna D, Rottoli D, Cavinato RA, Remuzzi G, Benigni A. Cyclin-dependent kinase inhibition limits glomerulonephritis and extends lifespan of mice with systemic lupus. *Arthritis Rheum*. 2007;56(5):1629-37.

[30] Melk A, Schmidt BM, Takeuchi O, Sawitzki B, Rayner DC, Halloran PF. Expression of p16INK4a and other cell cycle regulator and senescence associated genes in aging human kidney. *Kidney Int*. 2004;65(2):510-20.

[31] Tarry-Adkins JL, Ozanne SE, Norden A, Cherif H, Hales CN. Lower antioxidant capacity and elevated p53 and p21 may be a link between gender disparity in renal telomere shortening, albuminuria, and longevity. *Am J Physiol Renal Physiol*. 2006 ;290(2):F509-16.

[32] Bai X, Chen X, Hou K, Zhang P, Feng Z, Fu B. Effect of cell cycle inhibitor p19ARF on senescence of human diploid cell. *Sci China C Life Sci*. 2007;50(2):155-60.

[33] Zhang X, Chen X, Wu D, Liu W, Wang J, Feng Z, Cai G, Fu B, Hong Q, Du J. Downregulation of connexin 43 expression by high glucose induces senescence in glomerular mesangial cells. *J Am Soc Nephrol*. 2006;17(6):1532-42. Epub 2006 May 4.

[34] Yan QJ, Chen XM, Zhang YM, Xie Y, Shi SZ, Fu B, Hong Q, Xu GS, Zhang XG, Zhu HY, Wu D, Lu Y, Zhang YH. NANOG changes in mouse kidneys with age. *Rejuvenation Res*. 2005;8(4):248-53.

[35] Xi C, Chen X, Sun X, Shi S, Feng Z, Wang J, Hong Q, Lu Y, Lin S. Effects of alterations of glomerular fibrin deposition on renal inflammation in rats at different age stages. *J Gerontol A Biol Sci Med Sci*. 2005;60(9):1099-110.

[36] Qiao X, Chen X, Wu D, Ding R, Wang J, Hong Q, Shi S, Li J, Xie Y, Lu Y, Wang Z. Mitochondrial pathway is responsible for aging-related increase of tubular cell apoptosis in renal ischemia/reperfusion injury. *J Gerontol A Biol Sci Med Sci*. 2005;60(7):830-9.

[37] Chen X, Cao D, Wang J, Yuan L, Feng Z, Fu B, Hong Q, Zhang X, Bai X, Lu Y, Ding R. Effects of human Na(+)/dicarboxylate cotransporter 3 on the replicative senescence of human embryonic lung diploid fibroblasts. *J Gerontol A Biol Sci Med Sci*. 2005 Jun;60(6):709-14.

[38] Wang J, Chen X, Zhu H, Peng L, Hong Q. Relationship between aging and renal high-affinity sodium-dependent dicarboxylate cotransporter-3 expression characterized with antifusion protein antibody. *J Gerontol A Biol Sci Med Sci*. 2003;58(10):B879-88.

[39] Li Z, Chen X, Xie Y, Shi S, Feng Z, Fu B, Zhang X, Cai G, Wu C, Wu D, Gu Y. Expression and significance of integrin-linked kinase in cultured cells, normal tissue,

and diseased tissue of aging rat kidneys. *J Gerontol A Biol Sci Med Sci.* 2004;59(10):984-96.

[40] Wang J, Ouyang C, Chen X, Fu B, Lu Y, Hong Q. STAT3 Inhibits Apoptosis of Human Renal Tubular Epithelial Cells Induced by ATP Depletion/Recovery. *Nephron Exp Nephrol.* 2007;108(1):e11-e18.

[41] Lin H, Chen X, Wang J, Yu Z. Inhibition of apoptosis in rat mesangial cells by tissue inhibitor of metalloproteinase-1. *Kidney Int.* 2002;62(1):60-9.

[42] Molofsky AV, Slutsky SG, Joseph NM, He S, Pardal R, Krishnamurthy J, Sharpless NE, Morrison SJ. Increasing p16INK4a expression decreases forebrain progenitors and neurogenesis during ageing. *Nature*. 2006;443(7110):448-52.

[43] Zhu X, Rottkamp CA, Raina AK, Brewer GJ, Ghanbari HA, Boux H, Smith MA. Neuronal CDK7 in hippocampus is related to aging and Alzheimer disease. *Neurobiol Aging*. 2000;21(6):807-13.

[44] Kushner JA. Beta-cell growth: an unusual paradigm of organogenesis that is cyclin D2/CDK4 dependent. *Cell Cycle*. 2006;5(3):234-7.

[45] Krishnamurthy J, Ramsey MR, Ligon KL, Torrice C, Koh A, Bonner-Weir S, Sharpless NE. p16INK4a induces an age-dependent decline in islet regenerative potential. *Nature.* 2006;443(7110):453-7.

[46] Choi D, Yoon S, Lee E, Hwang S, Yoon B, Lee J. The expression of pseudogene cyclin D2 mRNA in the human ovary may be a novel marker for decreased ovarian function associated with the aging process. *J Assist Reprod Genet.* 2001;18(2):110-3.

[47] Salama M, Tsuji M, Tamura M, Kagawa S. Impact of aging and diabetes mellitus on the expression of the proliferating cell nuclear antigen in rat testicular tissue. *Arch Androl*. 1998;40(2):95-107.

[48] Rizk DE, Fahim MA, Hassan HA, Al-Marzouqi AH, Ramadan GA, Al-Kedrah SS, Al-Ghafri LS. The effect of ovariectomy on biomarkers of urogenital ageing in old versus young adult rats. *Int Urogynecol J Pelvic Floor Dysfunct*. 2007;18(9):1077-85.

[49] Kikuchi M, Zhu C, Senoo T, Obara Y, Joyce NC. p27kip1 siRNA induces proliferation in corneal endothelial cells from young but not older donors. *Invest Ophthalmol Vis Sci*. 2006;47(11):4803-9.

[50] Enomoto K, Mimura T, Harris DL, Joyce NC. Age differences in cyclin-dependent kinase inhibitor expression and rb hyperphosphorylation in human corneal endothelial cells. *Invest Ophthalmol Vis Sci*. 2006;47(10):4330-40.

[51] Witkowski JM, Bryl E. Paradoxical age-related cell cycle quickening of human CD4(+) lymphocytes: a role for cyclin D1 and calpain. *Exp Gerontol*. 2004 Apr;39(4):577-85.

[52] Hale TJ, Richardson BC, Sweet LI, McElligott DL, Riggs JE, Chu EB, Glynn JM, LaFrenz D, Ernst DN, Rochford R, Hobbs MV. Age-related changes in mature CD4+ T cells: cell cycle analysis. *Cell Immunol*. 2002;220(1):51-62.

[53] Hyland P, Barnett C, Pawelec G, Barnett Y. Age-related accumulation of oxidative DNA damage and alterations in levels of p16(INK4a/CDKN2a), p21(WAF1/CIP1/SDI1) and p27(KIP1) in human CD4+ T cell clones in vitro. *Mech Ageing Dev*. 2001;122(11):1151-67.

[54] Tamir A, Miller RA. Aging impairs induction of cyclin-dependent kinases and down-regulation of p27 in mouse CD4(+) cells. *Cell Immunol*. 1999;198(1):11-20.

[55] Arbogast A, Boutet S, Phelouzat MA, Plastre O, Quadri R, Proust JJ. Failure of T lymphocytes from elderly humans to enter the cell cycle is associated with low Cdk6 activity and impaired phosphorylation of Rb protein. *Cell Immunol.* 1999;197(1):46-54.

[56] Quadri RA, Arbogast A, Phelouzat MA, Boutet S, Plastre O, Proust JJ. Age-associated decline in cdk1 activity delays cell cycle progression of human T lymphocytes. *J Immunol.* 1998;161(10):5203-9.

[57] Matheu A, Maraver A, Klatt P, Flores I, Garcia-Cao I, Borras C, Flores JM, Viña J, Blasco MA, Serrano M. Delayed ageing through damage protection by the Arf/p53 pathway. *Nature.* 2007 Jul 19;448(7151):375-9.

[58] Choudhury AR, Ju Z, Djojosubroto MW, Schienke A, Lechel A, Schaetzlein S, Jiang H, Stepczynska A, Wang C, Buer J, Lee HW, von Zglinicki T, Ganser A, Schirmacher P, Nakauchi H, Rudolph KL. Cdkn1a deletion improves stem cell function and lifespan of mice with dysfunctional telomeres without accelerating cancer formation. *Nat Genet.* 2007;39(1):99-105.

[59] Chang BD, Watanabe K, Broude EV, Fang J, Poole JC, Kalinichenko TV, Roninson IB. Effects of p21Waf1/Cip1/Sdi1 on cellular gene expression: implications for carcinogenesis, senescence, and age-related diseases. *Proc Natl Acad Sci U S A.* 2000;97(8):4291-6.

[60] Melzer D, Frayling TM, Murray A, Hurst AJ, Harries LW, Song H, Khaw K, Luben R, Surtees PG, Bandinelli SS, Corsi AM, Ferrucci L, Guralnik JM, Wallace RB, Hattersley AT, Pharoah PD. A common variant of the p16(INK4a) genetic region is associated with physical function in older people. *Mech Ageing Dev.* 2007;128(5-6):370-7.

[61] Kim WY, Sharpless NE. The regulation of INK4/ARF in cancer and aging. *Cell.* 2006 ;127(2):265-75.

[62] Sharpless NE. Ink4a/Arf links senescence and aging. *Exp Gerontol.* 2004;39(11-12):1751-9.

[63] Krishnamurthy J, Torrice C, Ramsey MR, Kovalev GI, Al-Regaiey K, Su L, Sharpless NE. Ink4a/Arf expression is a biomarker of aging. J Clin Invest. 2004;114(9):1299-307.

[64] Keith C. Meyer. Aging. *The Proceedings of the American Thoracic Society* 2:433-439 (2005).

[65] Edward G. Lakatta and Steven J. Sollott. The "Heartbreak" of Older Age. *Molecular Interventions* 2:431-446 (2002).

[66] Sersté T, Bourgeois N. Ageing and the liver. *Acta Gastroenterol Belg.* 2006 Jul-Sep;69(3):296-8.

[67] JE Martin, MT Sheaff. Renal ageing. *J Pathol.* 211(2): 198 – 205.

[68] Janzen V, Forkert R, Fleming HE, Saito Y, Waring MT, Dombkowski DM, Cheng T, DePinho RA, Sharpless NE, Scadden DT. Stem-cell ageing modified by the cyclin-dependent kinase inhibitor p16INK4a. *Nature.* 2006 Sep 28;443(7110):421-6.

[69] Oguro H, Iwama A, Morita Y, Kamijo T, van Lohuizen M, Nakauchi H. Differential impact of Ink4a and Arf on hematopoietic stem cells and their bone marrow microenvironment in Bmi1-deficient mice. *J Exp Med.* 2006 Oct 2;203(10):2247-53.

[70] Ressler S, Bartkova J, Niederegger H, Bartek J, Scharffetter-Kochanek K, Jansen-Dürr P, Wlaschek M. p16INK4A is a robust in vivo biomarker of cellular aging in human skin. *Aging Cell.* 2006 Oct;5(5):379-89.

[71] Maurelli R, Zambruno G, Guerra L, Abbruzzese C, Dimri G, Gellini M, Bondanza S, Dellambra E. Inactivation of p16INK4a (inhibitor of cyclin-dependent kinase 4A) immortalizes primary human keratinocytes by maintaining cells in the stem cell compartment. *FASEB J.* 2006 Jul;20(9):1516-8.

[72] Yu L, Quinn DA, Garg HG, Hales CA. Gene expression of cyclin-dependent kinase inhibitors and effect of heparin on their expression in mice with hypoxia-induced pulmonary hypertension. *Biochem Biophys Res Commun.* 2006 Jul 14;345(4):1565-72.

[73] Natarajan E, Omobono JD 2nd, Guo Z, Hopkinson S, Lazar AJ, Brenn T, Jones JC, Rheinwald JG. A keratinocyte hypermotility/growth-arrest response involving laminin 5 and p16INK4A activated in wound healing and senescence. *Am J Pathol.* 2006 Jun;168(6):1821-37.

[74] Mark EB, Jonsson M, Asp J, Wennberg AM, Mölne L, Lindahl A. Expression of genes involved in the regulation of p16 in psoriatic involved skin. *Arch Dermatol Res.* 2006 Apr;297(10):459-67.

[75] Ramalho FS, Maestri C, Ramalho LN, Ribeiro-Silva A, Romão E. Expression of p63 and p16 in primary and recurrent pterygia. *Graefes Arch Clin Exp Ophthalmol.* 2006 Oct;244(10):1310-4.

[76] Chkhotua AB, Abendroth D, Froeba G, Schelzig H. Up-regulation of cell cycle regulatory genes after renal ischemia/reperfusion: differential expression of p16(INK4a), p21(WAF1/CIP1) and p27(Kip1) cyclin-dependent kinase inhibitor genes depending on reperfusion time. *Transpl Int.* 2006 Jan;19(1):72-7.

[77] Gizard F, Amant C, Barbier O, Bellosta S, Robillard R, Percevault F, Sevestre H, Krimpenfort P, Corsini A, Rochette J, Glineur C, Fruchart JC, Torpier G, Staels B. PPAR alpha inhibits vascular smooth muscle cell proliferation underlying intimal hyperplasia by inducing the tumor suppressor p16INK4a. *J Clin Invest.* 2005 Nov;115(11):3228-38.

[78] Atanasoski S, Boller D, De Ventura L, Koegel H, Boentert M, Young P, Werner S, Suter U. Cell cycle inhibitors p21 and p16 are required for the regulation of Schwann cell proliferation. *Glia.* 2006 Jan 15;53(2):147-57.

[79] Elias AN, Barr RJ, Nanda VS. p16 expression in psoriatic lesions following therapy with propylthiouracil, an antithyroid thioureylene. *Int J Dermatol.* 2004 Dec;43(12):889-92.

[80] Zhou HW, Lou SQ, Zhang K. Recovery of function in osteoarthritic chondrocytes induced by p16INK4a-specific siRNA in vitro. *Rheumatology (Oxford).* 2004 May;43(5):555-68.

[81] Min BM, Oh JE, Choi CM. Retinoic acid delays keratinocyte senescence by suppression of betaig-h3 and p16 expression and induction of telomerase activity. *Int J Mol Med.* 2004 Jan;13(1):25-31.

[82] Nakamura S, Nishioka K. Enhanced expression of p16 in seborrhoeic keratosis; a lesion of accumulated senescent epidermal cells in G1 arrest. *Br J Dermatol.* 2003 Sep;149(3):560-5.

[83] Beauséjour CM, Krtolica A, Galimi F, Narita M, Lowe SW, Yaswen P, Campisi J. Reversal of human cellular senescence: roles of the p53 and p16 pathways. *EMBO J.* 2003 Aug 15;22(16):4212-22.

[84] Oliva A, Borriello A, Zeppetelli S, Di Feo A, Cortellazzi P, Ventriglia V, Criscuolo M, Zappia V, Della Ragione F. Retinoic acid inhibits the growth of bone marrow mesenchymal stem cells and induces p27Kip1 and p16INK4A up-regulation. *Mol Cell Biochem.* 2003 May;247(1-2):55-60.

[85] Bayrak A, Oktay K. The expression of cyclin-dependent kinase inhibitors p15, p16, p21, and p27 during ovarian follicle growth initiation in the mouse. *Reprod Biol Endocrinol.* 2003 May 7;1:41.

[86] Minami R, Muta K, Umemura T, Motomura S, Abe Y, Nishimura J, Nawata H. p16(INK4a) induces differentiation and apoptosis in erythroid lineage cells. *Exp Hematol.* 2003 May;31(5):355-62.

[87] Shirin H, Hibshoosh H, Kawabata Y, Weinstein IB, Moss SF. p16Ink4a is overexpressed in H. pylori-associated gastritis and is correlated with increased epithelial apoptosis. *Helicobacter.* 2003 Feb;8(1):66-71.

[88] Itahana K, Zou Y, Itahana Y, Martinez JL, Beausejour C, Jacobs JJ, Van Lohuizen M, Band V, Campisi J, Dimri GP. Control of the replicative life span of human fibroblasts by p16 and the polycomb protein Bmi-1. *Mol Cell Biol.* 2003 Jan;23(1):389-401.

[89] Zygmunt A, Tedesco VC, Udho E, Krucher NA. Hypoxia stimulates p16 expression and association with cdk4. *Exp Cell Res.* 2002 Aug 1;278(1):53-60.

[90] Brookes S, Rowe J, Ruas M, Llanos S, Clark PA, Lomax M, James MC, Vatcheva R, Bates S, Vousden KH, Parry D, Gruis N, Smit N, Bergman W, Peters G. INK4a-deficient human diploid fibroblasts are resistant to RAS-induced senescence. *EMBO J.* 2002 Jun 17;21(12):2936-45.

[91] Sviderskaya EV, Hill SP, Evans-Whipp TJ, Chin L, Orlow SJ, Easty DJ, Cheong SC, Beach D, DePinho RA, Bennett DC. p16(Ink4a) in melanocyte senescence and differentiation. *J Natl Cancer Inst.* 2002 Mar 20;94(6):446-54.

[92] Lagresle C, Gardie B, Eyquem S, Fasseu M, Vieville JC, Pla M, Sigaux F, Bories JC. Transgenic expression of the p16(INK4a) cyclin-dependent kinase inhibitor leads to enhanced apoptosis and differentiation arrest of CD4-CD8- immature thymocytes. *J Immunol.* 2002 Mar 1;168(5):2325-31.

[93] Duan J, Zhang Z, Tong T. Senescence delay of human diploid fibroblast induced by anti-sense p16INK4a expression. *J Biol Chem.* 2001 Dec 21;276(51):48325-31.

[94] Tsui LV, Camrud A, Mondesire J, Carlson P, Zayek N, Camrud L, Donahue B, Bauer S, Lin A, Frey D, Rivkin M, Subramanian A, Falotico R, Gyuris J, Schwartz R, McArthur JG. p27-p16 fusion gene inhibits angioplasty-induced neointimal hyperplasia and coronary artery occlusion. *Circ Res.* 2001 Aug 17;89(4):323-8.

[95] Randle DH, Zindy F, Sherr CJ, Roussel MF. Differential effects of p19(Arf) and p16(Ink4a) loss on senescence of murine bone marrow-derived preB cells and macrophages. *Proc Natl Acad Sci U S A.* 2001 Aug 14;98(17):9654-9.

[96] Shapiro GI, Edwards CD, Rollins BJ. The physiology of p16(INK4A)-mediated G1 proliferative arrest. *Cell Biochem Biophys.* 2000;33(2):189-97.

[97] McArthur JG, Qian H, Citron D, Banik GG, Lamphere L, Gyuris J, Tsui L, George SE. p27-p16 Chimera: a superior antiproliferative for the prevention of neointimal hyperplasia. *Mol Ther*. 2001 Jan;3(1):8-13.

[98] Fuxe J, Akusjärvi G, Goike HM, Roos G, Collins VP, Pettersson RF. Adenovirus-mediated overexpression of p15INK4B inhibits human glioma cell growth, induces replicative senescence, and inhibits telomerase activity similarly to p16INK4A. *Cell Growth Differ*. 2000 Jul;11(7):373-84.

[99] Sandhu C, Peehl DM, Slingerland J. p16INK4A mediates cyclin dependent kinase 4 and 6 inhibition in senescent prostatic epithelial cells. *Cancer Res*. 2000 May 15;60(10):2616-22.

[100] Yang X, He Z, Xin B, Cao L. LMP1 of Epstein-Barr virus suppresses cellular senescence associated with the inhibition of p16INK4a expression. *Oncogene*. 2000 Apr 13;19(16):2002-13.

[101] Piepkorn M. The expression of p16(INK4a), the product of a tumor suppressor gene for melanoma, is upregulated in human melanocytes by UVB irradiation. *J Am Acad Dermatol*. 2000 May;42(5 Pt 1):741-5.

[102] Seki Y, Koizumi H, Koizuka I, Takakuwa T, Tsutsumi K. Nuclear expression of the p16CDKN2 gene product during senescence of human pharyngeal epithelial cells. *Auris Nasus Larynx*. 2000 Apr;27(2):147-51.

[103] You YO, Lee G, Min BM. Retinoic acid extends the in vitro life span of normal human oral keratinocytes by decreasing p16(INK4A) expression and maintaining telomerase activity. *Biochem Biophys Res Commun*. 2000 Feb 16;268(2):268-74.

[104] Wolf G, Wenzel U, Ziyadeh FN, Stahl RA. Angiotensin converting-enzyme inhibitor treatment reduces glomerular p16INK4 and p27Kip1 expression in diabetic BBdp rats. *Diabetologia*. 1999 Dec;42(12):1425-32.

[105] Morisaki H, Ando A, Nagata Y, Pereira-Smith O, Smith JR, Ikeda K, Nakanishi M. Complex mechanisms underlying impaired activation of Cdk4 and Cdk2 in replicative senescence: roles of p16, p21, and cyclin D1. *Exp Cell Res*. 1999 Dec 15;253(2):503-10.

[106] Nielsen GP, Stemmer-Rachamimov AO, Shaw J, Roy JE, Koh J, Louis DN. Immunohistochemical survey of p16INK4A expression in normal human adult and infant tissues. *Lab Invest*. 1999 Sep;79(9):1137-43.

[107] Taniguchi K, Kohsaka H, Inoue N, Terada Y, Ito H, Hirokawa K, Miyasaka N. Induction of the p16INK4a senescence gene as a new therapeutic strategy for the treatment of rheumatoid arthritis. *Nat Med*. 1999 Jul;5(7):760-7.

[108] Charles J. Sherr, and James M. Roberts. CDK inhibitors: positive and negative regulators of G1-phase progression. 1999; 13 (12): 1501-1512 .

[109] Huschtscha LI, Reddel RR. p16(INK4a) and the control of cellular proliferative life span. *Carcinogenesis*. 1999 Jun;20(6):921-6.

[110] Pavletich NP. Mechanisms of cyclin-dependent kinase regulation: structures of Cdks, their cyclin activators, and Cip and INK4 inhibitors. *J Mol Biol*. 1999 Apr 16;287(5):821-8.

[111] Arendt T, Holzer M, Gärtner U. Neuronal expression of cycline dependent kinase inhibitors of the INK4 family in Alzheimer's disease. *J Neural Transm*. 1998;105(8-9):949-60.

[112] Huschtscha LI, Noble JR, Neumann AA, Moy EL, Barry P, Melki JR, Clark SJ, Reddel RR. Loss of p16INK4 expression by methylation is associated with lifespan extension of human mammary epithelial cells. *Cancer Res*. 1998 Aug 15;58(16):3508-12.

[113] Erickson S, Sangfelt O, Heyman M, Castro J, Einhorn S, Grandér D. Involvement of the Ink4 proteins p16 and p15 in T-lymphocyte senescence. *Oncogene*. 1998 Aug 6;17(5):595-602.

[114] Robles SJ, Adami GR. Agents that cause DNA double strand breaks lead to p16INK4a enrichment and the premature senescence of normal fibroblasts. *Oncogene*. 1998 Mar 5;16(9):1113-23.

[115] Terada Y, Inoshita S, Nakashima O, Yamada T, Tamamori M, Ito H, Sasaki S, Marumo F. Cyclin D1, p16, and retinoblastoma gene regulate mitogenic signaling of endothelin in rat mesangial cells. *Kidney Int*. 1998 Jan;53(1):76-83.

[116] Shapiro GI, Edwards CD, Ewen ME, Rollins BJ. p16INK4A participates in a G1 arrest checkpoint in response to DNA damage. *Mol Cell Biol*. 1998 Jan;18(1):378-87.

[117] Narimatsu M, Nakajima K, Ichiba M, Hirano T. Association of Stat3-dependent transcriptional activation of p19INK4D with IL-6-induced growth arrest. *Biochem Biophys Res Commun*. 1997 Sep 29;238(3):764-8.

[118] Park JY, Schutzer WE, Lindsley JN, Bagby SP, Oyama TT, Anderson S, Weiss RH. p21 is decreased in polycystic kidney disease and leads to increased epithelial cell cycle progression: roscovitine augments p21 levels. *BMC Nephrol*. 2007 Aug 22;8:12.

[119] Li JM, Fan LM, George VT, Brooks G. Nox2 regulates endothelial cell cycle arrest and apoptosis via p21cip1 and p53. *Free Radic Biol Med*. 2007 Sep 15;43(6):976-86.

[120] Bhattacharya S, Guo H, Ray RM, Johnson LR. Basic helix-loop-helix protein E47-mediated p21Waf1/Cip1 gene expression regulates apoptosis of intestinal epithelial cells. *Biochem J*. 2007 Oct 15;407(2):243-54.

[121] Broude EV, Demidenko ZN, Vivo C, Swift ME, Davis BM, Blagosklonny MV, Roninson IB. p21 (CDKN1A) is a negative regulator of p53 stability. *Cell Cycle*. 2007 Jun 15;6(12):1468-71.

[122] Pajalunga D, Mazzola A, Puggioni E, Crescenzi M. Non-proliferation as an active state: conceptual and practical implications. *Cell Cycle*. 2007 Jun 15;6(12):1415-8. Epub 2007 Apr 30.

[123] Hwang CY, Kim IY, Kwon KS. Cytoplasmic localization and ubiquitination of p21(Cip1) by reactive oxygen species. *Biochem Biophys Res Commun*. 2007 Jun 22;358(1):219-25.

[124] Kim JA, Hong S, Lee B, Hong JW, Kwak JY, Cho S, Kim CC. The inhibition of T-cells proliferation by mouse mesenchymal stem cells through the induction of p16INK4A-cyclin D1/cdk4 and p21waf1, p27kip1-cyclin E/cdk2 pathways. *Cell Immunol*. 2007 Jan;245(1):16-23.

[125] Yi T, Baek JH, Kim HJ, Choi MH, Seo SB, Ryoo HM, Kim GS, Woo KM. Trichostatin A-mediated upregulation of p21(WAF1) contributes to osteoclast apoptosis. *Exp Mol Med*. 2007 Apr 30;39(2):213-21.

[126] Ju Z, Choudhury AR, Rudolph KL. A dual role of p21 in stem cell aging. *Ann N Y Acad Sci.* 2007 Apr;1100:333-44.

[127] Harms C, Albrecht K, Harms U, Seidel K, Hauck L, Baldinger T, Hübner D, Kronenberg G, An J, Ruscher K, Meisel A, Dirnagl U, von Harsdorf R, Endres M, Hörtnagl H. Phosphatidylinositol 3-Akt-kinase-dependent phosphorylation of p21(Wafl/Cipl) as a novel mechanism of neuroprotection by glucocorticoids. *J Neurosci.* 2007 Apr 25;27(17):4562-71.

[128] Pajalunga D, Mazzola A, Salzano AM, Biferi MG, De Luca G, Crescenzi M. Critical requirement for cell cycle inhibitors in sustaining nonproliferative states. *J Cell Biol.* 2007 Mar 12;176(6):807-18.

[129] Chuang TD, Guh JY, Chiou SJ, Chen HC, Huang JS, Yang YL, Chuang LY. Phosphoinositide 3-kinase is required for high glucose-induced hypertrophy and p21WAF1 expression in LLC-PK1 cells. *Kidney Int.* 2007 May;71(9):867-74.

[130] Jänicke RU, Sohn D, Essmann F, Schulze-Osthoff K. The multiple battles fought by anti-apoptotic p21. *Cell Cycle.* 2007 Feb 1;6(4):407-13.

[131] Arias CF, Ballesteros-Tato A, García MI, Martín-Caballero J, Flores JM, Martínez-A C, Balomenos D. p21CIP1/WAF1 controls proliferation of activated/memory T cells and affects homeostasis and memory T cell responses. *J Immunol.* 2007 Feb 15;178(4):2296-306.

[132] Zhang J, Scadden DT, Crumpacker CS. Primitive hematopoietic cells resist HIV-1 infection via p21. *J Clin Invest.* 2007 Feb;117(2):473-81.

[133] van Os R, Kamminga LM, Ausema A, Bystrykh LV, Draijer DP, van Pelt K, Dontje B, de Haan G. A Limited role for p21Cipl/Wafl in maintaining normal hematopoietic stem cell functioning. *Stem Cells.* 2007 Apr;25(4):836-43.

[134] Al-Mohanna MA, Al-Khalaf HH, Al-Yousef N, Aboussekhra A. The p16INK4a tumor suppressor controls p21WAF1 induction in response to ultraviolet light. *Nucleic Acids Res.* 2007;35(1):223-33.

[135] Sohn D, Essmann F, Schulze-Osthoff K, Jänicke RU.p21 blocks irradiation-induced apoptosis downstream of mitochondria by inhibition of cyclin-dependent kinase-mediated caspase-9 activation. *Cancer Res.* 2006 Dec 1;66(23):11254-62

[136] Cozar-Castellano I, Haught M, Stewart AF. The cell cycle inhibitory protein p21cip is not essential for maintaining beta-cell cycle arrest or beta-cell function in vivo. *Diabetes.* 2006 Dec;55(12):3271-8.

[137] Gehen SC, Vitiello PF, Bambara RA, Keng PC, O'Reilly MA. Downregulation of PCNA potentiates p21-mediated growth inhibition in response to hyperoxia. *Am J Physiol Lung Cell Mol Physiol.* 2007 Mar;292(3):L716-24.

[138] de Oliveira RM. Klotho RNAi induces premature senescence of human cells via a p53/p21 dependent pathway. *FEBS Lett.* 2006 Oct 16;580(24):5753-8.

[139] Lee YJ, Lee JH, Han HJ. Extracellular adenosine triphosphate protects oxidative stress-induced increase of p21(WAF1/Cip1) and p27(Kip1) expression in primary cultured renal proximal tubule cells: role of PI3K and Akt signaling. *J Cell Physiol.* 2006 Dec;209(3):802-10.

[140] Hochegger K, Koppelstaetter C, Tagwerker A, Huber JM, Heininger D, Mayer G, Rosenkranz AR. p21 and mTERT are novel markers for determining different ischemic

time periods in renal ischemia-reperfusion injury. *Am J Physiol Renal Physiol.* 2007 Feb;292(2):F762-8.

[141] Stepniak E, Ricci R, Eferl R, Sumara G, Sumara I, Rath M, Hui L, Wagner EF. c-Jun/AP-1 controls liver regeneration by repressing p53/p21 and p38 MAPK activity. *Genes Dev.* 2006 Aug 15;20(16):2306-14.

[142] Devgan V, Nguyen BC, Oh H, Dotto GP. p21WAF1/Cip1 suppresses keratinocyte differentiation independently of the cell cycle through transcriptional up-regulation of the IGF-I gene. *J Biol Chem.* 2006 Oct 13;281(41):30463-70.

[143] Okamoto H, Fujioka Y, Takahashi A, Takahashi T, Taniguchi T, Ishikawa Y, Yokoyama M. Trichostatin A, an inhibitor of histone deacetylase, inhibits smooth muscle cell proliferation via induction of p21(WAF1). *J Atheroscler Thromb.* 2006 Aug;13(4):183-91.

[144] McCullough CT, Tura BJ, Harrison DJ. Growth factor attenuation of IFNgamma-mediated hepatocyte apoptosis requires p21waf-1. *Int J Exp Pathol.* 2006 Aug;87(4):275-81.

[145] Woods JM, Klosowska K, Spoden DJ, Stumbo NG, Paige DJ, Scatizzi JC, Volin MV, Rao MS, Perlman H. A cell-cycle independent role for p21 in regulating synovial fibroblast migration in rheumatoid arthritis. *Arthritis Res Ther.* 2006;8(4):R113.

[146] Child ES, Mann DJ. The intricacies of p21 phosphorylation: protein/protein interactions, subcellular localization and stability. *Cell Cycle.* 2006 Jun;5(12):1313-9.

[147] Vitiello PF, Staversky RJ, Gehen SC, Johnston CJ, Finkelstein JN, Wright TW, O'Reilly MA. p21Cip1 protection against hyperoxia requires Bcl-XL and is uncoupled from its ability to suppress growth. *Am J Pathol.* 2006 Jun;168(6):1838-47.

[148] Zheng QH, Ma LW, Zhu WG, Zhang ZY, Tong TJ. p21Wafl/Cip1 plays a critical role in modulating senescence through changes of DNA methylation. *J Cell Biochem.* 2006 Aug 1;98(5):1230-48.

[149] Cozar-Castellano I, Weinstock M, Haught M, Velázquez-Garcia S, Sipula D, Stewart AF. Evaluation of beta-cell replication in mice transgenic for hepatocyte growth factor and placental lactogen: comprehensive characterization of the G1/S regulatory proteins reveals unique involvement of p21cip. *Diabetes.* 2006 Jan;55(1):70-7.

[150] Boncoeur E, Tabary O, Bonvin E, Muselet C, Fritah A, Lefait E, Redeuilh G, Clement A, Jacquot J, Henrion-Caude A. Oxidative stress response results in increased p21WAF1/CIP1 degradation in cystic fibrosis lung epithelial cells. *Free Radic Biol Med.* 2006 Jan 1;40(1):75-86.

[151] Sun R, Jaruga B, Kulkarni S, Sun H, Gao B. IL-6 modulates hepatocyte proliferation via induction of HGF/p21cip1: regulation by SOCS3. *Biochem Biophys Res Commun.* 2005 Dec 30;338(4):1943-9.

[152] Goulvestre C, Chéreau C, Nicco C, Mouthon L, Weill B, Batteux F. A mimic of p21WAF1/CIP1 ameliorates murine lupus. *J Immunol.* 2005 Nov 15;175(10):6959-67.

[153] Yu H, Yuan Y, Shen H, Cheng T. Hematopoietic stem cell exhaustion impacted by p18 INK4C and p21 Cip1/Wafl in opposite manners. *Blood.* 2006 Feb 1;107(3):1200-6.

[154] Nordborg C, Aman P, Persson M, Nordborg E. Cell-type-specific expression of p53 and p21 in giant cell arteritis. *APMIS.* 2005 Sep;113(9):594-9.

[155] Bond M, Sala-Newby GB, Wu YJ, Newby AC. Biphasic effect of p21Cip1 on smooth muscle cell proliferation: role of PI 3-kinase and Skp2-mediated degradation. *Cardiovasc Res*. 2006 Jan;69(1):198-206.

[156] Miyake N, Brun AC, Magnusson M, Miyake K, Scadden DT, Karlsson S. HOXB4-induced self-renewal of hematopoietic stem cells is significantly enhanced by p21 deficiency. *Stem Cells*. 2006 Mar;24(3):653-61.

[157] Tarry-Adkins JL, Ozanne SE, Norden A, Cherif H, Hales CN. Lower antioxidant capacity and elevated p53 and p21 may be a link between gender disparity in renal telomere shortening, albuminuria, and longevity. *Am J Physiol Renal Physiol*. 2006 Feb;290(2):F509-16.

[158] Seomun Y, Kim JT, Kim HS, Park JY, Joo CK. Induction of p21Cip1-mediated G2/M arrest in H2O2-treated lens epithelial cells. *Mol Vis*. 2005 Sep 16;11:764-74.

[159] Monkawa T, Pippin J, Yo Y, Kopp JB, Alpers CE, Shankland SJ. The cyclin-dependent kinase inhibitor p21 limits murine mesangial proliferative glomerulonephritis. *Nephron Exp Nephrol*. 2006;102(1):e8-18.

[160] Khanna AK. Reciprocal role of cyclins and cyclin kinase inhibitor p21WAF1/CIP1 on lymphocyte proliferation, allo-immune activation and inflammation. *BMC Immunol*. 2005 Sep 21;6:22.

[161] Gu D, Atencio I, Kang DW, Looper LD, Ahmed CM, Levy A, Maneval D, Zepeda ML. Recombinant adenovirus-p21 attenuates proliferative responses associated with excessive scarring. *Wound Repair Regen*. 2005 Sep-Oct;13(5):480-90.

[162] Wang J, Devgan V, Corrado M, Prabhu NS, El-Deiry WS, Riccardi C, Pandolfi PP, Missero C, Dotto GP. Glucocorticoid-induced tumor necrosis factor receptor is a p21Cip1/WAF1 transcriptional target conferring resistance of keratinocytes to UV light-induced apoptosis. *J Biol Chem*. 2005 Nov 11;280(45):37725-31.

[163] Granada JF, Ensenat D, Keswani AN, Kaluza GL, Raizner AE, Liu XM, Peyton KJ, Azam MA, Wang H, Durante W. Single perivascular delivery of mitomycin C stimulates p21 expression and inhibits neointima formation in rat arteries. *Arterioscler Thromb Vasc Biol*. 2005 Nov;25(11):2343-8.

[164] Kim HS, Yeo EJ, Park SH, Park JI, Park SC, Shin JY, Kim MJ, Oh SJ, Won MH, Kang TC, Park JB, Kim J, Kim JI, Lee HY, Lee JY. p21WAF/CIP1/SDI1 is upregulated due to increased mRNA stability during hydroxyurea-induced senescence of human fibroblasts. *Mech Ageing Dev*. 2005 Dec;126(12):1255-61.

[165] Ghanem L, Steinman R. A proapoptotic function of p21 in differentiating granulocytes. *Leuk Res*. 2005 Nov;29(11):1315-23.

[166] Kim HJ, Park SR, Park HJ, Choi BH, Min BH. Potential predictive markers for proliferative capacity of cultured human articular chondrocytes: PCNA and p21. *Artif Organs*. 2005 May;29(5):393-8.

[167] Terada Y, Inoshita S, Nakashima O, Tamamori M, Ito H, Kuwahara M, Sasaki S, Marumo F. Cell cycle inhibitors (p27Kip1 and p21CIP1) cause hypertrophy in LLC-PK1 cells. *Kidney Int*. 1999 Aug;56(2):494-501.

[168] Ueyama C, Akashiba H, Nakayama K, Nakayama KI, Nishiyama N, Matsuki N. Ablation of p27 enhance kainate-induced seizure and hippocampal degeneration. *Neuroreport*. 2007 Nov 19;18(17):1781-5.

[169] Koyanagi M, Fukada K, Uchiyama T, Yagi J, Arimura Y. Long-term exposure to superantigen induces p27Kip1 and Bcl-2 expression in effector memory CD4+ T cells. *Cell Immunol*. 2007 Aug;248(2):77-85 .

[170] Hinrichsen R, Haunsø S, Busk PK. Different regulation of p27 and Akt during cardiomyocyte proliferation and hypertrophy. *Growth Factors*. 2007 Apr;25(2):132-40.

[171] Zhong L, Georgia S, Tschen SI, Nakayama K, Nakayama K, Bhushan A. Essential role of Skp2-mediated p27 degradation in growth and adaptive expansion of pancreatic beta cells. *J Clin Invest*. 2007 Oct;117(10):2869-76.

[172] Minoda R, Izumikawa M, Kawamoto K, Zhang H, Raphael Y. Manipulating cell cycle regulation in the mature cochlea. *Hear Res*. 2007 Oct;232(1-2):44-51.

[173] Fischer BM, Zheng S, Fan R, Voynow JA. Neutrophil elastase inhibition of cell cycle progression in airway epithelial cells in vitro is mediated by p27kip1. *Am J Physiol Lung Cell Mol Physiol.* 2007 Sep;293(3):L762-8.

[174] Lopez-Sanchez E, Frances-Muñoz E, Chaques V, Sanchez-Benavent ML, Menezo JL. Optic nerve alterations in P27(Kip1) knockout mice. *Eur J Ophthalmol*. 2007 May-Jun;17(3):377-82.

[175] Borriello A, Cucciolla V, Oliva A, Zappia V, Della Ragione F. p27Kip1 metabolism: a fascinating labyrinth. *Cell Cycle*. 2007 May 2;6(9):1053-61.

[176] Defoe DM, Adams LB, Sun J, Wisecarver SN, Levine EM. Defects in retinal pigment epithelium cell proliferation and retinal attachment in mutant mice with p27(Kip1) gene ablation. *Mol Vis*. 2007 Feb 27;13:273-86.

[177] Kase S, Takahashi S, Sato I, Nakanishi K, Yoshida K, Ohno S. Expression of p27(KIP1) and cyclin D1, and cell proliferation in human pterygium. *Br J Ophthalmol*. 2007 Jul;91(7):958-61.

[178] Rachdi L, Balcazar N, Elghazi L, Barker DJ, Krits I, Kiyokawa H, Bernal-Mizrachi E. Differential effects of p27 in regulation of beta-cell mass during development, neonatal period, and adult life. *Diabetes*. 2006 Dec;55(12):3520-8.

[179] Kikuchi M, Zhu C, Senoo T, Obara Y, Joyce NC. p27kip1 siRNA induces proliferation in corneal endothelial cells from young but not older donors. *Invest Ophthalmol Vis Sci*. 2006 Nov;47(11):4803-9.

[180] Li L, Iwamoto Y, Berezovskaya A, Boussiotis VA. A pathway regulated by cell cycle inhibitor p27Kip1 and checkpoint inhibitor Smad3 is involved in the induction of T cell tolerance. *Nat Immunol*. 2006 Nov;7(11):1157-65.

[181] Akashiba H, Matsuki N, Nishiyama N. p27 small interfering RNA induces cell death through elevating cell cycle activity in cultured cortical neurons: a proof-of-concept study. *Cell Mol Life Sci*. 2006 Oct;63(19-20):2397-404.

[182] Soeiro I, Mohamedali A, Romanska HM, Lea NC, Child ES, Glassford J, Orr SJ, Roberts C, Naresh KN, Lalani el-N, Mann DJ, Watson RJ, Thomas NS, Lam EW. p27Kip1 and p130 cooperate to regulate hematopoietic cell proliferation in vivo. *Mol Cell Biol.* 2006 Aug;26(16):6170-84.

[183] Emons JA, Marino R, Nilsson O, Barnes KM, Even-Zohar N, Andrade AC, Chatterjee NA, Wit JM, Karperien M, Baron J. The role of p27Kip1 in the regulation of growth plate chondrocyte proliferation in mice. *Pediatr Res*. 2006 Sep;60(3):288-93.

[184] Kuczkowski J, Bakowska A, Pawelczyk T, Narozny W, Mikaszewski B. Cell cycle inhibitory protein p27 in human middle ear cholesteatoma. *ORL J Otorhinolaryngol Relat Spec*. 2006;68(5):296-301.

[185] Kobayashi H, Kambe F, Imai T, Hibi Y, Kikumori T, Ohmori S, Nakao A, Seo H. Differential expression of cyclin-dependent kinase inhibitors, p27Kip1 and p57Kip2, by corticotropin in rat adrenal cortex. *J Endocrinol*. 2006 Jun;189(3):671-9.

[186] Kanzaki S, Beyer LA, Swiderski DL, Izumikawa M, Stöver T, Kawamoto K, Raphael Y. p27(Kip1) deficiency causes organ of Corti pathology and hearing loss. *Hear Res*. 2006 Apr;214(1-2):28-36.

[187] Li WQ, Jiang Q, Aleem E, Kaldis P, Khaled AR, Durum SK. IL-7 promotes T cell proliferation through destabilization of p27Kip1. *J Exp Med*. 2006 Mar 20;203(3):573-82 .

[188] Nishijima K, Ando M, Sano S, Hayashi-Ozawa A, Kinoshita Y, Iijima S. Costimulation of T-cell proliferation by anti-L-selectin antibody is associated with the reduction of a cdk inhibitor p27. *Immunology*. 2005 Nov;116(3):347-53.

[189] Yu L, Quinn DA, Garg HG, Hales CA. Cyclin-dependent kinase inhibitor p27Kip1, but not p21WAF1/Cip1, is required for inhibition of hypoxia-induced pulmonary hypertension and remodeling by heparin in mice. *Circ Res*. 2005 Oct 28;97(9):937-45.

[190] Wolf G, Schanze A, Stahl RA, Shankland SJ, Amann K. p27(Kip1) Knockout mice are protected from diabetic nephropathy: evidence for p27(Kip1) haplotype insufficiency. *Kidney Int*. 2005 Oct;68(4):1583-9.

[191] Vlachos P, Nyman U, Hajji N, Joseph B. The cell cycle inhibitor p57(Kip2) promotes cell death via the mitochondrial apoptotic pathway. *Cell Death Differ*. 2007 Aug;14(8):1497-507.

[192] Buchold GM, Magyar PL, O'Brien DA. Mice lacking cyclin-dependent kinase inhibitor p19Ink4d show strain-specific effects on male reproduction. *Mol Reprod Dev*. 2007 Aug;74(8):1008-20.

[193] Buchold GM, Magyar PL, Arumugam R, Lee MM, O'Brien DA. p19Ink4d and p18Ink4c cyclin-dependent kinase inhibitors in the male reproductive axis. *Mol Reprod Dev*. 2007 Aug;74(8):997-1007.

[194] Scassa ME, Marazita MC, Ceruti JM, Carcagno AL, Sirkin PF, González-Cid M, Pignataro OP, Cánepa ET. Cell cycle inhibitor, p19INK4d, promotes cell survival and decreases chromosomal aberrations after genotoxic insult due to enhanced DNA repair. *DNA Repair (Amst)*. 2007 May 1;6(5):626-38.

[195] Kim ST, Park NC, Yi LS, Gye MC. Expression of p57kip2 in germ cells and Leydig cells in human testis. *Arch Androl*. 2006 Nov-Dec;52(6):463-9.

[196] Hofer TP, Frankenberger M, Staples KJ, Ziegler-Heitbrock L. Expression of p57-Kip2 in monocytes and macrophages. *Immunobiology*. 2006;211(6-8):455-62.

[197] Kim SH, Rowe J, Fujii H, Jones R, Schmierer B, Kong BW, Kuchler K, Foster D, Ish-Horowicz D, Peters G. Upregulation of chicken p15INK4b at senescence and in the developing brain. *J Cell Sci*. 2006 Jun 15;119(Pt 12):2435-43.

[198] den Besten W, Kuo ML, Tago K, Williams RT, Sherr CJ. Ubiquitination of, and sumoylation by, the Arf tumor suppressor. *Isr Med Assoc J*. 2006 Apr;8(4):249-51.

[199] Tavera-Mendoza LE, Wang TT, White JH. p19INK4D and cell death. *Cell Cycle*. 2006 Mar;5(6):596-8.

[200] Chen P, Zindy F, Abdala C, Liu F, Li X, Roussel MF, Segil N. Progressive hearing loss in mice lacking the cyclin-dependent kinase inhibitor Ink4d. *Nat Cell Bio*l. 2003 May;5(5):422-6.

[201] Zindy F, den Besten W, Chen B, Rehg JE, Latres E, Barbacid M, Pollard JW, Sherr CJ, Cohen PE, Roussel MF. Control of spermatogenesis in mice by the cyclin D-dependent kinase inhibitors p18(Ink4c) and p19(Ink4d). *Mol Cell Biol*. 2001 May;21(9):3244-55.

[202] O'Farrell AM, Parry DA, Zindy F, Roussel MF, Lees E, Moore KW, Mui AL. Stat3-dependent induction of p19INK4D by IL-10 contributes to inhibition of macrophage proliferation. *J Immunol*. 2000 May 1;164(9):4607-15.

[203] Zindy F, van Deursen J, Grosveld G, Sherr CJ, Roussel MF. INK4d-deficient mice are fertile despite testicular atrophy. *Mol Cell Biol*. 2000 Jan;20(1):372-8.

[204] Jiang H, Zhu H, Chen X, Peng Y, Wang J, Liu F, Shi S, Fu B, Lu Y, Hong Q, Feng Z, Hou K, Sun X, Cai G, Zhang X, Xie Y. TIMP-1 transgenic mice recover from diabetes induced by multiple low-dose streptozotocin. *Diabetes*. 2007 Jan;56(1):49-56.

In: Cell Cycle Control: New Research
Editors: N. H. Leroy, N. T. Fournier
ISBN: 978-1-60456-796-0

Chapter XI

The Role and Predictive Power of Cell Cycle Checkpoints in Aging, Cardiovascular, Cancer, and Neurodegenerative Diseases

Olga Golubnitschaja
Department of Radiology, Friedrich-Wilhelms-University of Bonn, D-53105 Germany.

Abstract

Maintenance of genomic integrity is critical for prevention of a wide variety of adverse cellular effects including apoptosis, cellular senescence, and malignant cell transformation. Under stress conditions and even during an unperturbed cell cycle, checkpoint proteins play the key role in genome maintenance by and mediating cellular response to DNA damage, and represent an essential part of the "cellular stress response proteome". Intact checkpoint signal transduction cascades check the presence of genome damage, trigger cell cycle arrest, and forward the information to the protein core of cell cycle machinery, replication apparatus, repair, and/or apoptotic protein cores. Genetic checkpoint defects lead to syndromes that demonstrate chromosomal instability, increased sensitivity to genotoxic stress, tissue degeneration, developmental retardation, premature aging, and cancer predisposition that is most extensively studied for the ATM-checkpoint mutated in *Ataxia telangiectasia*. Tissue specific epigenetic control over the function of cell cycle checkpoints can be, further, deregulated by aberrant DNA methylation status. The consequent checkpoint deregulation may result in tissue specific degenerative processes such as degeneration and calcification of heart aortic valves, diabetic cardiomyopathy, hyperhomocysteinemic cerebrovascular, peripheral vascular and coronary heart diseases, neurodegenerative disorders (Alzheimer and Parkinson diseases, amyotrophic lateral sclerosis, glaucoma), and premature aging frequently accompanied with cancer. This chapter focuses on the checkpoints shown to be crucial for unperturbed cell cycle regulation, deregulation of which might be considered as a potential molecular marker for early diagnosis of and therapy efficiency in neurodegenerative, cardiovascular and cancer diseases. An application of the most potent detection technologies such as “Clinical Proteomics” and “Disease-specific Transcriptomics” also considered here, allows a most specific selection of diagnostic markers.

1. Introduction

Stress response is the ability of each organism which decides over adaptation quality and survival under extreme as well as routine conditions. At the organismal level, stress response is controlled by hormones (Charmandary et al., 2005). Cellular stress response is frequently non-specific to the origin of stress factors causing the response; it is controlled by the quality/deformation of the essential cellular macromolecules as reviewed by Kultz (2005). The capacity of the cellular stress response depends on the pool of proteins expressed at the time point of stress application and on additionally activated ones in response to the stress. This pool of proteins is called “Stress Proteome”. It has been shown that proteins involved in key aspects of the cellular stress response are highly conserved in all organisms tested. The “Minimal Stress Proteome” of cellular organisms consists mainly of the following functional groups (Kultz 2003):

- “Energy metabolism and ATPases/ABC-transporters”,
- “Translation regulation, protein synthesis and degradation”,
- “Redox regulation”,
- “Nucleic acid metabolism”,
- “DNA damage sensing/repair and cell cycle control”,
- “Basic transcription, replication, and RNA processing”,
- “CoA metabolism”, and
- “Molecular chaperones”.

DNA damage and protein oxidation represent the most severe deformations of cellular macromolecules, and trigger plenty of intra- and extra-cellular signaling cascades. Thereby, genotoxicity is the most fatal consequence of any kind of stress factor. To the currently well-acknowledged and most ubiquitous stress factor belong emotion stress, smoking and alcohol consumption (Korystov 1997; Irie et al., 2001), hyperhomocysteinemia (Sachdev 2005), folate and taurine deficiency (Tomei et al., 1990; Chern et al., 2001; Golubnitschaja et al., 2003), UV- and ionizing irradiation (Brown et al., 2003), fluctuating blood glucose concentration (starvation, *Diabetes mellitus*, etc. (Massague 2004; Golubnitschaja et al., 2006a), asphyxia/ischemia/stroke/reperfusion (Dell´Anna et al., 1997; Didenko et al., 2002; Cherubini et al., 2005), genetic syndromes such as Down syndrome (Midorikawa and Kawanishi 2001) neurodegenerative diseases such as Huntington, Alzheimer, and Parkinson diseases, glaucoma (Chuang 2005; Migliore et al., 2005; Hedge et al., 2006; Moenkemann et al., 2005), etc.

Under stress conditions and even during an unperturbed cell cycle, checkpoint proteins play the key role in genome maintenance by and mediating cellular response to DNA damage, as it is schematically shown in Figure 1.

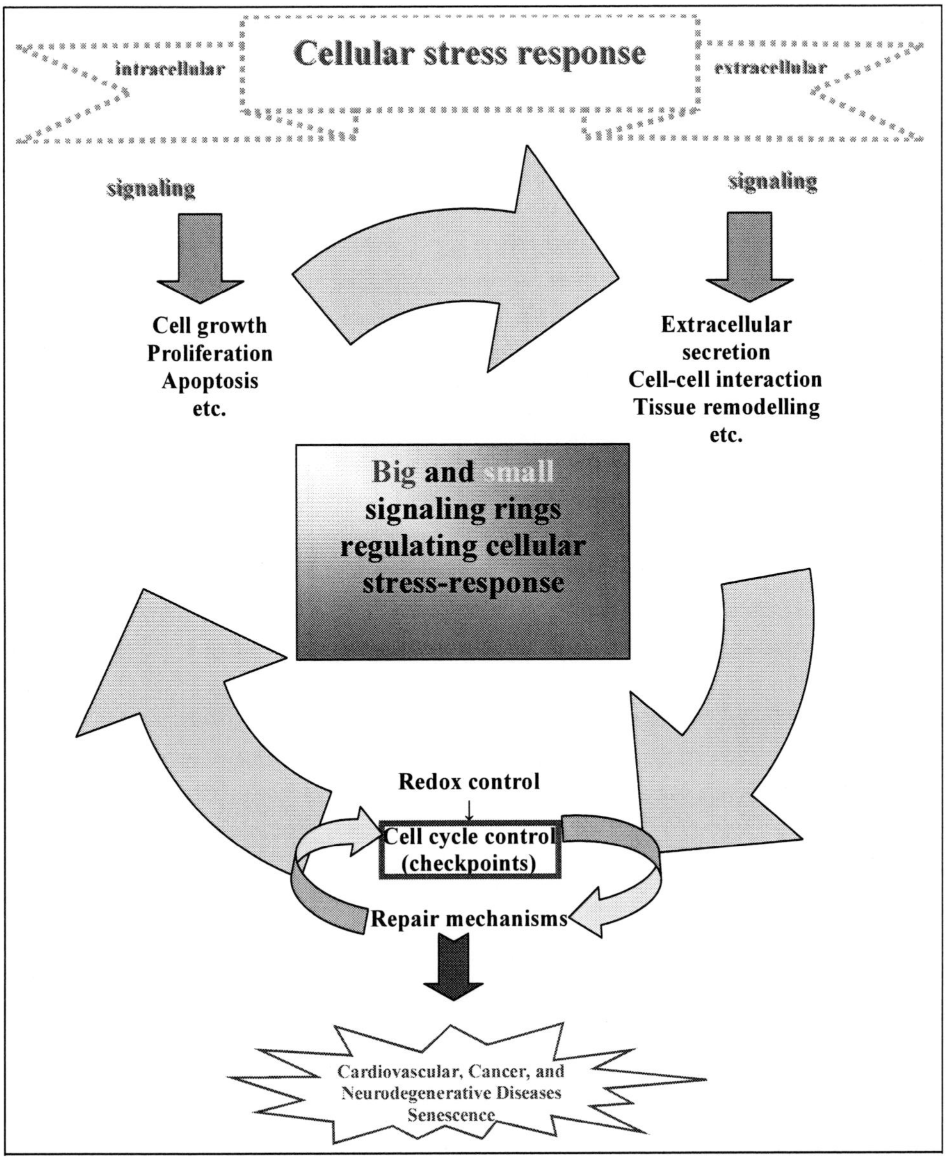

Figure 1. The role of cell cycle checkpoints in cellular stress response (Golubnitschaja, 2007).

Checkpoint signal transduction cascades check the presence of genome damage, trigger cell cycle arrest, and forward the information to the protein core of cell cycle machinery, replication apparatus, and either repair or apoptotic protein cores (Massague 2004). Damaged DNA in humans is detected by a cascade of sensor proteins (HUS1, RAD1, RAD9, RAD17, RAD26), which transmit signals via ATR to CHK1, or by the other cascade of sensor proteins (NBS1, BRCA1, RAD50, etc.), which transmit signals via the ataxia-telangiectasia mutated (ATM) protein kinase to CHK2 (Kawabe 2004). Double-stranded DNA (dsDNA)-breaks activate ATM which, in turn, modulates numerous signalling pathways via target phosphorylation that either enhances or represses activity of ATM-dependent proteins. ATM-mediated activation of cell-cycle checkpoints in response to (dsDNA)-breaks include following targets: P53, CDC25A, CDK2, CHK2, P21wafl/cipl (CDKN1A) in G_1/S phase transfer, and CHK1, CHK2, 14-3-3 proteins in G_2/M phase transfer of cell cycle (Xie et al., 1998; Shiloh 2003; Kawabe 2004; Pereg et al., 2006). Checkpoints response to single-stranded DNA (ssDNA)-lesions and DNA replication blockage is triggered by ATR-dependent pathways (Bakkenist and Kastan, 2004; McGowan and Russel, 2004; Levitt et al., 2005). By this way, checkpoint signal cascades response to a variety of stress conditions by coordinating of cell cycle arrest/promotion, various pro-apoptotic factors, the DNA-repair machinery, and oxidative-stress response genes (Massague 2004). Genetic defects affecting DNA damage response pathways lead to syndromes that demonstrate chromosomal instability, increased sensitivity to genotoxic stress, tissue degeneration, developmental retardation, premature aging, and cancer predisposition. The most prominent genomic instability syndromes are *Ataxia telangiectasia* and *Ataxia telangiectasia*-like disease, Bloom´s, Cockayne´s, Nijmegen breakage, Rothmund-Thompson, Werner´s syndromes, Fanconi´s anemia, Trichothiodystrophy, and *Xeroderma pigmentosum* as reviewed by Shilog (2003).

Recent works clearly showed that dysfunction of checkpoints uncoupled from cellular apoptotic/repair mechanisms leads to severe consequences such as human genetic syndromes, cardiovascular and cancer diseases, and senescence. Can we use cell cycle checkpoints as molecular markers for early diagnosis and prognosis of disorder severity?

2. Potential Application of Checkpoints Genes as Molecular Markers for Early Diagnosis of Cardiovascular Diseases

2.1. Checkpoints in Heart Diseases

Aortic valve degeneration is one of the leading causes of heart diseases in adults. The mechanisms underlying aortic valve degeneration are largely unknown. Cardiac tissue responds to a variety of stimuli by hypertrophic growth. Molecular mechanisms resulting in the hypertrophic response indicate similarity and overlap with those involved in both cell growth and death. This hypertrophic growth has been shown to be associated with the re-activation of the fetal gene program in cardiac cells (Liu and Olson, 2002; Bar et al., 2003), the clue of which is the positive regulation of the cell cycle progression. This switch in program might be crucial for the myocardial cell regulation. Such growth stimulation is responsible for the up-regulated activity of cyclin-dependent kinases (CDKs) that consist of a kinase core and an associated cyclin subunit acting as the positive regulator (Sherr, 1994). Thereby different CDK inhibitors keep the negative control over CDKs activities. CDK inhibitors are classified on the basis of their sequence homology and substrate specificity. Recently a novel cardiac helicase CHAMP that inhibits cell proliferation and cardiac hypertrophy was described (Liu and Olson, 2002). The CHAMP-dependent inhibition of cardiac hypertrophy is accompanied by an obligatory up-regulation of the cyclin-dependent protein-kinase inhibitor P21WAF1/CIP1, a 21-kDa protein, which is a member of the CIP/KIP family (Sherr, 1994). Furthermore, a targeted over-expression of P21WAF1/CIP1 prevents cell enlargement and suppresses a specific gene expression of the cardiac hypertrophy markers in the cell population *in vitro* (Tamamori et al., 1998) indicating the key role of p21WAF1/CIP1 in the regulation of the hypertrophic response. Recently clear correlations among

- the well coordinated regulation of $p21^{WAF1/CIP1}$ and 14-3-3 sigma - checkpoint genes arresting cell cycle in G_1 and G_2 phases, respectively,
- modulation of the density of valvular cell population, and
- progressive calcification of native aortic valve tissue

were reported using "Disease transcriptomics" technology (Golubnitschaja et al., 2006b).

Cell cycle control obviously plays a key role in the maintenance of cardiac cell population: high rates of cell death have been shown to be physiological in normal adult human hearts and those of mice and rats (Nadal-Ginard, 2001). The physiological expression of $p21^{WAF1/CIP1}$ shows a gradual increase during development in both rat and man, becoming maximal in adulthood (Burton et al., 1999). A direct link between the Bcl-2 dependent down-regulation of $p21^{WAF1/CIP1}$ and an increased myocyte density in the left ventricle has been shown recently in experimental work with transgenic mice (Limana et al., 2002). These findings are in agreement with the data demonstrating that down-regulation of both checkpoint genes $p21^{WAF1/CIP1}$ and 14-3-3 sigma correlates well with increasing cardiac cell density and calcification appearance in aortic valve tissue (Golubnitschaja et al., 2006b). Interestingly, the number of macrophages does not correlate with the cell density, whereas both the increased cell density and well coordinated down-regulation of $p21^{WAF1/CIP1}$ and 14-3-3 sigma gene expression were found to be characteristic for calcification. Therefore, the coordinated double-check of DNA quality and cell proliferation in valvular cells might be efficient only in non-calcified tissue, whereas in the calcified one this control becomes coordinately suppressed in both G_1 and G_2 dependent checkpoints. These findings give further evidence that the efficiency of cell cycle control in human non-calcified valvular tissue depends not only on the positive/negative CDKs regulation in the G_1 phase but also on the coordinated regulation of both G_1 and G_2 dependent checkpoints, that has not been considered from the view point of molecular pathomechanisms of valve calcification until now. Recent *in vitro* experiments on rat cardiac fibroblasts showed, that a target up-regulation of inhibitors for G_1 dependent CDKs effectively suppresses the DNA synthesis and may decrease a potential risk of cardiovascular diseases (Mercier et al., 2002).

The dissociation of $P21^{WAF1/CIP1}$ from the CDK complexes correlates well with the activation of CDK2, CDK4, CDK6, and the release from cell cycle arrest, whereby the number of cardiac cells in S phase rises considerably (von Harsdorf et al., 1999). Further, in contrast to P16 (a specific inhibitor of CDK4/6), the "universal" CDKs inhibitor $P21^{WAF1/CIP1}$ was shown to be able to block an E2F-1-induced G_1 exit completely (Akli et al., 1999). However, E1A binding activity to target protein complexes has effects on the cell cycle progression beyond those produced by E2F-1 alone, and can drive S-phase entry that is resistant to $P21^{WAF1/CIP1}$ (von Harsdorf et al., 1999). These facts explain the necessity of the coordinated regulation of both G_1 and G_2 dependent checkpoints, in order to keep the control over the cell population maintenance in cardiac tissue. Blockade of cell cycle progression results in a prolonged resistance to macrophage invasion and foam cell deposition (Mann et al., 1997). Therefore, it is likely that reduced cell cycle control in valvular tissue leads to the increased macrophage invasion that, in turn, can contribute to non-physiological calcification by both triggered unspecific inflammation and NO-toxicity (Schmidt and Walter, 1994; Zhuang and Wogan, 1997; Lutgens et al., 1999; Sanders et al., 2001; Frangogiannis et al., 2002). The coordinated activation of both G_1 and G_2 dependent checkpoint genes may be, therefore, an attribute of the valvular tissue resistance against the calcification processes. Molecular events leading to valve calcification are schematically shown in Figure 2, and should be taken into consideration, when designing therapeutic prevention of pro-calcification molecular pathways in human hearts.

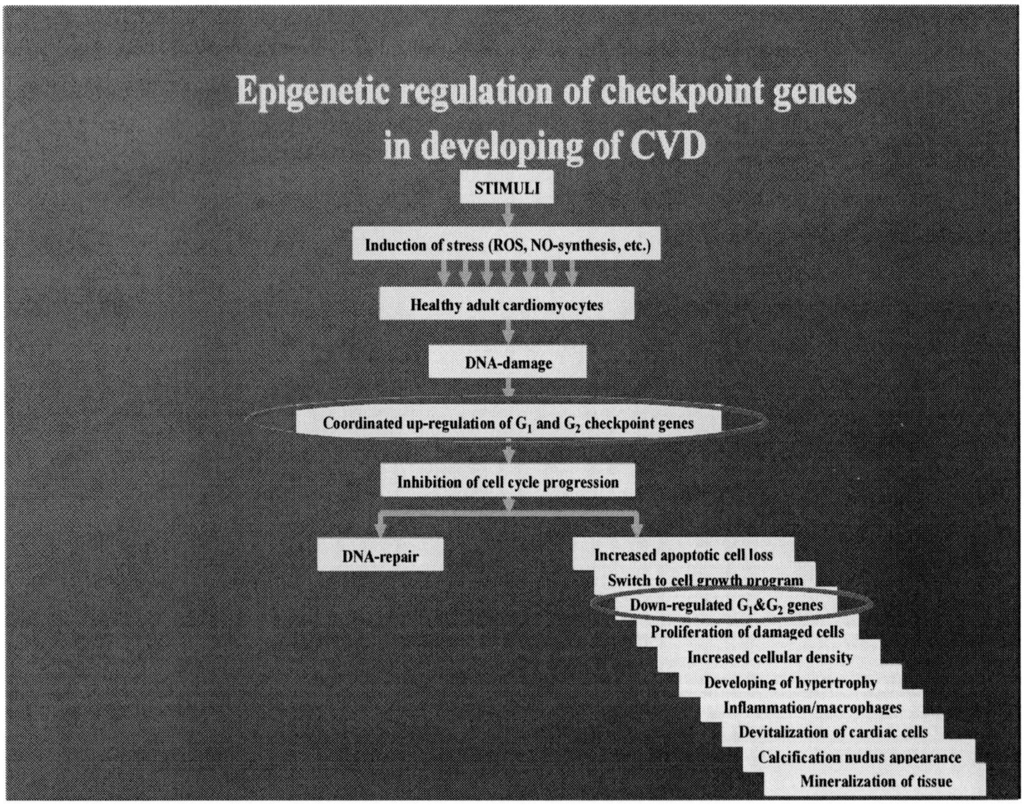

Figure 2. The cascade of molecular events leading to valve degeneration and mineralization of cardiac tissue. Coordinated up-regulation of G_1 and G_2 checkpoint genes was demonstrated in degenerated but non-calcified valves (Golubnitschaja et al., 2006b). In contrast, the expression of cell cycle checkpoints was completely suppressed in calcified tissue (Golubnitschaja, 2007).

2.2. Altered Expression of Checkpoints in Diabetic Cardiomyopathy

Many diabetic patients suffer from cardiomyopathy, resulting in heart failure and increased mortality, even in the absence of vascular disease (Liang et al., 2002). More than 50 % of patients with type 2 diabetes have coronary heart disease, related to silent ischemia and autonomic denervation of the heart. Little is known about molecular mechanisms leading to the degeneration of cardiac tissue under diabetic conditions. Hyperglycemia-induced oxidative stress and subsequent DNA damage is speculated to contribute to the pathogenesis of diabetic complications in both diabetes type 1 and 2 (Dandona et al., 1996; Hinokio et al., 1999; Sampson et al., 2001; Sardas et al., 2001). A repair capacity towards increased production and damage effects of reactive oxygen species (ROS) in diabetic patients or corresponding animal models is now under intensive investigation as a parameter or even biomarker of differential cell resistance towards diabetic complications (Hinokio et al., 1999; Sampson et al., 2001; Taniguchi et al., 1996; Leinonen et al., 1997; Krauss et al., 2003). Noteworthy, in the early phase of diabetes the heart is more resistant to ischemic damage than the non-diabetic heart (Tosaki et al., 1996; Paulson 1997). Therefore, it has been hypothesized that oxidative stress conditioned by hyperglycemia may be associated with an increased density of damaged cardiomyocytes followed by P53-dependent induction of cell cycle arrest in cardiomyocytes and consequent reduction in cell population with a concomitant remodeling of heart tissue (Golubnitschaja et al, 2006a). Indeed, in the early phase of STZ-induced insulin-dependent diabetes both cell cycle checkpoints $p21^{WAF1/CIP1}$ and 14-3-3 sigma become activated in cardiac tissue compared to the non-diabetic heart. Furthermore, a repair of damaged DNA is strongly promoted in diabetic cardiomyocytes, since the measured levels of repaired DNA molecules have been shown to be even higher than those of non-diabetic cardiomyocytes. These findings suggest that arrest of cardiomyocyte proliferation under diabetic conditions might be triggered in both G_1 and G_2 phases of the cell cycle, and accompanied by extensive DNA repair provided by a complex, both G_1- and G_2-dependent repair mashinery in the early phase of the disease. This conclusion is of high importance, since it could explain the observed but not yet understood increased resistance of the heart towards ischemia/reperfusion in early diabetes compared to the non-diabetic heart. Further observations of disappearing of this beneficial effect with the duration of diabetes proposes a dual role of cell cycle checkpoints under diabetic conditions: in early diabetes an increased expression of both $p21^{WAF1/CIP1}$ and 14-3-3 sigma genes has a beneficial effect activating DNA repair processes triggered by cell cycle arrest, and, therefore, preventing replication of damaged DNA. On the other hand, the double cell cycle arrest ultimately inhibits replication of cells which consequently accumulate in the G_1 and G_2 phases; this potentially might lead to the myocardial tissue degeneration well-documented in later phases of diabetes. The proposed molecular events leading to diabetic cardiomyopathy fit also well to the general scheme of cardiac tissue degeneration given in Figure 2. Indeed, in contrast to early stages, the advanced stages of the diabetic cardiac hypertrophy and fibrosis are accompanied by an inactivation of 14-3-3 (Gurusamy et al., 2005).

Further investigations of checkpoints expression in cardiomyocytes of diabetic rats demonstrated that the basic level of regulation - namely gene methylation status - is involved (Monkemann et al., 2002). There is a strong epigenetic control over gene expression of both $p21^{WAF1/CIP1}$ and 14-3-3 sigma: methylation status of abundant CpG islands located in a promoter region is known to have an important regulatory effect on their expression (Allan et al., 2000; Ferguson etal., 2000). The mechanisms of gene silencing by methylated cytosin are varied among promotors. The most generally reported mechanism is the repression of transcription by methyl-CpG-binding proteins (MeCP1 and MeCP2) that bind DNA in a sequence independent manner. Thereby, an activation of methyl-CpG-binding proteins results in alternating the chromatin structure and preventing the transcriptional factors like Sp1 from DNA binding (Lewis et al., 1992; Tate et al., 1996; Simmen et al., 1999).

Some therapeutic approaches to prevent both diabetic and non-diabetic cardiomyopathy are currently under consideration. Increasing evidence suggests that dietary taurine supplementation reverses cardiomyopathy in animals underwent taurine depletion (TD) and also confers beneficial effects in patients with heart failure (Schaffer et al., 2000a; Sole and Jeejeebhoy, 2002; Pion et al. 1987, Azuma et al., 1992). Several mechanisms have already been proposed to account for the beneficial effect of taurine in heart failure: improvement of tissue osmoregulatory properties and ion transport mechansims (Labudova et al., 1999a; Schaffer et al., 2000a; Suleiman 1994; Moenkemann et al., 1999) promotion of body fluid homeostasis by pronounced natriuresis and diuresis effects (Mozaffari and Schaffer, 2001), positive ionotropic effect by regulating $[Ca^{2+}]_i$, $[Na^{+}]_i$ and Na^{+}/Ca^{2+} exchanger activity (Azuma et al., 1992; Schaffer et al., 2000a; Schaffer et al., 2000b), and attenuation of angiotensin II mediated events such as induction of cardiac hypertrophy, myocardial remodelling and volume overload (Schaffer et al., 2000b). Oxidative stress is proposed to be the pathogenic mechanism linking TD to the ultimate manifestation of cardiomyopathy. Similarly to diabetic rats, the taurine depletion causes DNA damage accompanied by up-regulation of P53, $P21^{WAF1/CIP1}$ and 14-3-3 sigma expression in cardiomyocytes (Golubnitschaja et al, 2003). Since TD is associated with the development of cardiomyopathy, the relevance of these molecular events to the ultimate manifestation of cardiomyopathy caused by both *Diabetes mellitus* and TD represents a fertile area for further investigation.

2.3. Hyperhomocysteinemia as an Important Independent Risk Factor for Cardiovascular Diseases and Potential Moderator of Tumor Activity

Hyperhomocysteinemia (Hhcy) is associated with the development of cerebrovascular, peripheral vascular and coronary heart diseases (Wilcken et al., 1983; Brattstrom et al., 1984; Boers et al., 1985; Malinow et al., 1989; Coull et al., 1990; Clarke et al., 1991; Malinow et al., 1996; Graham et al., 1997). Elevated plasma levels of homocysteine (Hcys) can result from genetic (such as mutations in 5,10-Methylenetetrahydrofolate reductase – the key enzyme in both folate and homocysteine metabolic pathways) or nutrient-related disturbances (e.g. folate deficiency) in the transsulphuration or remethylation pathways for Hcys metabolism (Frosst et al., 1995; Akoglu et al., 2004). Hcys significantly stimulates DNA synthesis, promotes vascular smooth muscle cells (VSMCs) growth inhibiting endothelial cell growth and stimulating proliferation of arterial tissue, and consequently promotes the vessel wall thickening potentiating calcification of aortic smooth muscle cells (Lubec and Szekeres, 1994; Tsai et al., 1994; Li et al., 2003). What are the molecular mechanisms of this selective promotion of VSMCs proliferation? Lubec et al. demonstrated the ability of Hcys to activate the cell cycle progression by stimulation of aortic cyclin-dependent kinase at the transcriptional level in a dose-dependent manner (Lubec et al., 1996). This important finding has been later confirmed by numerous further experimental data demonstrating Hcys mitogenic effect on VSMCs with strong proliferative response on the luminal side of the vessel wall tapering, however, towards the adventitia, and showing also a significant inhibition of DNA synthesis and of proliferative effect by nutritional folate supplementation (Carmody et al., 1999; Chen et al., 2000; Buemi et al., 2001; Kartal et al., 2005). Further studies demonstrated induced B and T lymphocyte proliferation by increased Hcys concentration levels; the underlying cell cycle progression mechanisms involve PKC, p38 MAPK and NF-kappaB activity (Zhang et al, 2001; Zhang et al., 2002). This might be the further mechanism potentiating chronic inflammatory progression of atherosclerosis with Hhcy.

Recently achieved results represent a functional link of Hhcy associated cardiovascular and cancer diseases: 5,10-Methylenetetrahydrofolate being the key metabolite in both folate and Hcys metabolic pathways, is the main modulator of growth-promoting effects of Hcys as well as anti-proliferative ones of folate in colon cancer cells. (Akoglu et al., 2004). Hhcy-induced alterations of extracellular-matrix (ECM) metabolism can, further, contribute to the progression of both cardiovascular and cancer diseases: Hhcy increases the secretion of elastolytic Metalloproteinases 2 and 9 and Tissue inhibitor of metalloproteinases-1 (Yang and Zou, 2003; Chaussalet et al., 2004). These alterations can potentially trigger a non-physiologic deposition of ECM-elements and advance a tumor metastatic activity. Hhcy research provides us with a further important functional link between cardiovascular and cancer diseases: Hhcy is associated with altered overall DNA methylation status in different cell types (Hultberg et al., 2000; Castro et al., 2003) representing the basic mechanism of the epigenetic control, particularly, over cell cycle regulation. Further studies are required before the true role of Hcys and Hhcy in the regulation of overall DNA methylation status can be clearly evaluated. This can dramatically further our knowledge about the origin of cardiovascular and cancer diseases.

3. Checkpoints in Cancer: Differential Gene Expression, Epigenetic Control and Cell Cycle Progression, Potential Early Diagnosis and Therapy Monitoring

Mammalian cell cycle checkpoints activation has been proposed as the key tumor-suppressor mechanism that should prevent the accumulation of mutations that usually drives oncogenesis (Hartwell and Kastan, 1994). Mutations in and/or complete loss of even one checkpoint results in the pronounced predisposition to cancer diseases, as it has been well documented in the case of *Ataxia telangiectasia*, the human genomic instability syndrome characterized by a mutated ATM protein kinase - the cell cycle modulator (Shiloh, 2003). This checkpoints-associated increased tumor predisposition might be developed either during embryogenesis or can be, further, caused by some pathologic conditions in child- or adulthood such as highly increased stress and DNA modifying conditions such as altered DNA methylation status, DNA damage, viral infections etc. Inborn checkpoint mutations are also associated with diverse and usually well-detectable abnormalities like growth and mental retardation, facial dysmorphism, immune dysfunction, and others commonly found in syndromes associated with an impaired response to DNA damage. These abnormalities are well-documented in the human genetic disorders "Seckel syndrome" with inborn defect of *atm*-related *atr*-checkpoint gene (Alderton et al., 2004).

It is more complicate to predict a non-inborn tumor predisposition arose in child- and adulthood due to DNA modifying events. Similarly to inborn checkpoint mutations, the potential tumorigenesis selects for genetic and epigenetic changes, allowing evasion of anti-proliferative and apoptotic mechanisms, when the arising tumor acquires genetic instability. However, how early this occurs and whether it drives tumor development is currently unclear.

3.1. Predictive Power of Checkpoints in Breast Cancer

With one million new cases in the world each year, breast cancer is one of the most common malignancies (about 24 % of all cancer cases), the most frequent malignancy in women (10-12 per 100 women), and the leading cause of worldwide death for women 40-55 years of age (Baselga and Mendelsohn, 1994; Bange et al., 2001; Chopin et al., 2004a). The pathology is highly resistant to chemotherapy, and there is still no effective cure for the patients with advanced stages of breast cancer, especially in the case of hormone-independent disease (Bange et al., 2001; Chopin et al., 2004a). Most breast cancer patients succumb to the tumor due to the presence of undetected micrometastases; therefore, long-time survival is possible if the tumor becomes localized before metastasing occurs. Since the surgical removal of the primary tumor at very early stages is still the most effective treatment of breast cancer, an early diagnosis of the disease, therefore, would prevent the extremely high morbidity and mortality observed for breast cancer patients. Despite years of intense study, no molecular alterations common to the vast majority of breast cancer have been found till 2000, when Ferguson et al. has observed a silencing 14-3-3 sigma checkpoint gene, the DNA sequence of which is hypermethylated in more than 90 % of primary breast cancers tested (Ferguson et al., 2000). This checkpoint is responsible for instituting the cell cycle arrest in G_2 phase (Hermeking et al., 1997; Chan et al., 1999).

The mammalian family of 14-3-3 proteins consists of 7 members, encoded by separate genes (β, γ, ε, η, σ, τ, and ζ) with broad spectrum of functions in kinases conducted signaling pathways generally essential in the controlling of cell cycle progression, particularly during embryonic eye and brain development, as well as regulation of brain function, and general stress response (Aitken et al. 1995; Rittinger et al., 1999). 14-3-3 proteins possess a CpG-rich promoter sequence – the target for a regulation activity of DNA methyltransferases, which control the expression of 14-3-3 proteins on the basic level, by altering their gene methylation status. CpG island methylation is an epigenetic change that is largely responsible for silencing of the corresponding genes. In the case of 14-3-3 checkpoint genes, their hypermethylation may underlie an increased resistance of breast cancer towards genotoxic agents and may lead to a failure of chemo/radio therapies applied. Some other epigenetic alterations have been demonstrated in breast cancer (50 % of breast tumors exhibit a hypermethylation of E-cadherin and mammary derived growth factor inhibitor (Graff et al. 1995; Huynh et al., 1996)), which, however, are not as ubiquitous as it is in the case of the

14-3-3 sigma gene, indicating the importance of this checkpoint loss in breast cancer pathology. Also further checkpoints such as $P21^{WAF1/CIP1}$ and $P27^{KIP1}$ become considered as potent molecular targets for breast cancer therapy (Bange et al., 2001; Chopin et al., 2004a and 2004b).

Molecular mechanisms of the epigenetic control over the overall gene expression have been studied in more detail for the CHFR gene, encoding a mitotic stress checkpoint protein inactivated in plenty of human tumors such as primary colorectal, bone, head and neck cancers (Corn et al., 2003; Toyota et al., 2003). Expression of CHFR correlated well with the methylation status of the gene´s CpG-rich regulatory region. The gene hypermethylation and thus silencing of CHFR were strictly dependent on the activities of two DNA methyltransferases, DNMT1 and DNMT3b, inactivation of which restored CHFR expression in tumor cells. Taken together, these findings highlighted the key mechanism by which mitotic checkpoints can be bypassed in cancer.

As for therapy in breast cancer, the use of breast conserving lumpectomy followed by radiation therapy is common and has been shown to be a strategy competitive to mastectomy in preventing mortality caused by breast cancer. However, breast irradiation applied, particularly, after pre-irradiation chemotherapy frequently leads to serious short- and long-term side effects, the prediction of which is highly desirable in terms of individual therapy planning. Comparative protein profiling demonstrated highly individual stress-response in circulating leukocytes of treated patients. Characteristic expression patterns of stress-response proteins in circulating leukocytes might provide novel molecular targets for prediction of therapeutic side-effects and improve individual therapy planning for breast cancer patients, in order to avoid unnecessary and excessive treatment-related toxicity (Yeghiazaryan et al, 2007).

3.2. Checkpoints in Malignant Gliomas: "Disease Proteomics" Reveals a Failure of Current Therapy Approaches

Human gliomas are the most frequent malignant brain tumors and the most malignant of the astrocytomas. A catastrophic prognosis of the median survival in the range of one year can be only marginally improved by current therapy approaches (Stupp et al., 2005; Wild-Bode et al., 2001). The excessive proliferation, disseminated tumor growth, extremely rich neovascularization, and resistance towards apoptotic stimuli are the main features of malignant gliomas making their treatment especially complicated. The resistance of malignant glioma cells towards stress conditions results in sub-lethal effects under clinically relevant therapy approaches and, consequently, in an enhanced invasiveness of treated cells as revealed by “Clinical proteomics” and “Disease Transcriptomics” (Trog et al., 2006a, Trog et al., 2006b, Trog et al., 2007a). This scenario cannot lead to success in the therapy of malignant gliomas. In order to reconsider the current therapy approaches, some key mechanisms underlying the response of glioma cells to treatment conditions should be, further, clarified. The clue is obviously a complex molecular and resulting cellular response to sub-lethal genotoxic conditions. Recently achieved results indicate that not a lack of expression, but the deficits in response to genotoxic conditions in terms of P53 and

$P21^{WAF1/CIP1}$ expression regulation may play a crucial role in the resistance towards pro-apoptotic stimuli, uncontrolled cell growth and survival of malignant gliomas under clinically relevant chemo- or/and radiation-treatment (Trog et al, 2007b). In contrast, both P53 and $P21^{WAF1/CIP1}$ were found to be significantly up-regulated by high cell density at prolonged cultivation. The well-known specific feature of malignant gliomas is an obligatory dependence on the development of its own vascular system for adequate oxygen and nutrient delivery. Thus, malignant glioma cells of high density induce their own vasculature formation (Ader et al., 2003). It will be speculated, therefore, that the cell cycle control in glioma cells might have a function, which differs from maintaining the original phenotype of single cells: cell cycle attributing proteins are activated in concert with pro-angiogenic events, when the tissue becomes remodeled and the processes of neovascularization are triggered (Trog et al., 2007b). This study gives us an example for the application of checkpoints expression monitoring, in order to evaluate the cancer therapy efficiency and, if necessary, to suggest improved approaches.

4. The Key Role of Checkpoints in Longevity/Aging and Neurodegeneration

4.1. Longevity and Aging

This is one of the phenomena which being associated with organismal stress factors and cellular stress response becomes mostly discussed in literature. Plenty of studies have demonstrated that aging is a result of a complex interaction of genetic, epigenetic, and environmental factors. "The genetics of human longevity" undergoes currently an extensive investigation as reviewed by Capri et al., 2006. Different strategies have been proposed which should identify the gene-candidates contributing to either human longevity or aging. Thus, there will be looked for gender differences, since women live longer than men (Austad 2006), and for a positive association with "unsuccessful aging" by selection of genes possibly responsible for myocardial infarction, Alzheimer disease, and *Diabetes mellitus*, etc. (Capri et al., 2006). In this pool of candidates some inflammation-related and oxidative-stress-responsible genes are reported. Further, experiments carried out on long-lived organisms have proposed the decreased cellular production of reactive oxygen species (ROS) in parallel to an increased creation of antioxidants as the most optimal organismal strategy for longevity (Lutz et al., 2003; Corona et al., 2005). Concretely in mammalians, methionine (Met) metabolism has been shown to be strongly associated with a long life-span (Sanz et al., 2006):

- dietary Met-restriction and shortening of protein Met-content increases the longevity, since this amino acid is most susceptible to oxidation by ROS, and dietary Met-supplementation increases plasma Hcys – an independent risk factor for cardio-vascular disorders as reviewed above,
- knocking out methionine-sulfoxide-reductase (MSR) lowers longevity, whereas its over-expression increases life-span and delays aging (Moskovitz et al., 2001); MSR

reduces Met-sulfoxide – the main by-product of protein oxidation – back to Met using a reduction potential of the thioredoxin (TRX)-complex (Stadtman et al., 2003),
- overexpression of thioredoxin controls oxidative stress and life-span increasing considerably the longevity (Mitsui et al., 2002).

Thereby, TRX, an ancient low molecular weight oxido-reductase with highly conservative structure, acts in response to changes in cellular redox status as universal red-ox signaling transductor by "ping-pong" mechanism, and represents a functional link to the regulation of hormonal stress processing, DNA metabolism, replication and repair machinery, apoptosis and cell cycle checkpoints (Muller 1991; Golubnitchaya-Labudova et al., 1998; Turunen et al., 2004; Holmgren et al., 2005). Concerning the checkpoint regulation, recent evidence indicates direct control of the Cdc25s, and therefore the cell cycle, in response to changes in cellular redox status conducted by TRX (Rudolph 2005).

4.2. Functional Linkage Between the Regulation of Protein Redox Status and Cell Cycle Control: the Meaning for Senescence and Neurodegenerative Diseases

Oxidative damage to proteins, altered protein redox status, particularly of the key redox systems such as the thioredoxin-dependent methionine-sulfoxide-reductase and the thioredoxin-system, have been implicated in the pathology of several neurodegenerative conditions including Down Syndrome (DS), Alzheimer (AD) and Parkinson diseases (PD), and amyotrophic lateral sclerosis (Kim et al., 2001; Stadtman et al., 2003; Landino et al., 2004). Protein redox status is decisive for the signal transduction targeted in the control over cell cycle progression. The key redox-sensitive cell cycle regulators P53, NFkappaB and the TRX-system represent the functional link between a redox signal cascade and cell cycle machinery (Hinz et al., 1999; Rudolph 2005; Muller 1995). A deprivation of both NFkappaB and TRX has been demonstrated early during embryonic development in fetal brain of individuals with DS characterized by neurodgeneration, premature aging, early development of AD, and by an increased risk for cardiovascular and cancer diseases (Labudova et al., 1999b; Kitzmueller et al., 1999). Further analysis revealed strong neuroprotective effects of both NFkappaB and TRX indicating a functional link between altered protein redox status, cell cycle control, and neurodegenerative pathology of both DS and AD (Lovell et al., 2000; Piccioli et al., 2001; Chiueh et al., 2003). A growing body of evidence indicates that neurodegeneration involves the activation of cell cycle machinery in post-mitotic neurons: an accumulation of mitotic phosphoepitopes, re-activation of the M-phase regulator, Cdc2/cyclin B-complex, an activation of Cdc-2-activating Cdc25A-phosphatase as well as modulation of checkpoints activity in G_1/S have been described in neurones undergoing degeneration in AD (Ding et al., 2000; Lim and Qi, 2003). Misregulation of Cdks occurs in AD, PD, amyotrophic lateral sclerosis, and other neurological disorders affecting the following processes: proliferation, differentiation, senescence and apoptosis as reviewed by Nguyen et al. (2002).

Deregulation of cell-cycle-related Cdk2, Cgk4 and Cdk6 initiates death pathways by de-repressing E2F-1/Rb-dependent transcription at the neuronal G_1/S checkpoints, that, in concert with the activity of cell-cycle-unrelated Cdk5 possibly inducing the cell-cycle proteins, leads to neurodegeneration (Nguyen et al., 2002). Recent experimental data indicate, however, that only stress leading to DNA damage can initiate cell cycle activation linked to neuronal death; in contrast, apoptosis triggered by stimuli that do not target DNA, does not initiate cell cycle activation (Kruman et al., 2004). Therefore, genotoxic stress combined with misregulation of neuronal checkpoints seems to be the key molecular evens underlying neurodegenerative pathology. Inborn checkpoint misregulation is most extensively studied in ATM-deficiency which is accompanied with neuronal degeneration, premature aging and cancer predisposition in *Ataxia telangiectasia*. There is a growing body of evidence that a misregulation of other checkpoints can result in accelerated aging and senescence. Thus, recent animal experiments support the concept that hyper-activation of the checkpoint and tumor suppressor gene p53 may cause accelerated aging (Varela et al., 2005).

Whereas the function of checkpoint genes is relatively well understood in cell cycle progression, little is known about a role of checkpoint genes in non-proliferating cells. As mentioned above, the deregulated cell cycle activation under genotoxic conditions in non-proliferating cells might be the clue to neurodegeneration and accelerated aging. Thus, apart from the function of universal CDKs inhibitor originally described for $p21^{WAF1/CIP1}$, other functions, particularly, under stress conditions have been demonstrated for this gene as reviewed by R. H. Weiss, 2003: an induced $p21^{WAF1/CIP1}$ up-regulates a big set of genes implicated in senescence, age-related and neurodegenerative diseases. To this set belongs the β-amyloid precursor protein linked to Alzheimer disease, extracellular matrix components and receptors, cathepsin B, prosaposin, etc. (Chang et al., 2000). Compared to age-matched controls, the $p21^{WAF1/CIP1}$ gene was shown to be hypermethylated in circulating leukocytes of glaucoma patients - a neurodegenerative disease characterized by retinal ganglion cell loss and optic nerve head atrophy (Golubnitschaja, non-published data). At the moment we can only speculate about the role of $p21^{WAF1/CIP1}$ hypermethylation in glaucoma, whether this is due to anti-aging processes that act against oxidative stress damage implicated in the pathology of glaucoma or due to a deregulated epigenetic control initiating cell cycle activation linked to cellular death under genotoxic conditions, which are well documented in glaucoma (Moenkemann et al., 2005). However, a possible deregulation of epigenetic control over checkpoints expression should be, further, taken into account, if early diagnosis of neurodegenerative diseases and development of new targets for novel therapeutic approaches are considered.

Conclusion

There is no doubt about the key role of cell cycle control mechanisms in maintenance of genomic integrity that is absolutely decisive for adequate cellular stress response and appropriate tissue functionality. Inborn genetic checkpoint defects are usually well-detectable through morphological abnormalities like growth and mental retardation, facial dysmorphism, premature aging as well as immune dysfunction, etc., commonly found in human syndromes associated with an impaired response to DNA damage. Predicting a non-inborn tissue specific checkpoint deregulation associated with a predisposition to tissue specific pathologies such as neurodegeneration, diverse cardiovascular and tumor diseases, is more complicated. However, how early this occurs and whether it drives either tumor development or other pathologies is unclear. Therefore, novel diagnostic approaches should be developed, in order to estimate individual predispositions as early as possible. Molecular research technologies such as "Clinical Proteomics" and "Disease Genomics and Transcriptomics" represent the most powerful tools allowing a precise selection of disease-relevant gene-candidates that can be used as disease-specific molecular markers. Depending on the level of deregulation an appropriate diagnostic detection method can be chosen such as "Real-Time"-PCR for the level of transcription, and "methylation-specific PCR" for evaluation of epigenetic control quality. One of the most important aspects is the tissue specificity and accessibility. Some non-invasive molecular diagnostic approaches have been recently suggested for early diagnosis of glaucoma and diabetic retinopathy – two neurodegenerative eye disorders with a vascular component in the corresponding pathomechanisms involving pronounced checkpoints dysfunction. Both non-invasive diagnostic approaches are based on disease specific alterations in gene expression profiles of circulating leukocytes that can be evaluated *ex vivo* in individual blood samples (Golubnitschaja and Flammer, International Patent No. IB02/00648; Golubnitschaja, International Patent No. 06115230.2). Application of biotechnology in medicine furthers our knowledge about molecular pathomechanisms and opens new perspectives for early diagnosis and even prevention of severe and fatal diseases.

References

Ader, I., Delmas, C., Bonnet, J., Rochaix, P., Favre, G., Toulas, C. & Cohen-Jonathan-Moyal, E. (2003) Inhibition of Rho pathways induces radiosensitization and oxygenation in human glioblastoma xenografts. *Oncogene 22*, 8861-8869.

Aitken, A., Jones, D., Soneji, Y. & Howell, S. (1995) 14-3-3 proteins: biological function and domain structure. *Biochem Soc Trans 23*, 605-611.

Akli, S., Zhan, S., Abdellatif, M. & Schneider, M. D. (1999) E1A can provoke G1 exit that is refractory to p21 and independent of activating cdk2. *Circ Res 85*, 319-328.

Akoglu, B., Milovic, V., Caspary, W. F. & Faust, D. (2004) Hyperproliferation of homocysteine-treated colon cancer cells is reversed by folate and 5-methyltetrahydrofolate. *Eur J Nutr 43*, 93-99.

Alderton, G. K., Joenje, H., Varon, R., Borglum, A. D., Jeggo, P. A. & O'Driscoll, M. (2004) Seckel syndrome exhibits cellular features demonstrating defects in the ATR-signalling pathway. *Hum Mol Genet 13*, 3127-3138.

Allan, L. A., Duhig, T., Read, M. & Fried, M. (2000) The p21(WAF1/CIP1) promoter is methylated in Rat-1 cells: stable restoration of p53-dependent p21(WAF1/CIP1) expression after transfection of a genomic clone containing the p21(WAF1/CIP1) gene. *Mol Cell Biol 20*, 1291-1298.

Austad, S. N. (2006) Why women live longer than men: sex differences in longevity. *Gend Med 3*, 79-92.

Azuma, J., Sawamura, A. & Awata, N. (1992) Usefulness of taurine in chronic congestive heart failure and its prospective application. *Jpn Circ J 56*, 95-99.

Bakkenist, C. J. & Kastan, M. B. (2004) Initiating cellular stress responses. *Cell 118*, 9-17

Bange, J., Zwick, E. & Ullrich, A. (2001) Molecular targets for breast cancer therapy and prevention. *Nat Med 7*, 548-552.

Bar, M. D. H., Kreuzer, J., Cojoc, A. & Jahn, L. (2003) Upregulation of embryonic transcription factors in right ventricular hypertrophy. *Basic Res Cardiol 98*, 285-294.

Baselga, J. & Mendelsohn, J. (1994) The epidermal growth factor receptor as a target for therapy in breast carcinoma. *Breast Cancer Res Treat 29*, 127-138.

Boers, G. H., Smals, A. G., Trijbels, F. J., Fowler, B., Bakkeren, J. A., Schoonderwaldt, H. C., Kleijer, W. J. & Kloppenborg, P. W. (1985) Heterozygosity for homocystinuria in premature peripheral and cerebral occlusive arterial disease. *N Engl J Med 313*, 709-715.

Brattstrom, L. E., Hardebo, J. E. & Hultberg, B. L. (1984) Moderate homocysteinemia--a possible risk factor for arteriosclerotic cerebrovascular disease. *Stroke 15*, 1012-1016.

Brown, K. D., Rathi, A., Kamath, R., Beardsley, D. I., Zhan, Q., Mannino, J. L. & Baskaran, R. (2003) The mismatch repair system is required for S-phase checkpoint activation. *Nat Genet 33,* 80-84.

Buemi, M., Marino, D., Di Pasquale. G., Floccari, F., Ruello, A., Aloisi, C., Corica, F., Senatore, M., Romeo, A. & Frisina, N. (2001) Effects of homocysteine on proliferation, necrosis, and apoptosis of vascular smooth muscle cells in culture and influence of folic acid. *Thromb Res 104*, 207-213.

Burton, P. B., Yacoub, M. H. & Barton, P. J. (1999) Cyclin-dependent kinase inhibitor expression in human heart failure. A comparison with fetal development. *Eur Heart J 20*, 604-611.

Capri, M., Salvioli, S., Sevini, F., Valensin, S., Celani, L., Monti, D., Pawelec, G., De Benedictis, G., Gonos, E. S. & Franceschi, C. (2006) The genetics of human longevity. *Ann N Y Acad Sci 1067*, 252-263.

Carmody, B. J., Arora, S., Avena, R., Cosby, K. & Sidawy, A. N. (1999) Folic acid inhibits homocysteine-induced proliferation of human arterial smooth muscle cells. *J Vasc Surg 30*, 1121-1128.

Castro, R., Rivera, I., Struys, E. A., Jansen, E. E., Ravasco, P., Camilo, M. E., Blom, H. J., Jakobs, C. & Tavares de Almeida, I. (2003) Increased homocysteine and S-adenosylhomocysteine concentrations and DNA hypomethylation in vascular disease. *Clin Chem 49*, 1292-1296.

Chan, T. A., Hermeking, H., Lengauer, C., Kinzler, K. W. & Vogelstein, B. (1999) 14-3-3Sigma is required to prevent mitotic catastrophe after DNA damage. *Nature 401*, 616-620.

Chang, B. D., Watanabe, K., Broude, E. V., Fang, J., Poole, J. C., Kalinichenko, T. V. & Roninson, I. B. (2000) Effects of p21Waf1/Cip1/Sdi1 on cellular gene expression: implications for carcinogenesis, senescence, and age-related diseases. *Proc Natl Acad Sci U S A 97*: 4291-4296.

Charmandari, E., Tsigos, C. & Chrousos, G. (2005) Endocrinology of the stress response. *Annu Rev Physiol 67*, 259-284.

Chaussalet, M., Lamy, E., Foucault-Bertaud, A., Genovesio, C., Sabatier, F., Dignat-George, F. & Charpiot, P. (2004) Homocysteine modulates the proteolytic potential of human vascular endothelial cells. *Biochem Biophys Res Commun 316*, 170-176.

Chen, C., Halkos, M. E., Surowiec, S. M., Conklin, B. S., Lin, P. H. & Lumsden, A. B. (2000) Effects of homocysteine on smooth muscle cell proliferation in both cell culture and artery perfusion culture models. *J Surg Res 88*, 26-33.

Chern, C. L., Huang, R. F., Chen, Y. H., Cheng, J. T. & Liu, T. Z. (2001) Folate deficiency-induced oxidative stress and apoptosis are mediated via homocysteine-dependent overproduction of hydrogen peroxide and enhanced activation of NF-kappaB in human Hep G2 cells. *Biomed Pharmacother 55*, 434-442.

Cherubini, A., Ruggiero, C., Polidori, M. C. & Mecocci, P. (2005) Potential markers of oxidative stress in stroke. *Free Radic Biol Med 39*, 841-852.

Chiueh, C., Lee, S., Andoh, T. & Murphy, D. (2003) Induction of antioxidative and antiapoptotic thioredoxin supports neuroprotective hypothesis of estrogen. *Endocrine 21*, 27-31.

Chopin, V., Toillon, R. A., Jouy, N. & Le Bourhis, X. (2004a) P21(WAF1/CIP1) is dispensable for G1 arrest, but indispensable for apoptosis induced by sodium butyrate in MCF-7 breast cancer cells. *Oncogene 23*, 21-29.

Chopin, V., Slomianny, C., Hondermarck, H. & Le Bourhis, X. (2004b) Synergistic induction of apoptosis in breast cancer cells by cotreatment with butyrate and TNF-alpha, TRAIL, or anti-Fas agonist antibody involves enhancement of death receptors' signaling and requires P21(waf1). *Exp Cell Res 298*, 560-573.

Chuang, D. M. (2005) The antiapoptotic actions of mood stabilizers: molecular mechanisms and therapeutic potentials. *Ann N Y Acad Sci 1053*: 195-204.

Clarke, R., Daly, L., Robinson, K., Naughten, E., Cahalane, S., Fowler, B. & Graham, I. (1991) Hyperhomocysteinemia: an independent risk factor for vascular disease. *N Engl J Med 324*: 1149-1155.

Corn, P. G., Summers, M. K., Fogt, F., Virmani, A. K., Gazdar, A. F., Halazonetis, T. D. & El-Deiry, W. S. (2003) Frequent hypermethylation of the 5' CpG island of the mitotic stress checkpoint gene Chfr in colorectal and non-small cell lung cancer. *Carcinogenesis 24* 47-51.

Corona, M., Hughes, K. A., Weaver, D. B. & Robinson, G. E. (2005) Gene expression patterns associated with queen honey bee longevity. *Mech Ageing Dev 126*, 1230-1238.

Coull, B. M., Malinow, M. R., Beamer, N., Sexton, G., Nordt, F. & de Garmo, P. (1990) Elevated plasma homocyst(e)ine concentration as a possible independent risk factor for stroke. *Stroke 2*, 572-576.

Dandona, P., Thusu, K., Cook, S., Snyder, B., Makowski, J., Armstrong, D. & Nicotera, T. (1996) Oxidative damage to DNA in diabetes mellitus. *Lancet 347*, 444-445.

Dell'Anna, E., Chen, Y., Engidawork, E., Andersson, K., Lubec, G., Luthman, J. & Herrera-Marschitz, M. (1997) Delayed neuronal death following perinatal asphyxia in rat. *Exp Brain Res 115*, 105-115.

Didenko, V. V., Ngo, H., Minchew, C. L., Boudreaux, D. J., Widmayer, M. A. & Baskin, D. S. (2002) Visualization of irreparable ischemic damage in brain by selective labeling of double strand blunt-ended DNA breaks. *Mol Med 8*, 818-823.

Ding, X. L., Husseman, J., Tomashevski, A., Nochlin, D., Jin, L. W. & Vincent, I. (2000) The cell cycle Cdc25A tyrosine phosphatase is activated in degenerating postmitotic neurons in Alzheimer's disease. *Am J Pathol 157*, 1983-1990.

Ferguson, A. T., Evron, E., Umbricht, C. B., Pandita, T. K., Chan, T. A., Hermeking, H., Marks, J. R., Lambers, A. R., Futreal, P. A., Stampfer, M. R. & Sukumar, S. (2000) High frequency of hypermethylation at the 14-3-3 sigma locus leads to gene silencing in breast cancer. *Proc Natl Acad Sci U S A 97*, 6049-6054.

Frangogiannis, N. G., Smith, C. W. & Entman, M. L. (2002) The inflammatory response in myocardial infarction. *Cardiovasc Res 53*, 31-47.

Frosst, P., Blom, H. J., Milos, R., Goyette, P., Sheppard, C. A., Matthews, R. G., Boers, G. J., den Heijer, Kluijtmans, L. A., van den Heuve, L. P. & Rozen, R. (1995) A candidate genetic risk factor for vascular disease: a common mutation in methylenetetrahydrofolate reductase. *Nat Genet 10*, 111-113.

Golubnitschaja, O., Mönkemann, H., Kim, K. & Mozaffari, M.S. (2003) Depletion of taurine induces DNA damage and expression of checkpoint genes $p21^{WAF1/CIP1}$ and 14-3-3 σ in rat cardiomyocytes. *Biochem Pharmacol 66*, 511-517.

Golubnitschaja, O., Moenkemann, H., Trog, D. B., Blom, H. J. & De Vriese, A. S. (2006a) Activation of genes inducing cell-cycle arrest and of increased DNA repair in the hearts of rats with early streptozotocin-induced diabetes mellitus. *Med Sci Monit 12*, BR68-74.

Golubnitschaja, O., Yeghiazaryan, K., Skowasch, D., Schild, H. & Bauriedel, G. (2006b) p21(WAF1/CIP1) and 14-3-3 sigma gene expression in degenerated aortic valves: A link between cell cycle checkpoints and calcification. *Amino Acids 31*, 309-316.

Golubnitschaja, O. (2007) Cell cycle checkpoints: the role and evaluation for early diagnosis of senescence, cardiovascular, cancer and neurodegenerative diseases. *Amino acids 32*, 359-371.

Golubnitchaya-Labudova, O., Horecka, T., Kapalla, M., Perecko, D., Kutejova, E. & Lubec, G. (1998) Thioredoxin from Streptomyces aureofaciens controls coiling of plasmid DNA. *Life Sci 62*, 397-412.

Golubnitschaja, O. & Flammer, J.: International Patent No. IB02/00648: Diagnostic method for glaucoma.

Golubnitschaja, O.: International Patent No. 06115230.2: Method for early diagnosis of proliferative diabetic retinopathy.

Graff, J. R., Herman, J. G., Lapidus, R. G., Chopra, H., Xu, R., Jarrard, D. F., Isaacs, W. B., Pitha, P. M., Davidson, N. E. & Baylin, S. B. (1995) E-cadherin expression is silenced by DNA hypermethylation in human breast and prostate carcinomas. *Cancer Res 55*, 5195-5199.

Graham, I. M., Daly, L. E., Refsum, H. M., Robinson, K., Brattstrom, L. E., Ueland, P. M., Palma-Reis, R. J., Boers, G. H., Sheahan, R. G., Israelsson, B., Uiterwaal, C. S., Meleady, R., McMaster, D., Verhoef, P., Witteman, J., Rubba, P., Bellet, H., Wautrecht, J. C., de Valk, H. W., Sales, Luis, A. C., Parrot-Rouland, F. M., Tan, K. S., Higgins, I., Garcon, D., Andria, G., et al. (1997) Plasma homocysteine as a risk factor for vascular disease. The European Concerted Action Project. *JAMA 277*, 1775-1781.

Gurusamy, N., Watanabe, K., Ma, M., Zhang, S., Muslin, A. J., Kodama, M. & Aizawa, Y. (2005) Inactivation of 14-3-3 protein exacerbates cardiac hypertrophy and fibrosis through enhanced expression of protein kinase C beta 2 in experimental diabetes. *Biol Pharm Bull 28*, 957-962.

Hartwell, L. H. & Kastan, M. B. (1994) Cell cycle control and cancer. *Science 266*, 1821-1828.

Hegde, M. L., Gupta, V. B., Anitha, M., Harikrishna, T., Shankar, S. K., Muthane, U., Subba Rao, K. & Jagannatha Rao, K. S. (2006) Studies on genomic DNA topology and stability in brain regions of Parkinson's disease. *Arch Biochem Biophys 449*, 143-156.

Hermeking, H., Lengauer, C., Polyak, K., He, T. C., Zhang, L., Thiagalingam, S., Kinzler, K. W. & Vogelstein, B. (1997) 14-3-3 sigma is a p53-regulated inhibitor of G2/M progression. *Mol Cell 1*, 3-11.

Hinokio, Y., Suzuki, S., Hirai, M., Chiba, M., Hirai, A. & Toyota, T. (1999) Oxidative DNA damage in diabetes mellitus: its association with diabetic complications. *Diabetologia 42*: 995-998.

Hinz, M., Krappmann, D., Eichten, A., Heder, A., Scheidereit, C. & Strauss, M. (1999) NF-kappaB function in growth control: regulation of cyclin D1 expression and G0/G1-to-S-phase transition. *Mol Cell Biol 19*, 2690-2698.

Holmgren, A., Johansson, C., Berndt, C., Lonn, M. E., Hudemann, C. & Lillig, C. H. (2005) Thiol redox control via thioredoxin and glutaredoxin systems. *Biochem Soc Trans 33*, 1375-1377.

Hultberg, B., Andersson, A. & Isaksson, A. (2000) Hypomethylation as a cause of homocysteine-induced cell damage in human cell lines. *Toxicology 147*, 69-75

Huynh, H., Alpert, L. & Pollak, M. (1996) Silencing of the mammary-derived growth inhibitor (MDGI) gene in breast neoplasms is associated with epigenetic changes. *Cancer Res 56*, 4865-4870.

Irie, M., Asami, S., Nagata, S., Ikeda, M., Miyata, M. & Kasai, H. (2001) Psychosocial factors as a potential trigger of oxidative DNA damage in human leukocytes. *Jpn J Cancer Res 92*, 367-376.

Kartal Ozer, N., Taha, S. & Azzi, A. (2005) Homocysteine induces DNA synthesis and proliferation of vascular smooth muscle cells by interfering with MAPK kinase pathway. *Biofactors 24*, 193-199.

Kawabe, T. (2004) G2 checkpoint abrogators as anticancer drugs. *Mol Cancer Ther 3*, 513-519.

Kim, S. H., Fountoulakis, M., Cairns, N. & Lubec, G. (2001) Protein levels of human peroxiredoxin subtypes in brains of patients with Alzheimer's disease and Down syndrome. *J Neural Transm Suppl 61*, 223-235.

Kitzmueller, E., Labudova, O., Rink, H., Cairns, N. & Lubec, G. (1999) Altered gene expression in fetal Down syndrome brain as revealed by the gene hunting technique of subtractive hybridization. *J Neural Transm Suppl 57*, 99-124.

Korystov, IuN. (1997) Emotions, stress, smoking, alcohol consumption and cancer - the correlational and causal connections. *Zh Vyssh Nerv Deiat Im I P Pavlova. 47*, 627-657.

Krauss, H., Kozlik, J., Grzymislawski, M., Sosnowski, P., Mikrut, K., Piatek, J. & Paluszak, J. (2003) The influence of glimepiride on the oxidative state of rats with streptozotocin-induced hyperglycemia. *Med Sci Monit 9*, 389-393.

Kruman, I. I., Wersto, R. P., Cardozo-Pelaez, F., Smilenov, L., Chan, S. L., Chrest, F. J., Emokpae, R. Jr., Gorospe, M. & Mattson, M. P. (2004) Cell cycle activation linked to neuronal cell death initiated by DNA damage. *Neuron 41*, 549-561.

Kultz, D. (2003) Evolution of the cellular stress proteome: from monophyletic origin to ubiquitous function. *J Exp Biol 206*, 3119-3124.

Kultz, D. (2005) Molecular and evolutionary basis of the cellular stress response. *Annu Rev Physiol 67*, 225-257.

Labudova, O., Yeghiazarjan, C., Hoger, H. & Lubec, G. (1999a) Taurine modulates expression of transporters in rat brain and heart. *Amino Acids 17*, 301-313.

Labudova, O., Kitzmueller, E., Rink, H., Cairns, N. & Lubec, G. (1999b) Gene expression in fetal Down syndrome brain as revealed by subtractive hybridization. *J Neural Transm Suppl 57*, 125-136.

Landino, L. M., Skreslet, T. E. & Alston, J. A. (2004) Cysteine oxidation of tau and microtubule-associated protein-2 by peroxynitrite: modulation of microtubule assembly kinetics by the thioredoxin reductase system. *J Biol Chem 279*, 35101-35105.

Leinonen, J., Lehtimaki, T., Toyokuni, S., Okada, K., Tanaka, T., Hiai, H., Ochi, H., Laippala, P., Rantalaiho, V., Wirta, O., Pasternack, A. & Alho, H. (1997) New biomarker evidence of oxidative DNA damage in patients with non-insulin-dependent diabetes mellitus. *FEBS Lett 417*, 150-152.

Levitt, P. S., Liu, H., Manning, C. & Weiss, R. S. (2005) Conditional inactivation of the mouse Hus1 cell cycle checkpoint gene. *Genomics 86,* 212-224.

Lewis, J. D., Meehan, R. R., Henzel, W. J., Maurer-Fogy, I., Jeppesen, P., Klein, F. & Bird, A. (1992) Purification, sequence, and cellular localization of a novel chromosomal protein that binds to methylated DNA. *Cell 6,* 905-914.

Li, J., Chai, S., Tang, C. & Du, J. (2003) Homocysteine potentiates calcification of cultured rat aortic smooth muscle cells. *Life Sci 74*, 451-461.

Liang, Q., Carlson, E. C., Donthi, R. V., Kralik, P. M., Shen, X. & Epstein, P. N. (2002) Overexpression of metallothionein reduces diabetic cardiomyopathy. *Diabetes 51*, 174-181.

Lim, A. C. & Qi, R. Z. (2003) Cyclin-dependent kinases in neural development and degeneration. *J Alzheimers Dis 5*, 329-335.

Limana, F., Urbanek, K., Chimenti, S., Quaini, F., Leri, A., Kajstura, J., Nadal-Ginard, B., Izumo, S. & Anversa, P. (2002) bcl-2 overexpression promotes myocyte proliferation. *Proc Natl Acad Sci U S A 99*, 6257-6262.

Liu, Z. P. & Olson, E. N. (2002) Suppression of proliferation and cardiomyocyte hypertrophy by CHAMP, a cardiac-specific RNA helicase. *Proc Natl Acad Sci U S A 99*, 2043-2048

Lovell, M. A., Xie, C., Gabbita, S. P. & Markesbery, W. R. (2000) Decreased thioredoxin and increased thioredoxin reductase levels in Alzheimer's disease brain. *Free Radic Biol Med 28*, 418-427.

Lubec, B. & Szekeres, T. (1994) Growth promotion by homocysteine thiolactone and its alpha-alkylated derivative. *Amino Acids 7*, 109-111.

Lubec, B., Labudova, O., Hoeger, H., Muehl, A., Fang-Kircher, S., Marx, M., Mosgoeller, W. & Gialamas, J. (1996) Homocysteine increases cyclin-dependent kinase in aortic rat tissue. *Circulation 94*, 2620-2625.

Lutgens, E., de Muinck, E. D., Kitslaar, P. J., Tordoir, J. H., Wellens, H. J. & Daemen, M. J. (1999) Biphasic pattern of cell turnover characterizes the progression from fatty streaks to ruptured human atherosclerotic plaques. *Cardiovasc Res 41*, 473-479.

Lutz, P. L., Prentice, H. M. & Milton, S. L. (2003) Is turtle longevity linked to enhanced mechanisms for surviving brain anoxia and reoxygenation? *Exp Gerontol 38*, 797-800.

Malinow, M. R., Kang, S. S., Taylor, L. M., Wong, P. W., Coull, B., Inahara, T., Mukerjee, D., Sexton, G. & Upson, B. (1989) Prevalence of hyperhomocyst(e)inemia in patients with peripheral arterial occlusive disease. *Circulation 79*, 1180-1188.

Malinow, M. R., Ducimetiere, P., Luc, G., Evans, A. E., Arveiler, D., Cambien, F. & Upson, B. M. (1996) Plasma homocyst(e)ine levels and graded risk for myocardial infarction: findings in two populations at contrasting risk for coronary heart disease. *Atherosclerosis 126*, 27-34.

Mann, M. J., Gibbons, G. H., Tsao, P. S., von der Leyen, H. E., Cooke, J. P., Buitrago, R., Kernoff, R. & Dzau, V. J. (1997) Cell cycle inhibition preserves endothelial function in genetically engineered rabbit vein grafts. *J Clin Invest 99*, 1295-1301.

Massague, J. (2004) G1 cell-cycle control and cancer. *Nature 432*, 298-306.

McGowan, C. H. & Russell, P. (2004) The DNA damage response: sensing and signaling. *Curr Opin Cell Biol 16,* 629-633.

Mercier, I., Colombo, F., Mader, S. & Calderone, A. (2002) Ovarian hormones induce TGF-beta(3) and fibronectin mRNAs but exhibit a disparate action on cardiac fibroblast proliferation. *Cardiovasc Res 53*, 728-739.

Midorikawa, K. & Kawanishi, S. (2001) Superoxide dismutases enhance H2O2-induced DNA damage and alter its site specificity. *FEBS Lett 495,* 187-190.

Migliore, L., Fontana, I., Trippi, F., Colognato, R., Coppede, F., Tognoni, G., Nucciarone, B. & Siciliano, G. (2005) Oxidative DNA damage in peripheral leukocytes of mild cognitive impairment and AD patients. *Neurobiol Aging 26*, 567-573.

Mitsui, A., Hamuro, J., Nakamura, H., Kondo, N., Hirabayashi, Y., Ishizaki-Koizumi, S., Hirakawa, T., Inoue, T. & Yodoi, J. (2002) Overexpression of human thioredoxin in transgenic mice controls oxidative stress and life span. *Antioxid Redox Signal 4*, 693-696.

Moenkemann, H., Labudova, O., Yeghiazarian, K., Rink, H., Hoeger, H. & Lubec, G. (1999) Evidence that taurine modulates osmoregulation by modification of osmolarity sensor protein ENVZ-expression. *Amino Acids 17*, 347-355.

Monkemann, H., De Vriese, A. S., Blom, H. J., Kluijtmans, L. A., Heil, S. G., Schild, H. H. & Golubnitschaja, O. (2002) Early molecular events in the development of the diabetic cardiomyopathy. *Amino Acids 23*, 331-336.

Moenkemann, H., Flammer, J., Wunderlich, K., Breipohl, W., Schild, H. H. & Golubnitschaja, O. (2005) Increased DNA breaks and up-regulation of both G(1) and G(2) checkpoint genes p21(WAF1/CIP1) and 14-3-3 sigma in circulating leukocytes of glaucoma patients and vasospastic individuals. *Amino Acids 28*, 199-205.

Moskovitz, J., Bar-Noy, S., Williams, W. M., Requena, J., Berlett, B. S. & Stadtman, E. R. (2001) Methionine sulfoxide reductase (MsrA) is a regulator of antioxidant defense and lifespan in mammals. *Proc Natl Acad Sci U S A 98*, 12920-12925.

Mozaffari, M. S. & Schaffer, D. (2001) Taurine modulates arginine vasopressin-mediated regulation of renal function. *J Cardiovasc Pharmacol 37*, 742-750.

Muller, E. G. (1991) Thioredoxin deficiency in yeast prolongs S phase and shortens the G1 interval of the cell cycle. *J Biol Chem 266*, 9194-9202.

Muller, E. G. (1995) A redox-dependent function of thioredoxin is necessary to sustain a rapid rate of DNA synthesis in yeast. *Arch Biochem Biophys 318*, 356-361.

Nadal-Ginard, B. (2001) Generation of new cardiomyocytes in the adult heart: Prospects of myocardial regeneration as an alternative to cardiac transplantation. *Rev Esp Cardiol 54*, 543-550.

Nguyen, M. D., Mushynski, W. E. & Julien, J. P. (2002) Cycling at the interface between neurodevelopment and neurodegeneration. *Cell Death Differ 9*, 1294-1306.

Paulson, D. J. (1997) The diabetic heart is more sensitive to ischemic injury. *Cardiovasc Res 34*, 104-112.

Pereg, Y., Lam, S., Teunisse, A., Biton, S., Meulmeester, E., Mittelman, L., Buscemi, G., Okamoto, K., Taya, Y., Shiloh, Y. & Jochemsen, A. G. (2006) Differential roles of ATM- and Chk2-mediated phosphorylations of Hdmx in response to DNA damage. *Mol Cell Biol 26*, 6819-6831.

Piccioli, P., Porcile, C., Stanzione, S., Bisaglia, M., Bajetto, A., Bonavia, R., Florio, T. & Schettini, G. (2001) Inhibition of nuclear factor-kappaB activation induces apoptosis in cerebellar granule cells. *J Neurosci Res 66*, 1064-1073.

Pion, P. D., Kittleson, M. D., Rogers, Q. R. & Morris, J. G. (1987) Myocardial failure in cats associated with low plasma taurine: a reversible cardiomyopathy. *Science 237*, 764-768.

Rittinger, K., Budman, J., Xu, J., Volinia, S., Cantley, L. C., Smerdon, S. J., Gamblin, S. J. & Yaffe, M. B. (1999) Structural analysis of 14-3-3 phosphopeptide complexes identifies a dual role for the nuclear export signal of 14-3-3 in ligand binding. *Mol Cell 4*, 153-166

Rudolph, J. (2005) Redox regulation of the Cdc25 phosphatases. *Antioxid Redox Signal 7*, 761-767.

Sachdev, P. S. (2005) Homocysteine and brain atrophy. *Prog Neuropsychopharmacol Biol Psychiatry 29*, 1152-1161

Sampson, M. J., Astley, S., Richardson, T., Willis, G., Davies, I. R., Hughes, D. A. & Southon, S. (2001) Increased DNA oxidative susceptibility without increased plasma LDL oxidizability in Type II diabetes: effects of alpha-tocopherol supplementation. *Clin Sci (Lond) 101*, 235-241.

Sanders, D. B., Hunter, K., Wu, Y., Jablonowski, C., Bahl, J. J. & Larson, D. F. (2001) Modulation of the inflammatory response in the cardiomyocyte and macrophage. *J Extra Corpor Technol 33*, 167-174.

Sanz, A., Caro, P., Ayala, V., Portero-Otin, M., Pamplona, R. & Barja, G. (2006) Methionine restriction decreases mitochondrial oxygen radical generation and leak as well as oxidative damage to mitochondrial DNA and proteins. *FASEB J 20*, 1064-1073.

Sardas, S., Yilmaz, M., Oztok, U., Cakir, N. & Karakaya, A. E. (2001) Assessment of DNA strand breakage by comet assay in diabetic patients and the role of antioxidant supplementation. *Mutat Res 490*, 123-129.

Schaffer, S., Solodushko, V. & Azuma, J. (2000a) Taurine deficient cardiomyopathy: role of phospholipids, calcium and osmotic stress. *Adv Exp Med Biol 483*, 57-69

Schaffer, S. W., Lombardini, J. B. & Azuma, J. (2000b) Interaction between the actions of taurine and angiotensin II. *Amino Acids 18*, 305-318.

Schmidt, H. H. & Walter, U. (1994) NO at work. *Cell 78*, 919-925.

Sherr CJ (1994) G1 phase progression: cycling on cue. *Cell 79*, 551-555.

Shiloh Y (2003) ATM and related protein kinases: safeguarding genome integrity. *Nat Rev Cancer 3*, 155-168.

Simmen, M. W., Leitgeb, S., Charlton, J., Jones, S. J., Harris, B. R., Clark, V. H. & Bird, A. (1999) Nonmethylated transposable elements and methylated genes in a chordate genome. *Science 283*, 1164-1167.

Sole, M. J. & Jeejeebhoy, K. N. (2002) Conditioned nutritional requirements: therapeutic relevance to heart failure. *Herz 27*, 174-178.

Stadtman, E. R., Moskovitz, J. & Levine, R. L. (2003) Oxidation of methionine residues of proteins: biological consequences. *Antioxid Redox Signal 5*, 577-582.

Stupp, R., Mason, W. P., van den Bent, M. J., Weller, M., Fisher, B., Taphoorn, M. J., Belanger, K., Brandes, A. A., Marosi, C., Bogdahn, U., Curschmann, J., Janzer, R. C., Ludwin, S. K., Gorlia, T., Allgeier, A., Lacombe, D., Cairncross, J. G., Eisenhauer, E. & Mirimanoff, R. O. (2005) European Organisation for Research and Treatment of Cancer Brain Tumor and Radiotherapy Groups; National Cancer Institute of Canada Clinical Trials Group. Radiotherapy plus concomitant and adjuvant temozolomide for glioblastoma. *N Engl J Med 352*, 987-996.

Suleiman, M. S. (1994) New concepts in the cardioprotective action of magnesium and taurine during the calcium paradox and ischaemia of the heart. *Magnes Res 7*, 295-312.

Tamamori, M., Ito, H., Hiroe, M., Terada, Y., Marumo, F. & Ikeda, M. A. (1998) Essential roles for G1 cyclin-dependent kinase activity in development of cardiomyocyte hypertrophy. *Am J Physiol 275*, H2036-H2040.

Taniguchi, N., Kaneto, H., Asahi, M., Takahashi, M., Wenyi, C., Higashiyama, S., Fujii, J., Suzuki, K. & Kayanoki, Y. (1996) Involvement of glycation and oxidative stress in diabetic macroangiopathy. *Diabetes 45*, S81-S83.

Tate, P., Skarnes, W. & Bird, A. (1996) The methyl-CpG binding protein MeCP2 is essential for embryonic development in the mouse. *Nat Genet 12*, 205-208.

Tomei, L. D., Kiecolt-Glaser, J. K., Kennedy, S. & Glaser, R. (1990) Psychological stress and phorbol ester inhibition of radiation-induced apoptosis in human peripheral blood leukocytes. *Psychiatry Res 33*, 59-71.

Tosaki, A., Engelman, D. T., Engelman, R. M. & Das, D. K. (1996) The evolution of diabetic response to ischemia/reperfusion and preconditioning in isolated working rat hearts. *Cardiovasc Res 31*, 526-536.

Toyota, M., Sasaki, Y., Satoh, A., Ogi, K., Kikuchi, T., Suzuki, H., Mita, H., Tanaka, N., Itoh, F., Issa, J. P., Jair, K. W., Schuebel, K. E., Imai, K. & Tokino, T. (2003) Epigenetic inactivation of CHFR in human tumors. *Proc Natl Acad Sci U S A 100*, 7818-7823.

Trog, D., Fountoulakis, M., Friedlein, A. & Golubnitschaja, O. (2006a) Is current therapy of malignant gliomas beneficial for patients? Proteomics evidence of shifts in glioma cells expression patterns under clinically relevant treatment conditions. *Proteomics 6*, 2924-2930.

Trog, D., Yeghiazaryan, K., Fountoulakis, M., Friedlein, A., Moenkemann, H., Haertel, N., Schueller, H., Breipohl, W., Schild, H., Leppert, D. & Golubnitschaja, O. (2006b) Proinvasive gene regulating effect of irradiation and combined temozolomide-radiation treatment on surviving human malignant glioma cells. *Eur J Pharmacol 542*, 8-15.

Trog, D., Yeghiazaryan, K., Schild, H. H. & Golubnitschaja, O. (2007a) Up-regulation of vimentin expression in low-density malignant glioma cells as immediate and late effects under irradiation and temozolomide treatment. *Amino Acids*, PMID: 18046501.

Trog, D., Moenkemann, H., Breipohl, W., Schueller, H., Schild, H. & Golubnitschaja, O. (2007b) Non-sufficient cell cycle control as possible clue for the resistance of human malignant glioma cells to clinically relevant treatment conditions. *Amino Acids 32*, 373-379.

Tsai, J. C., Perrella, M. A., Yoshizumi, M., Hsieh, C. M., Haber, E., Schlegel, R. & Lee, M. E. (1994) Promotion of vascular smooth muscle cell growth by homocysteine: a link to atherosclerosis. *Proc Natl Acad Sci U S A 91*, 6369-6373.

Turunen, N., Karihtala, P., Mantyniemi, A., Sormunen, R., Holmgren, A., Kinnula, V. L. & Soini, Y. (2004) Thioredoxin is associated with proliferation, p53 expression and negative estrogen and progesterone receptor status in breast carcinoma. *APMIS 112*, 123-132.

Varela, I., Cadinanos, J., Pendas, A. M., Gutierrez-Fernandez, A., Folgueras, A. R., Sanchez, L. M., Zhou, Z., Rodriguez, F. J., Stewart, C. L., Vega, J. A., Tryggvason, K., Freije, J. M. & Lopez-Otin, C. (2005) Accelerated ageing in mice deficient in Zmpste24 protease is linked to p53 signalling activation. *Nature 437*, 564-568.

von Harsdorf, R., Hauck, L., Mehrhof, F., Wegenka, U., Cardoso, M. C. & Dietz, R. (1999) E2F-1 overexpression in cardiomyocytes induces downregulation of p21CIP1 and p27KIP1 and release of active cyclin-dependent kinases in the presence of insulin-like growth factor I. *Circ Res 85*, 128-136.

Weiss, R. H. (2003) p21Wafl/Cipl as a therapeutic target in breast and other cancers. *Cancer Cell 4*, 425-429.

Wilcken, D. E., Reddy, S. G. & Gupta, V. J. (1983) Homocysteinemia, ischemic heart disease, and the carrier state for homocystinuria. *Metabolism 32*, 363-370.

Wild-Bode, C., Weller, M., Rimner, A., Dichgans, J. & Wick, W. (2001) Sublethal irradiation promotes migration and invasiveness of glioma cells: implications for radiotherapy of human glioblastoma. *Cancer Res 61*, 2744-2750.

Xie, G., Habbersett, R. C., Jia, Y., Peterson, S. R., Lehnert, B. E., Bradbury, E. M. & D'Anna, J. A. (1998) Requirements for p53 and the ATM gene product in the regulation of G1/S and S phase checkpoints. *Oncogene 16*, 721-736.

Yang, Z. Z. & Zou, A. P. (2003) Homocysteine enhances TIMP-1 expression and cell proliferation associated with NADH oxidase in rat mesangial cells. *Kidney Int 63*, 1012-1020.

Yeghiazaryan, K., Mamlouk, S., Trog, D., Moenkemann, H., Braun, M., Kuhn, W., Schild, H. H. & Golubnitschaja, O. (2007) Irradiated breast cancer patients demonstrate subgroup specific regularities in protein expression patterns of circulating leukocytes. *Cancer Genomics Proteomics 4*, 411-418 .

Zhang, Q., Zeng, X., Guo, J. & Wang, X. (2001) Effects of homocysteine on murine splenic B lymphocyte proliferation and its signal transduction mechanism. *Cardiovasc Res 52*, 328-336.

Zhang, Q., Zeng, X., Guo, J. & Wang, X. (2002) Oxidant stress mechanism of homocysteine potentiating Con A-induced proliferation in murine splenic T lymphocytes. *Cardiovasc Res 53*, 1035-1042.

Zhuang, J. C. & Wogan, G. N. (1997) Growth and viability of macrophages continuously stimulated to produce nitric oxide. *Proc Natl Acad Sci U S A 94*, 11875-11880.

In: Cell Cycle Control: New Research
Editors: N. H. Leroy, N. T. Fournier
ISBN: 978-1-60456-796-0

Chapter XII

The Role of Cell Cycle Control Genes in the Pathogenesis of Human Hematological Neoplasms

Luís Vieira[1] and Maria Gomes da Silva[2]
[1]Centro de Genética Humana, Instituto Nacional de Saúde Dr. Ricardo Jorge, Av. Padre Cruz, 1649-016 Lisboa, Portugal;
[2]Serviço de Hematologia, Instituto Português de Oncologia de Francisco Gentil, R. Prof. Lima Basto, 1099-023 Lisboa, Portugal.

Abstract

Dysregulation of cell cycle control is of major importance in cancer pathogenesis. Consequently, the identification of genetic and epigenetic changes in genes encoding cell cycle control proteins is instrumental to understand how dysregulation of the cell cycle machinery contributes to the proliferation of tumour cells. Coiled-coil domain containing 94 (CCDC94) was recently characterized as a novel human nuclear protein that plays a role in cell cycle control. Ectopic expression of CCDC94 protein *in vitro* results in cell cycle arrest whereas suppression of *CCDC94* mRNA induces a dramatic increase in cell proliferation. *CCDC94* was originally identified as a target gene of an acquired chromosome translocation in a patient with acute myeloid leukemia (AML) that resulted in its haplo-insufficiency in leukemic cells. *CCDC94* adds to a number of other genes involved in the cell cycle machinery whose expression is frequently altered in hematological malignancies. Chromosome rearrangements and deletions involving D-type cyclins, cyclin-dependent kinases (CDKs) or cyclin-dependent kinase inhibitors (CDKIs) are frequent in lymphoid tumours. In particular, chromosome translocations cause overexpression of D-type cyclins in mature B-cell neoplasms such as mantle cell lymphoma and multiple myeloma whereas different types of rearrangements lead to overexpression of CDK6 in mature T-cell malignancies. In addition, deletion or promoter hypermethylation of *CDKN2A* and/or *CDKN2B* genes, which encode 2 members of the INK4A family of CDKIs, $p16^{INK4A}$ and $p15^{INK4B}$ respectively, are recurrent events in acute lymphoblastic leukemias. In contrast, patients with myelodysplastic syndrome and AML frequently present with hypermethylation of *CDKN2B* alone. This review summarizes the current knowledge on the molecular changes involving D-type cyclins, CDKs and CDKIs in several hematological neoplasms and addresses the most relevant effects of these alterations in the control of cell cycle.

Introduction

The mammalian cell cycle regulates crucial biological functions of cells. It is broadly divided into 4 consecutive and highly regulated phases known as first gap (G1), DNA synthesis (S), second gap (G2) and mitosis (M). When cells are not cycling, they enter a quiescent state known as G0. Cell cycle progression is typically initiated at the extracellular environment, with the binding of growth factors to their cognate membrane receptors, followed by the activation of multiple downstream signalling pathways. Stimulatory signals converge at the cell nucleus where a limited number of cell cycle control proteins are ultimately responsible for processing growth-promoting signals in a complex and cell type-specific interplay. Progression from one phase of the cycle to the next is carefully mediated by the sequential assembly of serine-threonine cyclin dependent kinases (CDKs) with their regulatory subunit cyclin proteins, thereby activating the kinase function. Cyclin-dependent kinase 4 (CDK4) and cyclin-dependent kinase 6 (CDK6) can be activated through their association with any of the 3 D-type cyclins (Cyclin D1, D2 and D3) during the early G1 phase, whereas CDK2 forms activated complexes with cyclins A and E at the late G1 phase and G1/S transition (Sherr and Roberts, 1999). Activated cyclin D-CDK4/6 complexes phosphorylate the retinoblastoma 1 (RB1) protein and a number of other substrates that play a role in cell cycle progression (Weinberg, 1995; Sherr and Roberts, 1999). Phosphorylation and concomitant inactivation of RB1 allows the release of sequestered E2F transcription factors, which subsequently drive the expression of several genes required for progression into S phase (Nevins, 1992). Progression along the cell cycle is negatively regulated by 2 families of cyclin-dependent kinase inhibitors (CDKIs). The members of the INK4A family ($p16^{INK4A}$, $p15^{INK4B}$, $p18^{INK4C}$ and $p19^{INK4D}$) arrest cell cycle progression at the G1 phase by binding to CDK4 or CDK6, thus inhibiting the activity of cyclin D proteins (Sherr and Roberts, 1999). The members of the CIP/KIP family ($p21^{CIP1}$, $p27^{KIP1}$ and $p57^{KIP2}$) are capable of inhibiting the activity of a broader number of CDKs during either G1, S or G2 phases and, on the opposite, to contribute to the assembly of cyclin D-CDK4/6 complexes at the G1/S transition (Sherr and Roberts, 1999). Dysregulation of normal cyclin, CDK or CDKI protein activities are important physiological alterations that contribute to the transformation of a normal cell into a tumour cell. This phenomenon frequently results from the acquisition of genetic or epigenetic alterations in their cognate genes, thereby affecting the normal regulation of cell growth and division. Here we review recurrent genetic and epigenetic changes that dysregulate cell cycle control in various hematological neoplasms including acute lymphoid and myeloid leukemia, and several subtypes of B- and T-cell lymphomas. We chose to focus on these disorders because they illustrate diverse genetic and epigenetic mechanisms by which dysregulation of cell cycle control leads to malignancy.

Mature B-Cell Neoplasms

Mantle Cell Lymphomas

Mantle cell lymphoma (MCL) is an aggressive B-cell malignant disorder genetically characterized by the presence of a reciprocal chromosome translocation t(11;14)(q13;q32) in approximately 95% of patients. The t(11;14) fuses the *CCND1* gene (11q13), which encodes cyclin D1 protein, to the immunoglobulin heavy chain (*IGH*) gene at 14q32 (Tsujimoto et al, 1994). The direct consequence of t(11;14) is overexpression of cyclin D1 due to the juxtaposition of the *CCND1* coding sequence to a strong *IGH* enhancer sequence (Bosch et al, 1994; de Boer et al, 1995). Full-length cyclin D1 is a critical effector protein in the control of G1/S transition of the cell cycle (Knudsen, 2006). Normal hematopoietic cells usually express low levels of cyclin D2/cyclin D3 but do not express cyclin D1 (Lukas et al, 1994; Lam et al, 2000; Sicinska et al, 2003). However, cyclin D1 is transcriptionally up-regulated upon mitogenic signalling and binds to either CDK4 or CDK6 in a cell type-specific manner, thereby leading to phosphorylation of RB1 and irreversibly promoting cell cycle progression into the S phase (Ewen et al, 1993; Dowdy et al, 1993; Bates et al, 1994; Weinberg, 1995). In MCL cells, the constitutive expression of cyclin D1 has been shown to lead to its association with CDK4 but not CDK6 (Marzec et al, 2006). In accordance, the RB1 protein was shown to be hyperphosphorylated in MCL tumours, indicating that in these cells the suppressive effect of RB1 at the G1/S transition is diminished as a consequence of the activity of cyclin D1/CDK4 complexes (Jares et al, 1996; Marzec et al, 2006). To exert its pro-proliferative action, however, cyclin D1 strongly depends on the existence of a functional RB1 protein. RB1-deficient tumour cell lines have very low levels of cyclin D1 and ectopic expression of RB1 in these cells specifically induces the expression of cyclin D1 at the transcriptional level (Müller et al, 1994). Moreover, proliferation of RB1-null cells is impaired even in the presence of active cyclin D1-CDK4 complexes (Lukas et al, 1995). However, several data indicate that the cell cycle-promoting activity of cyclin D1 in MCL tumours can, at least in part, be performed independently of the existence of a functional *RB1* gene or of its relative expression levels. Jares et al (1996) found that cyclin D1 mRNA levels in MCL cells were independent of RB1 protein levels or of the proliferative activity of the tumours. In addition, intragenic homozygous deletions of the *RB1* gene which result in total absence of the protein, have been shown to occur in a few patients with blastic variant of MCL (Pinyol et al, 2007). This finding indicates that loss of RB1 may be preferentially associated with disease progression.

It was initially suggested that the existence of a single A/G nucleotide polymorphism within the splice donor site of *CCND1* exon 4, facilitated the formation of a shorter cyclin D1 transcript (cyclin D1b) instead of the full-length cyclin D1a transcript (Betticher et al, 1995). The cyclin D1b transcript lacks 3′ untranslated (UTR) sequences that contain mRNA destabilizing elements and is highly potent in transformation activity (Lu et al, 2003; Solomon et al, 2003). However, transcription of cyclin D1b has recently been shown to occur independently of the nucleotide that is present at the splice site (Krieger et al, 2006; Wiestner et al, 2007). It was also shown that the variability in expression of cyclin D1 is preferentially due to genomic deletions in the 3′ UTR sequence of cyclin D1a mRNA, or to point mutations in the coding sequence of *CCND1* that lead to the removal of destabilizing elements from the mRNA by creating premature polyadenylation signals (Wiestner et al, 2007). Significantly, tumours showing a higher proliferation rate express truncated cyclin D1 isoforms whereas tumours with the lowest proliferation rate express only the full-length isoform (Wiestner et al, 2007). Microarray studies have also shown that the most proliferative and clinically aggressive MCL tumours are associated with expression of shorter cyclin D1 mRNA species (Rosenwald et al, 2003). Furthermore, the overall cyclin D1 mRNA abundance was higher in these tumours than in those where the full-length transcript was present (Rosenwald et al, 2003). Thus, it is likely that the greater stability of truncated cyclin D1 transcripts strengthens the role of cyclin D1 overexpression in t(11;14)-positive MCL cells by enhancing cell cycle progression through the G1/S transition.

Expression of either cyclin D2 or cyclin D3 mRNA, instead of cyclin D1 mRNA, was revealed by gene expression profiling of characteristic MCL cases that lacked t(11;14)(q13;q32) (Fu et al, 2005). However, chromosomal rearrangements involving the *CCND2* or *CCND3 loci*, respectively, were not detected in any of the t(11;14)-negative MCL cases by interphase fluorescence *in situ* hybridization (FISH) analysis (Fu et al, 2005). More recently, Gesk et al (2006) described two cyclin D1-negative MCL patients with a t(2;12)(p12;p13) that juxtaposes the *CCND2* gene (12p13) to the immunoglobulin κ light chain (*IGκ*) gene at 2p12. Immunohistochemical studies showed a strong nuclear expression of cyclin D2 in the tumour cells of both patients, indicating that *CCND2* was overexpressed as a consequence of proximity to *IGκ* sequences (Gesk et al, 2006). Although *CCND2* translocations are rare events, this study illustrates that cyclin D2 can functionally replace cyclin D1 overexpression in t(11;14)-negative MCL.

Homozygous deletions of the *CDKN2A* gene, which encodes p16^{INK4A}, were initially described in various malignant cell lines, including some of leukemic nature, suggesting that it could function as a tumour suppressor gene whose loss is implicated in diverse forms of cancer (Kamb et al, 1994; Nobori et al, 1994). Not surprisingly, hemi- or homozygous deletions of *CDKN2A* were also reported to occur in approximately half of patients with MCL (Dreyling et al, 1997). Although *CDKN2A* deletions are statistically independent of cyclin D1 mRNA expression, both alterations synergize to dictate tumour proliferation and survival (Rosenwald et al, 2003). Thus, deletion of *CDKN2A* and overexpression of cyclin D1 are cooperative genetic events that promote the G1/S transition by increasing the levels of active cyclin D1-CDK4/6 complexes.

The *CDKN1B* gene encodes p27^{KIP1}, an important CDKI in cell cycle regulation which levels are elevated during G0 but rapidly reduced upon entry of the cell into G1 phase (Nourse et al, 1994; Kato et al, 1994). The normal mantle zone and the interfollicular lymphocytes within the lymph nodes are mainly p27^{KIP1}-positive whereas the germinal centres tend to lack p27^{KIP1} expression, which is the opposite expression pattern of the proliferation-related Ki67 antigen (Quintanilla-Martinez et al, 1998). Accordingly, the expression of p27^{KIP1} is inversely proportional to that of Ki67 in various types of low-grade B-cell lymphomas (Quintanilla-Martinez et al, 1998; Chiarle et al, 2000). In contrast, expression of p27^{KIP1} in MCL tumours is independent of the proliferative rate and is not detected in approximately 80% of MCL patients due to an increased degradation of the protein via ubiquitin- and proteasome-dependent pathways (Quintanilla-Martinez et al, 1998;

Chiarle et al, 2000). Intriguingly, $p27^{KIP1}$ expression is detected in 80% of patients with the highly proliferative blastic variant of MCL, suggesting that in these cases $p27^{KIP1}$ is either sequestered by overexpressed cyclin D/CDK complexes (Cheng et al, 1998) or its expression tends to compensate for mutation or deletion of other cell cycle control genes such as *CDKN2A* and *CDKN1A* (encoding $p21^{CIP1}$) which are generally associated with blastic variant of MCL (Pinyol et al, 1997; Quintanilla-Martinez et al, 1998). In conclusion, the G1/S transition in MCL tumours results from a balance between different expression levels of various cell cycle control proteins which in turn are largely the consequence of rearrangements of genes that control this cell-cycle transition.

Multiple Myeloma

Multiple myeloma (MM) is a B-cell disorder characterized by bone marrow accumulation of antibody-secreting clonal plasma cells. Reciprocal chromosomal translocations involving *IGH* (or less often an *IGL locus*) and several different partner genes are frequently present in patients with MM or with a predisposing benign condition designated monoclonal gammopathy of undetermined significance (MGUS). The incidence of *IGH* translocations ranges from approximately 50% in MGUS to 80% in extramedullary MM although at present there is no unequivocal genetic marker that allows a distinction between these clinical conditions (Bergsagel and Kuehl, 2001). Five different recurrent chromosome translocations comprise approximately 40% of cases of MGUS/MM. These include t(11;14)(q13;q32), found in 10-15% of patients using conventional cytogenetics and interphase FISH analyses (Chesi et al, 1996; Avet-Loiseau et al, 1998; Fonseca et al, 1998). In addition to the cyclin D1-positive t(11;14), forty percent of patients ectopically express both alleles of *CCND1* in the absence of the translocation (Bergsagel et al, 2005). This finding indicates that biallelic dysregulation of cyclin D1 expression is a highly frequent alteration in MM/MGUS and suggests that the phenomenon is regulated by trans-acting transcription factors which have not yet been identified (Bergsagel et al, 2005). In cyclin D1-positive MM cells, cyclin D1a and cyclin D1b mRNA isoforms are co-expressed irrespectively of the presence of the translocation (Roué et al, 2003). However, cyclin D1b protein levels are almost undetectable in comparison to those of MCL tumours (Krieger et al, 2006), suggesting that this isoform may have a less prominent role in MM pathogenesis as compared to MCL.

In addition to t(11;14), three different recurrent translocations including t(4;14)(p16;q32), t(14;16)(q32;q23), and t(14;20)(q32;q11), are altogether observed in 20-25% of MGUS/MM patients. The t(14;16) and t(14;20) cause dysregulation of the *MAF* (16q23) and *MAFB* (20q11) transcription factor genes respectively (Chesi et al, 1998a; Hanamura et al, 2001; Boersma-Vreugdenhil et al, 2004), whereas the t(4;14) leads to overexpression (or less often to loss of expression) of the fibroblast growth factor receptor 3 (*FGFR3*) gene on derivative chromosome 14, and to the formation of a fusion product between the multiple myeloma SET domain (*MMSET*) gene and the *IGH locus* on the derivative chromosome 4 (Chesi et al, 1997; Richelda et al, 1997; Chesi et al, 1998b; Santra et al, 2002). Although these 3 translocations do not physically involve a cyclin D gene, high levels of cyclin D2 were found in cells harbouring either t(4;14), t(14;16) or t(14;20) (Bergsagel and Kuehl, 2003; Dring et al, 2004; Hurt et al, 2004), indicating that up-regulation of the *CCND2* gene is a downstream effect of gene fusion rearrangements. Finally, cyclin D3 expression may also be dysregulated as a consequence of the t(6;14)(p21;q32) chromosome translocation that juxtaposes the *CCND3* gene to *IGH* (Shaughnessy et al, 2001). However, this translocation is only found in approximately 3% of karyotypically-abnormal MM patients by cytogenetic and/or spectral karyotyping analyses (Sawyer et al, 2001). Taken together, these data shows that cyclin D overexpression occurs in at least two thirds of MGUS/MM patients, which led to the proposal that dysregulation of cyclin D activity is a unifying event in the early pathogenesis of these tumours (Bergsagel et al, 2005).

Initial investigation for alterations in members of the INK4A-family of CDKIs in patients with MM was reported by Tasaka et al (1997) who found one patient with both *CDKN2B* (encoding $p15^{INK4B}$) and *CDKN2C* (encoding $p18^{INK4C}$) homozygous deletions. Subsequently, hypermethylation of *CDKN2B* was reported to occur in 36-67% of MM patients (Ng et al, 1997; Guillerm et al, 2001). In addition, hypermethylation of the *CDKN2A* gene promoter was found in 19-75% of MM patients, and similarly to the findings in MCL, loss of *CDKN2A* expression is independent of cyclin D1 overexpression (Ng et al, 1997; Guillerm et al, 2001; Krämer et al, 2002). However, the impact of *CDKN2A* promoter methylation on survival of MM patients is controversial (Uchida et al, 2001; Krämer et al, 2002). The frequency of methylation for *CDKN2A*, *CDKN2B*, or both, is similar in patients with MGUS or MM, suggesting that epigenetic silencing of these genes also represents an early event in pathogenesis and is not associated with progression from MGUS to MM (Guillerm et al, 2001). More recently, deletions of the *CDKN2C* gene were described in 38% of cyclin D1-expressing and non-expressing MM cell lines (Kulkarni et al, 2002). Significantly, ectopic expression of $p18^{INK4C}$ in a cyclin D1-overexpressing MM cell line with biallelic *CDKN2C* deletion resulted in significant tumour regression through inhibition of the cyclin D-CDK-RB1 pathway and increased apoptotic cell death (Kulkarni et al, 2002).

Although hematopoietic cells have a strict requirement for cyclin D activity in order to proliferate (Park et al, 2003; Kozar et al, 2004), overexpression of D-type cyclins alone is not sufficient to induce cell cycle progression. In cyclin D1-overexpressing MM cells, high levels of CDK4, and the concomitant absence of CDK6, cyclin D2 and cyclin D3, are required to phosphorylate the RB1 protein and promote cell cycle progression (Ely et al, 2005). In contrast, cyclin D2 preferentially associates with CDK6 in cyclin D2-overexpressing cells and avoids inhibition by both $p18^{INK4C}$ and $p27^{KIP1}$, thereby inducing the inactivation of RB1 and promoting progression of cell cycle (Ely et al, 2005). Finally, normal B cell subpopulations require high levels of cyclin D3 and CDK6, and concomitant absence of $p27^{KIP1}$ for cytokine-induced growth of cyclin D3-expressing cells (Wagner et al, 1998).

Diffuse Large B-Cell Lymphoma

Diffuse large B-cell lymphoma (DLBCL) is the most common subtype of lymphoma worldwide and is strikingly heterogeneous at both the clinical and genetic levels. Although chromosome alterations are frequent in DLBCL patients, no single cytogenetic abnormality characterizes this disorder (Abramson and Shipp, 2005). Genome-scale microarray studies have been instrumental in dissecting the molecular basis of DLBCL by recognizing the existence of 3 different diseases (germinal centre B-cell like, activated B-cell like, and type 3) that are indistinguishable by current diagnostic methods (Rosenwald et al, 2002). Data obtained from microarray studies were further refined to build a quantitative-polymerase chain reaction (Q-PCR)-based multivariate model that predicts overall survival in DLBCL patients using the mRNA expression levels of only 6 genes (Lossos et al, 2004). The reduced expression of *LMO2*, *BCL6* and *FN1* genes correlates with prolonged survival whereas the increased expression of *BCL2*, *SCYA3* and *CCND2* correlates with shortened survival (Lossos et al, 2004). In agreement, absence of cyclin D2 expression correlated with improved disease-free survival (DFS) and overall survival (OS) in patients with the germinal centre B-cell like DLBCL type analysed by tissue microarrays (Hans et al, 2004; Amen et al, 2007). Thus, cyclin D2 overexpression is an adverse prognostic marker in DLBCL, reflecting a positive role of this cyclin in promoting cell cycle progression in tumour cells. In addition, cyclin D2 induces the nuclear export and ubiquitin-mediated cytoplasmic degradation of $p27^{KIP1}$ at the G0/G1 transition (Susaki et al, 2007). Lymphocytes that lack cyclin D2 do not proliferate in response to mitogenic stimuli mainly as a result of a defect in the rapid elimination of $p27^{KIP1}$ (Solvason et al, 2000), suggesting that cyclin D2 plays an important role in the down-regulation of this CDKI. Indeed, the abundance of cyclin D2 has been shown to increase in association with a decrease in $p27^{KIP1}$ expression in different hematopoietic cell types (Banerji et al, 2001; Parada et al, 2001).

Similarly to the findings in MCL, no correlation exists between $p27^{KIP1}$ expression levels and the proliferative index of DLBCL tumours. In at least half of the tumours, expression of $p27^{KIP1}$ is low or is not detected irrespectively of the proliferative index (Quintanilla-Martinez et al, 1998; Chiarle et al, 2000). However, $p27^{KIP1}$ is expressed at high levels in some DLBCL tumours that show a high proliferative rate, suggesting that in these cases the $p27^{KIP1}$ protein is not functional because of interaction with other cell cycle regulator proteins (Cheng et al, 1998; Sáez et al, 1999; Sánchez-Beato et al, 1999). This hypothesis is supported by immunohistochemistry studies of highly proliferating DLBCL tumours, which revealed that higher $p27^{KIP1}$ expression levels correlate with increased levels of cyclin D3 (Sánchez-Beato et al, 1999). Moreover, confocal microscopy analysis showed that cyclin D3 and $p27^{KIP1}$ colocalize in the nucleus of tumour cells (Sánchez-Beato et al, 1999), indicating that high levels of cyclin D3 appear to be responsible for sequestration of $p27^{KIP1}$ rendering it non-functional in its ability to restrain the activity of cyclin E-CDK2 complexes and, consequently, of inducing cell cycle arrest at the G1/S transition.

Chronic Lymphocytic Leukemia

Chronic lymphocytic leukemia (CLL) is a malignant B-cell disorder characterized by progressive accumulation of clonal, growth-arrested lymphocytes at the G0/G1 phases of the cell cycle. CLL is a major exception among B-cell tumours because activation of an oncogene by chromosome translocations is rarely observed (Hayette et al, 1998). However, dysregulation of cyclin D2 expression can be found in CLL patients at a high frequency and in the absence of detectable *CCND2* rearrangements. In particular, cyclin D2 overexpression was detected in approximately 85% of CLL patients although no correlation was found between increased expression of the *CCND2* gene, which localizes to chromosome band 12p13, and the presence of chromosome 12 trisomy in these patients (Delmer et al, 1995). In CLL cells, cyclin D2 preferentially associates with CDK4, instead of CDK6 as in normal B cells, and the activated cyclin D2-CDK4 complexes are also able to phosphorylate the RB1 protein (Wagner et al, 1998; Decker et al, 2002). $p27^{KIP1}$ is also consistently up-regulated in CLL cells and higher $p27^{KIP1}$ levels have been associated with a poorer patient outcome in different studies (Vrhovac et al, 1998; Wolowiec et al, 2001; Decker et al, 2002). Since $p27^{KIP1}$ is implicated in cell cycle arrest by inhibition of the kinase activity of cyclin E/CDK2 complexes, high levels of this CDKI may be responsible for accumulation of tumour cells at the G0/G1 transition and the low proliferative tumour rate that characterizes this disorder. In agreement, $p27^{KIP1}$ expression was found to be inversely proportional to the proliferative index of CLL cells (Sánchez-Beato et al, 1997; Quintanilla-Martinez et al, 1999; Chiarle et al, 2000). When CLL cells are stimulated to proliferate in the presence of interleukin-2, $p27^{KIP1}$ levels are rapidly reduced probably as a result of ubiquitin-dependent degradation (Decker et al, 2002). Accordingly, these cells undergo apoptosis after *in vitro* treatment with fludarabine by inducing a rapid, caspase-dependent cleavage of $p27^{KIP1}$ and a concomitant increase in CDK4 activity, thereby indicating an anti-apoptotic function for $p27^{KIP1}$ (Hiromura et al, 1999; Sanhes et al, 2003). Thus, these data indicate that $p27^{KIP1}$ plays an important role in CLL pathogenesis not only by arresting cell cycle progression at the G0/G1 transition but also by blocking normal B-cell apoptosis.

Mature T-Cell Neoplasms

Mature T-cell neoplasms are relatively uncommon disorders that are among the most clinically aggressive of all hematopoietic and lymphoid malignancies. In contrast with B-cell mature neoplasms, the *CDK6* gene localized at 7q21 is a recurrent target of chromosome rearrangements in mature T-cell malignancies. Using immunohistochemistry and flow cytometry analyses, Chilosi et al (1998) found that CDK6 was strongly expressed in the majority of cells obtained from T-cell lymphoblastic lymphoma/leukemia samples whereas only a very small percentage of cells derived from normal and hyperplastic thymuses expressed CDK6. Furthermore, with the exception of nasal angiocentric T-cell lymphomas and nasal natural killer/T-cell lymphomas, the expression of CDK6 was weak or not detected in other B- or T-cell lymphoma subtypes (Chilosi et al, 1998; Lien et al, 2000). Using Q-PCR and Western-blot analyses, Su et al (2004) confirmed the strong expression of the *CDK6* gene in T cell-acute lymphoblastic leukemia (ALL) patients with t(5;7)(q35;21) and its lack of expression in non-lymphoid leukemic cell lines. Recently, Nagel et al (2008) reported genomic amplification of the 7q22 region comprising the *CDK6* gene in an anaplastic large-cell lymphoma (ALCL) cell line. The extension of this analysis to primary T-cell lymphoma patient samples revealed that 23% of peripheral T-cell lymphomas had amplification of *CDK6* whereas this phenomenon was rarely observed among ALCL samples (Nagel et al, 2008). These data indicate that CDK6 overexpression is preferentially associated with T-cell lymphomas although it is not consistently associated with a specific disease subtype. Accordingly, increased CDK6 expression has been described in a few patients with mature B-cell lymphomas harbouring 7q21 translocations (Corcoran et al, 1999; Brito-Babapulle et al, 2002; Hayette et al, 2003). In addition, a 3-way translocation involving *MLL*, *AF4* and *CDK6* genes was described in an infant patient with B-cell ALL (Raffini et al, 2002). Although no functional studies have been performed, the finding of a *CDK6-MLL* fusion transcript indicated that *CDK6* was overexpressed as a consequence of the translocation. Taken together, the available data suggests that the overexpression of CDK6 likely contributes to cell proliferation by relieving the suppressive role of RB1-transcriptional repressor complexes, thus favoring progression through the G1 phase of the cell cycle.

Acute Lymphoblastic Leukemia

Deletion of the *CDKN2A* and/or *CDKN2B* genes was also suggested to occur in patients with ALL because both genes lie in close proximity within chromosome band 9p21, a frequent site for deletions and rearrangements in this disorder. Indeed, deletions of both *CDKN2A* and *CDKN2B* were found in 90% of pediatric ALL patients with cytogenetic abnormalities of chromosome 9p whereas only 22% of patients with apparently normal chromosomes 9 showed deletion of both genes (Okuda et al, 1995). Hebert et al (1994) reported the presence of homozygous deletions of *CDKN2A* and/or *CDKN2B* in 83% of T-lineage ALL patients, indicating that these are the most frequent genetic defects in T-cell ALL. In comparison, biallelic deletions of *CDKN2A* were only found in 6,5-8,6% of B-cell ALL patients (Hebert et al, 1994; Stranks et al, 1995). However, there was a predominance of deletions (21%) among patients with B-cell precursor ALL although no association was found between *CDKN2A* deletion and poor prognosis (Stranks et al, 1995). The majority of deletions are homozygous and in those cases with hemizygous deletions, sequence analysis did not reveal any mutations in the non-deleted allele (Okuda et al, 1995; Takeuchi et al, 1995). Although both genes may be deleted in the same patient, the *CDKN2A* gene is more frequently deleted than *CDKN2B* in either T-cell ALL or B-cell ALL (Takeuchi et al, 1995; Cayuela et al, 1996). In contrast, methylation of the CpG island promoter region of *CDKN2B* can be found in approximately half of all ALL patients but that of *CDKN2A* in only a small percentage of patients and concomitant to *CDKN2B* (Chim et al, 2001). Hypermethylation of *CDKN2B* alone showed no impact on complete remission, 5-year OS, or 5-year DFS (Chim et al, 2001). However, when the methylation status of *CDKN2A* and *CDKN2B* was evaluated in T-cell ALL patients in conjunction with that of other frequently methylated cancer-related genes, a positive correlation was found between higher numbers of methylated genes and a reduced OS or DFS (Roman-Gomez et al, 2005). The reconstitution of $p16^{INK4A}$ expression in $p16^{INK4A}$-deficient T-lineage ALL cells reinduces normal growth, differentiation and apoptosis of T-cells whilst causing cell cycle arrest by redistributing $p27^{KIP1}$ from CDK4/6 to CDK2, thereby inhibiting cyclin E/CDK2 complexes (Ausserlechner et al, 2005).

Aberrations of the short arm of chromosome 12 are frequently implicated in both lymphoid and myeloid hematological malignancies, and consist mainly of balanced or unbalanced translocations, insertions, and deletions (Vieira et al, 2005). The *CDKN1B* gene, which localizes to chromosome band 12p12-p13.1, was found to be inactivated by deletion in 83% of childhood ALL patients with 12p abnormalities (Komuro et al, 1999). With rare exceptions, homozygous inactivations of *CDKN1B* were not reported, suggesting that the primary consequence of deletions is haplo-insufficiency of $p27^{KIP1}$ (Cavé et al, 1995; Pietenpol et al, 1995; Komuro et al, 1999). Evidence obtained from mouse models showed that animals carrying only one functional allele of $p27^{KIP1}$ are more susceptible to the formation of tumours after exposure to ionising radiation or carcinogens than wild-type mice, thus providing a role for haploinsufficiency of $p27^{KIP1}$ in tumour formation (Fero et al, 1998). In the context of cell cycle regulation, expression of a single *CDKN1B* allele may not be sufficient to inhibit the activity of cyclin E-CDK2 complexes and induce cell cycle arrest at the late G1 phase. Moreover, since low levels of $p27^{KIP1}$ stabilize cyclin D-CDK4/6 complexes in the early G1 phase (Sherr and Roberts, 2004), the lower-than-normal levels of $p27^{KIP1}$ in ALL cells hemizygous for the *CDKN1B* gene may further facilitate cell cycle progression. Expression of the *CDKN1B* gene may also be altered as a result of dysregulated transcription factor activity. The hybrid protein produced by the *MLL-AF4* fusion gene, the result of t(4;11)(q21;q23) associated with 60% of infant B-lineage ALL (Gu et al, 1992; Rowley et al, 2001), is able to bind to the *CDKN1B* promoter in lymphoid cells and to upregulate the expression of $p27^{KIP1}$ (Xia et al, 2005). In this setting, increased levels of $p27^{KIP1}$ could change proliferation of *MLL-AF4* positive cells in favor of quiescence, thus rendering tumour cells less susceptible to chemotherapeutic treatment (Xia et al, 2005).

In approximately half of patients with 12p aberrations, deletions of *CDKN1B* were found to extend to the *CCND2* gene, which is located more telomeric than *CDKN1B* on chromosome 12p (Komuro et al, 1999), thus suggesting a minor role for *CCND2* allele loss comparatively to *CDKN1B* in ALL cells. Conversely, overexpression of cyclin D2 was demonstrated in rare cases of T-cell ALL with t(12;14)(p13;q11). In these cases the *CCND2* gene was found to be juxtaposed to the T-cell receptor (*TCR*) beta or *TCR* alpha/delta *loci* at 14q11 instead of *IGκ* as in MCL-associated t(2;12)(p12;p13) (Clappier et al, 2006; Karrman et al, 2006). Expression of *CCND2* but not that of other surrounding genes at 12p was shown in the t(12;14)-positive cases in comparison to T-cell ALL patients without the translocation (Clappier et al, 2006; Karrman et al, 2006), indicating that increased and sustained expression of cyclin D2 also favors proliferation of ALL cells. In contrast, overexpression of the *CCND2* gene was not found in acute leukemias of myeloid lineage (Delmer et al, 1995).

Acute Myeloid Leukemia

We have previously characterized a 3-way translocation t(9;11;19)(p22;q23;p13.3) in a childhood acute myeloid leukemia (AML) patient which disrupted the coiled-coil domain containing 94 (*CCDC94*) gene at 19p13.3, formerly known as full-length long Japan 10374 (Vieira et al, 2006). Molecular studies showed that the translocated *CCDC94* gene was rendered non-functional, thereby indicating that leukemic cells were haploinsufficient for CCDC94 protein. Bioinformatics analysis of the CCDC94 protein sequence indicated the presence of transcription factor features and identified cwf16 as an homologous protein in the fission yeast *Schizosaccharomyces pombe*. The cwf16 protein was originally identified as a member of the spliceosome complex, where it associates with the cdc5 protein involved in cell cycle regulation in yeast (Ohi et al, 2002). Ectopic expression of the CCDC94 protein in human and mouse cell lines confirmed that it localizes exclusively to the nucleus (Vieira et al, 2006). Furthermore, expression of exogenous CCDC94 represses cell cycle progression at the G1/S transition whereas RNA interference-mediated suppression of *CCDC94* strongly induces proliferation in a mouse fibroblast cell line. These data indicate that CCDC94 is a strong negative regulator of cell proliferation. In agreement, Hannemann et al (2006) identified *CCDC94* among a 35 gene-expression microarray signature which discriminated between all subtypes of ductal carcinoma *in situ* (DCIS) and invasive breast cancer. However, *CCDC94* was not included in gene-expression signatures that discriminated between each subtype of DCIS, suggesting that CCDC94 preferentially operates in a pathway that leads to proliferation and tissue invasion (Hannemann et al, 2006). However, additional studies are warranted to characterize the mechanisms by which CCDC94 exerts control over the mammalian cycle.

Whereas *CDKN2A* and *CDKN2B* are frequently inactivated in ALL patients, *CDKN2B* is the only INK4A family member that is recurrently inactivated in myeloid disorders (Drexler et al, 1998; Rosu-Myles and Wolff, 2007). In addition, inactivation by promoter hypermethylation is the predominant event, occuring in approximately 80% of AML patients and 50% of patients with pre-leukemic myelodysplastic syndrome (MDS), and is associated with poor prognosis in various disease subtypes (Wong et al, 2000; Teofili et al, 2003; Rosu-Myles and Wolff, 2007). The level of CpG methylation of *CDKN2B* increases as the disease progresses from the MDS phase to AML, indicating that methylation is a secondary event in tumorigenesis (Tien et al, 2001). Accordingly, transplantation of tumours without *CDKN2B* methylation into mice results in secondary tumours in which *CDKN2B* is methylated (Wolff et al, 2003). As an alternative mechanism to *CDKN2B* hypermethylation, transcription factor dysregulation by chromosome translocation has also been recently recognized as playing a role in the loss of $p15^{INK4B}$ expression in AML cells. The *CBFβ-MYH11* fusion transcript, the product of inv(16)(p13;q22) or t(16;16)(p13;q22) which is found in 6% of AML patients (Liu et al, 1993; Rowley, 2001), prevents the transcription of the *CDKN2B* gene by disrupting the normal assembly of heterodimeric transcription factors between *CBFβ* and RUNX-family proteins at the *CDKN2B* promoter (Markus et al, 2007).

Conclusion

Dysregulation of normal cell cycle progression is a recurrent pathogenic event in hematological malignancies and results in uncontrolled cell proliferation. This phenomenon is mainly dictated by genetic and epigenetic somatic changes within hematopoietic cells that affect the expression of several important cell cycle control proteins including D-type cyclins, CDKs and CDKIs. The end result of these changes is either the promotion of cell cycle progression in the absence of its normal repressor mechanisms or, alternatively, the forced entering of cells into a permanent quiescent state. An important notion that stems from the present review is that closely-related groups of hematological neoplasms favour different levels of the cyclin D-CDK-RB1 pathway to dysregulate the cell cycle apparatus. Thus, cyclin D overexpression is frequently associated with mature B-cell disorders whereas CDK overexpression is preferentially found in the mature T-cell neoplasms. In turn, loss of CDKI expression is the predominant event in acute leukemias. However, accumulating data indicates that cell cycle dysregulation can only be effective if changes in the expression of different members of the pathway are simultaneously present. Interestingly, both genetic and epigenetic alterations of cell cycle control genes are found in lymphoid malignancies whereas only epigenetic changes of a single gene (*CDKN2B*) are recurrently associated with acute myeloid neoplasms. Although the biological basis for this difference is not clear, recent data indicate that the cell cycle machinery in myeloid lineage cells may be frequently targeted through the indirect action of dysregulated transcription factor activity. The recent identification of the cell cycle regulator *CCDC94* gene, and of its haploinsufficiency in AML cells, further supports this hypothesis.

References

Abramson JS; Shipp MA. Advances in the biology and therapy of diffuse large B-cell lymphoma: moving toward a molecularly targeted approach. *Blood*, 2005 106, 1164-1174.

Amen F; Horncastle D; Elderfield K; Banham AH; Bower M; Macdonald D; Kanfer E; Naresh KN. Absence of cyclin-D2 and Bcl-2 expression within the germinal centre type of diffuse large B-cell lymphoma identifies a very good prognostic subgroup of patients. *Histopathol* 2007 51, 70-79.

Ausserlechner MJ; Obexer P; Geley S; Kofler R. G1 arrest by p16INK4A uncouples growth from cell cycle progression in leukemia cells with deregulated cyclin E and c-Myc expression. *Leukemia* 2005 19, 1051-1057.

Avet-Loiseau H; Li J-Y; Facon T; Brigaudeau C; Morineau N; Maloisel F; Rapp MJ; Talmant P; Trimoreau F; Jaccard A; Harousseau JL; Bataille R. High incidence of translocations t(11;14)(q13;q32) and t(4;14)(p16;q32) in patients with plasma cell malignancies. *Cancer Res* 1998 58, 5640–5645.

Banerji L; Glassford J; Lea NC; Thomas NS; Klaus GG; Lam EW. BCR signals target p27Kip1 and cyclin D2 via the PI3-K signalling pathway to mediate cell cycle arrest and apoptosis of WEHI 231 B cells. *Oncogene* 2001 20, 7352–7367.

Bates S; Bonetta L; MacAllan D; Parry D; Holder A; Dickson C; Peters G. CDK6 (PLSTIRE) and CDK4 (PSK-J3) are a distinct subset of the cyclin-dependent kinases that associate with cyclin D1. *Oncogene* 1994 9, 71–79.

Bergsagel PL; Kuehl WM. Chromosome translocations in multiple myeloma. *Oncogene* 2001 20, 5611-5622.

Bergsagel PL; Kuehl WM. Critical roles for immunoglobulin translocations and cyclin D dysregulation in multiple myeloma. *Immunol Rev* 2003 194, 96-104.

Bergsagel PL; Kuehl WM; Zhan F; Sawyer J; Barlogie B; Shaughnessy JJr. Cyclin D dysregulation: an early and unifying pathogenic event in multiple myeloma. *Blood* 2005 106, 296-303.

Betticher C; Thatcher N; Altermatt HJ; Hoban P; Ryder WD; Heighway J. Alternate splicing produces a novel cyclin D1 transcript. *Oncogene* 1995 11, 1005-1011.

Boersma-Vreugdenhil GR; Kuipers J; Van Stralen E; Peeters T; Michaux L; Hagemeijer A; Pearson PL; Clevers HC; Bast BJ. The recurrent translocation t(14;20)(q32;q12) in multiple myeloma results in aberrant expression of MAFB: a molecular and genetic analysis of the chromosomal breakpoints. *Br J Haematol* 2004 126, 355-363.

Bosch F; Jares P; Campo E; Lopez-Guillermo A; Piris MA; Villamor N; Tassies D; Jaffe ES; Monserrat E; Rozman C. Prad-1/cyclin D1 gene overexpression in chronic lymphoproliferative disorders: a highly specific marker for mantle cell lymphoma. *Blood* 1994 84, 2726–2732.

Brito-Babapulle V; Gruszka-Wetwood AM; Platt G; Anderson CL; Elnenaei MO; Matutes E; Wotherspoon AC; Weston-Smith SG; Catovsky D. Translocation t(2;7)(p12;q21-22) with dysregulation of the CDK6 gene mapping to 7q21-22 in a non-Hodgkin's lymphoma with leukemia. *Haematologica* 2002 87, 357-362.

Cavé H; Gérard B; Martin E; Guidal C; Devaux I; Weissenbach J; Elion J; Vilmer E; Grandchamp B. Loss of heterozygosity in the chromosomal region 12p12-13 is very common in childhood acute lymphoblastic leukemia and permits the precise localization of a tumour-suppressor gene distinct from p27KIP1. *Blood* 1995 86, 3869-3875.

Cayuela JM; Madani A; Sanhes L; Stern MH; Sigaux F. Multiple tumor-suppressor gene 1 inactivation is the most frequent genetic alteration in T-cell acute lymphoblastic leukaemia. *Blood* 1999 87, 2180-2186.

Cheng M; Sexl V; Sherr CJ; Roussel MF. Assembly of cyclin D-dependent kinase and titration of p27KIP1 regulated by mitogen activated protein kinase (MEK1). *Proc Natl Acad Sci USA* 1998 95, 1091-1096.

Chesi M; Bergsagel PL; Brents LA; Smith CM; Gerhard DS; Kuehl WM. Dysregulation of cyclin D1 by translocation into an IgH gamma switch region in two multiple myeloma cell lines *Blood* 1996 88, 674-681.

Chesi M; Bergsagel PL; Shonukan OO; Martelli ML; Brents LA; Chen T; Schröck E; Ried T; Kuehl WM. Frequent dysregulation of the c-*maf* proto-oncogene at 16q23 by translocation to an Ig locus in multiple myeloma. *Blood* 1998a 91, 4457-4463.

Chesi M; Nardini E; Brents LA; Schrock E; Ried T; Kuehl WE; Bergsagel PL. Frequent translocation t(4;14)(p16.3;q32.3) in multiple myeloma is associated with increased expression and activating mutations of fibroblast growth factor receptor 3. *Nat Genet* 1997 16, 260-264.

Chesi M; Nardini E; Lim RSC; Smith KD; Kuehl WM; Bergsagel PL. The t(4;14) translocation in myeloma dysregulates both FGFR3 and a novel gene, MMSET, resulting in IgH/MMSET hybrid transcripts. *Blood* 1998b 92, 3025-3034.

Chiarle R; Budel LM; Skolnik J; Frizzera G; Chilosi M; Corato A; Pizzolo G; Magidson J; Montagnoli A; Pagano M; Maes B; De Wolf-Peeters C; Inghirami G. Increased proteasome degradation of cyclin-dependent kinase inhibitor p27 is associated with a decreased overall survival in mantle cell lymphoma. *Blood* 2000 95, 619-626.

Chilosi M; Doglioni C; Yan Z; Lestani M; Menestrina F; Sorio C; Benedetti A; Vinante F; Pizzolo G; Inghirami G. Differential expression of cyclin-dependent kinase 6 in cortical thymocytes and T-cell lymphoblastic lymphoma/leukaemia. *Am J Pathol* 1998 152, 209-217.

Chim CS; Tam CY; Liang R; Kwong YL. Methylation of p15INK4B and p16INK4A genes in adult acute leukemia: lack of prognostic significance. *Cancer* 2001 91, 2222-2229.

Clappier E; Cuccuini W; Cayuela JM; Vecchione D; Baruchel A; Dombret H; Sigaux F; Soulier J. Cyclin D2 dysregulation by chromosomal translocations to TCR loci in T-cell acute lymphoblastic leukemias. *Leukemia* 2006 20, 82-86.

Corcoran MM; Mould SJ; Orchard JA; Ibbotson RE; Chapman RM; Boright AP; Platt C; Tsui L-C; Scherer SW; Oscier DG. Dysregulation of cyclin dependent kinase 6 expression in splenic marginal zone lymphoma through chromosome 7q translocations. *Oncogene* 1999 18, 6271-6277.

de Boer CJ; van Krieken JHM; Kluin-Nelemans HC; Kluin PM; Schuuring E. Cyclin D1 messenger RNA overexpression as a marker for mantle cell lymphoma. *Oncogene* 1995 10, 1833–1840.

Decker T; Schneller F; Hipp S; Miething C; Jahn T; Duyster J; Peschel C. Cell cycle progression of chronic lymphocytic leukemia cells is controlled by cyclin D2, cyclin D3, cyclin-dependent kinase (cdk) 4 and the cdk inhibitor p27kip1. *Leukemia* 2002 16, 327-334.

Delmer A; Ajchenbaum-Cymbalista F; Tang R; Ramond S; Faussat A-M; Marie J-P; Zittoun R. Overexpression of cyclin D2 in chronic B-cell malignancies. *Blood* 1995 85, 2870-2876.

Dowdy SF; Hinds PW; Louie K; Reed SI; Arnold A; Weinberg RA. Physical interaction of the retinoblastoma protein with human cyclins. *Cell* 1993 73, 499-511.

Drexler HG. Review of alterations of the cyclin-dependent kinase inhibitor INK4 family genes p15INK4B, p16INK4A, p18INK4C and p19INK4D in human leukemia-lymphoma cells. *Leukemia* 1998 12, 845-859.

Dreyling MH; Bullinger L; Ott G; Stilgenbauer S; Muller-Hermelink HK; Bentz M; Hiddemann W; Doöhner H. Alterations of the cyclin D1/p16INK4A-pRB pathway in mantle cell lymphoma. *Cancer Res* 1997 57, 4608–4614.

Dring AM; Davies FE; Fenton JAL; Roddam PL; Scott K; Gonzalez D; Rollinson S; Rawstron AC; Rees-Unwin KS; Li C; Munshi NC; Anderson KC; Morgan GJ. A global expression-based analysis of the consequences of the t(4;14) translocation in myeloma. *Clin Cancer Res* 2004 10, 5692–5701.

Ely S; Di Liberto M; Niesvizky R; Baughn LB; Cho HJ; Hatada EN; Knowles DM; Lane J; Chen-Kiang S. Mutually exclusive cyclin-dependent kinase 4/cyclin D1 and cyclin-dependent kinase 6/cyclin D2 pairing inactivates retinoblastoma protein and promotes cell cycle dysregulation in multiple myeloma. *Cancer Res* 2005 65, 11345-11353.

Ewen ME; Sluss HK; Scherr CJ; Matsushime H; Kato J; Livingston DM. Functional interactions of the retinoblastoma protein with mammalian D-type cyclins. *Cell* 1993 73, 487-497.

Fero ML; Randel E; Gurley KE; Roberts JM; Kemp CJ. The murine gene p27kip1 is haplo-insufficient for tumour suppression. *Nature* 1998 396, 177-180.

Fonseca R; Witzig TE; Gertz MA; Kyle RA; Hoyer JD; Jalal SM; Greipp PR. Multiple myeloma and the translocation t(11;14)(q13;q32): a report on 13 cases. *Br J Haematol* 1998 101, 296–301.

Fu K; Weisenburger DD; Greiner TC; Dave S; Wright G; Rosenwald A; Chiorazzi M; Iqbal J; Gesk S; Siebert R; De Jong D; Jaffe ES; Wilson WH; Delabie J; Ott G; Dave BJ; Sanger WG; Smith LM; Rimsza L; Braziel RM; Müller-Hermelink HK; Campo E; Gascoyne RD; Staudt LM; Chan WC and for the Lymphoma/Leukemia Molecular Profiling Project. Cyclin D1-negative mantle cell lymphoma: a clinicopathologic study based on gene expression profiling. *Blood* 2005 106, 4315-4321.

Gesk S; Klapper W; Martín-Subero JI; Nagel I; Harder L; Fu K; Bernd HW; Weisenburger DD; Parwaresch R; Siebert R. A chromosomal translocation in cyclin D1-negative/cyclin D2-positive mantle cell lymphoma fuses the CCND2 gene to the IGK locus. *Blood* 2006 108, 1109-1110.

Gu Y; Nakamura T; Alder H; Prasad R; Canaani O; Cimino G; Croce CM; Canaani E. The t(4;11) chromosome translocation of human acute leukemias fuses the ALL-1 gene, related to Drosophila trithorax, to the AF-4 gene. *Cell* 1992 71, 701–708.

Guillerm G; Gyan E; Wolowiec D; Facon T; Avet-Loiseau H; Kuliczkowski K; Bauters F; Fenaux P; Quesnel B. p16(INK4a) and p15(INK4b) gene methylations in plasma cells from monoclonal gammopathy of undetermined significance. *Blood* 2001 98, 244-246.

Hanamura I; Iida S; Akano Y; Hayami Y; Kato M; Miura K; Harada S; Banno S; Wakita A; Kiyoi H; Naoe T; Shimizu S; Sonta SI; Nitta M; Taniwaki M; Ueda R. Ectopic expression of MAFB gene in human myeloma cells carrying t(14;20)(q32;q11) chromosomal translocations. *Jpn J Cancer Res* 2001 92, 638-644.

Hannemann J; Velds A; Halfwerk JBG; Kreike B; Peterse JL; van de Vijver MJ. Classification of ductal carcinoma *in situ* by gene expression profiling. *Breast Cancer Res* 2006 8, R61.

Hans CP; Weisenburger DD; Greiner TC; Gascoyne RD; Delabie J; Ott G; Müller-Hermelink HK; Campo E; Braziel RM; Jaffe ES; Pan Z; Farinha P; Smith LM; Falini B; Banham AH; Rosenwald A; Staudt LM; Connors JM; Armitage JO; Chan WC. Confirmation of the molecular classification of diffuse large B-cell lymphoma by immunohistochemistry using a tissue microarray. *Blood* 2004 103, 275-282.

Hayette S; Gadoux M; Martel S; Bertrand S; Tigaud I; Magaud J-P; Rimokh R. FLRG (follistatin-related gene), a new target of chromosomal rearrangement in malignant blood disorders. *Oncogene* 1998 16, 2949-2954.

Hayette S; Tigaud I; Callet-Bauchu E; Ffrench M; Gazzo S; Wahbi K; Callanan M; Felman P; Dumontet C; Magaud J-P; Rimokh R. In B-cell chronic lymphocytic leukemias, 7q21 translocations lead to overexpression of the CDK6 gene. *Blood* 2003 102, 1549-1550.

Hebert J; Cayuela JM; Berkeley J; Sigaux F. Candidate tumor-suppressor genes MTS1 (p16INK4A) and MTS2 (p15INK4B) display frequent homozygous deletions in primary cells from T- but not from B-cell lineage acute lymphoblastic leukemias. *Blood* 1994 84, 4038-4044.

Hiromura K; Pippin JW; Fero ML; Roberts JM; Shankland SJ. Modulation of apoptosis by the cyclin-dependent kinase inhibitor p27(Kip1). *J Clin Invest* 1999 103, 597–604.

Hurt EM; Wiestner A; Rosenwald A; Shaffer AL; Campo E; Grogan T; Bergsagel PL; Kuehl WM; Staudt LM. Overexpression of c-maf is a frequent oncogenic event in multiple myeloma that promotes proliferation and pathological interactions with bone marrow stroma. *Cancer Cell* 2004 5, 191-199.

Jares P; Campo E; Pinyol M; Bosch F; Miquel R; Fernandez PL; Sanchez-Beato M; Soler F; Perez-Losada A; Nayach I; Mallofré C; Piris MA; Montserrat E; Cardesa A. Expression of retinoblastoma gene product (pRb) in mantle cell lymphomas. Correlation with cyclin D1 (PRAD1/CCND1) mRNA levels and proliferative activity. *Am J Pathol* 1996 148, 1591-1600.

Kamb A; Gruis NA; Weaver-Feldhaus J; Liu Q; Harshman K; Tavtigian SV; Stockert E; Day RS 3rd; Johnson BE; Skolnick MH. A cell cycle regulator potentially involved in genesis of many tumor types. *Science* 1994 264, 436-440.

Karrman K; Andersson A; Björgvinsdóttir H; Strömbeck B; Lassen C; Olofsson T; Nguyen-Khac F; Berger R; Bernard O; Fioretos T; Johansson B. Deregulation of cyclin D2 by juxtaposition with T-cell receptor alpha/delta locus in t(12;14)(p13;q11)-positive childhood T-cell acute lymphoblastic leukemia. *Eur J Haematol* 2006 77, 27-34.

Kato J; Matsuoka M; Polyak K; Massague J; Sherr CJ. Cyclic AMP induced G1 phase arrest mediated by an inhibitor (p27KIP1) of cyclin-dependent kinase 4 activation. *Cell* 1994 79, 487–496.

Knudsen KE. The cyclin D1b splice variant: an old oncogene learns new tricks. *Cell Div* 2006 1, 15.

Komuro H; Valentine MB; Rubnitz JE; Saito M; Raimondi SC; Carrol AJ; Yi T; Sherr CJ; Look AT. p27kip1 deletions in childhood acute lymphoblastic leukemia. *Neoplasia* 1999 1, 253-261.

Kozar K; Ciemerych MA; Rebel VI; Shigematsu H; Zagozdzon A; Sicinska E; Geng Y; Yu Q; Bhattacharya S; Bronson RT; Akashi K; Sicinski P. Mouse development and cell proliferation in the absence of D-cyclins. *Cell* 2004 118, 477–491.

Krämer A; Schultheis B; Bergmann J; Willer A; Hegenbart U; Ho AD; Goldschmidt H; Hehlmann R Alterations of the cyclin D1/pRb/p16INK4A pathway in multiple myeloma. *Leukemia* 2002 16, 1844–1851.

Krieger S; Gauduchon J; Roussel M; Troussard X; Sola B. Relevance of cyclin d1b expression and CCND1 polymorphism in the pathogenesis of multiple myeloma and mantle cell lymphoma. *BMC Cancer* 2006 6, 238.

Kulkarni MS; Daggett JL; Bender TP; Kuehl WM; Bergsagel PL; Williams ME. Frequent inactivation of the cyclin dependent kinase inhibitor p18 by homozygous deletion in multiple myeloma cell lines: ectopic p18 expression inhibits growth and induces apoptosis. *Leukemia* 2002 16, 127-134.

Lam EW; Glassford J; Banerji L; Thomas NS; Sicinski P; Klaus GG. Cyclin D3 compensates for loss of cyclin D2 in mouse B-lymphocytes activated via the antigen receptor and CD40. *J Biol Chem* 2000 275, 3479-3484.

Lien HC; Lin CW; Huang PH; Chang ML; Hsu SM. Expression of cyclin-dependent kinase 6 (cdk6) and frequent loss of CD44 in nasal-nasopharyngeal NK/T-cell lymphomas: comparison with CD56-negative peripheral T-cell lymphomas. *Lab Invest* 2000 80, 893–900.

Liu P; Tarlé SA; Hajra A; Claxton DF; Marlton P; Freedman M; Siciliano MJ; Collins FS. Fusion between transcription factor CBFβ/PEBP2β and a myosin heavy chain in acute myeloid leukemia. *Science* 1993 261, 1041-1044.

Lossos IS; Czerwinski DK; Alizadeh AA; Wechser MA; Tibshirani R; Botstein D; Levy R. Prediction of survival in diffuse large-B-cell lymphoma based on the expression of six genes. *N Engl J Med* 2004 350, 1828-1837.

Lu F; Gladden AB; Diehl JA. An alternative spliced cyclin D1 isoform, cyclin D1b, is a nuclear oncogene. *Cancer Res* 2003 63, 7056-7061.

Lukas J; Aagaard L; Strauss M; Bartek J. Oncogenic aberrations of p16INK4A/CDKN2 and cyclin Dl cooperate to deregulate G_1 control. *Cancer Res* 1995 55, 4818-4823.

Lukas J; Pagano M; Staskova Z; Draetta G; Bartek J. Cyclin D1 protein oscillates and is essential for cell cycle progression in human tumour cell lines. *Oncogene* 1994 9, 707-718.

Markus J; Garin MT; Bies J; Galili N; Raza A; Thirman MJ; Le Beau MM; Rowley JD; Liu PP; Wolff L. Methylation-independent silencing of the tumor suppressor INK4b (p15) by CBFbeta-SMMHC in acute myelogenous leukemia with inv(16). *Cancer Res* 2007 67, 992–1000.

Marzec M; Kasprzycka M; Lai R; Gladden AB; Wlodarski P; Tomczak E; Nowell P; DePrimo SE; Sadis S; Eck S; Schuster SJ; Diehl JA; Wasik MA. Mantle cell lymphoma cells express predominantly cyclin D1a isoform and are highly sensitive to selective inhibition of CDK4 kinase activity. *Blood* 2006 108, 1744-1750.

Müller H; Lukas J; Schneider A; Warthoe P; Bartek J; Eilers M; Strauss M. Cyclin D1 expression is regulated by the retinoblastoma protein. *Proc Natl Acad Sci USA* 1994 91, 2945-2949.

Nagel S; Leich E; Quentmeier H; Meyer C; Kaufmann M; Drexler HG; Zetti A; Rosenwald A; MacLeod RA. Amplification at 7q22 targets cyclin-dependent kinase 6 in T-cell lymphoma. *Leukemia* 2008 22, 387-392.

Nevins JR. E2F: a link between the Rb tumor suppressor protein and viral oncoproteins. *Science* 1992 258, 424-429.

Ng MH; Chung YF; Lo KW; Wickham NW; Lee JC; Huang DP. Frequent hypermethylation of p16INK4A and p15INK4B genes in multiple myeloma. *Blood* 1997 89, 2500–2506.

Nobori T; Miura K; Wu DJ; Lois A; Takabayashi K; Carson DA. Deletions of the cyclin-dependent kinase-4 inhibitor gene in multiple human cancers. *Nature* 1994 368, 753-756.

Nourse J; Firpo E; Flanagan WM; Coats S; Polyak K; Lee MH; Massague J; Crabtree GR; Roberts JM. Interleukin-2-mediated elimination of the p27Kip1 cyclin-dependent kinase inhibitor prevented by rapamycin. *Nature* 1994 372, 570–573.

Ohi MD; Link AJ; Ren L; Jennings JL; McDonald WH; Gould KL. Proteomics analysis reveals stable multiprotein complexes in both fission and budding yeasts containing Myb-related Cdc5p/Cef1p, novel pre-mRNA splicing factors, and snRNAs. *Mol Cell Biol* 2002 22, 2011–2024.

Okuda T, Shurtleff SA, Valentine MB; Raimondi SC, Head DR, Behm F, Curcio-Brint AM, Liu Q, Pui CH, Sherr CJ, Beach D, Look AT, Downing JR. Frequent deletion of p16INK4A/MTS1 and p15INK4B/MTS2 in pediatric acute lymphoblastic leukaemia. Blood *1995* 85, 2321-2330.

Parada Y; Banerji L; Glassford J; Lea NC; Collado M; Rivas C; Lewis JL; Gordon MY; Thomas NS; Lam EW. BCR-ABL and interleukin 3 promote haematopoietic cell proliferation and survival through modulation of cyclin D2 and p27Kip1 expression. *J Biol Chem* 2001 276, 23572–23580.

Park I; Qian D; Kiel M; Becker M; Pihalja M; Weissman I; Morrison S; Clarke M. Bmi-1 is required for maintenance of adult self-renewing haematopoietic stem cells. *Nature* 2003 423, 302–305.

Pietenpol JA; Bohlander SK; Sato Y; Papasopoulos N; Liu B; Friedman C; Trask BJ; Roberts JM; Kinzler KW; Rowley JD; Vogelstein B: Assignment of the human p27KIP-1 gene to 12p13 and its analysis in leukemias. *Cancer Res* 1995 55; 1206–1210.

Pinyol M; Bea S; Plà L; Ribrag V; Bosq J; Rosenwald A; Campo E; Jares P. Inactivation of RB1 in mantle-cell lymphoma detected by nonsense-mediated mRNA decay pathway inhibition and microarray analysis. *Blood* 2007 109; 5422-5429.

Pinyol M; Hernandez L; Cazorla M; Balbin M; Jares P; Fernandez PL; Montserrat E; Cardesa A; Lopez-Otin C; Campo E. Deletions and loss of expression of p16INK4a and P21Waf1 genes are associated with aggressive variants of mantle cell lymphomas. *Blood* 1997 89, 272–280.

Quintanilla-Martinez L; Thieblemont C; Fend F; Kumar S; Pinyol M; Campo E; Jaffe ES; Raffeld M. Mantle cell lymphomas lack expression of p27kip1, a cyclin-dependent kinase inhibitor. *Am J Pathol* 1998 153, 175-182.

Raffini LJ; Slater DJ; Rappaport EF; Lo Nigro L; Cheung NV; Biegel JA; Nowell PC; Lange BJ; Felix CA. Panhandle and reverse-panhandle PCR enable cloning of der(11) and der(other) genomic breakpoint junctions of MLL translocations and identify complex translocation of MLL, AF-4, and CDK6. *Proc Natl Acad Sci USA* 2002 99, 4568–4573.

Richelda R; Ronchetti D; Baldini L; Cro L; Viggiano L; Marzella R; Rocchi M; Otsuki T; Lombardi L; Maiolo AT; Neri A. A novel chromosomal translocation t(4;14)(p16.3;q32) in multiple myeloma involves the fibroblast growth-factor receptor 3 gene. *Blood* 1997 90; 4062-4070.

Roman-Gomez J; Jimenez-Velasco A; Agirre X; Prosper F; Heiniger A; Torres A. Lack of CpG island methylator phenotype defines a clinical subtype of T-cell acute lymphoblastic leukemia associated with good prognosis. *J Clin Oncol* 2005 23, 7043-7049.

Rosenwald A; Wright G; Chan WC; Connors JM; Campo E; Fisher RI; Gascoyne RD; Muller-Hermelink HK; Smeland EB; Giltnane JM; Hurt EM; Zhao H; Averett L; Yang L; Wilson WH; Jaffe ES; Simon R; Klausner RD; Powell J; Duffey PL; Longo DL; Greiner TC; Weisenburger DD; Sanger WG; Dave BJ; Lynch JC; Vose J; Armitage JO; Montserrat E; López-Guillermo A; Grogan TM; Miller TP; LeBlanc M; Ott G; Kvaloy S; Delabie J; Holte H; Krajci P; Stokke T; Staudt LM; Lymphoma/Leukemia Molecular Profiling Project. The use of molecular profiling to predict survival after chemotherapy for diffuse large B-cell lymphoma. *N Engl J Med* 2002 346, 1937-1947.

Rosenwald A; Wright G; Wiestner A; Chan WC; Connors JM; Campo E; Gascoyne RD; Grogan TM; Muller-Hermelink HK; Smeland EB; Chiorazzi M; Giltnane JM; Hurt EM; Zhao H; Averett L; Henrickson S; Yang L; Powell J; Wilson WH; Jaffe ES; Simon R; Klausner RD; Montserrat E; Bosch F; Greiner TC; Weisenburger DD; Sanger WG; Dave

BJ; Lynch JC; Vose J; Armitage JO; Fisher RI; Miller TP; LeBlanc M; Ott G; Kvaloy S; Holte H; Delabie J; Staudt LM. The proliferation gene signature is a quantitative integrator of oncogenic events that predicts survival in mantle cell lymphoma. *Cancer Cell* 2003 3, 185-197.

Rosu-Myles M; Wolff L. p15Ink4b: Dual function in myelopoiesis and inactivation in myeloid disease. *Blood Cells Mol Dis* 2007 Oct 26 (Epub ahead of print).

Roué G; Krieger S; Florent M; Roussel M; Duquesne F; Troussard X; Pichereau V; Sola B. Expression of the two alternative [a] and [b] transcripts of CCND1 gene in cyclin D1-expressing malignancies: relevance for the pathogenesis. *Leukemia* 2003 17, 652-655.

Rowley JD. Chromosome translocations: dangerous liaisons revisited. *Nat Rev Cancer* 2001 1, 245-250.

Sáez A; Sánchez E; Sánchez-Beato M; Cruz MA; Chacón I; Muñoz E; Camacho FI; Martínez-Montero JC; Mollejo M; García JF; Piris MA. p27KIP1 is abnormally expressed in diffuse large B-cell lymphomas and is associated with an adverse clinical outcome. *Br J Cancer* 1999 80, 1427-1434.

Sanchez-Beato M; Camacho FI; Martinez-Montero JC; Sáez AI; Villuendas R; Sánchez-Verde L; García JF; Piris A. Anomalous high p27kip1 expression in a subset of aggressive B-cell lymphomas is associated with cyclin D3 overexpression. p27kip1-cyclin D3 colocalization in tumor cells. *Blood* 1999 94, 765-772.

Sánchez-Beato M; Sáez AI; Martínez-Montero JC; Sol Mateo M; Sánchez-Verde L; Villuendas R; Troncone G; Piris MA. Cyclin-dependent kinase inhibitor p27kip1 in lymphoid tissue: p27kip1 expression is inversely proportional to the proliferative index. *Am J Pathol* 1997 151, 151–160.

Sanhes L; Tang R; Delmer A; DeCaprio JA; Ajchembaum-Cymbalista F. Fludarabine-induced apoptosis of B chronic lymphocytic leukemia cells includes early cleavage of p27kip1 by caspases. *Leukemia* 2003 17, 1104-1111.

Santra M; Zhan F; Tian E; Barlogie B; Shaughnessy J. A subset of multiple myeloma harboring the t(4;14)(p16;q32) translocation lack FGFR3 expression but maintain an IGH/MMSET fusion transcript. *Blood* 2002 101, 2374-2376.

Sawyer JR; Lukacs JL; Thomas EL; Swanson CM; Goosen LS; Sammartino G; Gilliland JC; Munshi NC; Tricot G; Shaughnessy JDJr; Barlogie B. Multicolour spectral karyotyping identifies new translocations and a recurring pathway for chromosome loss in multiple myeloma. *Br J Haematol* 2001 112, 167-174.

Shaughnessy JJr; Gabrea A; Qi Y; Brents L; Zhan F; Tian E; Sawyer J; Barlogie B; Bergsagel PL; Kuehl M. Cyclin D3 at 6p21 is dysregulated by recurrent chromosomal translocations to immunoglobulin loci in multiple myeloma. *Blood* 2001 98, 217-223.

Sherr CJ; Roberts JM. CDK inhibitors: positive and negative regulators of G1-phase progression. *Genes Dev* 1999 13, 1501-1512.

Sherr CJ; Roberts JM. Living with or without cyclins and cyclin-dependent kinases. *Genes Dev* 2004 18, 2699-2711.

Sicinska E; Aifantis I; Le Cam L; Swat W, Borowski C, Yu Q, Ferrando AA, Levin SD, Geng Y, von Boehmer H, Sicinski P. Requirement for cyclin D3 in lymphocyte development and T cell leukemias. *Cancer Cell* 2003 4, 451-461.

Solomon DA; Wang Y; Fox SR; Lambeck TC; Giesting S; Lan Z; Senderowicz AM; Conti CJ; Knudsen ES. Cyclin D1 splice variants. *J Biol Chem* 2003 278, 30339-30347.

Solvason N; Wu WW; Parry D; Mahony D; Lam EW; Glassford J; Klaus GG; Sicinski P; Weinberg R; Liu YJ; Howard M; Lees E. Cyclin D2 is essential for BCR-mediated proliferation and CD5 B cell development. *Int Immunol* 2000 12, 631–638.

Stranks G; Height SE; Mitchell P; Jadayel D; Yuille MA; De Lord C; Clutterbuck RD; Treleaven JG; Powles RL; Nacheva E; Oscier DG; Karpas A; Lenoir GM; Smith SD; Millar JL; Catovsky D; Dyer MJS. Deletions and rearrangements of CDKN2 in lymphoid malignancy. *Blood* 1995 85, 893-901.

Su XY; Busson M; Della Valle V; Ballerini P; Dastugue N; Talmant P; Ferrando AA; Baudry-Bluteau D; Romana S; Berger R; Bernard OA. Various types of rearrangements target TLX3 locus in T-cell acute lymphoblastic leukemia. *Genes Chromosomes Cancer* 2004 41, 243–249.

Susaki E; Nakayama K; Nakayama KI. Cyclin D2 translocates p27kip1 out of the nucleus and promotes its degradation at the G_0-G_1 transition. *Mol Cel Biol* 2007 27, 4626-4640.

Takeuchi S; Bartram CR; Seriu T; Millar CW; Tobler A; Janssen JW; Reiter A; Ludwig WD; Zimmermann M; Schwaller J; Lee E; Miyoshi I; Koeffler HP. Analysis of a family of cyclin-dependent kinase inhibitors: p15/MTS2/INK4B, p16/MTS1/INK4A, and p18 genes in acute lymphoblastic leukemia of childhood. *Blood* 1995 86, 755-760.

Tasaka T; Berenson J; Vescio R; Hirama T; Miller CW; Nagai M; Takahara J; Koeffler HP. Analysis of the p16INK4A, p15INK4B, and p18INK4C genes in multiple myeloma. *Br J Haematol* 1997 96, 98–102.

Teofili L ; Martini M; Luongo M; Diverio D; Capelli G; Breccia M; Lo Coco F; Leone G; Larocca LM. Hypermethylation of GpG islands in the promoter region of p15(INK4b) in acute promyelocytic leukemia represses p15(INK4b) expression and correlates with poor prognosis. *Leukemia* 2003 17, 919–924.

Tien HF ; Tang JH ; Tsay W; Liu MC; Lee FY; Wang CH; Chen YC; Shen MC. Methylation of the p15(INK4B) gene in myelodysplastic syndrome: it can be detected early at diagnosis or during disease progression and is highly associated with leukemic transformation. *Br J Haematol* 2001 112, 148–154.

Tsujimoto Y; Yunis J; Onarato-Showe L; Erikson J; Nowell PC; Croce CM. Molecular cloning of the chromosomal breakpoint of B-cell lymphomas and leukemias with the t(11;14) chromosome translocation. *Science* 1984 224, 1403–1406.

Uchida T; Kinoshita T; Ohno T; Ohashi H; Nagai H; Saito H. Hypermethylation of the p16INK4A gene promoter during the progression of plasma cell dyscrasia. *Leukemia* 2001 15, 157–165.

Vieira L; Marques B; Cavaleiro C; Ambrósio AP; Jorge M; Neto A; Costa JM; Júnior EC; Boavida MG. Molecular cytogenetic characterization of rearrangements involving 12p in leukemia. *Cancer Genet Cytogenet* 2005 157, 134-139.

Vieira L; Sousa AC; Matos P; Marques B; Alaiz H; Ribeiro MJ; Braga P; da Silva MG; Jordan P. Three-way translocation involves *MLL*, *MLLT3* and a novel cell cycle control gene, FLJ10374, in the pathogenesis of acute myeloid leukemia with t(9;11;19)(p22;q23;p13.3). *Genes Chromosomes Cancer* 2006 45, 455-469.

Vrhovac R; Delmer A; Tang R; Marie J-P; Zittoun R; Ajchenbaum-Cymbalista F. Prognostic significance of the cell cycle inhibitor p27kip1 in chronic B-cell lymphocytic leukemia. *Blood* 1998 91, 4694-4700.

Wagner EF; Hleb M; Hanna N; Sharma S. A pivotal role of cyclin D3 and cyclin-dependent kinase inhibitor p27kip1 in the regulation of IL-2-, IL-4-, or IL-10-mediated human B cell proliferation. *J Immunol* 1998 161, 1123-31.

Weinberg RB. The retinoblastoma protein and cell cycle control. *Cell* 1995 81, 323-330.

Wiestner A; Tehrani M; Chiorazzi M; Wright G; Gibellini F; Nakayama K; Liu H; Rosenwald A; Muller-Hermelink HK; Ott G; Chan WC; Greiner TC; Weisenburger DD; Vose J; Armitage JO; Gascoyne RD; Connors JM; Campo E; Montserrat E; Bosch F; Smeland EB; Kvaloy S; Holte H; Delabie J; Fisher RI; Grogan TM; Miller TP; Wilson WH; Jaffe ES; Staudt LM and for the Lymphoma/Leukemia Molecular Profiling Project. Point mutations and genomic deletions in CCND1 create stable truncated cyclin D1 mRNAs that are associated with increased proliferation rate and shorter survival. *Blood* 2007 109, 4599-4606.

Wolff L; Garin MT; Koller R; Bies J; Liao W; Malumbres M; Tessarollo L; Powell D; Perella C. Hypermethylation of the Ink4b locus in murine myeloid leukemia and increased susceptibility to leukemia in p15(Ink4b)-deficient mice. *Oncogene* 2003 22, 9265–9274.

Wolowiec D; Ciszac L; Kosmaczewska A; Bocko D; Teodorowska R; Frydecka I; Kuliczkowski K. Cell cycle regulatory proteins and apoptosis in B-cell chronic lymphocytic leukemia. *Haematologica* 2001 86, 1296-1304.

Wong IH; Ng MH; Huang DP; Lee JC. Aberrant p15 promoter methylation in adult and childhood acute leukemias of nearly all morphologic subtypes: potential prognostic implications. *Blood* 2000 95, 1942–1949.

Xia Z-B; Popovic R; Chen J; Theisler C; Stuart T; Santillan DA; Erfurth F; Diaz MO; Zeleznik-Le NJ. The MLL fusion gene, MLL-AF4, regulates cyclin-dependent kinase inhibitor CDKN1B (p27kip1) expression. *Proc Natl Acad Sci USA* 2005 102, 14028-14033.

In: Cell Cycle Control: New Research
Editors: N. H. Leroy, N. T. Fournier
ISBN: 978-1-60456-796-0

Chapter XIII

Cell Cycle Control in Drosophila

Zohra Rahmani*
CNRS UMR 7592, Institut Jacques Monod, 75251 Paris cedex 05, France.

Abstract

Cell cycle is an ordered set of processes by which one cell grows and divides into two daughter cells. Drosophila is an attractive organism that has been used as a model for the study of the mitotic cell cycle. The fruit fly has not only been extensively used for genetics studies but also its developmental biology is highly suitable for studies of the mitotic cycle. The mitotic division in the early embryos consists of rapid successions of Mitosis (M) and DNA synthesis (S) phases with no discernible G1 or G2 (Gap) phases as found at later stages of development. In the subsequent development of the organism, two different types of tissue arise, having either proliferating diploid cells or cells that do not divide but become highly polyploid. This review presents the various modes of cell cycling that occur during Drosophila development and discusses their regulation in light of the latest findings.

Overview of Drosophila Development

Five different cell cycle modes occur during the development of Drosophila: syncytial cycles, post blastoderm divisions, canonical mitotic cycles, endoreplication, and meiosis (Figure 1) (reviewed in Foe et al., 1993). Drosophila embryogenesis starts with 13 rapid synchronous nuclear divisions (less than 10 min) without cytokinesis in a common cytoplasm, called a syncytium. The first eight of these nuclear divisions occur within the embryo. During nuclear cycles 9 and 10, the majority of the nuclei migrate towards the cortex where they form a monolayer called the syncytial blastoderm. The duration of interphase is

* Correspondence concerning this article should be addressed to: Zohra Rahmani CNRS UMR 7592, Institut Jacques Monod, 2 Place Jussieu, Tour 43, 75251 Paris cedex 05, France. e-mail : rahmani.zohra@ijm.jussieu.fr.

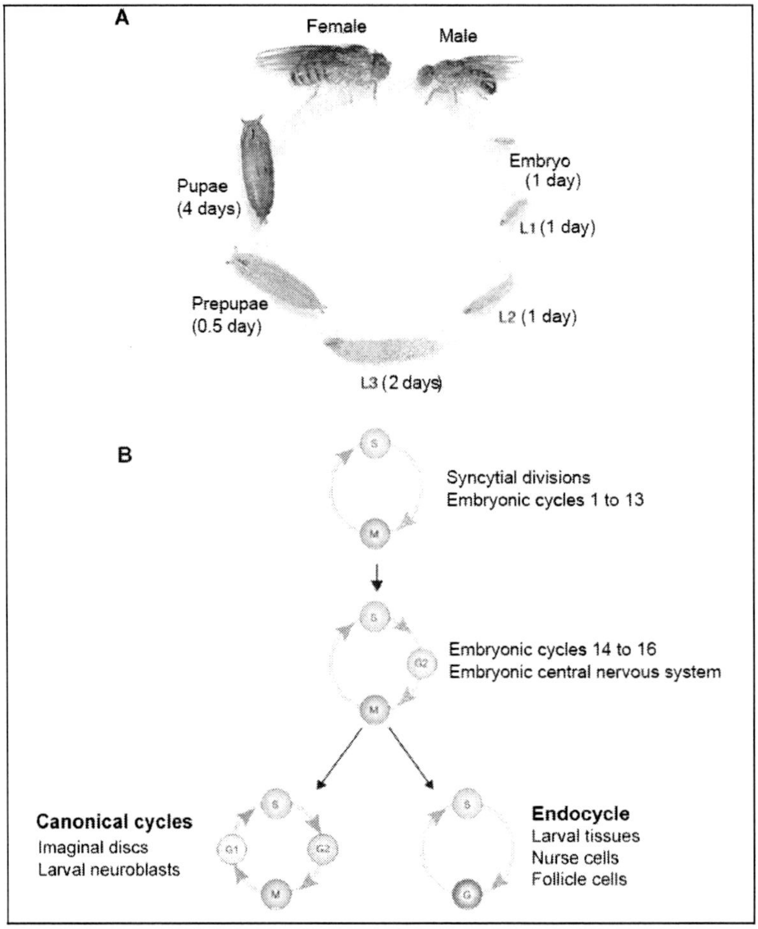

Figure 1. Different modes of cell cycle during development of Drosophila. (A) Timetable of Drosophila development from egg to adult. The period of embryonic development starts at fertilization and takes a day at the end of which the embryo hatches out of the egg shell to become a larva. The larva then goes through three stages L1, L2, L3 (or instars), separated by molts in which it sheds its old coat of cuticle and lays down a larger one. The end of the third instar is followed by pupation in which a radical remodeling of the body takes place. This process is called metamorphosis. An adult fly emerges about nine days after fertilization. (B) In the early embryo, the first 13 cycles consist of alternating phases of DNA synthesis (S) and mitosis (M) without intervening gap phases (G) in a syncytium. After cellularization, syncytial cycles are followed by G2-containing cycles. At cycle 17, cells from the central nervous system continue to divide with the same mode of cell cycle whereas most other cells from larval tissues, nurse cells and follicle cells enter endoreduplication cycles in which S phases alternate with G phases without mitosis. Cells of the larval central nervous system and imaginal lineages continue to divide throughout larval and pupal development with a canonical mitotic cycle.

lengthened from 6 to 14 min during the final syncytial divisions 11 to 13 (Sibon et al., 1997). Cellularization of the nuclei located on the embryo surface begins at the interphase of cycle 14. The syncytial cycle consists of alternating DNA synthesis (S) and mitosis (M) phases with no intervening gap (G) phases. This cycle is under the control of the maternal genes. Unlike syncytial cycles, blastoderm cellularization requires zygotic transcription and marks the midblastula transition. During divisions 14 to 16, cellular blastoderm embryos divide using S-G2-M cycles. At the end of mitosis 16, a G1 phase occurs for the first time, in which the epidermal cells arrest and exit the cell cycle. After that, the majority of cells exit the cell cycle, except cells in the central nervous system (CNS) and in the imaginal discs. During the larval stages, CNS cells and the imaginal discs undergo canonical mitotic cycles whereas the majority of the cells undergo endocycles without mitosis. During the metamorphosis, the polyploid larval cells degenerate and are replaced by a set of diploid imaginal cells. In the adult fly, both male and female germ line stem cells undergo meiosis in the apical region of the gonad.

Cell Cycles in Early Embryogenesis

The Drosophila embryo is a syncytium for the first two hours of its development. During this time, 13 rapid rounds of nuclear division occur. The first nine rounds of mitosis occur within the embryo and at telophase of nuclear cycle 9 the majority of the nuclei migrate to the cortex. Once at the surface, the nuclei undergo a further four cycles before cellularization occurs at interphase of cycle 14 (Zalokar and Erk, 1976; Foe and Alberts, 1983).

During this embryonic stage, the first 13 cell cycles in the Drosophila embryo are S-M cycles driven by maternal components that are essentially independent of zygotic transcription. After the 13th mitosis following cellularization of the embryo, the embryo arrests in an extended G2 phase due to the degradation of maternal stores of a protein phosphatase encoded by the Cdc25 homolog string (stg) (Edgar and Datar, 1996), which is required to activate mitotic Cyclin/Cdk1 complexes (Edgar et al., 1994b). Cdc25 is a dual specificity phosphatase that removes inhibitory phosphates from the adenosine triphosphate (ATP)-binding site of Cdk1. This activates Cyclin A-Cdk1 and Cyclin B-Cdk1 kinases, triggering mitosis. Initially, Cdc25 is expressed in large amounts from abundant maternal mRNA stored in the Drosophila egg, as are all other proteins required for cell cycle progression. These maternal regulators set the extreme pace of the initial embryonic cycles, which have a cycle time of less than 10 minutes.

Regulators of the S-M Cycles

Localized cyclin degradation has been shown to be required for mitotic exit during syncytial divisions (Su et al., 1998), suggesting that the S/M transition may be driven by increased mitotic cyclin levels as well. Alterations in maternal gene doses of mitotic cyclins affect cell cycle length (cyclin A) and microtubule dynamics (cyclin B) during syncytial divisions (Stiffler et al., 1999). However, neither cyclin B nor cyclin B3 is required for mitosis in Drosophila, but doubly heterozygous females produce embryo with abnormal syncytial divisions (Jacobs et al., 1998).

pan gu (png), plutonium (plu), and giant nuclei (gnu) have been shown to be required for the exit of the S phase and the entry into M phase. In *png*, *plu*, or *gnu* mutants, DNA replication occurs without mitosis and results in giant polyploid cells (Freeman et al., 1986, 1987; Shamanski and Orr-Weaver, 1991). Png, a serine-threonine kinase, forms a complex (the PNG kinase complex) with plu and gnu to promote mitotic entry. The downstream cell cycle target of the PNG kinase complex is Cyclin B. Mutations in any of the three genes encoding the PNG complex proteins result in decreased Cyclin B protein levels but does not affect the transcript levels (Fenger et al., 2000; Lee et al., 2003).

Regulation of G2/M Transition

During interphase 14, the cell cycles control becomes zygotic and regulated at G2/M transitions (Edgar and O'Farrell, 1990). These cycles have durations of 1-3 hours, and occur in invariant spatiotemporal proteins that are controlled by patterned transcription of the string gene (Edgar and O'Farrell, 1989; Edgar et al., 1994a). The decision to enter mitosis at the G2/M transition is largely governed by Cdk1 phosphorylation and mitotic cyclin levels. Evolutionarily conserved enzymes with opposing actions control Cdk1's phosphorylation state. Wee1 kinases negatively regulate Cdk1 by addition of an inhibitory phosphate that is removed by Cdc25 phosphatases, String (stg) and Twine in Drosophila, thereby promoting mitotic entry. A G2 phase is first utilized during post blastoderm divisions to permit gene expression and to coordinate division with morphogenetic movements during gastrulation. Maternally regulated S-M cycles of syncytial embryos switch to zygotically regulated S-G2-M post blastoderm cycles at the midblastula transition (MBT). Incomplete DNA replication is thought to trigger a checkpoint that delays mitotic entry via string downregulation, thereby setting the stage for introduction of a G2 phase (Sibon et al., 1997). Drosophila checkpoint genes that perform non essential functions in response to DNA damage or incomplete replication (Brodsky et al., 2000, Garner et al., 2001, Hari et al., 1995) have been shown to play essential developmental roles (Fogarty et al., 1997; Sibon et al., 1997, 1999). Embryos lacking *mei-41* or *grapes* (*grp*) undergo unusually short syncytial divisions, continue cleavage divisions after cycle 13, and fail to cellularize (Sibon et al., 1997, 1999). Drosophila *wee1* (*Dwee1*) mutants also fail to complete the nuclear division cycles (Price et al., 2000). A DNA damage checkpoint assay in Drosophila eye discs led to the identification of mus304, which encodes a coiled-coil cytoplasmic protein, as a new checkpoint component (Brodsky et al., 2000). In addition to a defective checkpoint, *mus304* mutants exhibit several developmental defects. The *mei-41* and *grp* phenotypes suggest a model in which a DNA replication checkpoint plays a key role in slowing the cell cycle and establishing the MBT (Sibon et al., 1997). During late syncytial divisions, an unidentified maternal cytoplasmic replication factor essential for nuclear division may be progressively depleted by exponentially increasing numbers of nuclei to slow down DNA replication, and the mei-41, grp checkpoint could then be triggered to ensure S-phase completion prior to mitosis. This is consistent with the observation that the nucleo-cytoplasmic ratio controls MBT timing (Edgar et al., 1986). Transcription of large genes required for the transition from maternal to zygotic cell cycle control may be possible by the lengthening of late syncytial interphases. Since inhibitory phosphorylation of Cdk1 is not detected during the late syncytial cycles (Edgar et al., 1994), mei-41 and grp may therefore affect cell cycle progression via other mechanisms. Indeed, studies of the early Drosophila embryonic syncytial cycles have revealed that these cycles seem to be sensitive to levels of Cyclin A. Changes in the levels of this mitotic cyclin alter the length of early syncytial cycles, but not of the later embryonic cell cycles (Stiffler et al., 1999). Interestingly, grp has been shown to promote Cyclin A degradation (Su et al., 1999), and the grp checkpoint has been shown to regulate chromosome condensation and the timing of nuclear envelope breakdown (Yu et al., 2000).

Degradation of maternal *string* mRNA by cycle 14 appears to be a prerequisite for the MBT; decreasing maternal string and twine advances the MBT, whereas increasing twine delays it (Edgar and Datar, 1996). string is required for the first zygotically controlled division, as *string* mutants arrest during interphase 14 (Edgar and O'Farrell, 1989). Ectopic expression of string in post blastoderm embryos drives G2-M in virtually all cells, confirming that string transcript levels are rate limiting for mitotic entry (Edgar and O'Farrell, 1990).

Regulation of G1/S Transition During late Embryogenesis

In the seventh hours of embryogenesis, cells start to exit the cell cycle and arrest in their first G1 phase (Edgar and O'Farrell, 1990). This is due to the transcriptional down-regulation of cyclin E (Knoblich et al., 1994), and induction of the cyclin E/Cdk2 inhibitor dacapo (de Nooij et al., 1996; Lane et al., 1996).

Cyclin E is the key regulator of the G1/S transition in Drosophila. cyclin E transcripts are constitutively present during syncytial and post blastoderm divisions (Knoblich et al., 1994). After mitosis 16, cyclin E transcripts are abruptly downregulated in the epidermis prior G1 arrest, but remain at high levels in mitotically dividing cells of the nervous system. Genetic studies indicate that the G1/S transition requires sufficient Cyclin E levels and that down regulation of cyclin E transcripts in embryos is necessary for developmental addition of a G1 phase. *Cyclin E* mutant embryos complete cycles 1-16, but DNA replication is blocked after cycle 16 coincident with addition of the first G1 phase and downregulation of maternal cyclin E pools (Knoblich et al., 1994). Cyclin E overexpression in the epidermis after mitosis 16 induces an ectopic S phase (Knoblich et al., 1994). Cyclin E activity is also negatively regulated by Dacapo (Dap), a homolog of a family of cdk inhibitors identified in mammalian cells. It is not clear how transcription of dacapo is regulated, although its tissue and developmental-stage-specific expression indicates the possible involvement of embryonic-patterning genes (Meyer et al., 2002).

The E2F transcription factor is a key regulator of G1/S in dividing mammalian cells. In Drosophila, there are two E2F subunits (E2F1 and E2F2), a single DP subunit, and two RB-related proteins (RBF1 and RBF2). E2F forms a heterodimer with DP to activate transcription of cyclin E and genes encoding replication proteins. When RBF forms a complex with the E2F/DP heterodimers, the transcription is repressed. Sequential phosphorylation of RBF by Cyclin D/Cdk4 and then by Cyclin E/Cdk2 leads to its dissociation from the E2F/DP heterodimer (reviewed in Coqueret, 2002). Duronio et al. showed that Dacapo is not required to shut off the transcription factor E2f1-mediated transcription, since E2F1 targets were still downregulated appropriately in *dacapo* mutants (Duronio et al., 1995, 1998). However, E2F1 protein abundance was found to be cell cycle regulated during the early embryonic cycles, and E2F1 was degraded during S phase, beginning in cycle 15 (Duronio et al., 1995, 1998; Royzman et al., 1997). Inactivation of Drosophila E2F1 (dE2F1) prevents DNA synthesis during later stages of embryogenesis (Duronio et al., 1995). Thus the destruction of E2F1 may provide a mechanism of terminating E2f1-dependent transcription, independently of Rbf1 (Shibutani et al., 2007).

Regulation of Cell Division in the Developing Imaginal Discs

Within Drosophila larvae, cell division occurs in groups of adult precursor cells called imaginal cells. The imaginal cells of appendages, such as the wings and legs, are contained in epithelial sheets called imaginal discs. Imaginal cells are also found in clusters in the gut, salivary glands, and abdominal epidermis, which are used in remodelling these structures during metamorphosis. Imaginal disc cells are G1-arrested when the embryo hatches, and enter S phase only after sustaining a substantial amount of cell growth, which increases their mass as much as six-fold (Madhavan and Schneiderman, 1977). After an exponential growth, cells gradually decrease until they differentiate five days later (Fain and Stevens, 1982; Neufeld et al., 1998). Each parcel of imaginal cells (a disc) has 10 to 50 cells before they differentiate into an adult structure such as a wing, a leg, or an eye. Since very little cell death is observed during disc growth (James and Bryant, 1981), it appears that cell number in these adult structures is dictated primarily by cell division. Proliferation occurs throughout the discs, with a cell cycle that averages 8 hours and includes substantial G1 and G2 periods (Madhavan and Schneiderman, 1977; Graves and Schubiger, 1982; Fain and Stevens, 1982; Adler and MacQueen, 1984). Although there is almost no apparent pattern of cell division until the onset of cell differentiation, clonal analysis has revealed regional differences in cell proliferation rates (Gonzàlez-Gaitàn et al., 1994; Resino et al., 2002). Thus, differential patterning of cell proliferation (zonal growth) is a plausible determinant of wing morphogenesis.

In the wing imaginal disc, a zone of non-proliferating cells (ZNC) is composed of cells surrounding the anterior-posterior subdomain and dorsal and ventral domains in the future wing blade. Cells of the ZNC stop proliferating 30 hours before most other cells in the disc but re-enter the cell cycle for two or three divisions after pupariation (O' Brochta and Bryant, 1985; Hartenstein and Posakony, 1989). Wingless in conjunction with Notch induces arrest in both the G1 and the G2 phases of the cell cycle in separate subdomains of the ZNC (Johnston and Edgar, 1998). The G2 arrest results from inhibition of String by Achaete and Scute transcription factors, which are activated by Wingless signalling (Johnston and Edgar, 1998). Notch signalling promotes G1 arrest in central cells by blocking the ability of Wingless to promote G2 arrest; as a result, these cells proceed through mitosis and subsequently arrest in G1 (Johnston and Edgar, 1998).

During the late third instar larvae stage, terminal differentiation is initiated in the posterior cells of the eye disc and progresses anteriorly (reviewed in Wolff and Ready, 1993). The boundary of this progression is called the morphogenetic furrow (MF), where rows of cells visibly constrict as they begin to differentiate. In the region of the disc anterior to the MF, cells proliferate asynchronously to constitute the "First Mitotic Wave". Posterior to the furrow, some cells go through an additional division, the "Second Mitotic Wave", which produces the necessary number of cells to form a normal eye. Cells complete mitosis anterior to the furrow and arrest in G1 in the furrow as photoreceptor differentiation begins. Functional characterization of the Drosophila gene roughex has indicated that entry into G1 requires inactivation of Cyclin A as well as Cyclin E. In *roughex* mutants cells, instead of arresting in G1, progress into S phase and continue proliferating, apparently because Cyclin A-dependent Cdk activity is not suppressed sufficiently during G1 (Dong et al., 1997; Thomas et al., 1994, 1997). Several signal transduction pathways impinge on G1-S during eye differentiation by controlling Cyclin E levels. The EGF receptor pathway has been shown to be necessary for G1 arrest in the furrow (Baker and Yu, 2001). The TGF-β pathway is also involved since Decapentaplegic is required for Cyclin E inhibition in the furrow (Horsfield et al., 1998). The Hedgehog pathway promotes cyclin E transcription through the Cubitus interruptus transcription factor in an E2F-independent manner (Duman-Scheel et al., 2002).

Switching from Mitotic Cell Cycles to Endocycles

Many cells in invertebrates do not stop cycling when they become post-mitotic, but switch to an endoreplication cycle in which repeated S phases occur without intervening mitoses. In the endocycle, DNA replication is uncoupled from mitosis allowing cells to increase dramatically their DNA content above diploid values. Following the cell-proliferative phase of Drosophila embryogenesis, many tissues initiate endocycles that lack all visible aspects of mitosis (Smith and Orr-Weaver, 1991). These tissues which include the gut, epidermis, fat body, malipighian tubules, trachea, and salivary glands continue to endocycle during larval development long after they have fully differentiated. Some adult tissues, including ovarian follicle and nurse cells, and sensory neurons in the wing, also employ endocycles. Endocycle parallels larval growth, and experimental inhibition of endocycle progression using DNA replication inhibitors, or mutations in genes essential for DNA replication, shows that increases in N values (N is the normal haploid chromosome number) are required for both cell and organism growth. Conversely, conditions that arrest growth invariably block endocycle progression, suggesting a tight regulatory linkage between these two processes (Galloni and Edgar, 1999; Zhang et al., 2000). Final DNA levels in the larval cells appear to be developmentally programmed ranging from 16N to 2048N depending on cell type.

Entry into the endocycle is accompanied by the transcriptional downregulation of many genes that promote mitotic progression including cyclin A, cyclin B, cyclin B3, cdk1 and string (Smith and Orr-Weaver, 1991; Sauer et al., 1995; Schaeffer et al., 2004). Several lines of evidence indicate that the proteolysis of the mitotic cyclins is essential for entry into a developmentally programmed endocycle. The mitotic cyclins are degraded by the highly conserved Anaphase Promoting Complex/ cyclosome (APC/C) (Sigrist and Lehner, 1997; Schaeffer et al., 2004). The APC/C is an E3 ubiquitin ligase that targets proteins for destruction by the 26S proteasome. Studies of Fzr/Cdk1, a regulatory subunit of the APC, show that APC activity is required for the switch from mitotic to endocycles. *Fzr* mutant embryos show inappropriate accumulation of G2 Cyclins in G1 cells and these cells never initiate their first endocycle S phases (Sigrist and Lehner, 1997). This is presumed to be due to the inhibitory action of persistent APC targets such as Cyclin A, Cyclin B, or perhap's Geminin (a negative regulator of DNA replication), on pre-replication Complex (pre-RC) formation. The pre-RC which contains ORC proteins (Origin of Replication Complex subunits), Cdc6, Double-Parked (Cdtl/Dup), a helicase which helps unwinding the DNA, and MCM onto DNA replication origins ensure that each genomic sequence is replicated only once per cell cycle (reviewed in Donaldson and Blow, 1999). Consistent with this idea, inhibition of Cdk1 activity in Drosophila can force cells that are normally mitotic to become endoreplicative (Hayashi, 1996). The observation that the accumulation of mitotic cyclins leads to a cell cycle arrest, and not re-entry into the mitotic cycle, suggests that the destruction of the mitotic cyclins through the upregulation of Fzr is only one of several mechanisms that act in parallel to inhibit mitosis at the mitotic/endocycle boundary.

Developmental changes in Drosophila's ovarian nurse cells have provided additional insights into the parameters of the endocycle. During the first five endocycles of nurse cell development, the chromosome are polytene, with tightly associated sister chromatids. In these cycles, genomic replication is complete and heterochromatin replicates late in S phase (Dej and Spradling, 1999). After the S phase of the fifth endocycle, the chromosomes condense and separate from each other. In subsequent cycles, the chromosomes are polyploid in appearance, although an association between replicated chromatids is maintained (Dej and Spradling, 1999). In addition to this cytological change, heterochromatin and some euchromatic regions such as the histone genes are endoreplicated from cycle 6 on (Hammond and Laird, 1985). morula (mr) encodes the APC subunit, APC2 (Kashevsky et al., 2002). In the hypomorphic female sterile mutant mr^1 and mr^2, the polyploid nurse cells of the ovary undergo four seemingly normal endocycles before arresting in a metaphase-like state during the fifth endocycles (Reed and Orr-Weaver, 1997). These observations reveal that during the first four endocycles, the nurse cells can retain some properties of the mitotic cycle and suggest that developmental controls can alter the extent of DNA replication.

One central regulator of all Drosophila endocycles is the S phase kinase, Cyclin E/Cdk2. The cyclin E gene is transcribed prior to the onset of endocycle S phases and is required for these cycles (Knoblich et al., 1994). The E2F1/DP transcription factor, thought to be important for cyclin E transcription, is also required for normal endocycle progression (Royzman et al., 1997). Ectopic induction of Cyclin E or E2F1 can trigger precocious DNA replication in endoreplicating tissues, even after their cells have been forced into a quiescent state by nutrient deprivation (Britton and Edgar, 1998). It is generally believed that Cyclin E/Cdk2 is the trigger that fires pre-RC assembly by phosphorylating as yet unidentified components therein (Duronio et al., 1998). However, one report on endoreplicating salivary glands showed that cyclin E induction caused the relocalization of an MCM protein to chromatin, and the authors suggested that Cyclin E might also be involved in pre-RC assembly (Hua et al., 1997; Su and O'Farrell, 1998).

In apparent contrast to its role in promoting endocycling, continuous overexpression of Cyclin E inhibits endocycle progression in Drosophila (Follette et al., 1998; Weiss et al., 1998). This suggests that oscillations of Cyclin E activity may be required for successive endocycling. In fact, oscillations of Cyclin E protein have been documented in endocycling ovarian nurse cells (Lilly and Spradling, 1996) and cyclin E mRNA fluctuates in the endocycling embryonic gut (Duronio and O'Farrell, 1995) as well as in endocycling larval tissues. It has been suggested that Cyclin E/Cdk2 activity may have to drop between S phases to allow pre-RC assembly and "relicensing" of the DNA for replication (Hua et al., 1997; Follette et al., 1998; Weiss et al., 1998). However, two apparently contrary observations from flies does not support the idea that Cdk activity inhibits pre-RC formation. First, all 16 mitotic cycles of Drosophila embryogenesis proceed in the presence of high, continuous Cyclin E/Cdk2 activity (Sauer et al., 1995). Second, although continuous overexpression of Cyclin E in mitotically proliferating wing imaginal cells does truncate G1 and elongate S phases, it fails to arrest cell cycle progression altogether. These apparent paradoxes suggest that pre-RC assembly can occur in mitotic Drosophila cells even in the presence of high Cyclin E activity. This might be explained by differences between mitotic and endocycle pre-RC components, and in fact one Drosophila MCM protein has been described that may be used in mitotic but not endocycle S phase (Feger et al., 1995). Another possibility is that nuclear envelope breakdown, which occurs in mitotic cycles but is bypassed in Drosophila's endocycles, may transiently lower nuclear levels of Cyclin E/Cdk2 and thus facilitate pre-RC assembly even while cytoplasmic Cdk2 activity remains high.

The mechanisms that account for the periodicity of Cyclin E in endocycling cells is still unresolved. However, the majority of the evidence points to the importance of the cycle accumulation and destruction of the Cyclin E protein. During the endocycle, the periodicity of Cyclin E expression is influenced by multiple inputs. The Skp1-Cul1-Fbox (SCF) protein complex, which functions as an E3 ubiquitin ligase, targets the Cyclin E protein for proteolysis (Koepp et al., 2001; Moberg et al., 2001). Archipelago (Ago), known as hCdc4 or Fbw7 in mammals, physically interacts with Cyclin E through an F-box which provides the substrate specificity. In *ago* mutants, Cyclin E protein accumulates in both mitotic and endocycling cells. Persistent Cyclin E accumulation has negative effects on endocycle progression, resulting in phenotypes ranging from a lengthening of the S phase in the polyploid nurse cells to a complete abrogation of endoreplication in the somatic follicle cells of the ovary (Doronkin et al., 2003; Shcherbata et al., 2004). The COP9 signalosome positively regulates the activity of the SCF complex and promotes the degradation of Cyclin E (Doronkin et al., 2003). Mutations in two components of the COP9 signalosome, *CSN4* and *CSN5*, alter the kinetics of Cyclin E oscillations as well as the dynamics of the S-G cycle in polyploid nurse cells in a manner similar to that observed in *ago* mutants. Analysis of the hypomorphic *cyclin* E^{01672} mutant shows that the oscillations of Cyclin E protein are severely dampened in *cyclin* E^{01672} polyploid nurse cells such that endocycling nuclei are rarely observed to have extremely high or extremely low levels of Cyclin E protein (Lilly and Spradling, 1996). While the exact cause of the reduced oscillations is not clear, one model is that in Cyclin E nurse cells, the reduced accumulation of Cyclin E^{01672} results in diminished ability to carry out the rapid destruction of the Cyclin E protein at the end of each endocycle S phase (Lilly and Spradling, 1996). Thus, an important component of the periodic accumulation of Cyclin E during the endocycle is likely to be an autoregulation loop, which couples Cyclin E/Cdk2 activity to the regulated proteolysis of Cyclin E protein via the SCF. However, it has also been proposed that the periodic inhibition of Cyclin E/Cdk2 activity by the $p27^{CIP/KIP}$ like cyclin-dependent kinase inhibitor (CKI) Dacapo (Dap) influences the kinetics of the S-G cycles during endoreplication (de Nooij et al., 2000; Hong et al., 2003). Specifically, Dap is proposed to inhibit transiently Cyclin E/Cdk2 activity to promote entry into the G phase. Dap binds and specifically inhibits Cyclin E/Cdk2 complexes in Drosophila (de Nooij et al., 1996; Lane et al., 2000). In the polyploid nurse cells of the ovary, Dap oscillations closely follow those of Cyclin E (de Nooij et al., 2000). In addition, in many tissues of Drosophila Cyclin E positively influences the accumulation of Dap (de Nooij et al., 2000). This suggests a feedback loop in which a rise in Cyclin E levels triggers DNA replication, as well as the accumulation of Dap. Moreover, in *dacapo* mutants, endocycling cells have inappropriately high Cyclin E/Cdk2 activity but a slower S phase, leading to double-strand DNA breaks (Hong et al., 2007). These replication problems are most likely due to the inhibition of origin licensing factor Double-parked (Dup, Cdt1), which function during the G phase to promote pre-RC formation. This finding is consistent with previous studies showing that high levels of Cyclin/Cdk2 promote the destruction of Dup (Cdt1) at the G1/S phase transition (Thomer et al., 2004).

Endocycle Control by Growth Regulatory Pathways

As embryogenesis and metamorphosis occur in a self-contained system, the egg and the pupae, respectively, the growth phase of Drosophila development occurs during the three larval stages. Egg hatching results in a first instar larvae that increases in mass 200-fold over an approximately 5-days period, culminating in pupation at the end of the third larval instar. Essentially, all dedicated larval tissues are polyploid with most of larval growth achieved via endocycles that increase cell size rather than cell number. The number of endocycles utilized varies dramatically between tissues as evidenced by the fact that midgut cells achieve a polyploidy of only 64N, whereas larval salivary gland cells become 1024-2048N (N is the normal haploid chromosome number). Because cell size and thus tissue size are proportional to ploidy, endocycle exit is likely to be actively regulated.

When a first instar larvae is deprived of food upon hatching, larval tissues do not initiate endocycles and remain quiescent (Britton and Edgar, 1998). When the starved larvae are provided nutrients, the endocycle resume. Thus, the endocycle responds to the larvae's environmental growth conditions. Similarly, endocycles in the polyploid nurse cells of the ovary respond to the nutritional status of the female (Drummond-Barbosa and Spradling, 2001). Adequate nutritional input is needed to drive endocycles during larval stages via the insulin receptor/PI3 kinase pathway, and differential responses to nutritional signals can explain variations in ploidy level attained in different tissues (Britton and Edgar, 1998; Britton et al., 2002). Genetic reduction of IR/PI3K signalling inhibits the endocycle and results in small cells with reduced ploidy, whereas stimulation of PI3K signaling induces endocycle progression and results in larger than normal cells with increased ploidy (Johnston and Gallant, 2002; Hafen and Stocker, 2003). One of the key effectors of the PI3K pathway is the protein kinase TOR, which controls ribosome biogenesis and cap-dependent translation in response to amino-acid availability by phosphorylating p70S6K and 4E-BP1, respectively. Genetic manipulation of TOR or its regulators (Tsc1/2 and Rheb) indicates that TOR activation stimulates growth and endocycle progression, whereas its inhibition has the opposite effect (Saucedo et al., 2003; Stocker et al., 2003; Zhang et al., 2003). Moreover, activation of IR/PI3K signalling or TOR activity can bypass the starvation arrest and induce endo-S phase but results in rapid death to the fly (Britton et al., 2002). This suggests that the downregulation of IR/PI3K signalling and cessation of the endocycle is a necessary response to conditions of nutrient limitation.

Neuroblast Proliferation and Cell Cycle Regulation

Neuroblasts segregate from the ectoderm in waves following the blastoderm stage in the early embryo. They proliferate during embryogenesis to give rise to neurons and glia which compose the larval CNS (Hartenstein and Campos-Ortega, 1984), arrest their cell cycle after completing this embryonic program, and remain quiescent until after larval hatching (Prokop and Technau, 1991). In the first instar larva, neuroblasts re-enter the cell cycle in a characteristic spatio-temporal pattern (White and Kankel, 1978; Truman and Bate, 1988; Ito and Hotta, 1992) and perform a second round of neurogenesis. The process of neuroblast cell cycle activation has been well studied in the ventral nerve cord (Truman and Bate, 1988) and in the brain (Ito and Hotta, 1992). Larval neuroblasts in the brain are the first to be reactivated then the thoracic ventral nerve cord, and finally those in the abdominal and terminal ventral nerve cord (reviewed in Truman et al., 1994). The small quiescent neuroblasts undergo a characteristic enlargement which begins before the initiation of their first S phase (Truman and Bate, 1988). This first G1/S transition, which appears to be growth dependent, is followed by a rapid series of cell cycles that average 1-2 hours. During this larval proliferation phase, each neuroblast gives rise to ~100 progeny cells. Proliferating neuroblasts constitutively express cyclin E but string is expressed only periodically, which suggest that their cell cycles may be G2 but not G1 regulated, as in the embryo. The presence of amino acids in the food appears to provide a physiological signal for the entire animal that triggers the start of the cell cycle (Britton and Edgar, 1998). The connection between such signals and cell cycle regulators is not known but the analysis of specific regulators has provided some information on the processes which are involved. Genetic studies have identified two genes that act together to regulate the timing of neuroblast cell cycle activation, anachronism (ana) and terribly reduced optic lobes (trol). ana encodes a secreted glycoprotein that is made in glial cells adjacent to larval neuroblasts. ana has been shown to inhibit premature neuroblast proliferation (Ebens et al., 1993). In starved larvae, *ana* mutant neuroblasts remain cell-cycle arrested (Britton and Edgar, 1988). Therefore, ana controls the developmental timing of neuroblast proliferation and acts downstream of the signals that activate the cell cycle in response to feeding. Unlike ana, the trol locus is required for neuroblast proliferation (Datta, 1995) and regulates G1 progression directly. The *trol* gene is required for normal levels of cyclin E expression in the larval CNS and overexpression of dE2F/dDP or Cyclin E bypasses trol-induced G1 arrest (Caldwell and Datta, 1998). Thus, ana and trol may provide a novel link between developmental control of CNS cell division and the G1-S cell cycle machinery.

Regulation of the Mitotic to Meiotic Cycle Transition

The male germ stem cells are flanked by somatic stem cells known as cyst progenitor cells and divide to produce new cyst progenitor cells and cyst cells (Figure 2) (Gönczy and DiNardo, 1996). Each Drosophila testis consists of a small number (16-18 in larvae, 5 to 9 in adults) of germ line stem cells arranged around a compact group of specialized somatic cells known as the hub in a germinal proliferation centre located at the apical tip of the testes (Hardy et al., 1979; reviewed in Fuller, 1993). After the fourth round of division, the 16 interconnected germ cells cease mitosis and initiate the meiotic program as primary spermatocytes. During the extended G2 phase that constitutes meiotic prophase, these primary spermatocytes increase in volume 25-fold in 3.5 days and undergo extensive gene expression (reviewed in Lindsley and Tokuyasu, 1980). At the end of primary spermatocyte growth, the growth period is followed by meiosis (G2/M transition) and spermatid differentiation, which occur with little or no transcription. The G2/M transition of meiosis I in males is regulated by the Cdc2 (Cdk1) kinase (Sigrist et al., 1995). cyclin A mRNA is expressed in primary spermatocytes in preparation for the G2/M transition of meiosis I. roughex regulates the level and the subcellular localization of Cyclin A during the extended G2 period of meiosis I in order to regulate the number of meiosis II divisions that occur subsequently (Gönczy et al., 1994). cyclin B is also expressed during the G2/M transition. The cdc25 homolog, twine, transcribed in primary spermatocytes is required to activate the G2/M transition of meiosis I (Alphey et al., 1992; Courtot et al., 1992).

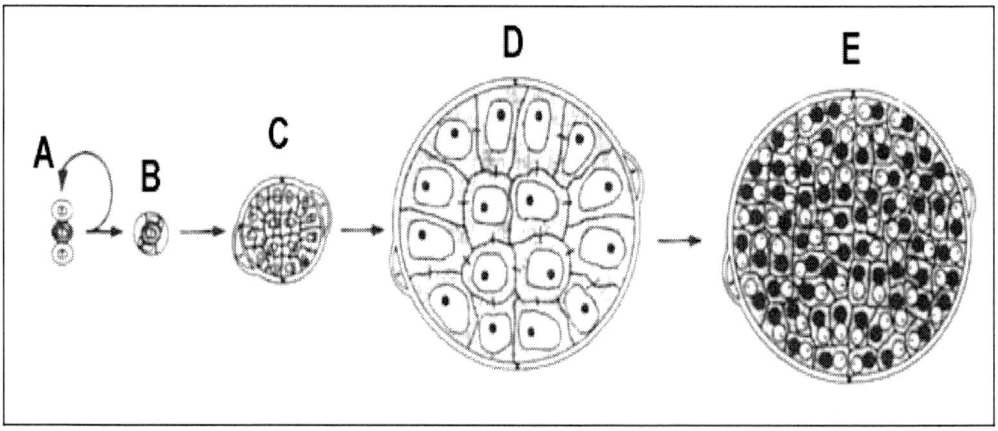

Figure 2. Early stages of spermatogenesis in Drosophila. (A) Asymmetric division of a stem cell produces another stem cell and a cyst progenitor cell. (B) The cyst progenitor cells undergo four rounds of mitosis to produce a cyst of 16 cells. (C) During a period of 90 hours, the 16 cells in the cyst express all the transcripts required for post meiotic development. (D) The cells then undergo two meiotic divisions to produce a 64 cell onion-stage cyst (E).

Mutations that block progression of spermatogenesis have revealed genes that act as key regulatory proteins in male germ cell differentiation. Function of the bag-of-marbles (bag) and benign gonial cell neoplasm (bgcn) genes in male germ cells is required for the transition from the spermatogonial program of mitotic amplification division to onset of the meiotic program and primary spermatocytes growth. In *bag* or *bgcn* mutant males, the spermatogonia continue mitotic divisions after the point at which they should have entered meiosis and their cysts commonly contain 32 or more early germ cells (Gönczy et al., 1997; McKearin and Spradling, 1990). Function of the meiotic arrest genes always early (aly), cannonball (can), meiosis I arrest (mia), and spermatocyte arrest (sa) is required for both meiotic cell cycle progression and for onset of spermatid differentiation. Male germ cell differentiation proceeds through the spermatocyte growth phase but arrests at the G2/M transition of meiosis I. The arrested cells fail to undergo spermatid differentiation and eventually degenerate (Lin et al., 1996). In *twine* mutants, spermatocytes skip certain events of meiosis, including spindle assembly, chromosome segregation and cytokinesis (Lin et al., 1996; White-Cooper et al., 1993). A similar phenotype is seen for cdc2 (cdk1) ts males shifted to a non-permissive temperature (Sigrist et al., 1995).

Oogenesis in Drosophila shows some similarities with spermatogenesis in that the gonial precursor cell undergo four mitotic divisions in the germarium of the ovary to produce a cyst of 16 cells interconnected by cytoplasmic bridges (Figure 3) (reviewed in Spradling, 1993). However, as the egg chamber develops only one cell becomes the oocyte while the other 15 cells develop into nurse cells. The nurse cells undertake up to ten rounds of endoreplication during which time twine and string are coexpressed together with many other maternally provided mRNAs that are required for early embryonic development. Prophase of meiosis is initiated in the oocyte nucleus in the germarium and continues during oogenesis. The meiotic spindle is not formed until the late stages of oogenesis, by the end of which the oocyte is arrested in metaphase of meiosis I, with the exchange chromosomes in a bundle at the metaphase plate and the non-exchange fourth chromosomes displaced from the plate towards the poles. The meiotic divisions are completed only after activation of the oocyte on passage through the oviduct (reviewed in Spradling, 1993). In contrast to the meiotic block seen in twine males, meiosis continues in females, but does so abnormally (Courtot et al., 1992; White-Cooper et al., 1993). Repeated abnormal meiotic divisions are seen in *twine* oocytes which appear to resemble the reductional division rather than the equational division. Repeated attempts at the reductional division would explain the dramatic non-disjunction that occurs during *twine* meiosis (Courtot et al., 1992; White-Cooper et al., 1993). This suggests that twine function is required to maintain meiotic arrest by keeping Cdk1 dephosphorylated and thereby active. In Drosophila, meiotic recombination only occurs in female. Therefore, the mechanisms regulating entry into the first meiotic division might differ between the sexes.

Myt1 is a conserved protein kinase that regulates the essential mitotic regulator Cdk1 by inhibitory phosphorylation (reviewed in Smits and Medema, 2001). A marked elevation of meiotic chromosome segregation defects is observed in the progeny of female *myt1* mutants (Jin et al., 2005). Myt1 has also been shown to promote proper cell cycle exit before the meiotic reductional divisions (Jin et al., 2005). These observations suggest that Myt1 serves regulatory functions that are important for meiotic cell cycles.

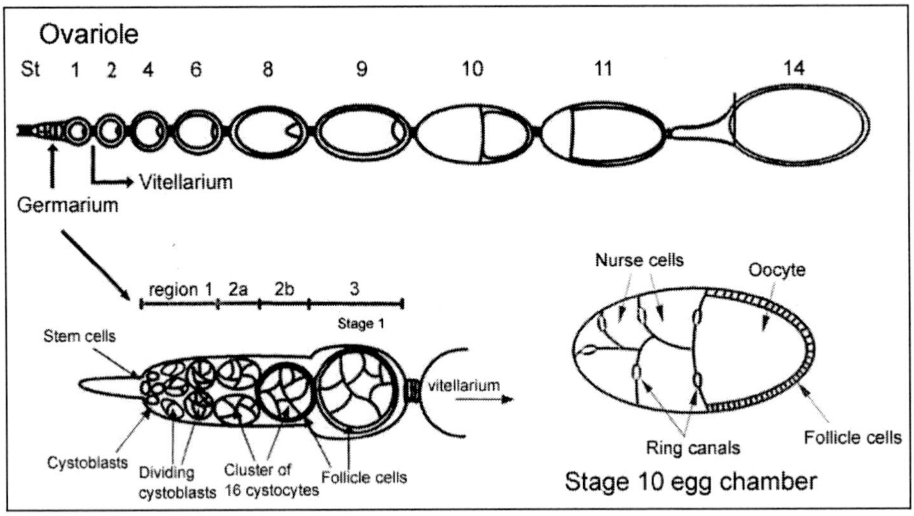

Figure 3. Oogenesis in Drosophila. The Drosophila ovary is divided into discrete units called ovarioles. In each ovariole, the germarium is localized at the anterior-most part, followed by a row of progressively older egg chambers. In region 1 of the germarium, stem cell daughters called cystoblasts undergo four rounds of mitosis to produce a 16-cell cyst. These 16 cells are interconnected by a series of cytoplasmic bridges called ring canals that develop from the incomplete cytokinesis of the contractile ring. In germarium region 2a/2b, the 16-cell cyst is surrounded by somatically derived follicle cells. In region 3, a round germ-line cyst completely surrounded by follicle cells buds off from the germarium as an egg chamber. The posterior-most cyst becomes the oocyte while the remaining 15 cells differentiate into nurse cells. Nurse cells supply the majority of the cytoplasmic components of the oocytes by transport through ring canals until stage 10. During stage 11, nurse cells extrude their cytoplasmic contents into the oocyte and degenerate. In the vitellarium, the maturation of the egg chamber is divided into various stages (St) until the mature stage 14.

Conclusion

Drosophila is an excellent system in which to investigate the developmental control of the cell cycle and its regulation. It has yielded many secrets about the basic machinery of the cell cycle in all cells. It has also led to the discovery of new molecular pathways and modes of regulation that have enhanced our understanding of the events that underlies cell-cycle progression during development. One of the most promising advances in Drosophila genetics is the RNAi (RNA interference) technology. High-throughput RNAi screens will facilitate the identification of novel regulators of a given cellular process. This should help to clarify the roles of many molecular networks that act together to regulate growth and cell cycle.

References

Adler, P. N. & MacQueen, M. (1984). Cell proliferation and DNA replication in the imaginal wing disc of Drosophila melanogaster. *Dev. Biol., 103*, 28-37.

Alphey, L., Jimenez, J., White-Cooper, H., Dawson, I., Nurse, P. & Glover, D. M. (1992). Twine, a cdc25 homologue that functions in the male and female germlines of Drosophila. *Cell, 69*, 977-988.

Baker, N. E. & Yu, S. Y. (2001). The EGF receptor defines domains of cell cycle progression and survival to regulate cell number in the developing Drosophila eye. *Cell, 104*, 699-708.

Britton, J. S. & Edgar, B. A. (1998). Environmental control of the cell cycle in Drosophila: nutrition activates mitotic and endoreplicative cells by distinct mechanisms. *Development, 125*, 2149-2158.

Britton, J. S., Lockwood, W. K., Li, L., Cohen, S. M. & Edgar, B. A. (2002). Drosophila's insulin/PI3-kinase pathway coordinates cellular metabolism with nutritional conditions. *Dev. Cell, 2*, 239-249.

Brodsky, M. H., Sekelsky, J. J., Tsang, G., Hawley, R. S. & Rubin, G. M. (2000). mus304 encodes a novel DNA damage checkpoint protein required during Drosophila development. *Genes Dev., 14*, 666-678.

Caldwell, M. C. & Datta, S. (1998). Expression of cyclin E or DP/E2F rescues the G1 arrest of trol mutant neuroblasts in the Drosophila larval central nervous system. *Mech. Dev., 79*, 121-130.

Coqueret, O. (2002). Linking cyclins to transcriptional control. *Gene, 299*, 35-55.

Courtot, C., Frankhauser, C., Simanis, V. & Lehner, C. (1992). The Drosophila cdc25 homologue twine is required for meiosis. *Development, 116*, 405-416.

Datta, S. (1995). Control of proliferation activation in quiescent neuroblasts of the Drosophila central nervous system. *Development, 121*, 1173-1182.

Dej, K. J. & Spradling, A. C. (1999). The endocycle controls nurse cell polytene chromosome structure during Drosophila oogenesis. *Development, 126*, 293-303.

De Nooij, J. C., Letendre, M. A. & Hariharan, I. K. (1996). A cyclin-dependent kinase inhibitor, Dacapo, is necessary for timely exit from the cell cycle during Drosophila embryogenesis. *Cell, 87*, 1237-1247.

De Nooij, J. C., Graber, K. H. & Hariharan, I. K. (2000). Expression of the cyclin-dependent kinase inhibitor Dacapo is regulated by Cyclin E. *Mech. Dev., 97*, 73-83.

Donaldson, A. D. & Blow, J. J. (1999). The regulation of replication origin activation. *Curr. Opin. Genet. Dev., 9*, 62-68.

Dong, X., Zavitz, K. H., Thomas, B. J., Lin, M., Campbell, S. & Zipursky, S. L. (1997). Control of G1 in the developing Drosophila eye: rca1 regulates Cyclin A. *Genes Dev., 11*, 94-105.

Doronkin, S., Djagaeva, I. & Beckendorf, S. K. (2003). The COP9 signalosome promotes degradation of Cyclin E during early Drosophila oogenesis. *Dev. Cell, 4*, 699-710.

Drummond-Barbosa, D. & Spradling, A. C. (2001). Stem cells and their progeny respond to nutritional changes during Drosophila oogenesis. *Dev. Biol., 231*, 265-278.

Duman-Scheel, M., Weng, L., Xin, S. J. & Du, W. (2002). Hedgehog regulates cell growth and proliferation by inducing cyclin D and cyclin E. *Nature, 417*, 299-304.

Duronio, R. J., O'Farrell, P. H., Xie, J. E., Brook, A. & Dyson, N. (1995). The transcription factor E2F is required for S phase during Drosophila embryogenesis. *Genes Dev., 9*, 1445-1455.

Duronio, R. J. & O'Farrell, P. H. (1995). Developmental control of the G1 to S transition in Drosophila: cyclin E is a limiting downstream target of E2F. *Genes Dev., 9*, 1456-1468.

Duronio, R. J., Bonnette, P. C. & O'Farrell, P. H. (1998). Mutations of the Drosophila dDP, dE2F, and cyclin E genes reveal distinct roles for the E2F-DP transcription factor and cyclin E during the G1-S transition. *Mol. Cell Biol., 18*, 141-151.

Ebens, A. J., Garren, H., Cheyette, B. N. & Zipursky, S. L. (1993). The Drosophila anachronism locus: a glycoprotein secreted by glia inhibits neuroblast proliferation. *Cell, 74*, 15-27.

Edgar, B. A., Kiehle, C. P. & Schubiger, G. (1986). Cell cycle control by the nucleo-cytoplasmic ratio in early Drosophila development. *Cell, 44*, 365-372.

Edgar, B. A. & O'Farrell, P. H. (1989). Genetic control of cell division patterns in the Drosophila embryo. *Cell, 57*, 177-187.

Edgar, B. A. & O'Farrell, P. H. (1990). The three postblastoderm cell cycles of Drosophila embryogenesis are regulated in G2 by string. *Cell, 62*, 469-480.

Edgar, B. A., Lehman, D. A. & O'Farrell, P. H. (1994a). Transcriptional regulation of string (cdc25): a link between developmental programming and the cell cycle. *Development, 120*, 3131-3143.

Edgar, B. A., Sprenger, F., Duronio, R. J., Leopold, P. & O'Farrell, P. H. (1994b). Distinct molecular mechanisms regulate cell cycle timing at successive stages of Drosophila embryogenesis. *Genes Dev., 8*, 440-452.

Edgar, B. A. & Datar, S. (1996). Zygotic degradation of two maternal Cdc25 mRNAs terminates Drosophila's early cell cycle program. *Genes Dev., 10*, 1966-1977.

Fain, M. J. & Stevens, B. (1982). Alterations in the cell cycle of Drosophila imaginal disc cells precede metamorphosis. *Dev. Biol., 92*, 247-258.

Feger, G., Vaessin, H., Su, T. T., Wolff, E., Jan, L. Y. & Jan, Y. N. (1995). Dpa, a member of the MCM family, is required for mitotic DNA replication but not endoreplication in Drosophila. *EMBO J., 14*, 5387-5398.

Fenger, D. D., Carminati, J. L., Burney-Sigman, D. L., Kashevsky, H., Dines, J. L., Elfring, L. K. & Orr-Weaver, T. L. (2000). PAN GU: a protein kinase that inhibits S phase and promotes mitosis in early Drosophila development. *Development, 127*, 4763-4774.

Foe, V. E. & Alberts, B. M. (1983). Studies of nuclear and cytoplasmic behaviour during the five mitotic cycles that precede gastrulation in Drosophila embryogenesis. *J. Cell Sci., 61*, 31-70.

Foe, V. E., Odell, G. M. & Edgar, B. A. (1993). *The Development of Drosophila melanogaster*, Cold Spring Harbor Laboratory Press, Cold Spring Harbor, New York.

Fogarty, P., Campbell, S. D., Abu-Shumays, R., Phalle, B. S., Yu, K. R., Uy, G. L., Goldberg, M. L. & Sullivan, W. (1997). The Drosophila grapes gene is related to checkpoint gene chk1/rad27 and is required for late syncytial division fidelity. *Curr. Biol., 7*, 418-426.

Follette, P. J., Duronio, R. J. & O'Farrell, P. H. (1998). Fluctuations in cyclin E levels are required for multiple rounds of endocycle S phase in Drosophila. *Curr. Biol., 8*, 235-238.

Freeman, M., Nusslein-Volhard, C. & Glover, D. (1986). The dissociation of nuclear and centrosomal division in *gnu*, a mutation causing giant nuclei in Drosophila. *Cell, 46*, 457-468.

Freeman, M. & Glover, D. (1987). The *gnu* mutation of Drosophila causes inappropriate DNA synthesis in unfertilized and fertilized eggs. *Genes Dev., 1*, 924-930.

Fuller, M. T. (1993). Spermatogenesis. In *Development in Drosophila* (A. Martinez-Arias & M. Bate, Eds.), pp. 71-147. Cold Spring Harbor Laboratory Press, New York.

Galloni, M. & Edgar, B. A. (1999). Cell-autonomous and non-autonomous growth-defective mutants of Drosophila melanogaster. *Development, 126*, 2365-2375.

Garner, M., van Kreeveld, S. & Su, T. T. (2001). mei-41 and bub1 block mitosis at two distinct steps in response to incomplete DNA replication in Drosophila embryos. *Curr. Biol., 11*, 1595-1599.

Gönczy, P., Thomas, B. J. & DiNardo, S. (1994). roughex is a dose-dependent regulator of the second meiotic division during Drosophila spermatogenesis. *Cell, 77*, 1015-1025.

Gönczy, P. & DiNardo, S. (1996). The germ line regulates somatic cyst cell proliferation and fate during Drosophila spermatogenesis. *Development, 122*, 2437-2447.

Gönczy, P., Matunis, E. & DiNardo, S. (1997). bag-of-marbles and benign gonial cell neoplasm act in the germline to restrict proliferation during Drosophila spermatogenesis. *Development, 124*, 4361-4371.

Gonzàlez-Gaitàn, M., Capdevila, M. P. & Garcia-Bellido, A. (1984). Cell proliferation patterns in the wing imaginal disc of Drosophila. *Mech. Dev., 40*, 183-200.

Graves, B. J. & Schubiger, G. (1982). Cell cycle changes during growth and differentiation of imaginal leg discs in Drosophila melanogaster. *Dev. Biol., 93*, 104-110.

Hafen, E. & Stocker, H. (2003). How are the sizes of cells, organs, and bodies controlled? *PLoS Biol., 1*, 319-323.

Hammond, M. P. & Laird, C. D. (1985). Chromosome structure and DNA replication in nurse and follicle cells of Drosophila melanogaster. *Chromosoma, 91*, 267-278.

Hardy, R. W., Tokuyasu, K. T., Lindsley, D. L. & Garavito, M. (1979). The germinal proliferation center in the testis of Drosophila melanogaster. *J. Ultrastruct. Res., 69*, 180-190.

Hari, K. L., Santerre, A., Sekelsky, J. J., McKim, K. S., Boyd, J. B. & Hawley, R. S. (1995). The mei-41 gene of D. melanogaster is a structural and functional homolog of the human ataxia telangictasia gene. *Cell, 82*, 815-821.

Hartenstein, V. & Campos-Ortega, J. A. (1984). Early neurogenesis in wild-type Drosophila melanogaster. *Roux's Arch. Dev. Biol., 193*, 308-325.

Hartenstein, V. & Posakony, J. W. (1989). Development of adult sensilla on the wing and notum of Drosophila melanogaster. *Development, 107*, 389-405.

Hayashi, S. (1996). A cdc2 dependent checkpoint maintains diploidy in drosophila. *Development, 122*, 1051-1058.

Hong, A., Lee-Kong, S., Iida, T., Sugimura, I. & Lilly, M. A. (2003). The p27cip/kip ortholog dacapo maintains the Drosophila oocyte in prophase of meiosis I. *Development, 130*, 1235-1242.

Hong, A., Narbonne-Reveau, K., Riesgo-Escovar, J., Fu, H., Aladjem, M. I. & Lilly, M. A. (2007). The cyclin-dependent kinase inhibitor Dacapo promotes replication licensing during Drosophila endocycles. *EMBO J., 26*, 2071-2082.

Horsfield, J., Penton, A., Secombe, J., Hoffman, F. M. & Richardson, H. (1998). decapentaplegic is required for arrest in G1 phase during Drosophila eye development. *Development, 125*, 5069-5078.

Hua, X. H., Yan, H. & Newport, J. (1997). A role for Cdk2 kinase in negatively regulating DNA replication during S phase of the cell cycle. *J. Cell. Biol., 137*, 183-192.

Ito, K. & Hotta, Y. (1992). Proliferation pattern of postembryonic neuroblasts in the brain of Drosophila melanogaster. *Dev. Biol., 149*, 134-148.

Jacobs, H. W., Knoblich, J. A. & Lehner, C. F. (1998). Drosophila Cyclin B3 is required for female fertility and is dispensable for mitosis like Cyclin B. *Genes Dev., 12*, 3741-3751.

James, A. A. & Bryant, P. J. (1981). A quantitative study of cell death and mitotic inhibition in gamma-irradiated imaginal wing discs of Drosophila melanogaster. *Radiat. Res., 87*, 552-564.

Jin, Z., Homola, E. M., Goldbach, P., Choi, Y., Brill, J. A. & Campbell, S. D. (2005). Drosophila Myt1 is a Cdk1 inhibitory kinase that regulates multiple aspects of cell cycle behaviour during gametogenesis. *Development, 132*, 4075-4085.

Johnston, L. A. & Edgar, B. A. (1998). Wingless and Notch regulate cell-cycle arrest in the developing Drosophila wing. *Nature, 394*, 82-84.

Johnston, L. A. & Gallant, P. (2002). Control of growth and organ size in Drosophila. *Bioessays, 24*, 54-64.

Kashevsky, H., Wallace, J. A., Reed, B. H., Lai, C., Hayashi-Hagihara, A. & Orr-Weaver, T. L. (2002). The anaphase promoting complex/cyclosome is required during development for modified cell cycles. *Proc. Natl. Acad. Sci. USA, 99*, 11217-11222.

Knoblich, J. A., Sauer, K., Jones, L., Richardson, H., Saint, R. & Lehner, C. F. (1994). Cyclin E controls S phase progression and its down-regulation during Drosophila embryogenesis is required for the arrest of cell proliferation. *Cell, 77*, 107-120.

Koepp, D. M., Schaefer, L. K., Ye, X., Keyomarsi, K., Chu, C., Harper, J. W. & Elledge, S. J. (2001). Phosphorylation-dependent ubiquitination of cyclin E by the SCFFbw7 ubiquitin ligase. *Science, 294*, 173-177.

Lane, M. E., Sauer, K., Wallace, K., Jan, Y. N., Lehner, C. F. & Vaessin, H. (1996). Dacapo, a cyclin-dependent kinase inhibitor, stops cell proliferation during Drosophila development. *Cell, 87*, 1225-1235.

Lane, M. E., Elend, M., Heidmannn, D., Herr, A., Marzodko, S., Herzig, A. & Lehner, C. F. (2000). A screen for modifiers of cyclin E function in Drosophila melanogaster identifies Cdk2 mutations, revealing the insignificance of putative phosphorylation sites in Cdk2. *Genetics, 155*, 233-244.

Lee, L. A., Van Hoewyk, D. & Orr-Weaver, T. L. (2003). The Drosophila cell cycle kinase PAN GU forms an active complex with PLUTONIUM and GNU to regulate embryonic divisions. *Genes Dev., 17*, 2979-2991.

Lilly, M. A. & Spradling, A. C. (1996). The Drosophila endocycle is controlled by Cyclin E and lacks a checkpoint ensuring S-phase completion. *Genes Dev., 10*, 2514-2526.

Lin, T. Y., Viswanathan, S., Wood, C., Wilson, P. G., Wolf, N. & Fuller, M. T. (1996). Coordinate developmental control of the meiotic cell cycle and spermatid differentiation in Drosophila males. *Development, 122*, 1331-1341.

Lindsley, D. & Tokuyasu, K. T. (1980). Spermatogenesis. In *genetics and biology of Drosophila* (M. Ashburner & T. R. F. Wright, Eds.), vol. 2, pp. 225-294. Academic Press, New York.

Madhavan, M. & Schneiderman, H. A. (1977). Histological analysis of the dynamics of growth of imaginal discs and histoblast nests during the larval development of Drosophila melanogaster. *Wilhelm Roux's Arch., 183*, 269-305.

McKearin, D. & Spradling, A. C. (1990). bag-of-marbles: a Drosophila gene required to initiate both male and female gametogenesis. *Genes Dev., 4*, 2242-2251.

Meyer, C. A., Kramer, I., Dittrich, R., Marzodko, S., Emmerich, J. & Lehner, C. F. (2002). Drosophila p27Dacapo expression during embryogenesis is controlled by a complex regulatory region independent of cell cycle progression. *Development, 129*, 319-328.

Moberg, K. H., Bell, D. W., Wahrer, D. C., Haber, D. A. & Hariharan, I. K. (2001). archipelago regulates Cyclin E levels in Drosophila and is mutated in human cancer cell lines. *Nature, 413*, 311-316.

Neufeld, T. P., de la Cruz, A. F., Johnston, L. A. & Edgar, B. A. (1998). Coordination of growth and cell division in the Drosophila wing. *Cell, 93*, 1183-1193.

O'Brochta, D. A. & Bryant, P. J. (1985). A zone of non-proliferating cells at a lineage restriction boundary in Drosophila. *Nature, 313*, 138-141.

Price, D., Rabinivitch, S., O'Farrell, P. H. & Campbell, S. D. (2000). Drosophila wee1 has an essential role in the nuclear divisions of early embryogenesis. *Genetics, 155*, 159-166.

Prokop, A. & Technau, G. M. (1991). The origin of postembryonic neuroblasts in the ventral nerve cord of Drosophila melanogaster. *Development, 111*, 79-88.

Reed, B. H. & Orr-Weaver, T. L. (1997). The Drosophila gene morula inhibits mitotic functions in the endo cell cycle and the mitotic cell cycle. *Development, 124*, 3543-3553.

Resino, J., Salama-Cohen, P. & Garcia-Bellido, A. (2002). Determining the role of patterned cell proliferation in the shape and size of the Drosophila wing. *Proc. Natl. Acad. Sci., 99*, 7502-7507.

Royzman, I., Whittaker, A. J. & Orr-Weaver, T. L. (1997). Mutations in Drosophila DP and E2F distinguish G1-S progression from an associated transcriptional program. *Genes Dev., 11*, 1999-2011.

Saucedo, L. J. & Edgar, B. A. (2002). Why size matters: altering cell size. *Curr. Opin. Genet. Dev., 12*, 565-571.

Sauer, K., Knoblich, J. A., Richardson, H. & Lehner, C. F. (1995). Distinct modes of cyclin E/cdc2c kinase regulation and S-phase control in mitotic and endoreduplication cycles of Drosophila embryogenesis. *Genes Dev., 9*, 1327-1339.

Schaeffer, V., Althauser, C., Shcherbata, H. R., Deng, W. M. & Ruohola-Baker, H. (2004). Notch-dependent Fizzy-related/Hec1/Cdh1 expression is required for the mitotic-to-endocycle transition in Drosophila follicle cells. *Curr. Biol., 14*, 630-636.

Shamanski, F. & Orr-Weaver, T. L. (1991). The Drosophila *plutonium* and *pan gu* genes regulate entry into S phase at fertilization. *Cell, 66*, 1289-1300.

Shcherbata, H. R., Althauser, C., Findley, S. D. & Ruohola-Baker, H. (2004). The mitotic-to-endocycle switch in Drosophila follicle cells is executed by Notch-dependent regulation of G1/S, G2/M and M/G1 cell-cycle transitions. *Development, 131*, 3169-3181.

Shibutani, S., Swanhart, L. M. & Duronio, R. J. (2007). Rbf1-independent termination of E2f1-target gene expression during early Drosophila embryogenesis. *Development, 134*, 467-78.

Sibon, O. C., Stevenson, V. A. & Theurkauf, W. E. (1997). DNA-replication checkpoint control at the the Drosophila midblastula transition. *Nature, 388*, 93-97.

Sibon, O. C., Laurençon, A., Hawley, R. & Theurkauf, W. E. (1999). The Drosophila ATM homologue Mei-41 has an essential checkpoint function at the midblastula transition. *Curr. Biol., 9*, 302-312.

Sigrist, S., Ried, G. & Lehner, C. F. (1995). Dmcdc2 kinase is required for both meiotic divisions during Drosophila spermatogenesis and is activated by the Twine/cdc25 phosphatase. *Mech. Dev., 53*, 247-260.

Sigrist, S. J. & Lehner, C. F. (1997). Drosophila fizzy-related down-regulates mitotic cyclins and is required for cell proliferation arrest and entry into endocycles. *Cell, 90*, 671-681.

Smith, A. V. & Orr-Weaver, T. L. (1991). The regulation of the cell cycle during Drosophila embryogenesis-the transition to polyteny. *Development, 112*, 997-1008.

Smits, V. A. & Medema, R. H. (2001). Checking out the G(2)/M transition. *Biochim. Biophys. Acta, 1519*, 1-12.

Spradling, A. C. (1993). Developmental genetics of oogenesis. In *Development in Drosophila* (A. Martinez-Arias & M. Bate, Eds.), pp. 1670. Cold Spring Harbor Laboratory Press, New York.

Stiffler, L. A., Ji, J. Y., Trautmann, S., Trusty, C. & Schubiger, G. (1999). Cyclin A and B functions in the early Drosophila embryo. *Development, 126*, 5505-5513.

Stocker, H., Radimerski, T., Schindelholz, B., Wittwer, F., Belawat, P., Daram, P, Breuer, S., Thomas, G. & Hafen, E. (2003). Rheb is an essential regulator of S6K in controlling cell growth in Drosophila. *Nat. Cell Biol., 5*, 559-565.

Su, T. T. & O'Farrell, P. H. (1998). Chromosome association of minichromosome maintenance proteins in Drosophila endoreplication cycles. *J. Cell Biol., 140*, 451-460.

Su, T. T., Sprenger, F., DiGregorio, P. J., Campbell, S. D. & O'Farrell, P. H. (1998). Exit from mitosis in Drosophila syncytial embryos requires proteolysis and cyclin degradation, and is associated with localized dephosphorylation. *Genes Dev., 12*, 495-503.

Su, T. T., Campbell, S. D. & O'Farrell, P. H. (1999). Drosophila grapes/CHK1 mutants are defective in cyclin proteolysis and coordination of mitotic events. *Curr. Biol., 9*, 919-922.

Thomas, B. J., Gunning, D. A., Cho, J. & Zipursky, S. L. (1994). Cell cycle progression in the developing Drosophila eye: roughex encodes a novel protein required for the establishment of G1. *Cell, 77*, 1003-1014.

Thomas, B. J., Zavitz, K. H., Dong, X., Lane, M. E., Weigmann, K., Finley Jr., R. L., Brent, R., Lehner, C. F. & Zipursky, S. L. (1997). *roughex* down-regulates G2 cyclins in G1. *Genes Dev*., 1289-1298.

Thomer, M., May, N. R., Aggarwal, B. D., Kwok, G. & Calvi, B. R. (2004). Drosophila double-parked is sufficient to induce re-replication during development and is regulated by cyclin E/CDK2. *Development, 131*, 4807-4818.

Truman, J. W. & Bate, M. (1988). Spatial and temporal patterns of neurogenesis in the central nervous system of Drosophila melanogaster. *Dev. Biol., 125*, 145-157.

Truman, J. W., Taylor, B. J. & Awad, T. A. (1994). Formation of the adult nervous system. In *The Development of Drosophila melanogaster* (M. Bate, M. & A. Martinez Arias, Eds.), vol. 2, pp. 1245-1275. Cold Spring Harbor Laboratory Press, New York.

Weiss, A., Herzig, A., Jacobs, H. & Lehner, C. F. (1998). Continuous Cyclin E expression inhibits progression through endoreduplication cycles in Drosophila. *Curr. Biol., 8*, 239-242.

White, K. & Kankel, D. R. (1978). Patterns of cell division and cell movement in the formation of the imaginal nervous system in Drosophila melanogaster. *Dev. Biol., 65*, 296-321.

White-Cooper, H., Alphey, L. & Glover, D. M. (1993). The cdc25 homologue twine is required for only some aspects of the entry into meiosis in Drosophila. *J. Cell Sci., 106*, 1035-1044.

Wolff, T. & Ready, D. F. (1993). Pattern formation in the Drosophila retina. In *The Development of Drosophila melanogaster* (M. Bate & A. Martinez Arias, Eds.), vol. 2, pp. 1277-1325. Cold Spring Harbor Laboratory Press, New York.

Yu, K. R., Saint, R. B. & Sullivan, W. (2000). The Grapes checkpoint coordinates nuclear envelope breakdown and chromosome condensation. *Nat. Cell Biol., 2*, 609-615.

Zalokar, M. & Erk, I. (1976). Division and migration of nuclei during early embryogenesis of Drosophila melanogaster. *J. microbiol. Cell, 25*, 97-106.

Zhang, H. B., Stallock, J. P., Ng, J. C., Reinhard, C. & Neufeld, T. P. (2000). Regulation of cellular growth by the Drosophila target of rapamycin dTOR. *Genes Dev., 14*, 2712-2724.

Zhang, Y., Gao, X., Saucedo, L. J., Ru, B., Edgar, B. A. & Pan, D. (2003). Rheb is a direct target of the tuberous sclerosis tumour suppressor proteins. *Nat. Cell Biol., 5*, 578-581.

In: Cell Cycle Control: New Research
Editors: N. H. Leroy, N. T. Fournier
ISBN: 978-1-60456-796-0

Chapter XIV

Cell Cycle Control of Mammals During the Embryonic Period and Recent Research

***Paula Lumy Takeuchi*[1], *Christina Vretos*[2] *and Luis Garcia Alonso*[2]**
[1]Genetics Department of Medicine Faculty of Ribeirão Preto – São Paulo University – Brazil;
[2]Morphology and Genetics Department, São Paulo Federal University-Paulista Medical School, Disciplina de Anatomia Descritiva e Topográfica, Brazil.

1. Cell Cycle Control: Historical

The Cell Theory was proposed by Theodor Schleiden and Jacob Schwann in 1838.This theory predict that every living organism is composed of one or more cells and that new cells can arise only by the division of preexisting cells. To reproduce, cells must replicate their chromosomes and then segregate them into two daughter cells. By the late 1950s the demonstration that DNA was the genetic material and the elucidation of its structure enabled biochemists to ask questions about DNA replication. Advances in cell biology during the 1960s identified the molecules that form the cytoskeleton that gives cells their shape and organizes the segregation of chromosomes and the division of cells. Thus by 1970, scientists understood the basic composition of cells and could frame questions about the molecular details of individual processes in the cell cycle (Murray and Hunt, 1993).

In 2001 Leland Hartwell, Paul Nurse and Timothy Hunt had been awardees with the Nobel Prize in Physiology or Medicine for their discoveries of key regulators of the cell cycle (http://nobelprize.org/nobel_prizes/medicine/laureates/2001/press.html).

Leland Hartwell (Fred Hutchinson Cancer Research Center, Seattle, USA) realized already at the end of the 1960s the possibility of studying the cell cycle with genetic methods. He used baker's yeast, *Saccharomyces cerevisiae*, as a model system, which proved to be highly suitable for cell cycle studies. In an elegant series of experiments 1970-71, he isolated yeast cells in which genes controlling the cell cycle were altered. By this approach he succeeded to identify more than one hundred genes specifically involved in cell cycle control, so called CDC-genes (Cell Division Cycle genes). One of these genes, designated *CDC28* by Hartwell, controls the first step in the progression through the G_1 phase of the cell cycle, and was therefore also called "Start".

In addition, Hartwell studied the sensitivity of yeast cells to irradiation. On the basis of his findings he introduced the checkpoint concept, which means that the cell cycle is arrested when DNA is damaged. The purpose of this is to allow time for DNA repair before the cell continues to the next phase of the cycle. Later Hartwell extended the checkpoint concept to include also controls ensuring a correct order between the cell cycle phases (http://nobelprize.org/nobel_prizes/medicine/laureates/2001/press.html).

Paul Nurse (Imperial Cancer Research Fund, London) used a different type of yeast, *Schizosaccharomyces pombe,* as a model organism. In the middle of the 1970s, Paul Nurse discovered the gene CDC2 in *S. pombe,* he showed that this gene had a key function in the control of cell division (transition from G_2 to mitosis, M). Later he found that *CDC2* had a more general function. It was identical to the gene (Start) that Hartwell earlier had identified in baker's yeast, controlling the transition from G_1 to S.

CDC2 was thus found to regulate different phases of the cell cycle. In 1987 Paul Nurse isolated the corresponding gene in humans, and it was later given the name *CDK1* (Cyclin Dependent Kinase 1). The gene encodes a protein that is a member of a family called cyclin dependent kinases, CDK. Nurse showed that activation of CDK is dependent on reversible phosphorylation, i.e. that phosphate groups are linked to or removed from proteins. On the basis of these findings, half a dozen different CDK molecules have been found in humans (http://nobelprize.org/nobel_prizes/medicine/laureates/2001/press.html).

Timothy Hunt (Imperial Cancer Research Fund, London) discovered the first cyclin molecule in the early 1980s. Cyclins are proteins formed and degraded during each cell cycle. They were named cyclins because the levels of these proteins vary periodically during the cell cycle. The cyclins bind to the CDK molecules, thereby regulating the CDK activity and selecting the proteins to be phosphorylated.

The discovery of cyclin, which was made using sea urchins, *Arbacia,* as a model system, was the result of Hunt's finding that this protein was degraded periodically in the cell cycle. Tim Hunt later discovered cyclins in other species and found that also the cyclins were conserved during evolution. Today around ten different cyclins have been found in humans (http://nobelprize.org/nobel_prizes/medicine/laureates/2001/press.html)

1.1. Cyclins and CDKs

Cyclins are found in all eukaryotes that have been examined, including yeasts, coelenterates, flies, echinoderms, mollusk, amphibians, mammals, and plants. Actually, we know that there are many different cyclins, which form a large family of related proteins with different functions. Mitotic cyclins are components of MPF, S phase cyclins play role in the control of DNA replication, and G_1 cyclins are important in catalyzing the events that move the somatic cell cycle from G_1 into S phase.

The cell cycle is divided into two fundamental parts: interphase, which occupies the majority of the cell cycle, and mitosis, which lasts about 30 minutes, ending with the division of the cell. During interphase the DNA is diffusely distributed within the nucleus, and individual chromosomes cannot be distinguished. Continuous processes occur throughout interphase and are referred to collectively as growth. These processes include the synthesis of new ribosomes, membranes, mitochondria, endoplasmic reticulum, and most cellular proteins (Alberts et al., 1997).

Stepwise processes occur once per cell cycle. For example, chromosomes replication is restricted to a specific part of interphase called *S phase* (DNA synthesis). S phase occurs in the middle of interphase, preceded by a gap G_1 and followed by a gap called G_2 (Figure 1). After each chromosome has replicated, the two daughter chromosomes remain attached to each other at multiple points along their length and are referred as sister chromatids. In a typical animal cell cycle, G_1 lasts 12 hours, S phase 6 hours, G_2 6 hours, and mitosis about 30 minutes. Cells in G_1, if still not compromised with DNA synthesis, could interrupt their growth and pass to resting state called G_0, where cells can remain for days, weeks or years before return to replicate (Alberts et al., 1997).

The embryonic cell cycles in early phases are shorter than the bacterial cell cycle. During embryonic cell cycles G_1 and G_2 phases are shortened and the cells take around 8 to 60 minutes to divide (Murray and Hunt, 1993).

CDKs transfer gamma-phosphate from ATP to a specific protein substrate (Morgan 1995). However, the kinase subunit alone is inactive because the bound ATP is not properly oriented, and access of the protein substrate is blocked by a section of CDK called the T-loop (De Bondt et al. 1993). These impediments are removed by association with cyclin and T-loop phosphorylation by the multicomponent CAK (CDK-Activating Kinase) (Russo et al. 1996). The key to ordering and controlling cell cycle progression is thought to lie in periodic expression of different cyclins, which associate with CDKs at defined intervals and determine their specificity (Murray and Hunt, 1993). These unique cyclin–CDK complexes must phosphorylate specific substrates at the proper time to drive controlled progression through the cell cycle. After completing their task, complexes are disassembled and cyclin degraded as a prerequisite for subsequent steps (Murray et al. 1989).

Temporal order is achieved and maintained by linking cyclin expression to completion of previous events, then regulating activity of the resulting cyclin–CDK complex. Controlling complexes can be accomplished by removing activating modifications, inhibitory phosphorylation of the CDK subunit, or tight binding of CKIs (CDK Inhibitory Proteins) (Morgan 1995). CDK activity can also be modulated by altering its location and/or accessibility to substrates, although these regulatory mechanisms are less well characterized (Murray 2004). Together, this molecular circuitry provides a mechanistic explanation of cell cycle progression during G_1.

G_1 phase of the cell cycle is organized around the concept of a restriction point (R point; called Start in yeast) (Hartwell et al. 1973; Pardee 1974; Blagosklonny and Pardee 2002). Before this G_1 checkpoint, the cell receives and interprets information from a variety of internal and external sources. A decision is then made whether or not to continue with the cycle and initiate another round of cell division. If conditions are not appropriate for proliferation, or the cell receives orders to adopt an alternative fate, it withdraws from the cycle into a G_0 resting state. It can remain in this position until proliferative conditions are re-established, or initiate an alternative program resulting in differentiation, senescence, or apoptosis (Alberts et al., 1997).

The idea of a R point arose from analyzing how newly generated mammalian fibroblasts respond to nutrient and growth factor starvation (Zetterberg et al. 1982). If serum is removed up to an experimentally determined point, cells halt cell cycle progression in G_1 phase. Upon serum re-addition, completion of the cell division cycle is significantly extended compared with continually fed cells. Thus, starvation not only blocks cell cycle progression, but causes cells to exit the cycle and enter G_0. However, if serum is removed after this point, cells continue through the cycle unhindered (Zetterberg and Larsson 1985). Subsequent analysis identified other criteria that differ before and after this period in G_1. Up until the R point, cells stop cycling in response to low concentrations of cyclohexamide (a protein synthesis inhibitor), while after the R point they are resistant (Pardee 1989). These observations suggested a molecular switch (such as an unstable protein) might define R point control (Zetterberg and Larsson 1991).

2. The First Organisms Studied

Since the 1940s, the efforts of many scientists have developed techniques for genetically analyzing budding yeast to the point where it is the organism of choice for genetic studies of the cell cycle (Murray and Hunt, 1993).

Budding yeast introduces us to the somatic cell cycle, whose organization differs from that of the early embryonic cell cycle. In early embryonic cell cycles, the inactivation of MPF and exit from mitosis commit cells to replicating their DNA and microtubule organizing center. In the somatic cell cycle, however, and additional transition, which takes place during G_1, must occur before the cell can initiate replication. This transition is called Start in unicellular eukaryotes and the R point in multicellular ones, although we believe that it is fundamentally the same in both types of cell (Murray and Hunt, 1997).

Start divides interphase into two periods. Before Start, cells are in G_1; they have unreplicated chromosomes and a microtubule organizing center. Passage through Start induces replication of DNA and the microtubule organizing center, leading to a new state that corresponds to G_2 in the conventional view of the cell cycle. Start act as a checkpoint at which the progress of the cell cycle engine can be regulated by events inside and outside the cell (Alberts et al., 1997).

Much of what we know about the cell cycle was discovered through the isolation of temperature sensitive (*ts*), cell division cycle (*cdc*) mutants. In 1970 Hartwell et al. discovered that a number of these *ts* mutants, upon shifting to the restrictive temperature, arrested the cell population with the same morphology, suggesting that the

mutant product was required only at a specific point in the cell cycle. Approximately 60 different *CDC* mutants have been isolated in budding yeast, and approximately 30 have been isolated in fission yeasts. In addition to *CDC* genes, a large number of new cell cycle genes have been identified on the basis of interactions with preexisting cell cycle genes (MacNeill and Nurse, 1997; Lew et al., 1997).

Detailed analysis of all the cdc^{ts} mutants produced a logical map of the cell cycle. Start induces three independent pathways, of which one leads to budding, another to spindle pole body duplication, and the third to the initiation of DNA replication. The genetic analysis of budding yeast cdc^{ts} mutants presented a very different picture of the cell cycle from the one derived from studies on frog eggs. At first glance, the complex interdependencies in the yeast cell cycle bore no resemblance to the free-running cell cycle engine of the frog egg (Murray and Hunt, 1993).

By 1984 research on yeast and frogs had generated two very different views of the cell cycle. Experiments on frog eggs suggested that the cell cycle was driven by a free-running cell cycle engine that produced regular oscillations in the activity of MPF (Maturation Promotor Factor). In frog experiments the transfer of cytoplasm from eggs to oocytes demonstrated that meiosis is dominant to interphase. Isolated oocytes were treated with progesterone to induce them to mature into unfertilized eggs. A hollow microneedle was used to remove about 5% of the cytoplasm from an egg and inject it into an oocyte. The recipient oocyte matured as if it had been treated with progesterone. The injected oocytes that had matured were then used as cytoplasmic donors to inject fresh oocytes. This second round of recipients also matured, and they could in turn act as donors of cytoplasm that induced yet another round of recipient oocytes to mature (Masui and Markert, 1971; Newport and Kirschner, 1984).

Fertilization leads rapidly to inactivation of MPF and interphase of the first mitotic cell cycle. After fertilization, MPF activity rises each time the embryo enters mitosis and then falls as it enters the next interphase. When fertilized frog eggs were treated with drugs that inhibit DNA synthesis, the regular rise and fall of MPF activity continued unabated. As MPF activity increased, the nuclei broke down, and as it decreased, the nuclei re-formed, even though DNA replication was completely blocked. Treatment with drugs that inhibit spindle assembly also failed to prevent the regular oscillation of MPF activity or the response of the nuclei to its rise and fall. Even in enucleated eggs MPF rose and fell normally. Together these experiments show that, in frog embryos, purely cytoplasmic reactions activate and inactivate MPF and drive the nucleus, whether ready or not, into and out of mitosis (Murray and Hunt, 1993).

The regular oscillations in MPF activity drive the cell cycle but did not reveal the molecular nature of MPF or the enzymes that turn it on and off. The only clue was that protein synthesis was required to activate MPF in every cell cycle. Perhaps the newly made proteins were components, or activators, of MPF. It now seems obvious that in either case the critical proteins would have to be used up or destroyed by passage through mitosis. Many unsuccessful attempts were made to identify periodic protein, whose periodic synthesis had been proposed to induce cell cycle transitions (Gehart and Kirschner, 1984).

The breakthrough came from studies of proteins synthesis in sea urchin eggs. Newly fertilized eggs were incubated with radioactive amino acid and sample every 10 minutes to analyze the pattern of radioactively labeled proteins. The amount of radioactivity in most proteins increased continuously throughout the experiment, but one protein behaved quite differently. It disappeared abruptly at the end of each mitosis and then gradually reappeared during the next interphase. Additional experiments showed that cyclin, as it is now called, is a periodic protein synthesized throughout the cell cycle but degraded at the end of each mitosis (Evans et al., 1983).

The behavior of cyclin suggested a simple model of the cell cycle engine, in which the activation of MPF drives cells into mitosis and the decline of MPF activity leads to the next interphase. The model rests on three postulates: the accumulation of cyclin during interphase activates MPF; active MPF induces the destruction of cyclin, and the destruction cyclin leads to inactivation of MPF. In this model the cell cycle engine flip-flops periodically between mitosis and interphase. The accumulation of cyclin in interphase leads to the activation of MPF and the induction of mitosis, and the ability of MPF to induce the degradation of cyclin leads to the inactivation of MPF and the next interphase (Murray and Hunt, 1993).

In most animal cells, the two stages of M phase (mitosis and cytokinesis) occur in rapid succession before entry into the following G_1 phase. In some cell types, however, cytokinesis is a relatively slow process that is not completed until long after mitosis, and can even occur after S phase of the following cell cycle. In the early embryos of sea urchins, for example, DNA synthesis begins in the last stage of mitosis and is completed shortly thereafter, while cytokinesis takes longer and is not finished until after the next S phase. Similarly, in rapidly dividing fission yeast, S phase occurs soon after nuclear division but the final separation of daughter cells occurs after S phase is complete (Alberts et al., 1997).

The early embryos of the *Drosophila melanogaster* provide an extreme example of the dissociation of cell division from nuclear division. The remarkably rapid cell cycles of these embryos (about 8 minutes each) occur without cytokinesis, resulting in the formation of a multinucleate cell, or syncytium, with thousands of nuclei in a shared cytoplasm. Only after about 13 rounds of nuclear division does a cytoplasm membrane from around each nucleus, resulting in the division of the syncytial cytoplasm into individuals cells (Murray and Hunt, 1993).

In the majority of cells, M phase does not begin until after end of S phase. The budding yeast, *Saccharomyces cerevisiae*, is an exception to this general rule. In this organism the entry into mitosis is not as extensively regulated as in most higher eukaryotes, and preparations for mitosis, such as the assembly of the mitotic spindle, can sometimes begin before DNA synthesis is completed. These cells therefore lack a clearly defined G_2 between S and M phases (Alberts et al., 1997).

3. Cell Cycle in Mammalian

As for other eukaryotic cells, the mammalian cell cycle is regulated by the sequential formation, activation, and inactivation of a series of cell cycle regulatory molecules that include the cyclins and the CDKs (Alberts et al.,1997).

Specific cyclin/CDK complexes become activated and thereby modulate a distinct phase(s) of the cell cycle. The regulation of cyclin synthesis and degradation, in addition to CDK activity levels, are tightly controlled and is key to ordered progression through the mammalian cell cycle (Murray and Hunt, 1993).

The mammalian cell cycle consists of a number of checkpoints that exist to ensure normal cell cycle progression. The primary checkpoint acts late in G_1 and it is the R point.

Once cells have passed this point, they normally are committed to a round of cell division. Other checkpoints exist in S phase to activate DNA repair mechanisms if necessary and at the G_2/M transition to ensure that cells have fully replicated their DNA and that it is undamaged before they enter mitosis. Finally, there are checkpoint control mechanisms within mitosis to ensure that conditions remain suitable for the cell to complete cell division. The length of time for a mammalian cell to progress around the cell cycle and undergo division varies depending on the cell type but on average it takes approximately 24 h (Alberts et al., 1997).

3.1. The G_0/G_1 Transition

The mammalian cell cycle is influenced by external signals during the G_0 and G_1 phases. The MAPK (Mitogen-Activated Protein Kinase) cascade is one of the most ubiquitous signal transduction pathways; it regulates several biological processes including progression of the cell cycle. The MAPK cascade consists of three evolutionarily conserved protein kinases: MAPKKK (MAPK Kinase Kinase), MAPKK (MAPK kinase), and MAPK, which are activated sequentially in a Ras-dependent manner (Marshall, 1995).

The MAPK cascade influences cellular proliferation by targeting the cyclin D-dependent kinases (Lavoie et al., 1996; Cheng et al., 1998; Balmanno and Cook, 1999). Evidence for this comes from the fact that cells that proliferates in the absence of mitogens, for example during embryogenesis, have very little cyclin D-dependent kinase activity (Meyer et al., 2002).

3.2. The G_1/S Transition

One of the most extensively studied substrates of the cyclin/CDKs is the retinoblastoma (Rb) family of pocket proteins. The Rb family of pocket proteins comprises a group of tumor suppressor proteins consisting of three members; pRb, p107, and p130. These proteins contain a pocket region that binds cellular targets. This region also is capable of binding a number of viral oncoproteins such as the adenovirus E1A protein, SV40 large T antigen, and the human papillomavirus 16E7 protein (Vousden, 1995), demonstrating one mechanism by which tumor viruses can interfere with cell cycle progression in mammalian cells. In addition to phosphorylation events, the functions of different Rb family members are also regulated by changes in expression. During G_1, the Rb pocket proteins are found in a hypophosphorylated state in which they bind to members of the E2F transcription factor family. As cells progress through the cell cycle, these proteins become hyperphosphorylated as a result of phosphorylation by cyclin D/CDK4(CDK6) and cyclin E/CDK2 complexes. Each family member also displays differential expression throughout the cell cycle. Thus, pRb is expressed throughout the cell cycle but is hyperphosphorylated and therefore inactivated in late G_1, although by mitosis it becomes dephosphorylated; p130 is highly expressed in G_0, whereas levels diminish as cells progress into S phase, consistent with a role for p130 in maintaining quiescence (Grana and Mayol, 1998); and p107 shows a reciprocal expression pattern to p130 such that low levels are found in G_0, which then increase as cells progress through G_1 into S.

During very early development cells make a decision whether to become part of the inner cell mass that eventually forms the embryo, or commit to become extraembryonic cells (such as trophoblasts) that help establish the placenta needed for proper development (Alberts et al., 1997). Rb loss causes hyperproliferation of these extraembryonic trophoblast cells, resulting in severe disruption of placenta architecture required for viability. Remarkably, Rb-/- embryos supplied with a wild-type placenta were carried to term and only died after birth. The animals exhibited no defects in the hematopoietic or nervous systems, suggesting that earlier phenotypes might be the result of non-cell autonomous effects. These results are quite similar in outcome to partial rescue of placenta defects in the cyclin E knockouts. Rb is therefore largely dispensable during embryogenesis; cells still maintain control of G_1 progression and respond appropriately to extracellular signaling. Given the presumed importance of controlling E2F activity during the G_1 phase, it will be interesting to see if Rb family members p107 or p130 assume this burden (Wu et al., 2003).

Rb-/- cells displayed increased levels of E2F dependent transcripts and premature synthesis of cyclin E, resulting in a shorter G_1 phase and smaller cells that grew faster than wild-type cells. They were less responsive to mitogen removal and failed to G_1 arrest in response to cyclohexamide, consistent with the idea that Rb makes an important contribution to R point control and hence G_1 progression. However, it now seems possible that these earlier results are non-cell autonomous and arise from placenta defects. Given the ability of embryogenesis to proceed in the absence of cyclin D-CDK4/6 and Rb, it seems reasonable to ask whether E2F is a critical component in this process. E2F/Dp complexes activate or repress transcription of target genes depending on presence or absence of Rb (Frolov and Dyson, 2004). Thus, loss of E2F transcription would be expected to severely compromise a cell's ability to proceed through G_1. The E2F family has seven members, six of which have been individually ablated in the mouse (Trimarchi and Lees 2002).

Phenotypes are tissue specific and indicate extensive functional overlap amongst family members. However, the E2F1-3 triple knockout prevents proliferation of primary mouse embryonic fibroblasts, consistent with a requirement for E2F activity during normal development (Wu et al., 2001). These observations indicate E2F transcriptional activity is required for development both in the mouse and in cultured cells. However, there appear to be fundamental differences in how its activity is positively and negatively regulated in different systems.

3.2.1. E2F Transcription Factor

To date, seven E2F members have been described (Dyson, 1998) and these molecules exist as heterodimers paired with a DP subunit. Two mammalian DP genes have so far been identified (DP-1 and -2) (Nevins, 1992). E2F and DP proteins contain highly conserved DNA-binding and dimerization domains. The E2F and DP proteins activate transcription in a synergistic manner, and DP proteins appear to act indirectly by enhancing the activity of E2F (Helin et al., 1993).

The Rb pocket proteins bind to, and sterically hinder transcriptional activity of, the E2F/DP complex, thereby enabling the E2F transcription factors to act as repressors of gene transcription (Dyson, 1998). Phosphorylation of the pocket protein component of the E2F/pocket protein complex by cyclin D/CDK4(CDK6) complexes in the G_1 phase of the cycle leads to dissociation of the phosphorylated pocket protein and E2F, followed by E2F-mediated transactivation of promoters of genes necessary for S phase progression (Kaldis, 2006).

E2F transcription factors are divided into three main groups: E2Fs 1–3, which play a role in progression from G_1 into S phase of the cell cycle and possess a pRb binding site within their transactivation domain; E2F-4 and -5, which bind to p107 or p130 members of the pocket protein family, and E2F-7—these three play a role in differentiation and proliferation; and E2F-6, which is unique since it lacks both the transactivation and the pocket protein binding domains. E2F-6 (also known as EMA) is thought to regulate cell cycle progression via its role as a transcriptional repressor (Morkel et al., 1997; Trimarchi et al., 1998; Gaubatz et al., 1998; Trimarchi et al., 2001). Although overexpression of E2F-6 did not block cycling NIH3T3 (mouse fibroblast cell line) fibroblasts from entry into S phase of the cycle, there was a significant decrease in S phase entry when G_0 arrested E2F-6 overexpressing cells were stimulated to reenter the cell cycle with serum (Gaubatz et al., 1998).

3.2.2. Cyclin E/CDK2 in G_1/S Transition

As cells approach the G_1 /S border, control of the cell cycle becomes dominated by cyclin E/CDK2 complexes. It has been demonstrated that overexpression of cyclin E/ CDK2 promotes S phase entry and that blocking the kinase activity of this complex inhibits progression into S phase (Ohtsubo and Roberts, 1993; Tsai et al., 1993; Resnitzky et al., 1994).

Cyclin E/CDK2 has other G_1 targets since it is still required under these conditions. In addition to helping remove negative impediments such as Rb and p27, cyclin E/CDK2 carries out various tasks required for upcoming S and M phases. It helps license replication origins to ensure DNA is only duplicated once per cell cycle. This process involves assembly of a pre-replicative complex (PRC) after mitosis, which then recruits MCM (minichromosome maintenance) helicases onto the DNA (Coverley et al.,, 2002; Diffley and Labib 2002). Cyclin E/CDK2 may then participate in subsequent origin firing during S phase (Woo and Poon 2003).

Cyclin E/CDK2 has also been implicated in modulating chromatin structure. Histone H1 continues to serve as a substrate for purified cyclin E/CDK2, but the physiological relevance of this reaction remained enigmatic. Recent evidence suggests cyclin E/CDK2 phosphorylates histone H1 in cells, destabilizing its interaction with chromatin (Contreras et al.,, 2003). In addition, cyclin E/CDK2 influences chromatin structure by phosphorylating Rb and altering its association with histone deacetylases (HDAC) at E2F promoters (Takaki et al. 2004).

CDK2 and CDK3 are related kinases associating with E and A type cyclins to drive G_1 progression and initiate S phase (Sherr, 1994; Morgan 1995). Work in cultured cells showed that dominant negative CDK2, microinjection of CDK2 antibodies, CDK2 antisense, and CDK2 inhibitors (protein and small molecule) all block DNA synthesis and cell proliferation (Sherr and Roberts 1999).

CDK2 is not essential for mitotic cell division of most if not all cell types, it is necessary for completion of prophase 1 during meiotic cell division in male and female germcells. This requirement explains sterility of both male and female knockouts.

Cyclins E1 and E2 are closely related and associate with CDK2/3 to drive G_1 progression and prepare for S phase (Sherr 1994; Geng et al.,, 2001). Cyclin E/CDK2 activity is thought to be required because blocking its function with microinjected cyclin E1 antibodies, cyclin E1 antisense, inhibitory peptides derived from p27 or p21, and small drug inhibitors all result in G_1 arrest (Sherr and Roberts 1999). Cyclin E/CDK2 activity also appears at least partially rate limiting for G_1 progression because cyclin E overexpression shortens G_1.

3.2.3. Cyclin D-CDK4/6

The main roles of cyclin D-CDK4/6 are Rb inactivation and sequestration of the cyclin E/CDK2 inhibitor p27 (Sherr and Roberts 1999). Cyclin D-CDK4/6 has also been implicated in phosphorylation of SMAD3 to inhibit transcriptional complexes responsible for cell growth inhibition by the TGF beta family (Matsuura et al. 2004). In each of these cases cyclin D-CDK4/6 removes an impediment to cell cycle progression. Similarly, cyclin E/CDK2 inactivates Rb and its own inhibitor p27 to alleviate inhibitory thresholds in G_1. Therefore, ablating either of these impediments in the mouse was predicted to cause inappropriate proliferation detrimental to development and/or survival.

CDK4 and CDK6 are closely related kinases associating with D-type cyclins to initiate G_1 progression in response to proliferative signals (Sherr 1994). This idea arose from extensive work in cultured cells showing: 1) overexpressed cyclin D1 shortens G_1 phase, suggesting activity of cyclin D1/CDK4 is rate limiting for G_1 progression, 2) cyclin D1 overexpression overcomes a G_1 arrest caused by DNA damage or the unfolded protein response, 3) microinjected cyclin D1 antibodies, cyclin D1 antisense, inhibitory peptides derived from p16INK4a, and small drug inhibitors all result in G_1 arrest (Sherr and Roberts 1999).

Mice lacking CDK4 are viable and display proliferative defects in a limited range of endocrine cell types, indicating CDK4 is dispensable for proliferation in most situations (Tsutsui et al.,, 1999; Moons et al. 2002). Likewise, CDK6-/- mice are also viable and for the most part develop normally (Malumbres et al.,, 2004). CDK6 is preferentially expressed in hematopoietic cells, and its absence leads to delayed G_1 progression in lymphocytes but not mouse embryo fibroblasts (MEFs) (Meyerson et al. 1992).

During mouse development D-type cyclins and CDK4/6 are only necessary in a few select compartments. While surprising, these results are consistent with earlier observations that cyclin D or CDK4 ablation affects postembryonic growth but not embryonic development in *Caenorhabditis elegans* and *Drosophila* (Datar et al.,, 2000; Meyer et al.,, 2002). Cells derived from knockout mice lacking CDK4/6 or all forms of cyclin D can clearly proliferate and respond to mitogens, in apparent contradiction to previous work suggesting that cyclin D-CDK4/6 is ubiquitously required (Sherr and Roberts 1995). It therefore remains unclear how mitogen responsiveness is linked to cell cycle progression. The cell culture derived model of G_1 progression posits that cyclin D generates cyclin E in order to connect extracellular signals and control of S phase entry. In support of this interpretation, all cyclin D1-/- mice phenotypes were rescued by inserting cyclin E into the cyclin D1 gene locus

(Geng et al.,, 1999). This explanation is now called into question because cells lacking D type cyclins or CDK4/6 still express cyclin E and enter S phase (Kozar et al,. 2004). Based on these surprising results, it is necessary to reconsider whether a linear series of interdependent events initiated by cyclin D-CDK4/6 lies at the heart of G_1 progression.

3.2.4. Cyclin-dependent kinase inhibitors

The cyclins and CDKs often are referred to as positive regulators of the eukaryotic cell cycle, but there are the negative regulators called CDKIs (Cyclin- Dependent Kinase Inhibitors).The CDKIs comprise the INK4 (inhibitor of CDK4) and CIP (CDK-Interacting Protein)/KIP (Kinase Inhibitor Protein) (Humphrey and Brooks, 2006).

The INK4 includes p14, p15, p16, p18 and p19, which specifically inhibit cyclin D/CDK4 and cyclin D/CDK6 at G_1 phase (Humphrey and Brooks, 2006).

The CIP/KIP includes p21, p27, and p57 which interact with, and inhibit the kinase activities of, cyclin E/CDK2, cyclin D/CDK4, cyclin D/CDK6, cyclcin A/CDK2, and cyclin B/CDC2 complexes and also function throughout the cell cycle (Alberts et al.,, 1997).

p27 is thought to establish a critical threshold of CDK inhibition that must be overcome before cells can proceed through G_1 and initiate DNA synthesis (Sherr and Roberts 1995). The p27 inhibitory threshold is overcome in part by cyclin E/CDK2 phosphorylation of p27 at T187, which marks it for ubiquitination and degradation by the proteasome (Sherr and Roberts 1999).

Current understanding of p27's role in the cell is based largely on its ability to inhibit cyclin/CDK complexes. However, increasing evidence suggests p27 performs CDK independent functions that may better explain its effects on cell fate determination and tumor suppression, as well as phenotypes of the p27 knockout mouse (Moeller et al. 2003).Members of CIP/KIP bind to, and inhibit the activity of, the entire cyclin/CDK complex, whereas INK4 family members block cyclin binding indirectly by causing allosteric changes in the CDK and hence altering the cyclin binding site; they also act by distorting the ATP binding site that leads to reduce affinity for ATP (Kaldis, 2006).

3.3. S Phase

S phase is the point at which a cell duplicates its chromosomes in readiness for mitosis and cell division. A number of checkpoints exist to ensure that DNA is replicated only once per cycle, that it is fully and correctly replicated, and that replication occurs before cell division. Another important event during S phase is centrosome duplication. Failure of cells to coordinate centrosome duplication with DNA replication leads to abnormal segregation of chromosomes, causing genomic instability that can lead to cancer (Alberts et al., 1997).

DNA synthesis is known to occur at specific sites on the DNA known as origins of replication. The best characterized origins of replication are those found in *Saccharomyces cerevisiae* and are known as ARS (Autonomous Replication Sequences) (Marahens and Stillman, 1992). The ARS contains a highly conserved region of 100–200 bp known as the ARS consensus sequence (ACS), and this is an essential component of the origin of replication to which the origin recognition complex (ORC) binds. The ORC is conserved in all eukaryotes (Bell and Stillman, 1992).

ORC is highly conserved, with homologues identified in *A. thaliana, S. pombe, D. melanogaster, X. laevis, M. musculus, H. sapiens*, and others (Carpenter et al., 1996; Dhar and Dutta, 2000; Gavin et al., 1995; Gossen et al., 1995; Leatherwood et al., 1996; Masuda et al., 2004; Muzi-Falconi and Kelly, 1995; Pinto et al., 1999; Quintana et al., 1997; Quintana et al., 1998; Tugal et al., 1998). The ORC subunits have been shown to form a functional complex in *D. melanogaster* (Chesnokov et al., 1999), *S. pombe* (Moon et al., 1999), *X. laevis* (Gillespie et al., 2001), and *H. sapiens* (Dhar et al., 2001a; Vashee et al., 2001).

Three ORC subunits have been identified in humans, HsORC 1, 2, and 4 (Gavin et al., 1995), all of which are involved in the initiation of DNA replication by recruitment of specific factors to the DNA. Human ORC has been shown to interact with a HAT, and this may be involved in making the initiation site accessible, thereby facilitating replication (Iizuka and Stillman, 1999).

Human ORC shows no preference for known origins compared to random DNA sequences, but does show a slight preference for AT rich DNA While ORC may be responsible for DNA binding, the mechanism of such binding becomes unclear with increasing organism complexity, perhaps due to the increasing complexity of factors affecting origin selection. Although the intricacies of the ORC-DNA interaction are not fully understood, the general role of ORC in replication is now accepted (Vashee et al., 2003).

As a replicative *initiator*, ORC should be able to recognize the *replicator* sequences. ORC has been shown by many to bind DNA, and this binding is dependent on ATP and the ATP binding functions of ORC (Bell and Stillman, 1992; Chesnokov et al., 2001; Gillespie et al., 2001; Seki and Diffley, 2000). Specific *replicator* sequence association has been observed in *S. cerevisiae* and *S. pombe.* Work in the latter organism has revealed that Orc4 dictates the specificity via a newly defined AT hook region (Kong and De-Pamphilis, 2001; Lee et al., 2001). ORC from higher eukaryotes, however, does not seem to have the same sequence specificity.

As ORC is a key player in defining and recruiting a replication complex to an origin, it is an important potential target for controlling replication. In yeasts, the ORC remains bound to chromatin throughout the cell cycle. However, in *S. cerevisiae*, Orc2 and Orc6 are phosphorylated by S phase cyclin/CDK1 during the G_1 /S transition. This ORC phosphorylation was found to be part of a mechanism to limit origin firing activity to only once per cell cycle; when Orc2 and Orc6 phosphorylation site mutants were introduced with constitutively active CDC6 and MCM proteins, rereplication was observed (Nguyen et al.,, 2001; Wilmes et al.,, 2004). Similarly, *S. pombe* Orc2 is phosphorylated, which may be due to its similar interaction with CDK1/cyclin B in the G_2 phase. This interaction serves to prevent rereplication without an intervening mitosis, again ensuring only one replication event per cell cycle takes place (Wuarin et al., 2002).

3.3.1. CDC6

A key regulator of DNA replication in mammalian cells is CDC6. The CDC6 was originally identified in *S. cerevisiae* as a protein essential for cell cycle progression (Hartwell et al., 1974), and was thereafter shown to have an early DNA synthesis defect (Hartwell,

1976). CDC6 interacts with ORC, forming a complex with an extended nuclease protected DNA footprint (Cocker et al., 1996; Liang et al., 1995).

Immunodepletion of CDC6 in human cells blocks S phase entry (Hateboer et al., 1998; Yan et al., 1998) and has been shown to affect the interaction of ORC with MCM (Minichromosome Maintenance) proteins but not its interactions with DNA (Coleman et al., 1996; Cook et al., 2002). These data suggest that CDC6 may act as an adaptor protein for interactions of the ORC with other proteins (e.g., MCM proteins). Levels of CDC6 in cycling human cells remain fairly stable during S phase, G_2, and mitosis (Saha et al., 1998; Jiang et al., 1999), but lower amounts are present in early G_1 when CDC6 is degraded by proteolysis (Mendez and Stillman, 2000; Petersen et al., 2000).

Phosphorylation of CDC6 is carried out by cyclin A/CDK2.By phosphorylating CDC6 in S and G_2/M, its activity was limited, preventing inappropriate origin firing in mitosis. CDC6 was also found to be marked for degradation at the G_1/S transition by Clb/CDC28 and the CDC4/CDC34/CDC53 ubiquitination machinery, adding a further layer of regulation (Drury et al., 1997; Elsasser et al., 1999).

Human CDCD6 is partially cell cycle regulated; it is under the control of the E2F transcription factor, which is responsible for promoting expression of numerous genes required for proliferation (Ohtani et al., 1998; Yan et al., 1998). Human CDC6 may not be degraded at the G_1 /S transition, but has been found to be exported from the nucleus in a phosphorylation dependent manner (Delmolino et al., 2001; Fujita et al., 1999; Jiang et al., 1999; Saha et al., 1998). This phosphorylation is controlled by cyclin A/CDK2, which allows for export of CDC6 soon after replication has initiated (Petersen et al., 1999). Some evidence exists that human CDC6 is degraded in Sphase, just as in yeast, a point that will need resolution (Coverley et al., 2000; Mendez and Stillman, 2000). Recently it was also reported that although exogenous CDC6 is exported from the nucleus in S phase, endogenous CDC6 is not (Alexandrow and Hamlin, 2004). One further study demonstrates that although CDC6 is released from chromatin and then degraded, it is constantly being resynthesized and immediately binds chromatin, replacing molecules which were displaced (Biermann et al., 2002). This observation may reconcile earlier observed differences, allowing CDC6 to bind chromatin in S phase in a tenuous manner so that rapid regulation can be achieved when needed.

3.3.2. MCM

In humans, MCMs seem to dissociate during S phase, and reassociate in late G_2/M early G_1 (Mendez and Stillman, 2000). This dissociation could be a critical step to prevent rereplication. Once MCM complex is removed, it cannot reassociate until an intervening mitosis occurs. Once it is removed from chromatin, it requires the activity of ORC and CDC6 to be reloaded. However, these proteins are unable to reload MCM in S phase; they are held inactive by CDK activity through various mechanisms described above. Therefore, in order to reload MCM, a period of low CDK activity must be reached (in G_1) to activate these proteins, allowing them to reload the MCM complex (Seki and Diffley, 2000).

3.3.3. Checkpoints

During S phase of the cell cycle, the entire genome is duplicated by DNA replication forks. In addition to duplicating the genome, replication forks also function to scan the genome for DNA damage or other types of interference, and to translate these replication stresses into DNA structures that can be recognized by the checkpoint sensors. DNA repair also plays important roles in the processing of DNA lesions and the activation of checkpoint. The activation of checkpoint signaling eventually leads to regulation of many cellular processes that can help cells to remove or overcome the stresses on their genomes and to maintain genomic stability. Furthermore, the checkpoint seems to regulate different cellular processes in a coordinated manner, giving the cells the greatest probability to survive under stressed conditions (Kaldis, 2006).

It is important to remember that many processes regulated by the checkpoint can crosstalk to each other, and they might also affect how checkpoint signals flow in cells. It should also be noted that although checkpoint activation can result in responses at the cellular level (e.g. arrest of the cell cycle), many other checkpoint-mediated responses might take place more locally at the sites of DNA damage. Instead of being a on-or-off switch, the checkpoint control on many cellular processes might be quantitative. In many cases, the checkpoint regulates multiple effectors involved in the same cellular process important for the stability of genome, ensuring that these processes are under fine-tuned controls (Humphrey and Brooks, 2006).

The tumor suppressor protein p53 also plays an important role in cell cycle arrest at the G_1 and G_2 checkpoints subsequent to inducing apoptosis (Prives, 1993). The p53 protein has a central sequence-specific DNA binding domain and a transcriptional activation domain at its amino-terminus; in response to DNA damage, it can induce the transcription of the CDKI p21, which inhibits the activation of various G_1 cyclin/CDK complexes (Ewen, 1996; Gartel et al., 1996). Antiproliferative signals such as contact inhibition, senescence (Alcorta et al., 1996), extracellular antimitogenic factors (Reynisdottir et al., 1995), and cell cycle checkpoints, such as p53 (el-Deiry et al., 1993), induce expression of p27, p16, p15, and p21, respectively. The role of cell cycle molecules in regulating proliferation is highlighted by the fact that a number of these molecules are found to be mutated or deregulated in numerous tumors. Indeed, it has been suggested that most human tumours result from a mutation or deletion in one or more cell cycle regulators, especially those that control G_1 /S progression. For example, *p16* is mutated in approximately one-third of all human cancers (Serrano, 1997; Kamb et al., 1994; Nobori et al., 1994), and *p53* is the most frequently mutated gene identified in human tumors (Levine, 1997). Also, many types of tumors show low expression levels of *p27*, which is associated with a poor prognosis (Porter et al., 1997), and cyclin D1 (Sherr, 1996; Hunter and Pines, 1994) and B1 (Yu et al., 2002) are often found at increased levels in breast cancer. Furthermore, there is a direct correlation between inactivation of p53 function and cyclin B1 overexpression in many tumors (Yu et al., 2002), although no direct interaction between these two molecules has been shown. However, as a direct consequence of this correlation, it has been proposed that cyclin B1 could be used as a tumor antigen and a cancer vaccine in some instances (Finn, 2003). CDKs have also been found to be deregulated in some tumors; for example, CDK4 is mutated in melanoma (Wolfel et al., 1995; Zuo et al., 1996), and CDK2 expression is increased in some breast cancer cells. Indeed, targeting

CDK2 expression with antisense oligonucleotides and RNA interference technologies reduces cellular proliferation in breast tumor cells (Payton et al., 2004).

3.4. The G_2/M Transition

Cyclin B/CDC2 is the key mitotic regulator of the G_2/M transition and was originally identified as the MPF, a factor capable of inducing M phase in immature *Xenopus* oocytes (Masui and Markert, 1971; Dunphy et al., 1988; Gautier et al., 1988). As is the case with other cyclin/CDK complexes, activation of the cyclin B/CDC2 complex is tightly regulated by phosphorylation and dephosphorylation events and also changes in subcellular localization (Smits and Medema, 2001; Takizawa and Morgan, 2000).

3.4.1. Cyclin B/CDC2 in G_2/M Transition

Cyclin B synthesis begins at the end of S phase (Pines and Hunter, 1989). Two cyclin B isoforms exist in mammalian cells, cyclin B1 and B2. Cyclin B/CDC2 complexes are regulated both positively and negatively by phosphorylation. Phosphorylation of CDC2 is required for activation, as is the case with all CDKs, and this phosphorylation event is mediated by CAK. During G_2, cyclin B/CDC2 complexes are held in an inactive state by phosphorylation of CDC2 Thr14 and Tyr15. Phosphorylation on Thr14 prevents ATP binding (Edicot et al., 1994), whereas that on Tyr15 interferes with phosphate transfer to the substrate owing to its positioning in the ATP binding site on CDC2 (Altherton-Fessler et al., 1993). These inhibitory phosphorylation events are carried out by the kinases Weel and Mytl; Weel specifically phosphorylates Tyr15, and Mytl phosphorylates both Tyr15 and Thr14, with a stronger affinity for Thr14 (Lundgre et al., 1991; Parker and Piwnica-Worms, 1992; Li et al., 1995). Cyclin B/CDC2 becomes fully activated following dephosphorylation of these sites by the protein phosphatase CDC25C.

3.5. M Phase and Cytokinesis

Mitosis and cytokinesis constitute the shortest phase of the eukaryotic cell cycle. Mitosis itself comprises five distinct phases as follows:

1. *Prophase*: where begins chromosomes condensation in the nucleus and ends with breakdown of the nuclear envelope.
2. *Prometaphase*: at this stage the mitotic spindle forms.
3. *Metaphase:* all chromosomes are bioriented and positioned near the spindle equator. All chromosomes align themselves along the metaphase plate.
4. *Anaphase*: the sister chromatids that comprise each chromosome separate to form two independent chromosomes.
5. *Telophase*: this is the final stage of mitosis, in which the chromosomes decondense and a nuclear envelope forms around each set of chromatids. The contractile ring

begins to form in readiness for the cell to split into two daughter cells, each with one nucleus.

Cytokinesis is the process that ensure to cell division is taken to completion. Following chromosome segregation, the microtubules bundle in the midregion of the spindle, forming the spindle midzone. As the contractile ring contracts, it creates a membrane barrier between each cell. The spindle midzone remains connected, forming a cytoplasmic bridge until this is finally cut during abscission (Cao and Wang, 1996). Specific CDKs also play a role in cytokinesis, and it has been shown that mammalian cells injected with a nondestructible form of cyclin B undergo anaphase and chromosome segregation but do not form a spindle midzone and fail to undergo cytokinesis suggesting a role for cyclin B in inhibiting cytokinesis (Wheatley et al., 1997).

4. Research in Cell Cycle Control in Embryonic Period

4.1. Bovine Embryos in Preimplantation Period as Animal Model

Some studies showed the Y-specific genes expression in preimplantation embryos. Increased mitotic rates in male embryos with one X chromosome and one Y chromosome were attributed to precocious expression of Y genes like *TSPY* (*Testis-Specific Protein, Y-encoded*) and *SRY* (*Sex-determining Region on Y chromosome*) which show mitogenic function (Mittwoch, 1993; Gutiérrez-Adán et al., 1997; Bernardi et al., 1996).

TSPY was discovered 20 years ago (Arnemann et al. 1987) and could be involved with spermatogenesis regulation (Lardone et al., 2007). Oram et al. (2006) studied *TSPY* and observed this gene promoted cellular proliferation *in vitro* and tumorigenesis *in vivo*, a time that its ectopic expression was able to mediate G_2/M transition in HeLa cells.

The *SRY* is a gene with just one exon and codified a protein with a domain HMG (*High Mobility Group*) which act like a chromatin modeling factor. This gene belongs to *SOX* (*Sry-related HMG box*) family and it acts on the DNA by HMG domain link to a sequence of 6bp of the DNA that can promote the association of regulators elements linked to separated DNA regions forming a complex that control another genes activities (Sinclair et al., 1990; Haqq et al., 1993).

Zwingman et al. (1993) and Cao et al. (1995) related that *SRY* was expressed during mice embryo preimplantational development since from 2 cells stage until blastocyst.

In the experiments carried out by Gutiérrez-Adán et al. (1997) was observed othe *SRY* expression in bovine embryos in the 4-8 e 8-16 cells, morulae and blastocyst stages.

The differences in the embryonic development rate between males and females in the preimplantation period can be attibuted to double expression of X-genes from female embryos before X inactivation occurs (Kocchar et al., 2001).

Many studies about gene expression of the preimplantaion embryos analyzed X-genes, mainly *G6PD* (*Glucose-6-Phosphate Dehydrogenase*), *HPRT* (*Hypoxanthine Phosphoribosyl*

Transferase), *XIST* (*X-inactive Specific Transcript*) and *ZFX* (*Zinc Finger Protein X-linked*) (Gutiérrez-Adán et al.,2000; Peippo et al., 2002; Nino-Soto et al.,2007).

Experiments carried out by Gutiérrez-Adán et al. (2000) demonstrated that female bovine blastocyts expressed more *G6PD* and *HPRT* than males. This results can explain the differences in the development rates between male and female embryos because G6PD e HPRT enzymes participate of detoxification process of oxygen radicals and in spite of their deleterious effects they stimulate cell profieration too (Riegger, 1992).

Decreased levels of oxygen radicals in females due to double expression of *G6PD* and *HPRT* in preimplantation period could be responsible for more slow development of these embryos what it would explain the more male embryos produced *in vitro* in blastocyst stage (Gutiérrez-Adán et al., 2000).

Some groups have identified the same sequence homologue to *TSPY* present in X chromosome knew like: *CDA1* (*Cell Division Autoantigen 1*), *HRIHFB2216,* clone *se20-4*, *DENTT* (*Differentially Expressed Nucleolar TGFβ_1 Target*) and *TSPX* (Chai et al., 2001; Eichmuller et al., 2001; Ozbun et al., 2001, 2003; Delbridge et al., 2004).

Identification in TSPX protein of a SET/NAP conserved domain which define the superfamily TTSN (*TSPY, TSPY-like, SET, NAP-1*) suggest that TSPX can be a member from this family (Ozbun et al., 2001).

The *TSPX* gene play a role in cell cycle because it can be envolved in trancription, mRNA processing, replication and chromosomal organization (Chai et al., 2001; Ozbun et al., 2001).

Chai et al. (2001) verified that TSPX late the HeLa cells growth. This autors observed that TSPX can be phospholated by AMPc, GMPc - dependente kinases, kinase type C, thimidine kinase, casein kinase II e by D1/CDK4 , A/CDK2 and B/CDK1.

According to Tu et al. (2007) the *TSPX* overexpression induced the expression of others proteins inhibitors of CDKs including p27, p57, p16 e p15, point to the negative regulator function in cell cycle. *TSPX* is induced by TGF-β1 and it does not promote the colonies formation of NCI-H157, NCI-H727 and NCI-H1299 (lung cancer) lineage (Ozbun et al., 2001).

The *TSPX* expression is being investigated in embryos bovine in preimplantation period to know if this gene late the cell cycle of this animals.

4.2. Post Implantation Period and Perspectives of Research

Genetic epidemiology of human malformations has a high priority area for future research. Since 1999, The National Institute of Child Health and Human Development (NICHD), the National Institute of Dental and Craniofacial Research (NIDCR), the National Institute of Environmental Health Sciences (NIEHS) of the National Institutes of Health (NIH) and the Environmental Protection Agency (EPA) has been studying the genetic susceptibility and mechanisms of human congenital malformations, including research on the contribution of potential genetic and environmental factors, identified at the molecular level, to the etiology, distribution and prevention of disease within families and across populations.

Only in the United States, birth defects are the leading cause of infant mortality, accounting for one in five infant deaths. It is estimated that more than 120,000 babies in the United States (about four percent of all live births) are born with major birth defects each year. These birth defects are involved in about half of all pediatric hospital admissions and, next to accidents, are the leading cause of death in children.

Focusing the research, Israeli and German scientists using computational biology techniques discovered 480 genes that play a role in human cell division and identified more than 100 of those genes that have an abnormal pattern of activation in cancer cells (http://grants.nih.gov/grants/guide/rfa-files/RFA-HD-99-002.html).

In the case of fetal malformation, a great part of it can be caused by problems in the development of the tissue. We had special attempt for the cardiac muscle development.

The congenital cardiopathies has great frequent fetal malformation and present incidence of 3,5: 12:1000 children born alive. Considering the fetal losses in the first trimester of the gestation and the stillborn ones, the incidence of the fetal congenital cardiopathies is up to five times bigger that the values described. Approximately half of these malformations is complex and needs interventionist treatment after the birth, presenting in its natural history, and mortality of 20 40% in the period neonatal. The congenital cardiopathies are of multifactorial etiology and the majority occurs in gestations without risk factor, constituting the ultrasonografic examination the best form of detention of these malformations before birth (Carvalho et al., 2006).

Cardiac muscle development exhibit two related modes of growth, highly regulated during development and disease, rapidly proliferating during fetal life. Much research has focused on identifying the candidate mitogens, hypertrophic agonists, and signaling pathways that mediate these processes in isolated cells, what drives the proliferative growth of embryonic myocardium. Technologies for gain- and loss-of-function permits a mutational analysis of these growth control pathways (MacLellan & Schneider, 2000).

In our service, one aim of our research is the hipoplastic left heart syndrome (HLHS) and the mechanisms that involve this disease. The HLHS (Figure 1) is a spectrum of syndromes displaying underdevelopment of the left-sided heart structure, characterized by aortic valve atresia or severe stenosis with a small ascending aorta and hypoplasia of the left ventricle (Sano et al., 2005). The natural history of HLHS is obscure, generally fatal within the first month of life without treatment [Silva et al., 2007].

The HLHS (OMIM 241550) can be caused by mutation in the GJA1 gene (http://www.ncbi.nlm.nih.gov/entrez/dispomim.cgi?cmd=entry&id=241550). This gene is responsible for the production of one type of gap junction protein, called connexin 43 or CX43 heart connexin (http://www.ncbi.nlm.nih.gov/Omim/getmap.cgi?l121014).

The gap junctional intercellular communication (GJIC) plays an important role in the maintenance of tissue independence and homeostasis in multicellular organisms. The GJIC are specializations of membrane that form channels between two connected cells. The structural proteins comprising these channels, collectively called connexins, are members of a highly related multigene family, consisting of at least 13 members expressed in mammals. Each channel is composed by two hemichannels (connexons), each one yielded by each cell of the pair. Each connexon is one hexameric assemblies of the connexin transmembrane protein (Yamasaki et al., 1999) (see Figure 2).

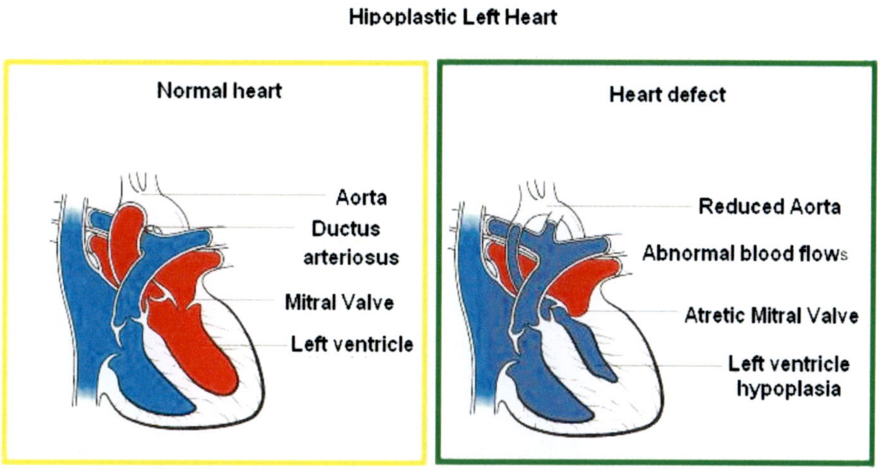

Figure 1. Schematic comparison between the normal heart and the hipoplastic left heart.

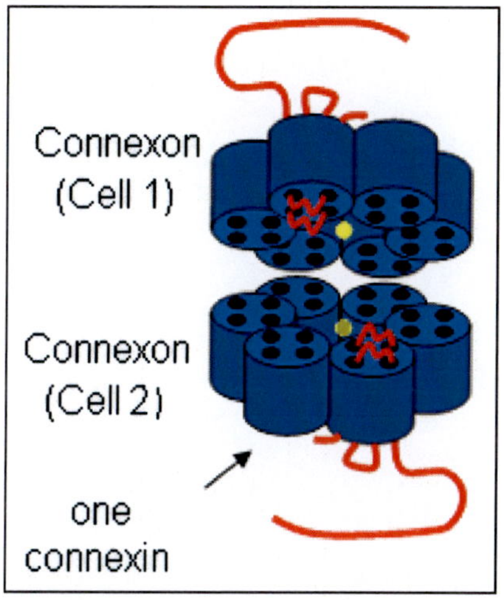

Figure 2. Schematic vision of connexin composed by hemichannels of connexons.

This intercellular communication permits coordinated cellular activity, a critical feature for organ homeostasis during development and adult life of multicellular organisms. The possible involvement of aberrant GJIC in abnormal cell growth and carcinogenesis has long been postulated in recent studies and confirmed that expression and function of connexin genes play an important role in cell growth control. In their studies, Yamasaki et al (1999) noticed that connexin expression per se, is more closely related to growth control specific cell types. Furthermore, the importance of GJIC in various physiological functions is verified by deletion of different connexin genes that results in various disorders including heart malformation, cataract, cancers and others.

Since the cloning of the first connexin in 1986, considerable progress has been made in our understanding of its function. The connexins are proteins components of gap junctions which allow cytoplasmic water-soluble molecules (Bruzone et al., 1996). All members of the connexin protein family share a very similar tertiary structure (Yamasaki et al., 1999). The GJIC was postulated to contribute to the maintenance of tissue homeostasis and coordinated cell growth, controlling the development of small cell populations which manifest aberrant cell growth and thus helping to suppress tumor development (Loewenstein, 1981).

Our research is focused in the study of the mutations of the GJA1 gene in the patients attempt at the São Paulo Hospital of the São Paulo Federal University-Paulista Medical School, and the mechanism involving the connexin 43 in the cell cycle control during the development of cardiacle muscle.

References

Books

Alberts, A; Bray, D; Lewis, J; Raff, M; Roberts, K; Watson, JD (1997). *Molecular Biology of the Cell* (3rd edition). New York. Garland.

Humphrey, T & Brooks, G (2006). *Cell Cycle Control: Mechanisms and Protocols*. New Jersey. Humana Press.

Kaldis, P (2006).*Cell Cycle Regulation: Results and Problems in Cell Differentiation*. New York. Springer-Verlag Berlin Heidelberg.

Lew, D. J., Weinert, T., and Pringle, J. R. (1997) *Cell cycle control in Saccharomyces cerevisiae, in Molecular and Cellular Biology of the Yeast Saccharomyces* (Pringle, J. R., Roach, J. R., and Jones, E. W., eds.), Cold Spring Harbor Laboratory Press, Cold Spring Harbor, NY, pp. 607–695.

MacNeill, S. A. & Nurse, P. (1997) *Cell cycle control in fission yeast, in Yeast III* (Pringle, J. R., Broach, J. and Jones, E. W., eds.), Cold Spring Harbor Labroratory Press, Cold Spring Harbor, New York.697–763.

Murray, A & Hunt, T (1993). *The Cell Cycle An Introduction*. New York. Oxford University Press.

Articles in Journals

Alcorta, D. A., Xiong, Y., Phelps, D., Hannon, G., Beach, D., and Barrett, J. C. (1996) Involvement of the cyclin-dependent kinase inhibitor p16 (INK4a) in replicative senescence of normal human fibroblasts. *Proc. Natl. Acad. Sci. USA 93*, 13742–13747.

Alexandrow MG, Hamlin JL (2004) Cdc6 chromatin affinity is unaffected by serine-54 phosphorylation, S-phase progression, and overexpression of cyclin A. *Mol Cell Biol 24*:1614–1627.

Arnemann J, Epplen JT, Cooke HJ, Sauermann U, Engel W, Schmidtke J. (1987) A human Y-chromosomal DNA sequence expressed in testicular tissue. *Nucleic Acids Res, 15(21)*:8713-24.

Atherton-Fessler, S., Parker, L. L., Geahlen, R. L., and Piwnica-Worms, H. (1993) Mechanisms of p34cdc2 regulation. *Mol. Cell. Biol. 13*, 1675–1685.

Balmanno, K& Cook, S. J. (1999) Sustained MAP kinase activation is required for the expression of cyclin D1, p21Cip1 and a subset of AP-1 proteins in CCL39 cells. *Oncogene 18*, 3085–3097.

Bell SP & Stillman B (1992) ATP-dependent recognition of eukaryotic origins of DNA replication by a multiprotein complex. *Nature 357*:128–134.

Bernardi ML, Cotinot C, Payen E, Delouis C. (1996) Transcription of Y- and X-linked genes in preimplantation ovine embryos. *Mol Reprod Dev 45*(2):132-8.

Biermann E, Baack M, Kreitz S, Knippers R (2002) Synthesis and turn-over of the replicative Cdc6 protein during the HeLa cell cycle. *Eur J Biochem 269*:1040–1046.

Blagosklonny MV, Pardee AB. (2002). The restriction point of the cell cycle. *Cell Cycle. 1*(2):103-10.

Bruzzone,R., White,T.W. and Paul,D.L. (1996) Connections with connexins: the molecular basis of direct intercellular signaling. *European Journal of Biochemistry 238*, 1–27.

Cao, L. G. and Wang, Y. L. (1996) Signals from the spindle midzone are required for the stimulation of cytokinesis in cultured epithelial cells. *Mol. Biol. Cell. 7*, 225–232.

Cao QP, Gaudette MF, Robinson DH, Crain WR. (1995) Expression of the mouse testis-determining gene Sry in male preimplantation embryos. *Mol Reprod Dev, 40* (2):196-204.

Carpenter PB, Mueller PR, DunphyWG (1996) Role for a Xenopus Orc2-related protein in controlling DNA replication. Nature 379:357–360.

Carvalho SR, Mendes MA, Cavalli RC *et al.* (2006). Rastreamento e diagnóstico ecocardiográfico das arritmias e cardiopatias congênitas fetais. Revista Brasileira de *Ginecologia e Obstetrícia 28* (5): 304-309.

Chai Z, Sarcevic B, Mawson A, Toh BH. (2001) SET-related cell division autoantigen-1 (CDA1) arrests cell growth. *J Biol Chem, 276*(36):33665-74.

Cheng, M., Sexl, V., Sherr, C. J., and Roussel, M. F. (1998) Assembly of cyclin Ddependent kinase and titration of p27Kip1 regulated by mitogen-activated protein kinase kinase (MEK1). *Proc. Natl. Acad. Sci. USA 95*, 1091–1096.

Chesnokov I, Gossen M, Remus D, Botchan M (1999) Assembly of functionally active Drosophila origin recognition complex from recombinant proteins. *Genes Dev 13*:1289–1296.

Chesnokov I, Remus D, Botchan M (2001) Functional analysis of mutant and wild-type Drosophila origin recognition complex. *Proc Natl Acad Sci USA 98*:11997–12002

Cocker JH, Piatti S, Santocanale C, Nasmyth K, Diffley JF (1996) An essential role for the Cdc6 protein in forming the pre-replicative complexes of budding yeast. *Nature 379*:180–182.

Coleman, T. R., Carpenter, P. B., and Dunphy, W. G. (1996) The Xenopus Cdc6 protein is essential for the initiation of a single round of DNA replication in cell-free extracts. *Cell 87*, 53–63.

Contreras A, Hale TK, Stenoien DL, Rosen JM, Mancini MA, Herrera RE (2003) The dynamic mobility of histone H1 is regulated by cyclin/Cdk phosphorylation. *Mol Cell Biol 23*:8626–8636.

Cook, J. G., Park, C. H., Burke, T. W., et al. (2002) Analysis of Cdc6 function in the assembly of mammalian prereplication complexes. *Proc. Natl. Acad. Sci. USA 99*, 1347–1352.

Coverley D, Pelizon C, Trewick S, Laskey RA (2000) Chromatin-bound Cdc6 persists in S and G2 phases in human cells, while soluble Cdc6 is destroyed in a cyclin A-cdk2 dependent process. *J Cell Sci 113*:1929–1938.

Datar SA, Jacobs HW, de la Cruz AF, Lehner CF, Edgar BA (2000) The *Drosophila* cyclin DCdk4 complex promotes cellular growth. *Embo J 19*:4543–4554.

De Bondt HL, Rosenblatt J, Jancarik J, Jones HD, Morgan DO, Kim SH. (1993) Crystal structure of cyclin-dependent kinase 2. *Nature. 363*(6430):595-602.

Delbridge ML, Longepied G, Depetris D, Mattei MG, Disteche CM, Marshall Graves JA, *et al.*(2004) TSPY, the candidate gonadoblastoma gene on the human Y chromosome, has a widely expressed homologue on the X - implications for Y chromosome evolution. *Chromosome Res, 12*(4):345-56.

Delmolino LM, Saha P, Dutta A (2001) Multiple mechanisms regulate subcellular localization of human CDC6. *J Biol Chem 276*:26947–26954.

Dhar SK, Delmolino L, Dutta A (2001a) Architecture of the human origin recognition complex. *J Biol Chem 276*:29067–29071.

Drury LS, Perkins G, Diffley JF (1997) The Cdc4/34/53 pathway targets Cdc6p for proteolysis in budding yeast. *Embo J 16*:5966–5976.

Dunphy, W. G., Brizuela, L., Beach, D., and Newport, J. (1988) The Xenopus cdc2 protein is a component of MPF, a cytoplasmic regulator of mitosis. *Cell 54*, 423–431.

Dyson, N. (1998) The regulation of E2F by pRB-family proteins. Genes Dev. 12, 2245–2262.

el-Deiry, W. S., Tokino, T., Velculescu, V. E., et al. (1993) WAF1, a potential mediator of p53 tumor suppression. *Cell 75*, 817–825.

Eichmuller S, Usener D, Dummer R, Stein A, Thiel D, Schadendorf D. (2001) Serological detection of cutaneous T-cell lymphoma-associated antigens. *Proc Natl Acad Sci U S A., 98*(2):629-34.

Elsasser S, Chi Y, Yang P, Campbell JL (1999) Phosphorylation controls timing of Cdc6p destruction: A biochemical analysis. *Mol Biol Cell 10*:3263–3277.

Endicott, J. A., Nurse, P., and Johnson, L. N. (1994) Mutational analysis supports a structural model for the cell cycle protein kinase p34. *Protein Eng. 7*, 243–253.

Evans, T; Rosenthal, ET; Youngblom, J; Distel, D & Hunt T. (1983). Cyclin: A protein specified by maternal mRNA in sea urchin eggs that is destroyed at each cleavage division. *Cell 33*, 389-396.

Ewen, M. E. (1996) p53-dependent repression of cdk4 synthesis in transforming growth factor-b-induced G1 cell cycle arrest. *J. Lab. Clin. Med. 128*, 355–360.

Finn, O. J. (2003) Cancer vaccines: between the idea and the reality. *Nat. Rev. Immunol. 3*, 630–641.

Frolov MV, Dyson NJ (2004) Molecular mechanisms of E2F-dependent activation and pRB-mediated repression. *J Cell Sci 117*:2173–2181.

Fujita M, Yamada C, Goto H, Yokoyama N, Kuzushima K, Inagaki M, Tsurumi T (1999) Cell cycle regulation of human CDC6 protein. Intracellular localization, interaction with the human mcm complex, and CDC2 kinase-mediated hyperphosphorylation. *J Biol Chem 274*:25927–25932.

Gartel, A. L., Serfas, M. S., and Tyner, A. L. (1996) p21—negative regulator of the cell cycle. *Proc. Soc. Exp. Biol. Med. 213*, 138–149.

Gaubatz, S., Wood, J. G., and Livingston, D. M. (1998) Unusual proliferation arrest and transcriptional control properties of a newly discovered E2F family member, E2F-6. *Proc. Natl. Acad. Sci. USA 95*, 9190–9195.

Gautier, J., Norbury, C., Lohka, M., Nurse, P., and Maller, J. (1988) Purified maturation-promoting factor contains the product of a Xenopus homolog of the fission yeast cell cycle control gene cdc2+. *Cell 54*, 433–439.

Gavin KA, Hidaka M, Stillman B (1995) Conserved initiator proteins in eukaryotes. *Science 270*:1667–1671.

Gehart, J; Wu, M & Kirschner, M. (1984). Cell cycle dynamics of an M-phase-specific cytoplasmic factors in Xenopus laevis oocytes and eggs. *J. Cell Biol. 98*, 1247-1255.

Geng Y, Whoriskey W, Park MY, Bronson RT, Medema RH, Li T, Weinberg RA, Sicinski P(1999) Rescue of cyclin D1 deficiency by knockin cyclin E. *Cell 97*:767–777.

Geng Y, Yu Q, Whoriskey W, Dick F, Tsai KY, Ford HL, Biswas DK, Pardee AB, Amati B,Jacks T, Richardson A, Dyson N, Sicinski P (2001) Expression of cyclins E1 and E2 during mouse development and in neoplasia. *Proc Natl Acad Sci USA 98*:13138–13143.

Gillespie PJ, Li A, Blow JJ (2001) Reconstitution of licensed replication origins on Xenopus sperm nuclei using purified proteins. *BMC Biochem 2*:15.

Gossen M, Pak DT, Hansen SK, Acharya JK, Botchan MR (1995) A Drosophila homolog of the yeast origin recognition complex. *Science 270*:1674–1677.

Grana, X., Garriga, J., and Mayol, X. (1998) Role of the retinoblastoma protein family, pRB, p107 and p130 in the negative control of cell growth. *Oncogene 17*, 3365–3383.

Gutiérrez-Adán A, Oter M, Martinez-Madrid B, Pintado B, De La Fuente J. Differential expression of two genes located on teh X chromosome between male and female *in vitro*-produced bovine embryos at the blastocyst stage. *Mol Reprod Dev* 2000, 55:146-51.

Gutiérrez-Adán A, Behboodi E, Murray JD, Anderson, GB. Early transcription of the SRY gene by bovine preimplantation embryos. *Mol Reprod Dev* 1997, 48:246-50.

Haqq CM, King CY, Donahoe PK, Weiss MA. SRY recognizes conserved DNA sites in sex-specific promoters. *Proc Natl Acad Sci U S A* 1993, 90(3):1097-101.

Hartwell, LH; Cullotti, J & Reid, B (1970) Genetic control of cell division in yeast, I. Detection of mutants. *Proc Natl Acad Sci USA 66*, 352-359.

Hartwell LH, Mortimer RK, Culotti J, Culotti M. (1973) Genetic Control of the Cell Division Cycle in Yeast: V. Genetic Analysis of cdc Mutants. *Genetics 74*(2):267-286.

Hartwell LH, Culotti J, Pringle JR, Reid BJ (1974) Genetic control of the cell division cycle in yeast. *Science 183*:46–51.

Hartwell LH (1976) Sequential function of gene products relative to DNA synthesis in the yeast cell cycle. *J Mol Biol 104*:803–817.

Hateboer, G., Wobst, A., Petersen, B. O., et al. (1998) Cell cycle-regulated expression of mammalian CDC6 is dependent on E2F. *Mol. Cell. Biol. 18*, 6679–6697.

Helin, K., Wu, C. L., Fattaey, A. R., et al. (1993) Heterodimerization of the transcription factors E2F-1 and DP-1 leads to cooperative trans-activation. *Genes Dev. 7*,1850–1861.

Hunter, T. and Pines, J. (1994) Cyclins and cancer. II: Cyclin D and CDK inhibitors come of age. *Cell 79*, 573–582.

Iizuka, M. and Stillman, B. (1999) Histone acetyltransferase HBO1 interacts with the ORC1 subunit of the human initiator protein. *J. Biol. Chem. 274*, 23027–23034.

Jiang W, Wells NJ, Hunter T (1999) Multistep regulation of DNA replication by Cdk phosphorylation of HsCdc6. *Proc Natl Acad Sci USA 96*:6193–6198.

Kamb, A., Gruis, N. A., Weaver-Feldhaus, J., et al. (1994) A cell cycle regulator potentially involved in genesis of many tumor types. *Science 264*, 436–440.

Kocchar HPS, Peippo J, King WA. Sex related embryo development. *Theriogenology* 2001, 55:3-14.

Kong D, DePamphilis ML (2001) Site-specific DNA binding of the Schizosaccharomyces pombe origin recognition complex is determined by the Orc4 subunit. *Mol Cell Biol 21*:8095–8103.

Kozar K, Ciemerych MA, Rebel VI, Shigematsu H, Zagozdzon A, Sicinska E, Geng Y, Yu Q, Bhattacharya S, Bronson RT, Akashi K, Sicinski P (2004) Mouse development and cell proliferation in the absence of D-cyclins. *Cell 118*:477–491.

Lardone MC, Parodi DA, Valdevenito R, Ebensperger M, Piottante A, Madariaga M, *et al* Quantification of DDX3Y, RBMY1, DAZ and TSPY mRNAs in testes of patients with severe impairment of spermatogenesis. *Mol Hum Reprod* 2007, 13(10):705-12.

Lavoie, J. N., L'Allemain, G., Brunet, A., Muller, R., and Pouyssegur, J. (1996) Cyclin D1 expression is regulated positively by the p42/p44MAPK and negatively by the p38/HOGMAPK pathway. J. Biol. Chem. 271, 20608–20616. *The Mammalian Cell Cycle* 141.

Lee JK, Moon KY, Jiang Y, Hurwitz J (2001) The Schizosaccharomyces pombe origin recognition complex interacts with multiple AT-rich regions of the replication origin DNA by means of the AT-hook domains of the spOrc4 protein. *Proc Natl Acad Sci USA 98*:13589–13594.

Leatherwood J, Lopez-Girona A, Russell P (1996) Interaction of Cdc2 and Cdc18 with a fission yeast ORC2-like protein. *Nature 379*:360–363.

Levine, A. J. (1997) p53, the cellular gatekeeper for growth and division. *Cell 88*, 323–331.

Li, J., Meyer, A. N., and Donoghue, D. J. (1995) Requirement for phosphorylation of cyclin B1 for Xenopus oocyte maturation. *Mol. Biol. Cell. 6*, 1111–1124.

Liang C, Weinreich M, Stillman B (1995) ORC and Cdc6p interact and determine the frequency of initiation of DNA replication in the genome. *Cell 81*:667–676.

Loewenstein, W.R. (1981). Junctional intercellular communication: the cell-to-cell membrane channel. *Physiological Reviews 61*, 829–913.

Lundgren, K., Walworth, N., Booher, R., Dembski, M., Kirschner, M., and Beach, D. (1991) mik1 and wee1 cooperate in the inhibitory tyrosine phosphorylation of cdc2. *Cell 64*, 1111–1122.

MacLellan WR & Schneider MD. (2000). Genetic dissection of cardiac growth control pathways. *Annual Review of Physiology 62*, 289-320.

Malumbres M, Sotillo R, Santamaria D, Galan J, Cerezo A, Ortega S, Dubus P, Barbacid M (2004) Mammalian cells cycle without the D-type cyclin-dependent kinases Cdk4 and Cdk6. *Cell 118*:493–504.

Marahrens, Y. and Stillman, B. (1992) A yeast chromosomal origin of DNA replication defined by multiple functional elements. *Science 255*, 817–823.

Marshall, M. S. (1995) Ras target proteins in eukaryotic cells. *FASEB J. 9*, 1311–1318.

MasudaHP, RamosGB, de Almeida-Engler J, Cabral LM, CoqueiroVM,Macrini CM, Ferreira PC, Hemerly AS (2004) Genome based identification and analysis of the prereplicative complex of Arabidopsis thaliana. *FEBS Lett 574*:192–202.

Masui, Y & Markert, CL. (1971). Cytoplasm control of nuclear behavior during meiotic maturation of frog oocytes. *J. Exp. Zool. 177*, 129-145.

Matsuura I, Denissova NG, Wang G, He D, Long J, Liu F (2004) Cyclin-dependent kinases regulate the antiproliferative function of Smads. *Nature 430*:226–231.

Mendez J, Stillman B (2000) Chromatin association of human origin recognition complex, cdc6, and minichromosome maintenance proteins during the cell cycle: assembly of prereplication complexes in late mitosis. *Mol Cell Biol 20*:8602–8612.

Meyer CA, Jacobs HW, Lehner CF (2002) Cyclin D-Cdk4 is not a master regulator of cell multiplication in *Drosophila* embryos. *Curr Biol 12*:661–666.

Meyerson M, Enders GH, Wu C-L, Su L-K, Gorka C, Nelson C, Harlow E, Tsai L-H (1992) A family of human cdc2-related protein kinases. *EMBO J 11*:2909–2917.

Mittwoch U. Blastocyst prepare for the race to be male. *Hum Reprod* 1993, 8:1550-5.

Moeller SJ, Head ED, Sheaff RJ (2003). p27Kipl inhibition of GRB2-SOS formation can regulate Ras activation. *Mol Cell Biol 23*:3735–3752.

Moon KY, Kong D, Lee JK, Raychaudhuri S, Hurwitz J (1999) Identification and reconstitution of the origin recognition complex from Schizosaccharomyces pombe. *Proc Natl Acad Sci USA 96*:12367–12372.

Moons DS, Jirawatnotai S, Parlow AF, Gibori G, Kineman RD, Kiyokawa H (2002) Pituitary hypoplasia and lactotroph dysfunction in mice deficient for cyclin-dependent kinase-4. *Endocrinology 143*:3001–3008.

Morgan DO (1995). Principles of CDK regulation. *Nature 374*(6518):131-4.

Morgan DO (1997) Cyclin-dependent kinases: engines, clocks, and microprocessors. *Annu Rev Cell Dev Biol 13*:261–291.

Morkel, M., Wenkel, J., Bannister, A. J., Kouzarides, T., and Hagemeier, C. (1997) An E2F-like repressor of transcription. *Nature 390*, 567–568.

Murray AW, Kirschner MW.(1989). Cyclin synthesis drives the early embryonic cell cycle. Nature 339(6222):275-80.

Muzi-Falconi M, Kelly TJ (1995) Orpl, a member of the Cdc18/Cdc6 family of S-phase regulators, is homologous to a component of the origin recognition complex. *Proc Natl Acad Sci USA 92*:12475–12479.

Nevins, J. R. (1992) E2F: a link between the Rb tumor suppressor protein and viral oncoproteins. *Science 258*, 424–429.

Newport, JW & Kirschner, MW. (1984). Regulation of the cell cycle during early Xenopus development. *Cell 37*, 731-742.

Nguyen VQ, Co C, Li JJ (2001) Cyclin-dependent kinases prevent DNA re-replication through multiple mechanisms. *Nature 411*:1068–1073.

Nino-Soto MI, Basrur PK, King WA. Impact of in vitro production techniques on the expression of X-linked genes in bovine (bos taurus) oocytes and pre-attachment embryos. *Mol Reprod Dev* 2007, 74(2):144-53.

Nobori, T., Miura, K., Wu, D. J., Lois, A., Takabayashi, K., and Carson, D. A. (1994) Deletions of the cyclin-dependent kinase-4 inhibitor gene in multiple human cancers. *Nature 368*, 753–756.

Ohtani K, Tsujimoto A, Ikeda M, NakamuraM (1998) Regulation of cell growth-dependent expression of mammalian CDC6 gene by the cell cycle transcription factor E2F. *Oncogene 17*:1777–1785.

Ohtsubo, M. and Roberts, J. M. (1993) Cyclin-dependent regulation of G1 in mammalian fibroblasts. *Science 259*, 1908–1912.

Oram SW, Liu XX, Lee TL, Chan WY, Lau YF. TSPY potentiates cell proliferation and tumorigenesis by promoting cell cycle progression in HeLa and NIH3T3 cells. *BMC Cancer*. 2006, 9;6:154.

Ozbun LL, You L, Kiang S, Angdisen J, Martinez A, Jakowlew SB. Identification of differentially expressed nucleolar TGF-beta1 target (DENTT) in human lung cancer cells that is a new member of the TSPY/SET/NAP-1 superfamily. *Genomics*. 2001, 73(2):179-93.

Pardee AB (1974) A restriction point for control of normal animal cell proliferation. *Proc Natl Acad Sci USA 71*:1286–1290.

Pardee AB.(1989) G1 events and regulation of cell proliferation. *Science. 246*(4930):603-8.

Parker, L. L., and Piwnica-Worms, H. (1992) Inactivation of the p34cdc2-cyclin B complex by the human WEE1 tyrosine kinase. *Science 257*, 1955–1957.

Payton, M., Chung, G., Yakowec, P., et al. (2004) in The Cell Cycle, Chromosomes and Cancer, vol. 15. (Deutscher, M. P., Black, S., Boehmer, P. E., et al., eds.), Miami Nature Biotechnology Winter Short Reports, Miami Beach, FLA, pp. 59–60.

Peippo J, Farazmand A, Kurkilahti M, Markkula M, Basrur PK, King WA. Sex-chromosome linked gene expression in in-vitro produced bovine embryos. *Mol Hum Reprod* 2002, 8(10):923-9.

Petersen BO, Lukas J, Sorensen CS, Bartek J, Helin K (1999) Phosphorylation of mammalian CDC6 by cyclin A/CDK2 regulates its subcellular localization. *Embo J 18*:396–410.

Pines, J. and Hunter, T. (1989) Isolation of a human cyclin cDNA: evidence for cyclin mRNA and protein regulation in the cell cycle and for interaction with p34cdc2. *Cell 58*, 833–846.

Pinto S, Quintana DG, Smith P, Mihalek RM, Hou ZH, Boynton S, Jones CJ, Hendricks M, Velinzon K, Wohlschlegel JA et al. (1999) latheo encodes a subunit of the origin recognition complex and disrupts neuronal proliferation and adult olfactory memory when mutant. *Neuron 23*:45–54.

Porter, P. L., Malone, K. E., Heagerty, P. J., et al. (1997) Expression of cell-cycle regulators p27Kip1 and cyclin E, alone and in combination, correlate with survival in young breast cancer patients. *Nat. Med. 3*, 222–225.

Prives, C. (1993) Doing the right thing: feedback control and p53. *Curr. Opin. Cell Biol. 5*, 214–218.

Quintana DG, Thome KC, Hou ZH, Ligon AH, Morton CC, Dutta A (1998) ORC5L, a new member of the human origin recognition complex, is deleted in uterine leiomyomas and malignant myeloid diseases. *J Biol Chem 273*:27137–27145.

Reynisdottir, I., Polyak, K., Iavarone, A., and Massague, J. (1995) Kip/Cip and Ink4 Cdk inhibitors cooperate to induce cell cycle arrest in response to TGF-b. *Genes Dev. 9*, 1831–1845.

Resnitzky, D., Gossen, M., Bujard, H., and Reed, S. I. (1994) Acceleration of the G1/S phase transition by expression of cyclins D1 and E with an inducible system. *Mol. Cell. Biol. 14*, 1669–1679.

Rieger D. Relationship between energy metabolism and development of early mammalian embryos. *Theriogenology* 1992, 37:75-93.

Russo AA, Jeffrey PD, Pavletich NP (1996). Structural basis of cyclin-dependent kinase activation by phosphorylation. *Nat Struct Biol 3*(8):696-700.

Saha P, Chen J, Thome KC, Lawlis SJ, Hou ZH, Hendricks M, Parvin JD, Dutta A (1998) HumanCDC6/Cdc18 associates with Orc1 and cyclin-cdk and is selectively eliminated from the nucleus at the onset of S phase. *Mol Cell Biol 18*:2758–2767.

Sano S, Ishino K, Kawada M, Honjo O. (2005). Right Ventricle-Pulmonary Artery Shunt in First-Stage Palliation of Hypoplastic Left Heart Syndrome. Operative Techniques in Thoracic and Cardiovascular Surgery: A Comparative Atlas 10(2): 141-151.

Seki T & Diffley JF (2000) Stepwise assembly of initiation proteins at budding yeast replication origins in vitro. *Proc Natl Acad Sci USA 97*:14115–14120.

Serrano, M. (1997) The tumor suppressor protein p16INK4a. *Exp. Cell Res. 237*, 7–13.

Sherr CJ (1994) G1 phase progression: cycling on cue. *Cell 79*:551–555.

Sherr, C. J. (1996) Cancer cell cycles. *Science 274*, 1672–1627.

Sherr CJ, Roberts JM (1995) Inhibitors of mammalian G1 cyclin-dependent kinases. *Genes Dev 9*:1149–1163.

Sherr CJ, Roberts JM (1999) CDK inhibitors: positive and negative regulators of G1-phaseprogression. *Genes Dev 13*:1501–1512.

Silva, José Pedro da et al (2007). Hypoplastic left heart syndrome: the report of a surgical strategy and comparative results of Norwood x Norwood-Sano approach. *Revista Brasileira de Cirurgia Cardiovascular 22*(2): 160-168.

Sinclair AH, Berta P, Palmer MS, Hawkins JR, Griffiths BL, Smith MJ, *et al.* A gene from human sex-determining region encodes a protein with homology to a conserved DNA-binding motif. *Nature* 1990, 346:240-4.

Smits, V. A. and Medema, R. H. (2001) Checking out the G(2)/M transition. *Biochim. Biophys. Acta 1519*, 1–12.

Takaki T, Fukasawa K, Suzuki-Takahashi I, Hirai H (2004) Cdk-mediated phosphorylation of pRB regulates HDAC binding in vitro. *Biochem Biophys Res Commun 316*:252–255.

Takizawa, C. G. and Morgan, D. O. (2000) Control of mitosis by changes in the subcellular location of cyclin-B1-Cdk1 and Cdc25C. *Curr. Opin. Cell Biol. 12*, 658–665.

Trimarchi, J. M., Fairchild, B., Verona, R., Moberg, K., Andon, N., and Lees, J. A. (1998) E2F-6, a member of the E2F family that can behave as a transcriptional repressor. *Proc. Natl. Acad. Sci. USA 95*, 2850–2855.

Trimarchi JM, Lees JA (2002) Sibling rivalry in the E2F family. *Nat RevMol Cell Biol 3*:11–20.

Tsai, L. H., Lees, E., Faha, B., Harlow, E., and Riabowol, K. (1993) The cdk2 kinase is required for the G1-to-S transition in mammalian cells. *Oncogene 8*, 1593–1602.

Tsutsui T, Hesabi B, Moons DS, Pandolfi PP, Hansel KS, Koff A, Kiyokawa H (1999) Targeteddisruption of CDK4 delays cell cycle entry with enhanced p27Kip1 activity. *Mol Cell Biol 19*:7011–7019.

Tugal T, Zou-Yang XH, Gavin K, Pappin D, Canas B, Kobayashi R, Hunt T, Stillman B (1998) The Orc4p and Orc5p subunits of the Xenopus and human origin recognition complex are related to Orc1p and Cdc6p. *J Biol Chem 273*:32421–32429.

Tu Y, Wu W, Wu T, Cao Z, Wilkins R, Toh BH, *et al.* Antiproliferative autoantigen CDA1 transcriptionally up-regulates p21(Waf1/Cip1) by activating p53 and MEK/ERK1/2 MAPK pathways. *J Biol Chem*. 2007, 282(16):11722-31.

Vashee S, Simancek P, ChallbergMD, Kelly TJ (2001) Assembly of the human origin recognition complex. *J Biol Chem 276*:26666–26673.

Vashee S, Cvetic C, Lu W, Simancek P, Kelly TJ, Walter JC (2003) Sequence-independentDNA binding replication initiation by the human origin recognition complex. *Genes Dev 17*:1894–1908.

Vousden, K. H. (1995) Regulation of the cell cycle by viral oncoproteins. *Semin. Cancer Biol. 6*, 109–116.142 Harper and Brooks.

Weinberg, R. A. (1995) The retinoblastoma protein and cell cycle control. *Cell 81*, 323–330.

Wheatley, S. P., Hinchcliffe, E. H., Glotzer, M., Hyman, A. A., Sluder, G., and Wang, Y. (1997) CDK1 inactivation regulates anaphase spindle dynamics and cytokinesis in vivo. *J. Cell Biol. 138*, 385–393.

Wilmes GM, Archambault V, Austin RJ, Jacobson MD, Bell SP, Cross FR (2004) Interactionof the S-phase cyclin Clb5 with an "RXL" docking sequence in the initiator protein Orc6 provides an origin-localized replication control switch. *Genes Dev 18*:981–991.

Wolfel, T., Hauer, M., Schneider, J., et al. (1995) A p16INK4a-insensitive CDK4 mutant targeted by cytolytic T lymphocytes in a human melanoma. *Science 269*, 1281–1284.

Woo RA, Poon RY (2003) Cyclin-dependent kinases and S phase control in mammalian cells. *Cell Cycle 2*:316–324.

Wuarin J, Buck V, Nurse P, Millar JB (2002) Stable association of mitotic cyclin B/Cdc2 to replication origins prevents endoreduplication. *Cell 111*:419–431

Wu L, Timmers C, Maiti B, Saavedra HI, Sang L, Chong GT, Nuckolls F, Giangrande P, Wright FA, Field SJ, Greenberg ME, Orkin S, Nevins JR, Robinson ML, Leone G (2001) The E2F1-3 transcription factors are essential for cellular proliferation. *Nature 414*:457–462.

Wu L, de Bruin A, Saavedra HI, Starovic M, Trimboli A, Yang Y, Opavska J, Wilson P, Thompson JC, Ostrowski MC, Rosol TJ, Woollett LA, Weinstein M, Cross JC, Robinson ML, Leone G (2003) Extra-embryonic function of Rb is essential for embryonic development and viability. *Nature 421*:942–947.

Yan Z, DeGregori J, Shohet R, Leone G, Stillman B, Nevins JR, Williams RS (1998) Cdc6 is regulated by E2F and is essential for DNA replication in mammalian cells. *Proc Natl Acad Sci USA 95*:3603–3608.

Yamasaki H, Krutovskikh V, Mesnil M, Tanaka T, Zaidan-Dagli ML, Omori Y. (1999). Role of connexin (gap junction) genes in cell growth control and carcinogenesis. *Comptes Rendus de l Academie des Sciences*. Serie III, Sciences de la Vie (Paris) 322(2-3):151-9.

Yu, M., Zhan, Q., and Finn, O. J. (2002) Immune recognition of cyclin B1 as a tumor antigen is a result of its overexpression in human tumors that is caused by non functional p53. *Mol. Immunol. 38*, 981–987.

Zetterberg A, Engström W, Larsson O (1982) Growth activation of resting cells: induction of balanced and imbalanced growth. *Ann N Y Acad Sci. 397*:130-47.

Zetterberg A, Larsson O (1985) Kinetic analysis of regulatory events in G1 leading to proliferation or quiescence of Swiss 3T3 cells. *Proc Natl Acad Sci U S A. 82*(16):5365-9.

Zetterberg A, Larsson O.(1991) Coordination between cell growth and cell cycle transit in animal cells. *Cold Spring Harb Symp Quant Biol 56*:137-47.

Zuo, L., Weger, J., Yang, Q., et al. (1996) Germline mutations in the p16INK4a binding domain of CDK4 in familial melanoma. *Nat. Genet. 12*, 97–9.

Zwingman, T, Erickson RP, Boyer T. Transcription of the sex-determining region genes SRY and Zfy in the mouse preimplantation embryo. *Proc Natl Acad Sci USA* 1993, 90:814-7.

In: Cell Cycle Control: New Research
Editors: N. H. Leroy, N. T. Fournier

ISBN: 978-1-60456-796-0

Chapter XV

Cell Cycle Arrest by Dietary Chemopreventive Agents

Fazlul H. Sarkar*[*] *and Yiwei Li
Department of Pathology, Karmanos Cancer Institute, Wayne State University School of Medicine, Detroit, MI, USA.

Abstract

In recent years, several dietary compounds or nutritional agents, i.e. isoflavone genistein, indole-3-carbinol (I3C), 3,3'-diindolylmethane (DIM), curcumin, (-)-epigallocatechin-3-gallate (EGCG), resveratrol, lycopene, etc, have been recognized as cancer chemopreventive agents because of their anti-carcinogenic activity. Moreover, these compounds also exert their anti-tumor activities through the regulation of different cell signaling pathways, resulting in the cell cycle arrest and subsequent cell growth inhibition. The cell cycle is known to be tightly regulated by different cyclins and cyclin dependent kinases (CDKs), which form key regulatory complex, in different phases of the cell cycle. The activity of cyclins/CDKs complex is negatively regulated by several CDK inhibitors (CDKIs) including $p21^{WAF1}$, $p27^{KIP1}$, $p16^{INK4a}$, etc. In addition, the molecules which interact with the key regulatory complex and CDKIs also influence the progression of cell cycle. The data from our laboratory and others have demonstrated that dietary chemopreventive agents inhibit the kinase activity of CDKs and regulate the expression of cyclins, CDKs, and CDKIs, leading to the cell cycle arrest at different phases of cell cycle. A number of molecules, which are critical for the cell cycle control, have been found to be altered by chemopreventive agents in the cell cycle synchronization assay. Moreover, we and other investigators have utilized gene expression profiling showing alteration in the expression of genes that are related to cell cycle progression by chemopreventive agents. In summary, emerging studies using genomics and proteomics is likely to advance our knowledge regarding the molecular

[*] Correspondence concerning this article should be addressed to: Fazlul H. Sarkar, Ph.D., Department of Pathology, Karmanos Cancer Institute, Wayne State University School of Medicine, 740 Hudson Webber Cancer Research Center, 4100 John R, Detroit, MI 48201. Phone: 313-576-8327; Fax: 313-576-8389; E-mail: fsarkar@med.wayne.edu.

mechanism of cell cycle regulation as a target for the prevention and/or treatment of human cancers by nutritional agents.

Introduction

Cell cycle and its control are very important physiological events in maintaining the homeostasis of cell growth control in human health and diseases. Progression of cell cycle from G_1 to M phase leads to cell division and proliferation. Cell cycle and its phases were first described by Howard and Pele in 1951. Since then, more and more evidences from experimental studies have demonstrated that the cell cycle is a highly conserved and ordered set of events which control cell growth and division [1,2]. It has been known that cell cycle is tightly controlled by many regulatory proteins. The main families of regulatory proteins that play key roles in controlling cell cycle progression are cyclins, cyclin dependent kinases (CDKs), CDK inhibitors (CDKI), and tumor suppressor gene products such as p53 and pRb [1,3-5]. These families comprise the basic regulatory machinery which permit or stop cell cycle transition, checkpoint progression, and cell proliferation.

Abnormality of cell cycle control plays important roles in carcinogenesis. Uncontrolled proliferation is a characteristic of cancer cells. The abnormality of cell cycle contributes largely to the uncontrolled proliferation in cancer cells. Most of the cell cycle regulators, which control the entry and progression in the different phases of the cell cycle, are altered in cancer cells, resulting in the uncontrolled cell proliferation and cell division. Therefore, the strategy to stop cell cycle progression and let cells stay quiescent is needed for cancer prevention and treatment.

In recent years, many dietary compounds or classically known nutritional agents, i.e. isoflavone genistein, indole-3-carbinol (I3C), 3,3'-diindolylmethane (DIM), curcumin, (-)-epigallocatechin-3-gallate (EGCG), resveratrol, lycopene, etc, have been recognized as cancer chemopreventive agents because of their anti-carcinogenic activity [6,7]. Moreover, these compounds also exert their anti-tumor activities through the regulation of different cell signaling pathways, resulting in the cell cycle arrest and subsequent cell growth inhibition [6-9]. These dietary compounds have been found to arrest cancer cells in the different phases of the cell cycle and here we are presenting evidence on the regulation of cell cycle by nutritional agents.

Cell Cycle Control

The cell cycle consists of a series of tightly controlled steps that allow cells to reproduce DNA, synthesize proteins, and divide [5]. Whole cell cycle contains four distinct phases: G_1 phase, S phase, G_2 phase, and M phase (Figure 1). G_1 phase is from the end of the previous M phase till the beginning of DNA synthesis. This phase is marked by synthesis of various enzymes that are required in S phase, mainly those needed for DNA replication. Then, the cells enter S phase to synthesize DNA. When S phase is complete, the chromosomes have been replicated. The cells then enter G_2 phase, which lasts until the cells enter mitosis.

During G_2 phase, significant proteins are synthesized. In M phase, nuclear and cytoplasmic divisions occur to form two new cells. Non-proliferative cells generally enter the quiescent G_0 state from G_1 and may remain quiescent for a long period of time. G_0 is a biochemically distinct state from which cells can re-enter the cell cycle and go on to DNA replication and mitosis.

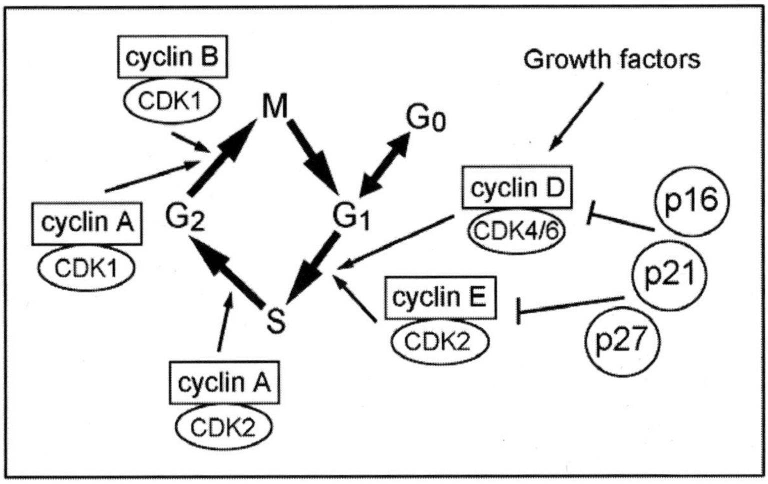

Figure 1. Cell cycle and its control.

The transitions between different phases are regulated by cyclins and CDKs. The cyclins are a family of 56 kDa proteins which bind to specific CDK and form a protein complex. The specific cyclin/CDK complexes control cell cycle progression through cell cycle checkpoints by phosphorylating and inactivating target substrates [3-5]. Cyclin D/CDK4, cyclin D/CDK6, and cyclin E/CDK2 complexes control G_1 transition to S phase while cyclin A/CDK1 and cyclin B/CDK1 complexes control the transition G_2 to M phase. The predominant cyclin and CDK complex in S phase is cyclin A/CDK2. CDK proteins generally remain at constant levels throughout the cell cycle. However, the sequential accumulation of different cyclins allows the formation of specific cyclin and CDK complexes which control the transitions between the cell cycle phases.

In addition to the basic control of cyclin/CDK activity by the alteration in the concentration of cyclins in the different phases of the cell cycle, the activity of cyclin/CDK complex is also regulated by many hormones and growth factors that modify cyclin activity and influence cellular growth through regulation of cell signal transduction pathways. More importantly, CDK inhibitors such as $p21^{WAF1}$, $p27^{KIP1}$, $p16^{INK4a}$, etc, and tumor suppressor gene product such as p53 and pRb also form regulatory network with cyclins and CDKs to regulate the activity of cyclin/CDK complex and influence the transition of cell cycle phases.

G_1/S and G_2/M transitions are two important checkpoints in cell cycle. Cyclin D1 is a critical modulator of the G_1/S transition through pRb phosphorylation and $p21^{WAF1}/p27^{KIP1}$ titration [10]. Mitogenic factors can modulate cyclin D1 by regulation of its transcription, translation, stability, localization, and ability to form complex with CDK. During G_1, cyclin D1/CDK4 and cyclin D1/CDK6 can be activated. The activated cyclin D1/CDK6 can phosphorylate pRb which inhibits E2F, leading to the release of E2F, the subsequent E2F-mediated transcription of genes essential for DNA synthesis in S phase, and the cells entering S phase. However, both $p21^{WAF1}$ and $p27^{KIP1}$ are able to inhibit the activity of cyclin D1/CDK complexes. Moreover, another CDK inhibitor, $p16^{INK4a}$, can bind specifically to CDK4 and CDK6, inhibiting CDK4 and CDK6 activity and disrupting cyclin D association [11].

Cyclin E is another important modulator of the G_1/S transition [12]. During G_1 phase, cyclin E binds to CDK2 and activates CDK2 which then phosphorylates target proteins such as pRb, facilitating the first steps of S phase. In addition to pRb phosphorylation, cyclin E/CDK2 also phosphorylates a number of proteins which play important roles in DNA replication [12]. The activity of cyclin E/CDK2 is mainly regulated by the expression level of cyclin E and the cyclin E degradation. During G_1 phase, as the cyclin E/CDK2 complex is activated it phosphorylates pRb and the pRb phosphorylation allows the release of E2F which, in turn, drives cyclin E expression, forming a positive feedback loop for cyclin E transcription [12,13]. In the same time, cyclin E is also regulated by ubiquitin-mediated degradation. Therefore, the positive feedback loop activating cyclin E transcription is counterbalanced by the rapid degradation of cyclin E protein. Importantly, CDK inhibitors, $p21^{WAF1}$ and $p27^{KIP1}$ can bind to cyclin E/CDK2 and inhibit CDK2 kinase activity [12]. Moreover, $p57^{KIP2}$ also exerts negative effects on the activity of cyclin/CDK2 complexes during G_1 phase and cyclin/CDK1 complexes during G_2 phase of the cell cycle. These are typical features in maintaining cellular homeostasis and the deregulation of these processes could lead to carcinogenesis.

Cyclin A is an interesting cyclin because it can activate two different CDKs, CDK2 and CDK1, in S phase and G_2/M transition [14]. Cyclin A starts to accumulate during S phase and is abruptly destroyed during M phase in advance of cyclin B. In S phase, phosphorylation of components in DNA replication machinery such as CDC6 by cyclin A/CDK2 is believed to be important for initiation of DNA replication. In M phase, cyclin A may contribute to the control of cyclin B stability.

Cyclin B is a mitotic cyclin. The level of cyclin B expression and the activity of cyclin B/CDK1 complex rise through the cell cycle until M phase, where they fall abruptly due to degradation [15]. Under the control of cell cycle regulatory machinery, cyclin B is expressed predominantly during G2/M phase. The cyclin B/CDK1 complex is also known as mitosis promoting factor (MPF), which controls G_2/M transition.

It is important to note that cell cycle is also controlled by hormones, growth factors, and the molecules which regulate CDK inhibitors, CDKs, and cyclins [3]. It has been known that tumor suppressor p53 can promote the expression of $p21^{WAF1}$, resulting in the cell cycle arrest. Akt can inhibit the expression of $p27^{KIP1}$ through the phosphorylation of FOXO3a, leading to cell cycle progression and cell proliferation. Many hormones and growth factors modify the activity of cyclins while an important transcription factor, NF-κB, regulates the activity of CDK [16].

In cancer cells, the cell cycle remains uncontrolled because of the abnormal levels of cyclins, CDKs, and CDKIs [17,18]. The abnormalities of cyclins, CDKs, and CDKIs could be caused by the abnormal level of hormones, growth factors, oncogene products, or tumor suppressor gene products. Moreover, NF-κB and Akt have been described as the major culprits in cancer. Most of these molecules, which control the entry and progression in the different phases, are altered in cancer cells, resulting in the uncontrolled cell proliferation and cell division. Therefore, the strategy to stop cell cycle progression and uncontrolled cell proliferation is a key step for cancer prevention and treatment.

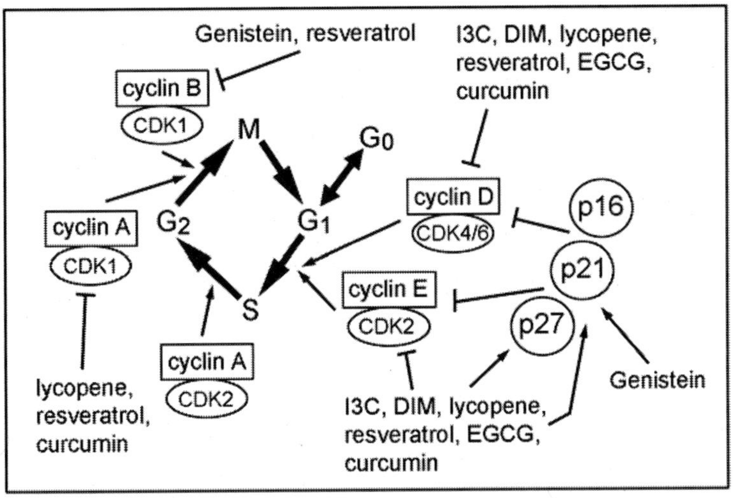

Figure 2. The effects of dietary compounds on cell cycle control.

Cell Cycle Arrested by Dietary Compounds

In recent years, dietary components as chemopreventive agents have received much attention among the public and the medical community. The experimental *in vitro* and *in vivo* studies have demonstrated that some dietary components or known nutritional agents such as soy isoflavones, indole-3-carbinol (I3C), 3,3'-diindolylmethane (DIM), curcumin, (-)-epigallocatechin-3-gallate (EGCG), resveratrol, lycopene, have inhibitory effects on human cancers [19-26], suggesting that they may serve as chemopreventive agents. Soy isoflavones mainly derived from soybean have been found to inhibit cancer growth *in vivo* and *in vitro* [27-29]. I3C and its *in vivo* dimeric product DIM, produced from naturally occurring glucosinolates in the family Cruciferae, have shown the inhibition of cancer cell growth through the modulation of genes that are related to the control of cell proliferation, cell cycle, apoptosis, signal transduction, oncogenesis, and transcription regulation [22,25]. Curcumin, a natural compound present in turmeric and possessing both anti-inflammatory and antioxidant activity, has been studied as a chemopreventive agent in several cancer models [26,30]. Beneficial effects attributed to green tea, such as its anticancer and antioxidant properties, are believed to be mainly mediated by EGCG [31,32]. Resveratrol, a phytoalexin present in a wide variety of plant species including grapes, has been found to inhibit the growth of various cancer cells and induce apoptotic cell death [33-36]. Lycopene from tomato and its products inhibits cell growth in breast, prostate and endometrial cancer cells mediated by the regulation of cell cycle-related genes [37,38]. It is becoming clear that these dietary components exert their effects on the regulation of the molecules involved in cell cycle control, resulting in the cell cycle arrest at different phases and subsequent cell growth inhibition (Figure 2).

Soy Isoflavones

Soy isoflavones mainly from soybean and soy product include genistein, daidzein, and glycitein. It has been reported that soy isoflavones, in particular genistein, behave as a cell growth inhibitor [39,40]. Experiments from our laboratory have shown that isoflavone genistein inhibits the growth of MDA-MB-231, MDA-MB-435 and MCF-7 breast cancer cells, PC-3 and LNCaP prostate cancer cells, H460 and H322 non-small cell lung cancer cells, and HN4 head and neck squamous carcinoma cells, regardless of the status of estrogen (ER), androgen receptor (AR), and p53 [41-45]. The growth inhibition of cancer cells could be due to the cell cycle arrest, which ultimately results in ceasing cell proliferation. Our data showed that genistein induces a G_2/M cell cycle arrest in PC-3 and LNCaP prostate cancer cells [43], H460 and H322 non-small cell lung cancer cells [41], MDA-MB-231 and MCF10CA1a breast cancer cells [46], and these results are consistent with reports by other investigators showing that genistein induced a G_2/M cell cycle arrest in gastric adenocarcinoma, melanoma, and prostate cancer cells [47-49].

In order to explore the molecular mechanisms by which isoflavone genistein induces cell cycle arrest, we have examined the expression of cell cycle-related genes including cyclins, CDKs, and CDKIs. The data from our laboratory showed no alternations in the expression of cyclins A, D and E and their associated CDKs in the PC-3 and LNCaP prostate cancer cells, and H460 and H322 non-small cell lung cancer cells treated with isoflavone genistein [41,43]. However, we found that compared to the untreated control cells, the cancer cells treated with different concentrations of genistein showed a dose-dependent decrease in the expression of cyclin B [41,43], which is known to play an important role in the positive regulation of CDK1 activity by forming cyclin B/CDK1 complex during the G_2/M phase progression of the cell cycle. Moreover, isoflavone induced G_2/M arrest has been associated with the inhibition of CDK1 kinase activity [50,51]. Our laboratory have also examined whether isoflavone genistein alters the expression of $p21^{WAF1}$ in MDA-MB-231, MDA-MB-435 and MCF-7 breast cancer cells, PC-3 and LNCaP prostate cancer cells, H460 and H322 non-small cell lung cancer cells, and HN4 head and neck squamous carcinoma cells by Western blot and/or Northern blot analysis. Our data revealed a significant dose dependent up-regulation of $p21^{WAF1}$ expression in isoflavone genistein treated cancer cells compared to control cells [41-45]. This finding was consistent with the results on the inhibition of cancer cell growth, G_2/M cell cycle arrest, and down-regulation of cyclin B expression. We also found that isoflavone genistein inhibited the activation of Akt and NF-κB which could influence cell cycle progression [52,53]. Moreover, our microarray data showed that genistein inhibited cell growth through down-regulation of cell proliferation and cell cycle related genes such as cyclin B, CDC25A, TGF-β, and ki67 [24]. These results suggest that down-regulation of cyclin B_1, CDK1, CDC25A, TGF-β, and ki67, and up-regulation of $p21^{WAF1}$ could be one of the mechanisms by which isoflavone genistein arrests cancer cells in G_2/M phase and inhibits cancer cell growth.

In addition, it has been reported that genistein could also arrest mouse fibroblast and melanoma cells at G_0/G_1 phase of cell cycle [54]. These data suggest that isoflavone genistein induces either G_2/M or G_0/G_1 cell cycle arrest, depending on the origin of cultured cell lines.

I3C and DIM

I3C is produced from naturally occurring glucosinolates contained in a wide variety of plants including members of the family Cruciferae, and particularly members of the genus Brassica [55]. I3C is biologically active and it is easily converted *in vivo* under acid condition to its dimeric product DIM, which is also biologically active. It has been shown that I3C and DIM inhibit growth of cancer cells including breast, prostate, colon, and cervical cancer cells [56-63]. Our laboratory has experimentally studied the effects of I3C and DIM on the growth of various cancer cells including PC-3 prostate cancer cells, MDA-MB-231, MDA-MB-435, and MCF10CA1a breast cancer cells [22,25,64,65]. The growth of these cancer cells was inhibited by treatment with 30 to 100 μM I3C or 20 to 60 μM DIM for a period of 24-72 hours. The inhibition of cell growth was found to be dose- and time-dependent. I3C at 60 μM and DIM at 30 μM significantly inhibited cancer cell growth. By flow cytometry analysis, we found that I3C induced a G_1 cell cycle arrest in PC-3 prostate cancer cells [22], and this result is consistent with results published by other investigators showing that I3C induces G_1 cell cycle arrest in breast cancer cells [66]. By flow cytometry, other investigators also found that DIM induced a G_1 arrest in DU 145 prostate cancer cells [67] and in actively proliferating human umbilical vein endothelial cells (HUVECs) [68].

We have further examined the status of cyclins, CDKs and CDKIs in I3C treated PC-3 prostate cancer cells. By Western Blot analysis, we found that I3C down-regulated the expression of CDK6 and up-regulated the expression of $p21^{WAF1}$ and $p27^{KIP1}$ in a dose-dependent manner [22]. Furthermore, we immunoprecipitated cyclin D1 and cyclin E complexes, and detected CDK6, $p21^{WAF1}$, and $p27^{KIP1}$ in the complexes. The results showed that I3C decreased CDK6 binding to cyclin D1 and increased $p21^{WAF1}$ and $p27^{KIP1}$ binding to cyclin D1 and cyclin E complexes [22], suggesting the inhibition of cell cycle progression at G_1 phase. Moreover, we found that I3C inhibited the CDK6 kinase activity [22], which plays important roles in the regulation of G_1 phase of the cell cycle. These findings were consistent with the results showing the cell growth inhibition and cell cycle arrest induced by I3C, suggesting that I3C inhibits the growth of prostate cancer cells through regulation of genes that are related to the control of cell proliferation and cell cycle. The reports from other laboratories also showed that I3C inhibited the activity of CDK6 promoter, the expression of CDK6, and the CDK2 kinase activity for its substrate pRb protein, and induced a G_1 cell cycle arrest in breast cancer cells [66,69,70]. Moiseeva et al. also reported that in MCF7 and MDA-MB-231 breast cancer cells, I3C increased the expression of $p21^{WAF1}$ and $p27^{KIP1}$, and decreased levels of cyclin E, cyclin D1 and CDK6 [71]. In LNCaP prostate cancer cells, I3C has been found to down-regulate CDK2 activity and up-regulate mRNA and protein levels of the CDK2 inhibitor $p21^{WAF1}$ through regulation of p53 [72]. I3C could stimulate p53 phosphorylation and disrupt p53-MDM2 interaction which releases p53 to induce $p21^{WAF1}$ [73]. Similarly, DIM can inhibit cyclin D1, CDK2, CDK4, and CDK6 in prostate cancer cells

[67] and increase the expression of $p27^{KIP1}$ in HUVECs [68]. Therefore, the up-regulation of $p21^{WAF1}$ and $p27^{KIP1}$ and down-regulation of cyclins and CDKs may be the molecular mechanisms by which I3C and DIM inhibit cancer cell growth and induces cell cycle arrest. In addition, we also found that I3C and DIM inhibited the activation of Akt and NF-κB which could regulate cell cycle control [74,75].

We have also tested the effect of DIM on the molecules which are critical for the cell cycle control in the cell cycle synchronization assay. We synchronized LNCaP cells by isoleucine-deprivation for 36 hours and the cells were found to be arrested in G_0/G_1 phase. Then, the isoleucine-deprived cells were released to enter the cell cycle by incubating the cells in complete media containing 10% FCS, 10 nM testosterone and 50 μM DIM. Cell cycle assay and Western blot analysis were conducted. We found that DIM inhibited the transition of synchronized prostate cancer cells from G_1 to S phase. We observed that androgen receptor (AR) level in control cells increased gradually from 8 to 20 hours after releasing from isoleucine-block and the expression of cyclin A also increased at 16 hours after releasing from isoleucine-block, suggesting the entry of cells into S phase. However, in DIM treated cells, the increase of AR levels during 8 to 20 hours after release from isoleucine-block was completely suppressed, and the cyclin A expression was undetectable, suggesting that DIM-treated cells failed to enter S phase. Interestingly, in DIM treated cells there was also a noticeable increase in $p27^{KIP1}$ level starting at 4 hours after release from isoleucine-block. Collectively, these results suggest that DIM arrests prostate cancer cells at G_1 phase of the cell cycle and thus inhibits cell proliferation and cell growth.

Curcumin

Curcumin is a compound from *Curcuma longa* (tumeric), a plant widely cultivated in tropical regions of Asia and Central America. Turmeric extract from the rhizomes, commonly called curcuminoids, is mainly composed of curcumin. It has been known that curcumin inhibits the growth of cancer cells, induces apoptosis, reduces cell survival signal protein Akt and NF-κB, and regulates the expression of genes related to cell cycle control, apoptosis, and anti-invasion [76-79]. The reported cell cycle arrest by curcumin involves different phases such as G_0/G_1, G_1/S, or G_2/M, dependent on cell types and curcumin concentration. In vascular smooth muscle cells treated with curcumin, flow cytometry analysis showed a G_0/G_1 arrest and a reduction in the percentage of cells in S phase [80]. However, in androgen-sensitive LNCaP and androgen-insensitive PC-3 prostate cancer cells, curcumin treatment resulted in cell cycle arrest at G_1/S phase and this cell cycle arrest ultimately led to the induction of apoptosis [81]. In human melanoma, glioma, and ovarian cancer cells, curcumin was able to induce a cell cycle arrest at G_2/M phase [82-84]. Moreover, high dose of curcumin not only enhanced G_2/M cell cycle arrest, but also caused S phase of cell cycle arrest [83].

Curcumin has been found to regulate cell cycle and NF-κB related genes. In LNCaP and PC-3 prostate cancer cells, curcumin induced the expression of CDKIs such as $p16^{INK4a}$, $p21^{WAF1}$ and $p27^{KIP1}$, and inhibited the expression of cyclin E and cyclin D1, and hyperphosphorylation of Rb protein [81,85]. Curcumin also inhibited both the constitutive and TNF-α-induced NF-κB activation [85] and the NF-κB-regulated gene products, leading to the suppression of proliferation and the cell cycle arrest at the G_1/S phase [86]. In human melanoma cells, curcumin induced melanoma cell apoptosis and cell cycle arrest at G_2/M, which was associated with the down-regulation of NF-κB activation, iNOS and DNA-dependent protein kinase catalytic subunit expression, and up-regulation of p53, $p21^{WAF1}$, $p27^{KIP1}$ and checkpoint kinase 2 [82]. In human glioma U251 cells, curcumin induced cell cycle arrest at G_2/M phase with up-regulated p53 expression followed by induction of $p21^{WAF1}$ and ING4 [87]. Curcumin also induced G_2/M arrest and apoptosis in cisplatin-resistant human ovarian cancer cells by modulating Akt and p38 MAPK [84], suggesting its effect on reducing drug resistance and sensitizing cancer cells to chemotherapeutic agents.

EGCG

Consumption of green tea has been associated with human health including the prevention of cancers. Green tea and its constituents have been studied both *in vitro* and *in vivo*. Green tea contains several catechins including epicatechin (EC), epigallocatechin (EGC), epicatechin-3-gallate (ECG), and epigallocatechin-3-gallate (EGCG). However, EGCG has been believed to be the most potent for inhibition of carcinogenesis [32,88]. There are growing evidences showing that EGCG inhibits the proliferation of various cancer cells and induces cell cycle arrest and apoptotic processes in cancer cells [32,88], suggesting its inhibitory effects on cancers. Similar to curcumin, the reported cell cycle arrests by EGCG involve G_1 and G_2/M phases. DNA flow cytometric analysis indicated that 30 μM EGCG blocked cell cycle progression at G_1 phase in asynchronous MCF-7 cells [89]. However, another report showed that treatment with 50 μM and 100 μM EGCG increased the percentages of lung cancer cells in the G_2/M phase from 13.8% to 15.6% and 24.1%, respectively [90].

It has been found that EGCG treatment resulted in a significant dose-dependent decrease in cyclin D1 and CDK2 protein levels and induction of $p16^{INK4a}$, $p21^{WAF1}$ and $p27^{KIP1}$ in human melanoma cells [91]. In asynchronous MCF-7 breast cancer cells, 30 μM EGCG treatment caused cell cycle arrest at G_1 phase with a loss of both CDK2 and CDK4 kinase activities. EGCG also induced the expression of $p21^{WAF1}$ protein through the increase in p53 levels in breast cancer cells [89]. In prostate cancer cells, EGCG treatment resulted in significant dose- and time-dependent up-regulation of the protein expression of $p16^{INK4a}$, $p21^{WAF1}$ and $p27^{KIP1}$, down-regulation of the protein expression of cyclin D1, cyclin E, CDK4 and CDK6, inhibition of the kinase activities associated with cyclin E, cyclin D1, CDK2, CDK4 and CDK6, increase in the binding of cyclin D1 to $p21^{WAF1}$ and $p27^{KIP1}$, and decrease in the binding of cyclin E to CDK2, leading to the cell cycle arrest at G_1 phase [23,92,93]. In head and neck cancer cells, treatment with EGCG increased the proportion of cells in the G_1 phase of the cell cycle and induced apoptosis with a decrease in the cyclin D1 protein, an increase in the $p21^{WAF1}$ and $p27^{KIP1}$ proteins, and a reduction in the hyperphosphorylated form of pRB [94].

Resveratrol

Resveratrol is a phytoalexin present in a wide variety of plant species including grapes, mulberries, and peanuts. Relatively high quantities of resveratrol are found in grapes. The concentration of resveratrol in red wine and grape juice is also high. Resveratrol has been shown to have beneficial effects on the prevention of heart diseases and cancers [95,96]. Experimental studies have shown that resveratrol inhibits the growth of various cancer cells and induces cell cycle arrest and apoptotic cell death [33-36]. Ragione et al. reported that analysis by flow cytometry in resveratrol treated cells showed the absence of G_2/M peak and the accumulation of cells in G_1 and S phases. They performed a detailed biochemical analysis which demonstrated that the G_1 phase of the cell cycle engine was completely unmodified by resveratrol addition, thus indicating that the G_1/S transition occurred normally. They observed a significant increase in cyclins A and E with the accumulation of CDK2 after 24 hours of treatment. These data demonstrate that resveratrol causes a cell cycle arrest at the S phase checkpoint [97]. Several other reports also showed that resveratrol induced S phase arrest in leukemia and melanoma cells [98-100]. However, it has also been found that resveratrol modulated cyclin/CDK/CDKI machinery, resulting in a G_1 phase arrest of the cell cycle followed by apoptosis in human epidermoid carcinoma A431 cells [101].

Resveratrol treatment has been found to regulate cyclins, CDKs and CDKIs. Treatment of human epidermoid carcinoma A431 cells and colon cancer cells by resveratrol caused an increase in the expression of $p21^{WAF1}$ and a decrease in the protein expressions of cyclin D1, cyclin D2, cyclin E, CDK2, CDK4, and CDK6 and the kinase activities associated with CDKs, leading to the cell cycle arrest at G_1 phase [101,102]. Resveratrol has also been reported to cause cell cycle arrest at G1 and G1/S phases by inducing the expression of $p21^{WAF1}$ and $p27^{KIP1}$ [103]. However, treatment with high concentration of resveratrol (300 μM) arrested cancer cells in S phase with decreased levels of cyclin D1, cyclin A, cyclin B1, and β-catenin protein in cancer cells [104]. Resveratrol also caused CDK1-tyr15 phosphorylation which could be a mechanism for S phase arrest in ovarian cancer cells [105].

Lycopene

Tomatoes are rich in lycopene, which is the pigment principally responsible for the deep-red color of tomato and its products. Tomato products including ketchup, tomato juice, and pizza sauce, are the richest sources of lycopene in the US diet. The consumption of tomatoes and tomato products containing lycopene have been associated with decreased risk of cardiovascular diseases and cancers [106]. Experimental studies have shown that lycopene inhibits cell growth in breast, prostate and endometrial cancer cells with regulation of cell cycle-related genes [37,38]. Flow cytometry analysis showed a G_1/S phase cell cycle arrest after treatment of MCF-7, HBL-100, and MDA-MB-231 breast cancer cell lines and fibrocystic breast cell line MCF-10a with 10 μM lycopene for 48 hours [107]. However, in LNCaP prostate cancer cells, lycopene at 5 μM increased the number of cells in the G_2/M phase of cell cycle from 13% to 28% and decreased S phase cells from 45% to 29%, suggesting a G_2/M arrest [108]. In PrEC prostate epithelial cells and HT-29 and HCT-116 colon cancer cells, lycopene arrested cell cycle at G_0/G_1 phase with inhibition of cyclin D1, Bcl-2 and Bcl-xL protein expression [109,110].

Gene Expression Profiles Showing Alteration in the Expression of Genes that are Related to Cell Cycle Control by Chemopreventive Agents

With the development of microarray and proteomic technologies in recent years, nutrigenomics and nutriproteomics have received increased attention in the research field of cancer prevention and treatment. Using these new technologies, the molecular effects of dietary compounds on cancer cells can be easily and rapidly accessed. We and other investigators have utilized gene expression profiling to investigate the alteration in the expression of genes that are related to cell cycle control by chemopreventive agents. We have conducted microarray analysis to interrogate 12,558 known genes to determine the gene expression profiles altered by genistein treatment in PC-3 prostate cancer cells [24]. We found a total of 832 genes which showed >2 fold change after genistein treatment from two independent experiments with high degree of concordance. Among these genes, 774 genes were down-regulated and 58 genes were up-regulated with genistein treatment. These genes were also subjected to cluster analysis according to their biological functions. We found that genistein regulated the expression of genes, which are critically involved in the regulation of cell growth, cell cycle, apoptosis, cell signaling transduction, angiogenesis, tumor cell invasion and metastasis. Genistein inhibited the expression of several genes such as cyclin B, cyclin A, cdc25A, survivin, TGF-β, ki67, pescadillo, which are involved in the regulation of cell cycle. Several genes including $p57^{KIP2}$, cyclin G2, growth arrest and DNA-damage-inducible protein, and elafin, which are related to the control of cell growth, were found to be up-regulated. These results from microarray analysis are consistent with our earlier results showing the down-regulation of cyclin B and the G_2/M arrest in genistein treated prostate cancer cells [43].

We have also utilized Affymetrix Human Genome U133A Array to interrogate 22,215 known genes altered by I3C and DIM treatment in PC-3 prostate cancer cells [25]. We found a total of 738 genes which showed >2 fold change after 24 hours of DIM treatment. Among these genes, 677 genes were down-regulated and 61 genes were up-regulated. Similarly, 727 genes showed >2 fold changes in the expression with down-regulation of 685 genes and up-regulation of 42 genes in I3C treated cells. The altered expressions of genes were observed as early as 6 hours and were more evident with longer treatment. Upon cluster analysis, we found that both I3C and DIM regulated the expression of genes, which are critically involved in the regulation of cell growth, cell cycle, apoptosis, signal transduction, Pol II transcription factor, and oncogenesis. We observed that I3C and DIM treatment down-regulated EGFR, TGF-β2, and FGF which promote cell cycle progression, demonstrating the inhibitory effects of I3C and DIM on the growth of PC-3 prostate cancer cells. Our results also showed that I3C and DIM inhibited the expression of cyclin E2, ATF5, MIG-2, and BCL2, and induced the expression of $p57^{KIP2}$, leading to the induction of cell cycle arrest at G_1 phase.

Microarray analysis has also been used to explore the gene expression profiles altered by resveratrol in LNCaP prostate cancer cells [111]. Time-dependent expression changes in >1,600 transcripts have been observed as early as 6 hours after treatment with resveratrol. Resveratrol down-regulated the expression of cell cycle and proliferation-specific genes involved in all phases of the cell cycle and induced negative regulators of proliferation, causing accumulation of cells at the G_1/S phases. Resveratrol inhibited the mRNA expression of genes critical for the cell cycle control, such as cyclins D, cyclin E, cyclin A, and cyclin B, while negative regulators of proliferation, such as cyclins G_1 and cyclin G_2, were induced after resveratrol treatment, consistent with G_1/S phase cell cycle arrest. These results suggest that resveratrol-induced growth arrest seems to be mediated by deregulation of a complex network of cell cycle regulatory genes [111].

Gene expression profiles altered by lycopene have also been accessed in breast cancer cells [112]. Total of 391 genes were found to be differentially modulated by lycopene between estrogen positive MCF-7 cells and estrogen negative MDA-MB-231 cells. Hierarchical clustering revealed 726 discriminatory genes between MCF-7 and MDA-MB-231 breast cancer cell lines and MCF-10a fibrocystic breast cell line. Lycopene regulated the expression of genes in various molecular pathways such as apoptosis, cell communication, MAPK and cell cycle. Interestingly, cell cycle related genes were modulated by lycopene treatment with up-regulation of CDK6, Cyclin D1, Cyclin B3 and Cyclin E1 [112], which is not consistent with the cell cycle arrest at G_1/S or G_2/M phase reported by others [107,108,110]. Further research is still ongoing which is likely to bring more mechanistic sight as to the regulation of cell cycle and cell cycle regulatory genes by chemopreventive (nutritional) agents.

Conclusion and Perspective

In summary, cell cycle control is an important physiological event in the maintenance of homeostasis in normal human cells and the deregulation of cell cycle is known to contribute in disturbing this homeostasis during carcinogenesis. Thus, abnormality of cell cycle and subsequent uncontrolled proliferation of cells result in the onset of cancer. Targeting altered cyclin/CDK/CDKI network in cancer cells by dietary compounds, therefore, could be a novel strategy for the prevention and treatment of cancers. Emerging studies using genomics and proteomics is also likely to advance our knowledge regarding the molecular mechanism of cell cycle regulation as a target for the prevention and/or treatment of human cancers using dietary compounds.

The authors' work cited in this review was funded by grants from the National Cancer Institute, NIH (5R01CA083695, 5R01CA101870, and 5R01CA108535 awarded to FHS), a sub-contract award to FHS from the University of Texas MD Anderson Cancer Center through a SPORE grant (5P20-CA101936) on pancreatic cancer awarded to James Abbruzzese, and a grant from the Department of Defense (DOD Prostate Cancer Research Program DAMD17-03-1-0042 awarded to FHS).

References

[1] Genovese, C; Trani, D; Caputi, M; Claudio, PP. Cell cycle control and beyond: emerging roles for the retinoblastoma gene family. *Oncogene,* 2006;25:5201-5209.

[2] Cam, H; Dynlacht, BD. Emerging roles for E2F: beyond the G1/S transition and DNA replication. *Cancer Cell,* 2003;3:311-316.

[3] Grana, X; Reddy, EP. Cell cycle control in mammalian cells: role of cyclins, cyclin dependent kinases (CDKs), growth suppressor genes and cyclin-dependent kinase inhibitors (CKIs). *Oncogene,* 1995;11:211-219.

[4] van den, HS; Harlow, E. Distinct roles for cyclin-dependent kinases in cell cycle control. *Science,* 1993;262:2050-2054.

[5] Nurse, P. Cyclin dependent kinases and cell cycle control (nobel lecture). *Chembiochem,* 2002;3:596-603.

[6] Khan, N; Afaq, F; Mukhtar, H. Apoptosis by dietary factors: the suicide solution for delaying cancer growth. *Carcinogenesis,* 2007;28:233-239.

[7] Sarkar, FH; Li, Y. Cell signaling pathways altered by natural chemopreventive agents. *Mutat Res,* 2004;555:53-64.

[8] Sarkar, FH; Li, Y. Using chemopreventive agents to enhance the efficacy of cancer therapy. *Cancer Res,* 2006;66:3347-3350.

[9] Dorai, T; Aggarwal, BB. Role of chemopreventive agents in cancer therapy. *Cancer Lett,* 2004;215:129-140.

[10] Hunter, T; Pines, J. Cyclins and cancer. II: Cyclin D and CDK inhibitors come of age. *Cell,* 1994;79:573-582.

[11] Pei, XH; Xiong, Y. Biochemical and cellular mechanisms of mammalian CDK inhibitors: a few unresolved issues. *Oncogene,* 2005;24:2787-2795.

[12] Hwang, HC; Clurman, BE. Cyclin E in normal and neoplastic cell cycles. *Oncogene,* 2005;24:2776-2786.

[13] Sherr, CJ; Roberts, JM. Living with or without cyclins and cyclin-dependent kinases. *Genes Dev,* 2004;18:2699-2711.

[14] Yam, CH; Fung, TK; Poon, RY. Cyclin A in cell cycle control and cancer. *Cell Mol Life Sci,* 2002;59:1317-1326.

[15] John, PC; Mews, M; Moore, R. Cyclin/Cdk complexes: their involvement in cell cycle progression and mitotic division. *Protoplasma,* 2001;216:119-142.

[16] Joyce, D; Albanese, C; Steer, J; Fu, M; Bouzahzah, B; Pestell, RG. NF-kappaB and cell-cycle regulation: the cyclin connection. *Cytokine Growth Factor Rev,* 2001;12:73-90.

[17] Schwartz, GK; Shah, MA. Targeting the cell cycle: a new approach to cancer therapy. *J Clin Oncol,* 2005;23:9408-9421.

[18] Kastan, MB; Bartek, J. Cell-cycle checkpoints and cancer. *Nature,* 2004;432:316-323.

[19] Lamartiniere, CA; Cotroneo, MS; Fritz, WA; Wang, J; Mentor-Marcel, R; Elgavish, A. Genistein chemoprevention: timing and mechanisms of action in murine mammary and prostate. *J Nutr,* 2002;132:552S-558S.

[20] Choudhuri, T; Pal, S; Agwarwal, ML; Das, T; Sa, G. Curcumin induces apoptosis in human breast cancer cells through p53-dependent Bax induction. *FEBS Lett,* 2002;512:334-340.

[21] Hastak, K; Gupta, S; Ahmad, N; Agarwal, MK; Agarwal, ML; Mukhtar, H. Role of p53 and NF-kappaB in epigallocatechin-3-gallate-induced apoptosis of LNCaP cells. *Oncogene,* 2003;22:4851-4859.

[22] Chinni, SR; Li, Y; Upadhyay, S; Koppolu, PK; Sarkar, FH. Indole-3-carbinol (I3C) induced cell growth inhibition, G1 cell cycle arrest and apoptosis in prostate cancer cells. *Oncogene,* 2001;20:2927-2936.

[23] Gupta, S; Hussain, T; Mukhtar, H. Molecular pathway for (-)-epigallocatechin-3-gallate-induced cell cycle arrest and apoptosis of human prostate carcinoma cells. *Arch Biochem Biophys,* 2003;410:177-185.

[24] Li, Y; Sarkar, FH. Gene expression profiles of genistein-treated PC3 prostate cancer cells. *J Nutr,* 2002;132:3623-3631.

[25] Li, Y; Li, X; Sarkar, FH. Gene expression profiles of I3C- and DIM-treated PC3 human prostate cancer cells determined by cDNA microarray analysis. *J Nutr,* 2003;133:1011-1019.

[26] Mukhopadhyay, A; Bueso-Ramos, C; Chatterjee, D; Pantazis, P; Aggarwal, BB. Curcumin downregulates cell survival mechanisms in human prostate cancer cell lines. *Oncogene,* 2001;20:7597-7609.

[27] Barnes, S. The chemopreventive properties of soy isoflavonoids in animal models of breast cancer. *Breast Cancer Res Treat,* 1997;46:169-179.

[28] Dixon, RA; Ferreira, D. Genistein. *Phytochemistry,* 2002;60:205-211.

[29] Li, Y; Sarkar, FH. Down-regulation of invasion and angiogenesis-related genes identified by cDNA microarray analysis of PC3 prostate cancer cells treated with genistein. *Cancer Lett,* 2002;186:157-164.

[30] Shao, ZM; Shen, ZZ; Liu, CH; Sartippour, MR; Go, VL; Heber, D; Nguyen, M. Curcumin exerts multiple suppressive effects on human breast carcinoma cells. *Int J Cancer,* 2002;98:234-240.

[31] Katiyar, SK; Afaq, F; Perez, A; Mukhtar, H. Green tea polyphenol (-)-epigallocatechin-3-gallate treatment of human skin inhibits ultraviolet radiation-induced oxidative stress. *Carcinogenesis,* 2001;22:287-294.

[32] Mukhtar, H; Ahmad, N. Green tea in chemoprevention of cancer. *Toxicol Sci,* 1999;52:111-117.

[33] Delmas, D; Lancon, A; Colin, D; Jannin, B; Latruffe, N. Resveratrol as a chemopreventive agent: a promising molecule for fighting cancer. *Curr Drug Targets,* 2006;7:423-442.

[34] Laux, MT; Aregullin, M; Berry, JP; Flanders, JA; Rodriguez, E. Identification of a p53-dependent pathway in the induction of apoptosis of human breast cancer cells by the natural product, resveratrol. *J Altern Complement Med,* 2004;10:235-239.

[35] Scarlatti, F; Sala, G; Somenzi, G; Signorelli, P; Sacchi, N; Ghidoni, R. Resveratrol induces growth inhibition and apoptosis in metastatic breast cancer cells via de novo ceramide signaling. *FASEB J,* 2003;17:2339-2341.

[36] Shih, A; Davis, FB; Lin, HY; Davis, PJ. Resveratrol induces apoptosis in thyroid cancer cell lines via a. *J Clin Endocrinol Metab,* 2002;87:1223-1232.

[37] Kim, L; Rao, AV; Rao, LG. Effect of lycopene on prostate LNCaP cancer cells in culture. *J Med Food,* 2002;5:181-187.

[38] Nahum, A; Hirsch, K; Danilenko, M; Watts, CK; Prall, OW; Levy, J; Sharoni, Y. Lycopene inhibition of cell cycle progression in breast and endometrial cancer cells is associated with reduction in cyclin D levels and retention of p27(Kip1) in the cyclin E-cdk2 complexes. *Oncogene,* 2001;20:3428-3436.

[39] Barnes, S; Peterson, TG. Biochemical targets of the isoflavone genistein in tumor cell lines. *Proc Soc Exp Biol Med,* 1995;208:103-108.

[40] Barnes, S. Effect of genistein on in vitro and in vivo models of cancer. *J Nutr,* 1995;125:777S-783S.

[41] Lian, F; Bhuiyan, M; Li, YW; Wall, N; Kraut, M; Sarkar, FH. Genistein-induced G2-M arrest, p21WAF1 upregulation, and apoptosis in a non-small-cell lung cancer cell line. *Nutr Cancer,* 1998;31:184-191.

[42] Alhasan, SA; Pietrasczkiwicz, H; Alonso, MD; Ensley, J; Sarkar, FH. Genistein-induced cell cycle arrest and apoptosis in a head and neck squamous cell carcinoma cell line. *Nutr Cancer,* 1999;34:12-19.

[43] Davis, JN; Singh, B; Bhuiyan, M; Sarkar, FH. Genistein-induced upregulation of p21WAF1, downregulation of cyclin B, and induction of apoptosis in prostate cancer cells. *Nutr Cancer,* 1998;32:123-131.

[44] Li, Y; Bhuiyan, M; Sarkar, FH. Induction of apoptosis and inhibition of c-erbB-2 in MDA-MB-435 cells by genistein. *Int J Oncol,* 1999;15:525-533.

[45] Li, Y; Upadhyay, S; Bhuiyan, M; Sarkar, FH. Induction of apoptosis in breast cancer cells MDA-MB-231 by genistein. *Oncogene,* 1999;18:3166-3172.

[46] Upadhyay, S; Neburi, M; Chinni, SR; Alhasan, S; Miller, F; Sarkar, FH. Differential sensitivity of normal and malignant breast epithelial cells to genistein is partly mediated by p21(WAF1). *Clin Cancer Res,* 2001;7:1782-1789.

[47] Matsukawa, Y; Marui, N; Sakai, T; Satomi, Y; Yoshida, M; Matsumoto, K; Nishino, H; Aoike, A. Genistein arrests cell cycle progression at G2-M. *Cancer Res,* 1993;53:1328-1331.

[48] Casagrande, F; Darbon, JM. p21CIP1 is dispensable for the G2 arrest caused by genistein in human melanoma cells. *Exp Cell Res,* 2000;258:101-108.

[49] Shenouda, NS; Zhou, C; Browning, JD; Ansell, PJ; Sakla, MS; Lubahn, DB; Macdonald, RS. Phytoestrogens in common herbs regulate prostate cancer cell growth in vitro. *Nutr Cancer,* 2004;49:200-208.

[50] Su, SJ; Chow, NH; Kung, ML; Hung, TC; Chang, KL. Effects of soy isoflavones on apoptosis induction and G2-M arrest in human hepatoma cells involvement of caspase-3 activation, Bcl-2 and Bcl-XL downregulation, and Cdc2 kinase activity. *Nutr Cancer,* 2003;45:113-123.

[51] Frey, RS; Li, J; Singletary, KW. Effects of genistein on cell proliferation and cell cycle arrest in nonneoplastic human mammary epithelial cells: involvement of Cdc2, p21(waf/cip1), p27(kip1), and Cdc25C expression. *Biochem Pharmacol,* 2001;61:979-989.

[52] Gong, L; Li, Y; Nedeljkovic-Kurepa, A; Sarkar, FH. Inactivation of NF-kappaB by genistein is mediated via Akt signaling pathway in breast cancer cells. *Oncogene,* 2003;22:4702-4709.

[53] Li, Y; Sarkar, FH. Inhibition of nuclear factor kappaB activation in PC3 cells by genistein is mediated via Akt signaling pathway. *Clin Cancer Res,* 2002;8:2369-2377.

[54] Kuzumaki, T; Kobayashi, T; Ishikawa, K. Genistein induces p21(Cipl/WAF1) expression and blocks the G1 to S phase transition in mouse fibroblast and melanoma cells. *Biochem Biophys Res Commun,* 1998;251:291-295.

[55] Broadbent, TA; Broadbent, HS. 1. The chemistry and pharmacology of indole-3-carbinol (indole-3-methanol) and 3-(methoxymethyl)indole. [Part II]. *Curr Med Chem,* 1998;5:469-491.

[56] Bonnesen, C; Eggleston, IM; Hayes, JD. Dietary indoles and isothiocyanates that are generated from cruciferous vegetables can both stimulate apoptosis and confer protection against DNA damage in human colon cell lines. *Cancer Res,* 2001;61:6120-6130.

[57] Chen, DZ; Qi, M; Auborn, KJ; Carter, TH. Indole-3-carbinol and diindolylmethane induce apoptosis of human cervical cancer cells and in murine HPV16-transgenic preneoplastic cervical epithelium. *J Nutr,* 2001;131:3294-3302.

[58] Frydoonfar, HR; McGrath, DR; Spigelman, AD. The effect of indole-3-carbinol and sulforaphane on a prostate cancer cell line. *ANZ J Surg,* 2003;73:154-156.

[59] Hong, C; Kim, HA; Firestone, GL; Bjeldanes, LF. 3,3'-Diindolylmethane (DIM) induces a G(1) cell cycle arrest in human breast cancer cells that is accompanied by Sp1-mediated activation of p21(WAF1/CIP1) expression. *Carcinogenesis,* 2002;23:1297-1305.

[60] Hong, C; Firestone, GL; Bjeldanes, LF. Bcl-2 family-mediated apoptotic effects of 3,3'-diindolylmethane (DIM) in human breast cancer cells. *Biochem Pharmacol,* 2002;63:1085-1097.

[61] Jin, L; Qi, M; Chen, DZ; Anderson, A; Yang, GY; Arbeit, JM; Auborn, KJ. Indole-3-carbinol prevents cervical cancer in human papilloma virus type 16 (HPV16) transgenic mice. *Cancer Res,* 1999;59:3991-3997.

[62] Kim, DJ; Shin, DH; Ahn, B; Kang, JS; Nam, KT; Park, CB; Kim, CK; Hong, JT; Kim, YB; Yun, YW; Jang, DD; Yang, KH. Chemoprevention of colon cancer by Korean food plant components. *Mutat Res,* 2003;523-524:99-107.

[63] Nachshon-Kedmi, M; Yannai, S; Haj, A; Fares, FA. Indole-3-carbinol and 3,3'-diindolylmethane induce apoptosis in human prostate cancer cells. *Food Chem Toxicol,* 2003;41:745-752.

[64] Rahman, KM; Aranha, O; Glazyrin, A; Chinni, SR; Sarkar, FH. Translocation of Bax to mitochondria induces apoptotic cell death in indole-3-carbinol (I3C) treated breast cancer cells. *Oncogene,* 2000;19:5764-5771.

[65] Rahman, KW; Aranha, O; Sarkar, FH. Indole-3-Carbinol (I3C) Induces Apoptosis in Tumorigenic but Not in Nontumorigenic Breast Epithelial Cells. *Nutr Cancer,* 2003;45:101-112.

[66] Cover, CM; Hsieh, SJ; Tran, SH; Hallden, G; Kim, GS; Bjeldanes, LF; Firestone, GL. Indole-3-carbinol inhibits the expression of cyclin-dependent kinase-6 and induces a

G1 cell cycle arrest of human breast cancer cells independent of estrogen receptor signaling. *J Biol Chem,* 1998;273:3838-3847.

[67] Garikapaty, VP; Ashok, BT; Tadi, K; Mittelman, A; Tiwari, RK. 3,3'-Diindolylmethane downregulates pro-survival pathway in hormone independent prostate cancer. *Biochem Biophys Res Commun,* 2006;340:718-725.

[68] Chang, X; Tou, JC; Hong, C; Kim, HA; Riby, JE; Firestone, GL; Bjeldanes, LF. 3,3'-Diindolylmethane inhibits angiogenesis and the growth of transplantable human breast carcinoma in athymic mice. *Carcinogenesis,* 2005;26:771-778.

[69] Cover, CM; Hsieh, SJ; Cram, EJ; Hong, C; Riby, JE; Bjeldanes, LF; Firestone, GL. Indole-3-carbinol and tamoxifen cooperate to arrest the cell cycle of MCF-7 human breast cancer cells. *Cancer Res,* 1999;59:1244-1251.

[70] Jump, SM; Kung, J; Staub, R; Kinseth, MA; Cram, EJ; Yudina, LN; Preobrazhenskaya, MN; Bjeldanes, LF; Firestone, GL. N-Alkoxy derivatization of indole-3-carbinol increases the efficacy of the G1 cell cycle arrest and of I3C-specific regulation of cell cycle gene transcription and activity in human breast cancer cells. *Biochem Pharmacol,* 2008;75:713-724.

[71] Moiseeva, EP; Heukers, R; Manson, MM. EGFR and Src are involved in indole-3-carbinol-induced death and cell cycle arrest of human breast cancer cells. *Carcinogenesis,* 2007;28:435-445.

[72] Hsu, JC; Dev, A; Wing, A; Brew, CT; Bjeldanes, LF; Firestone, GL. Indole-3-carbinol mediated cell cycle arrest of LNCaP human prostate cancer cells requires the induced production of activated p53 tumor suppressor protein. *Biochem Pharmacol,* 2006;72:1714-1723.

[73] Brew, CT; Aronchik, I; Hsu, JC; Sheen, JH; Dickson, RB; Bjeldanes, LF; Firestone, GL. Indole-3-carbinol activates the ATM signaling pathway independent of DNA damage to stabilize p53 and induce G1 arrest of human mammary epithelial cells. *Int J Cancer,* 2006;118:857-868.

[74] Bhuiyan, MM; Li, Y; Banerjee, S; Ahmed, F; Wang, Z; Ali, S; Sarkar, FH. Down-regulation of androgen receptor by 3,3'-diindolylmethane contributes to inhibition of cell proliferation and induction of apoptosis in both hormone-sensitive LNCaP and insensitive C4-2B prostate cancer cells. *Cancer Res,* 2006;66:10064-10072.

[75] Li, Y; Chinni, SR; Sarkar, FH. Selective growth regulatory and pro-apoptotic effects of DIM is mediated by AKT and NF-kappaB pathways in prostate cancer cells. *Front Biosci,* 2005;10:236-243.

[76] Radhakrishna, PG; Srivastava, AS; Hassanein, TI; Chauhan, DP; Carrier, E. Induction of apoptosis in human lung cancer cells by curcumin. *Cancer Lett,* 2004;208:163-170.

[77] Chen, HW; Yu, SL; Chen, JJ; Li, HN; Lin, YC; Yao, PL; Chou, HY; Chien, CT; Chen, WJ; Lee, YT; Yang, PC. Anti-invasive gene expression profile of curcumin in lung adenocarcinoma based on a high throughput microarray analysis. *Mol Pharmacol,* 2004;65:99-110.

[78] Chaudhary, LR; Hruska, KA. Inhibition of cell survival signal protein kinase B/Akt by curcumin in human prostate cancer cells. *J Cell Biochem,* 2003;89:1-5.

[79] Aggarwal, BB; Kumar, A; Bharti, AC. Anticancer potential of curcumin: preclinical and clinical studies. *Anticancer Res,* 2003;23:363-398.

[80] Chen, HW; Huang, HC. Effect of curcumin on cell cycle progression and apoptosis in vascular smooth muscle cells. *Br J Pharmacol,* 1998;124:1029-1040.

[81] Srivastava, RK; Chen, Q; Siddiqui, I; Sarva, K; Shankar, S. Linkage of curcumin-induced cell cycle arrest and apoptosis by cyclin-dependent kinase inhibitor p21(/WAF1/CIP1). *Cell Cycle,* 2007;6:2953-2961.

[82] Zheng, M; Ekmekcioglu, S; Walch, ET; Tang, CH; Grimm, EA. Inhibition of nuclear factor-kappaB and nitric oxide by curcumin induces G2/M cell cycle arrest and apoptosis in human melanoma cells. *Melanoma Res,* 2004;14:165-171.

[83] Bharti, AC; Donato, N; Singh, S; Aggarwal, BB. Curcumin (diferuloylmethane) down-regulates the constitutive activation of nuclear factor-kappa B and IkappaBalpha kinase in human multiple myeloma cells, leading to suppression of proliferation and induction of apoptosis. *Blood,* 2003;101:1053-1062.

[84] Weir, NM; Selvendiran, K; Kutala, VK; Tong, L; Vishwanath, S; Rajaram, M; Tridandapani, S; Anant, S; Kuppusamy, P. Curcumin induces G2/M arrest and apoptosis in cisplatin-resistant human ovarian cancer cells by modulating Akt and p38 MAPK. *Cancer Biol Ther,* 2007;6:178-184.

[85] Hour, TC; Chen, J; Huang, CY; Guan, JY; Lu, SH; Pu, YS. Curcumin enhances cytotoxicity of chemotherapeutic agents in prostate cancer cells by inducing p21(WAF1/CIP1) and C/EBPbeta expressions and suppressing NF-kappaB activation. *Prostate,* 2002;51:211-218.

[86] Shishodia, S; Amin, HM; Lai, R; Aggarwal, BB. Curcumin (diferuloylmethane) inhibits constitutive NF-kappaB activation, induces G1/S arrest, suppresses proliferation, and induces apoptosis in mantle cell lymphoma. *Biochem Pharmacol,* 2005;70:700-713.

[87] Liu, E; Wu, J; Cao, W; Zhang, J; Liu, W; Jiang, X; Zhang, X. Curcumin induces G2/M cell cycle arrest in a p53-dependent manner and upregulates ING4 expression in human glioma. *J Neurooncol,* 2007;85:263-270.

[88] Lambert, JD; Yang, CS. Mechanisms of cancer prevention by tea constituents. *J Nutr,* 2003;133:3262S-3267S.

[89] Liang, YC; Lin-Shiau, SY; Chen, CF; Lin, JK. Inhibition of cyclin-dependent kinases 2 and 4 activities as well as induction of Cdk inhibitors p21 and p27 during growth arrest of human breast carcinoma cells by (-)-epigallocatechin-3-gallate. *J Cell Biochem,* 1999;75:1-12.

[90] Okabe, S; Suganuma, M; Hayashi, M; Sueoka, E; Komori, A; Fujiki, H. Mechanisms of growth inhibition of human lung cancer cell line, PC-9, by tea polyphenols. *Jpn J Cancer Res,* 1997;88:639-643.

[91] Nihal, M; Ahmad, N; Mukhtar, H; Wood, GS. Anti-proliferative and proapoptotic effects of (-)-epigallocatechin-3-gallate on human melanoma: possible implications for the chemoprevention of melanoma. *Int J Cancer,* 2005;114:513-521.

[92] Ahmad, N; Cheng, P; Mukhtar, H. Cell cycle dysregulation by green tea polyphenol epigallocatechin-3-gallate. *Biochem Biophys Res Commun,* 2000;275:328-334.

[93] Bhatia, N; Agarwal, R. Detrimental effect of cancer preventive phytochemicals silymarin, genistein and epigallocatechin 3-gallate on epigenetic events in human prostate carcinoma DU145 cells. *Prostate,* 2001;46:98-107.

[94] Masuda, M; Suzui, M; Weinstein, IB. Effects of epigallocatechin-3-gallate on growth, epidermal growth factor receptor signaling pathways, gene expression, and chemosensitivity in human head and neck squamous cell carcinoma cell lines. *Clin Cancer Res,* 2001;7:4220-4229.

[95] Dong, Z. Molecular mechanism of the chemopreventive effect of resveratrol. *Mutat Res,* 2003;523-524:145-150.

[96] Fremont, L. Biological effects of resveratrol. *Life Sci,* 2000;66:663-673.

[97] Ragione, FD; Cucciolla, V; Borriello, A; Pietra, VD; Racioppi, L; Soldati, G; Manna, C; Galletti, P; Zappia, V. Resveratrol arrests the cell division cycle at S/G2 phase transition. *Biochem Biophys Res Commun,* 1998;250:53-58.

[98] Bernhard, D; Tinhofer, I; Tonko, M; Hubl, H; Ausserlechner, MJ; Greil, R; Kofler, R; Csordas, A. Resveratrol causes arrest in the S-phase prior to Fas-independent apoptosis in CEM-C7H2 acute leukemia cells. *Cell Death Differ,* 2000;7:834-842.

[99] Estrov, Z; Shishodia, S; Faderl, S; Harris, D; Van, Q; Kantarjian, HM; Talpaz, M; Aggarwal, BB. Resveratrol blocks interleukin-1beta-induced activation of the nuclear transcription factor NF-kappaB, inhibits proliferation, causes S-phase arrest, and induces apoptosis of acute myeloid leukemia cells. *Blood,* 2003;102:987-995.

[100] Larrosa, M; Tomas-Barberan, FA; Espin, JC. Grape polyphenol resveratrol and the related molecule 4-hydroxystilbene induce growth inhibition, apoptosis, S-phase arrest, and upregulation of cyclins A, E, and B1 in human SK-Mel-28 melanoma cells. *J Agric Food Chem,* 2003;51:4576-4584.

[101] Ahmad, N; Adhami, VM; Afaq, F; Feyes, DK; Mukhtar, H. Resveratrol causes WAF-1/p21-mediated G(1)-phase arrest of cell cycle and induction of apoptosis in human epidermoid carcinoma A431 cells. *Clin Cancer Res,* 2001;7:1466-1473.

[102] Wolter, F; Akoglu, B; Clausnitzer, A; Stein, J. Downregulation of the cyclin D1/Cdk4 complex occurs during resveratrol-induced cell cycle arrest in colon cancer cell lines. *J Nutr,* 2001;131:2197-2203.

[103] Shankar, S; Singh, G; Srivastava, RK. Chemoprevention by resveratrol: molecular mechanisms and therapeutic potential. *Front Biosci,* 2007;12:4839-4854.

[104] Joe, AK; Liu, H; Suzui, M; Vural, ME; Xiao, D; Weinstein, IB. Resveratrol induces growth inhibition, S-phase arrest, apoptosis, and changes in biomarker expression in several human cancer cell lines. *Clin Cancer Res,* 2002;8:893-903.

[105] Tyagi, A; Singh, RP; Agarwal, C; Siriwardana, S; Sclafani, RA; Agarwal, R. Resveratrol causes Cdc2-tyr15 phosphorylation via ATM/ATR-Chk1/2-Cdc25C pathway as a central mechanism for S phase arrest in human ovarian carcinoma Ovcar-3 cells. *Carcinogenesis,* 2005;26:1978-1987.

[106] Heber, D; Lu, QY. Overview of mechanisms of action of lycopene. *Exp Biol Med (Maywood),* 2002;227:920-923.

[107] Chalabi, N; Le, CL; Maurizis, JC; Bignon, YJ; Bernard-Gallon, DJ. The effects of lycopene on the proliferation of human breast cells and BRCA1 and BRCA2 gene expression. *Eur J Cancer,* 2004;40:1768-1775.

[108] Hwang, ES; Bowen, PE. Cell cycle arrest and induction of apoptosis by lycopene in LNCaP human prostate cancer cells. *J Med Food,* 2004;7:284-289.

[109] Obermuller-Jevic, UC; Olano-Martin, E; Corbacho, AM; Eiserich, JP; van, d, V; Valacchi, G; Cross, CE; Packer, L. Lycopene inhibits the growth of normal human prostate epithelial cells in vitro. *J Nutr,* 2003;133:3356-3360.
[110] Palozza, P; Serini, S; Boninsegna, A; Bellovino, D; Lucarini, M; Monastra, G; Gaetani, S. The growth-inhibitory effects of tomatoes digested in vitro in colon adenocarcinoma cells occur through down regulation of cyclin D1, Bcl-2 and Bcl-xL. *Br J Nutr,* 2007;98:789-795.
[111] Jones, SB; DePrimo, SE; Whitfield, ML; Brooks, JD. Resveratrol-induced gene expression profiles in human prostate cancer cells. *Cancer Epidemiol Biomarkers Prev,* 2005;14:596-604.
[112] Chalabi, N; Satih, S; Delort, L; Bignon, YJ; Bernard-Gallon, DJ. Expression profiling by whole-genome microarray hybridization reveals differential gene expression in breast cancer cell lines after lycopene exposure. *Biochim Biophys Acta,* 2007;1769:124-130.

In: Cell Cycle Control: New Research
Editors: N. H. Leroy, N. T. Fournier

ISBN: 978-1-60456-796-0

Chapter XVI

Chromatin: The Biophysical Approach to Elucidating Structure and Function

James M. Nicholson[1], Colin D. Reynolds[2], Mark J. Dickman[3], Christopher M. Wood[2] and John P. Baldwin[1]

[1]Photon Science Department, Science & Technology Facilities Council, Daresbury Laboratory, Warrington, Cheshire WA4 4AD, United Kingdom;

[2]School of Biomolecular Sciences, Liverpool John Moores University, Byrom Street, Liverpool L3 3AF, United Kingdom;

[3]Department of Chemical and Process Engineering, University of Sheffield, Mappin Street, Sheffield S1 3JD, United Kingdom.

Abstract

Chromatin is the intricate complex of proteins and DNA that is involved in the control of the cell cycle and the modulation of gene expression in higher organisms. Methods in biochemistry and molecular biology have made it possible to identify and isolate many of the chromatin-related molecules vital to these processes. A wealth of information on the structure and function of these molecules and larger complexes is increasingly being elucidated by biophysical techniques, such as: mass spectrometry for highly accurate protein identification; electron microscopy (EM), nuclear magnetic resonance (NMR), small-angle X-ray and neutron scattering (SAXS & SANS) and in particular X-ray crystallography at third generation synchrotron radiation sources for solving complete three-dimensional static structures of macromolecules to atomic resolution.

In this review a non-exhaustive overview of the current status and recent developments in these and other biophysical techniques in relation to the structure and function of chromatin-related molecules is given. Some of the possibilities for this crucial area of research with single-molecule methods and future fourth-generation light sources are also discussed.

Introduction

Chromatin is the complex of gene-carrying DNA and proteins responsible for the impressive feat of DNA-packaging in higher eukaryotes, whereby more than two metres of DNA is compressed into a cell nucleus that is only a few micrometres in diameter. Modifications to the structure of chromatin (euchromatin and heterochromatin) control the accessibility of the DNA within the nucleus during the different stages of the cell cycle. A vastly complex array of cross-talking chemical modifications to chromatin at the atomic level modulate gene expression in all higher organisms via multiple signalling pathways. For more in-depth reviews of chromatin structure and function see Nicholson & Wood (2006), Turner (2002) and Wolffe (1999).

Nucleosomes, the repeat units of chromatin, are formed by the interaction of the small, positively-charged histones with DNA. Each nucleosome core particle consists of 146 base pairs of DNA wrapped around the histone octamer, containing two copies each of the highly-conserved core histones H2A, H2B, H3 and H4. In the full nucleosome, a fifth, more variable, histone, H1 (H5 in birds), known as linker histone, interacts with the DNA at the entry and exit points from the nucleosome. In addition there is a variable (species-dependent) 0 to 60 base-pair section of linker DNA, which interconnects the nucleosomes in a zigzag pattern. Each core histone contains a C-terminal globular domain as well as an N-terminal tail rich in arginine and lysine residues. These tails are the main sites of chemical, or post-translational modifications (PTMs) on the histone proteins, which together with DNA methylation, modulate gene expression. The histone tails extend beyond the DNA in the nucleosome to interact with the modifying protein complexes. The PTMs include acetylation (associated with cancer), methylation (marking transcriptionally active/inactive chromatin), phosphorylation (marking damaged DNA; cancer), ADP-ribosylation (DNA-repair; histone chaperoning during nucleosome formation) and ubiquitination (regulation of gene expression) (reviewed by Reynolds *et al*, 2008).

Rather than act as simple genetic on/off switches, several studies have shown that the PTMs function synergistically to provide combinations of chemical markers responsible for specific gene regulation and a broad spectrum of up and down regulated states. Hence in the analogy with electrical on/off switches, the function of PTMs in gene regulation can be compared more to an electrical dimmer switch.

Biophysical techniques are becoming increasingly important in studying chromatin structure and function. In particular, state of the art mass spectrometry that can resolve protein masses to better than one percent of one Dalton unit is crucial in identifying the histone proteins in their myriad combination of chemically modified states. It is also proving invaluable in accurately identifying the modifying enzymes and large molecular complexes responsible for triggering nucleosome remodelling, as well as other proteins relevant to chromatin cell-cycle regulation, such as cyclins and their catalytic partners cyclin-dependent kinases (CDKs), apoptosis inhibitors and high-mobility group proteins (HMGs), which also interact with histones bound to DNA to modify chromatin structure.

X-ray crystallography at synchrotron radiation sources has been established for the past twenty years as the most powerful technique for resolving macromolecular structures at atomic resolution and deriving functional mechanisms. Newer, brighter third-generation sources are increasing both the throughput and achievable resolution for these studies. Other techniques, such as nuclear magnetic resonance (NMR), electron microscopy (EM), electron diffraction (ED), small-angle X-ray (SAXS) and neutron (SANS) scattering and synchrotron radiation circular dichroism (SRCD) provide complementary, lower resolution information concerning the dynamics of functioning molecules, higher-order chromatin structures and protein folding.

Identifying Molecules by Mass Spectrometry

Mass spectrometry is a powerful technique that enables the identification and characterisation of proteins and their associated post translational modifications (PTMs) (Witze *et al*, 2007). Mass spectrometry has been widely used to study chromatin and its associated modifications (Su *et al*, 2007). The analysis of PTMs of chromatin has been performed at both the intact protein level and *via* proteolytic digestion of the proteins and the subsequent analysis of the peptides generated. These approaches have been termed 'top-down' and 'bottom-up' respectively. Bottom-up mass spectrometry has been more widely used to study histone PTMs. However, such approaches create a number of challenges in the analysis of PTMs. Although identification of the protein may be performed with the identification of a small number of peptide sequences, the analysis of PTMs requires a more rigorous approach. High sequence coverage of the protein or in the case of the histones, high sequence coverage of all the protein isoforms is required. In addition, due to the basic nature of the proteins (lysine- and arginine-rich sequences are present in a number of domains in the core and linker histone proteins) a number of alternative proteolytic enzymes, time course reactions and chemical modifications prior to MS analysis are often required to generate high sequence coverage of the proteins. Furthermore PTMs are often present in sub-stoichiometric quantities *in vivo* and therefore the presence of a modified peptide is usually at a much lower concentration than the unmodified peptide. Therefore the application of a variety of separation and enrichment technologies have been employed to study the post translational modifications of proteins (Witze *et al*, 2007; Su *et al*, 2007; Burlingame *et al*, 2005).

Mass Spectrometry Approaches

Peptide mass fingerprinting (PMF) is an approach used to identify proteins following proteolytic digestion of the intact protein by comparing the masses of peptides generated with those of a theoretical protein digest. PMF in combination with Fourier transfer-ion cyclotron resonance (FT-ICR) has been widely used in the study of histone PTMs (Zhang *et al*, 2002 & Ren *et al*, 2005). FT-ICR offers high mass resolving power ($>10^6$) and mass accuracy (<1 ppm) (Comisarov & Marshall, 1974) therefore can be used to distinguish lysine acetylation from tri-methylation (mass shifts of 42.0106 and 42.0470 Da, respectively).

However, PMF has a number of limitations; for example it is unable to differentiate overlapping isobaric peptides of different sequences. Tandem mass spectrometry (MS/MS) is an alternative approach which can be used to unambiguously identify a wide range of PTMs.

Tandem MS involves the initial isolation of the ions of interest with subsequent fragmentation of the selected ion. The peptide/protein sequence can then be inferred from the fragmentation pattern obtained. Fragmentation of the peptides can be performed using collision-induced dissociation (CID) or collision-activated dissociation (CAD) and electron capture/transfer dissociation (ECD/ETD) (Zubarev *et al*, 1998; McLafferty *et al,* 2001; Zubarev *et al*, 2000; Syka *et al*, 2004). Both ETD and ECD offer the advantage of retaining PTMs such as phosphorylation, which are often lost during CID.

More recently top-down proteomic approaches (MS analysis of the intact protein) have been used to identify histone PTMs (reviewed by Siuti & Kelleher, 2007). Different gene families of histone H2A, H2B and H3 variants were identified by use of high mass-accuracy FT-ICR-MS along with their changes in relative expression and modifications during cell cycling (Siuti *et al*, 2006; Thomas *et al*, 2006; Boyne *et al*, 2006). The same group employed FT-ICR-MS to quantify the modified forms or positional isomers of histone H4 using the relative abundances of their fragment ions following HPLC and hydrophilic interaction liquid chromatography (HILIC) fractionation (Pesavento *et al*, 2006). Top-down MS/MS of intact proteins enables the determination of modifications as they occur in combination and generates additional information regarding the PTM hierarchies. However, top-down proteomic approaches require purified histone fractions (such as H4, H2A, H2B and H3) in large amounts. Therefore analysis where sample concentration is limited, such as clinical samples, is not currently practical.

Structure and Function at Atomic Resolution

X-ray crystallography is without doubt the most powerful method for determining static three-dimensional macromolecular structures to atomic resolution. The protein structure determination pipeline (Figure 1) involves stages of protein identification by mass spectrometry, as discussed above, protein production and purification, crystallisation, data collection, phasing, molecular modelling and validation using a variety of software packages and finally deposition in the globally and freely accessible Protein Data Bank (PDB: Berman *et al*, 2000).

Traditionally proteins were purified by biochemical techniques from native source material but, increasingly researchers are producing high yields of recombinant proteins through cloning techniques. The bottleneck of this structure determination process has been for many years and remains the crystallisation process. Crystallisation of any biological molecule occurs in supersaturated conditions at precisely the right pH and salt conditions. Automated robotic crystal-screening systems have become commonplace at many laboratories around the world and have greatly increased the throughput of successful protein crystallisations. However many important molecules, such as membrane proteins, are notoriously difficult to crystallise due in this case to the fact that membrane proteins require detergents or other means to solubilise them in isolation, and such detergents often interfere with crystallization. Membrane proteins form a large proportion of the genome and include many proteins of great physiological importance, such as ion channels and receptors.

Perhaps a statistic that best illustrates the problems associated with crystallising macromolecules is that there are currently (as of end of 2007) 9318 unique protein families identified from sequenced genomes in the Pfam database (Finn *et al*, 2006), however currently only 3551 (38%) contain a member with a known structure (deposited in the PDB), which allows the folds of all other members of the family to be inferred.

While vast amounts of protein sequence data are produced by modern large-scale DNA sequencing efforts, such as the Human Genome Project, the output of experimentally determined protein structures through structural genomics initiatives (Chandonia & Brenner, 2006) continues to lag behind largely due to the difficulties in obtaining suitable ordered crystals.

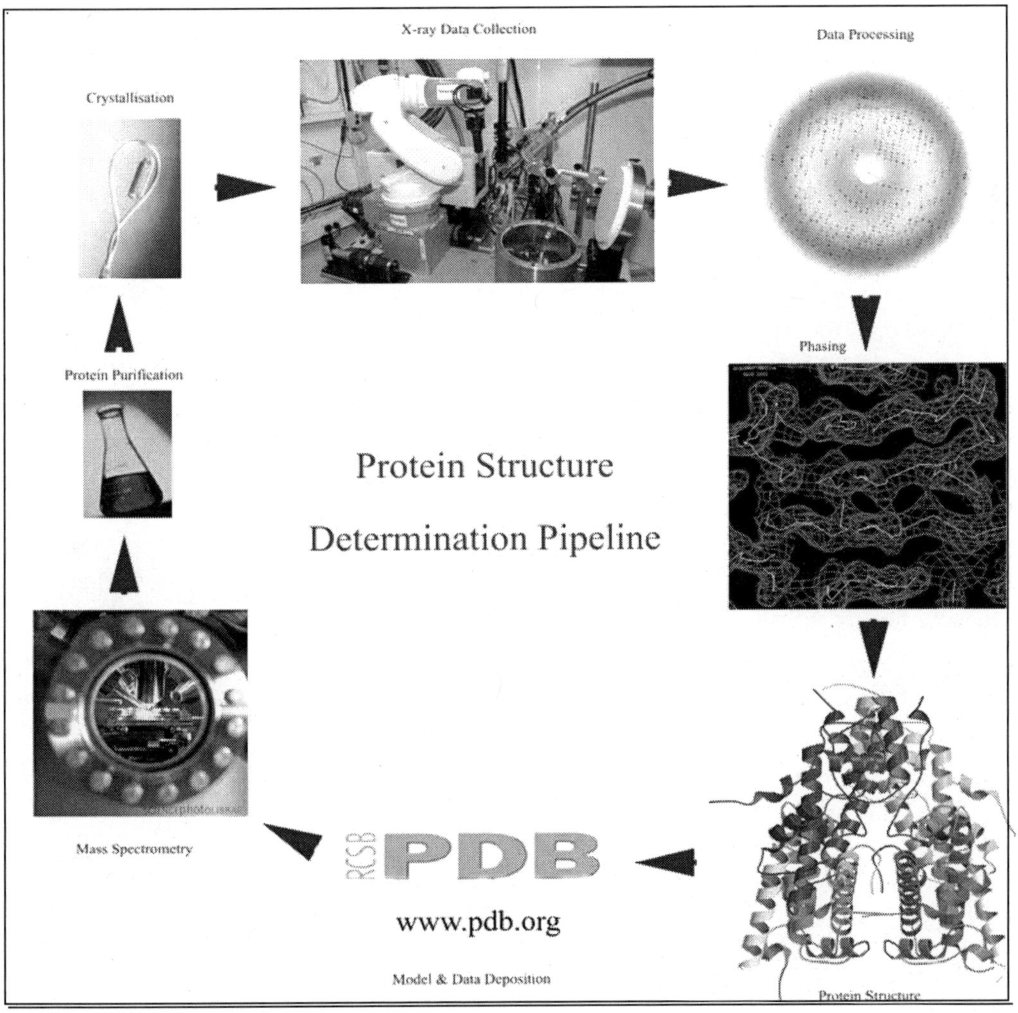

Figure 1. The macromolecular structure determination pipeline: target identification and selection using mass spectrometry and bioinformatics; protein production and purification using biochemical and/or molecular biology techniques; crystallisation; data collection at synchrotron radiation sources; phasing, refinement and model building to solve the structure; ultimate deposition of the molecular coordinates and X-ray data sets in the Protein Data Bank (PDB: Berman *et al*, 2000).

In comparison the X-ray data collection, phasing and molecular modeling stages of the pipeline have become relatively straightforward. Structure determination by macromolecular crystallography is now a mature, well-developed technique that is now at least semi-automatic at most synchrotron radiation facilities. The only slight drawback of the high intensity third-generation synchrotron sources is the rapid onset of radiation damage to protein crystals due to beam-heating effects and secondary damage caused by free radicals damaging protein bonds (Ravelli & Garman, 2006). The primary heating damage is largely negated by sample cooling with liquid nitrogen or helium jets. However, the secondary damage continues to be an issue and it is vital to minimise the X-ray exposure plus dead time for each crystal, and to collect the data before the damage becomes significant.

The generic weakness of the crystallographic method is generally known as the phase problem of crystallography. While it is possible to record the diffracted intensities from protein crystals, usually using large area, fast readout charge-couple devices (CCDs), the phase differences between the diffracted rays that are vital for interpreting the diffraction patterns by inverse fast Fourier transform, are lost. Fortunately there are several methods for increasing the phase information within a diffraction dataset and large-scale tunable synchrotron sources are ideal for optimising the data collection parameters to maximise the phase information obtained. In the past, crystallographers would attempt to soak their crystals with as many different heavy-atom derivatives as possible in order to extract phase information. However, tunable synchrotron beamlines have made it possible to incorporate a single heavy-atom type and vary the wavelength around the absorption edge of that atom and collect the anomalous X-ray scattering intensities that are characteristic for each atom, in order to maximise the phase differences obtained. This method is known as Multi-wavelength Anomalous Dispersion (MAD) and has the distinct advantage of only requiring a single crystal hence minimising statistical errors in the data collected. Selenium (Se) has become the heavy atom of choice as the amino acid methionine (Met), which is common in most proteins, can be replaced by selenium-methionine (Se-Met) with minimum disruption to the protein structure. In addition the absorption edge of selenium is at a favourable wavelength (0.979 Å) for most synchrotron X-ray beamline energy outputs. The MAD method has been refined and in many cases it is now possible to solve the phase problem with just one dataset in Single-wavelength Anomalous Dispersion (SAD), or by using the naturally-occuring sulphur atoms in protein disulphide bridges as the heavy atom with longer wavelength (or softer energy) X-rays in the Softer Single Wavelength Anomalous Technique (S-SWAT) (Olczak *et al*, 2003).

The development and increasing sophistication of software packages for data processing, phasing and model refinement, such as MOSFLM and the CCP4 (Collaborative Computing Project Number 4; www.ccp4.ac.uk) suite have greatly simplified these processes.

Structure determination by X-ray crystallography has an increasingly significant role in chromatin research with many relevant structures solved and functional significance derived. Both the recombinant nucleosome core particle (Luger *et al*, 1997; Davey *et al*, 2002) and the native intact histone octamer (Chantalat *et al*, 2003; Wood *et al*, 2005) structures have been solved to better than 2.0 Å resolution, as well as the globular domain of the avian-specific linker histone H5 (Ramakrishnan *et al*, 1993). These studies have yielded vital information regarding the interactions that hold the DNA to the histones, keep the histone octamer intact and the possible mechanisms for nucleosome remodeling (Nicholson *et al*, 2003).

While the sites of PTMs on the histone tails are shown to protrude beyond the DNA, their precise position is unclear as they are disordered in all structures to date. However, several structures of the modifying enzymes in complex with short histone-tail peptides have been reported, such as a histone acetyl-transferase where the lysine-acetylated recognition bromodomain is bound to the histone H4 tail with acetylated lysine residue 16 (K16) (Owen *et al*, 2000). Bromodomains are the only protein constituents known to recognise the acetylated lysines of histones. There are several protein domains identified as binding methylated lysine or arginine residues but the most extensively studied is the chromodomain. The HP1 chromodomain bound to the methylated histone H3 tail has been solved by crystallography and shows clear selectivity for methylated lysine residue 9 (K9) over lysine 4 (K4) (Jacobs & Khorasanizadeh, 2002). Several other histone-tail binding domains have been solved to near-atomic resolution using X-ray crystallography, including: histone deactylases (Somoza *et al*, 2004); histone chaperones (Namboodiri *et al*, 2004) and chromatin remodeling factors (Thoma *et al*, 2005). A review of chromatin associated molecules solved by X-ray crystallography and NMR to date is available (Nicholson & Wood, 2006).

Visualising the Dynamic Complexes

As in all biological systems the interacting molecules of chromatin, which are responsible for the PTMs and nucleosome formation and remodeling during the different stages of the cell-cycle and gene regulation, are not simply the static bundles of atoms depicted in X-ray structures. These models provide the most accurate end-state structures of proteins and macromolecular complexes but they only provide snapshots of the functioning molecular machinery inside the nucleus. To obtain a more complete image of these processes it is necessary to use other complementary techniques that probe biological molecules in a more natural state than the constraints of the crystal lattice.

Nuclear magnetic resonance (NMR) can be used to solve static protein structures as well as provide important data regarding molecular motion, which gives NMR an advantage over X-ray crystallography in studying the *in vivo* dynamics of macromolecules, although at a significantly reduced resolution. An additional advantage is that NMR is extremely useful for analyzing biological samples non-destructively in contrast to X-ray diffraction. Protein samples can be studied using NMR for weeks or months before conducting more destructive biochemical experiments.

In chromatin research NMR has been utilised to non-destructively study the intact globular domains of linker histone H1 (Ali *et al*, 2004), which have to date not been successfully crystallised. In addition there are a small number of histone-modifying enzyme structures (methyltransferases, acetyltransferases, deacetylases) solved by NMR. Examples of dynamic NMR studies in chromatin include the bromodomain of human brahma-related gene 1 (Brg1), which is a core protein in the human SWI/SNF chromatin-remodeling complex (Shen *et al*, 2007). NMR perturbation studies showed that the Brg1 bromodomain binds acetylated lysine residues in the histone tails, with no comparable affinity for unacetylated peptides. NMR studies have demonstrated the conformational flexibility upon binding of the various subunits to form the complex co-repressor Sin3 (van Ingen *et al*, 2006). Sin3 is the essential scaffold protein of the Sin3/HDAC corepressor complex, which is recruited to DNA by transcriptional repressors and targets genes involved in the regulation of the cell cycle, proliferation and differentiation.

Small-angle X-ray scattering (SAXS) is an X-ray diffraction method applied widely in both the laboratory and at synchrotron facilities. SAXS can be used to analyse proteins and other macromolecules in solution without the requirement for crystals and allowing useful data to be collected regarding the dynamic functioning state of biological molecules. SAXS utilises hard energy X-rays, similar to those of X-ray crystallography scattered from molecules freely tumbling in solution. This method gives overall molecular size and shape information, to resolutions ranging from thousands of angstroms to the order of 10 Å, such as the conformation of the surface areas of binding sites and gross molecular shape changes during the formation of protein complexes (Grossmann, 2007).

The higher-order structure of chromatin compacts the nucleosomes into a fibre, 30 nm in diameter. The chromatin fibre has been studied by many groups, using techniques including SAXS (Lasters *et al*, 1985), small-angle neutron scattering (SANS) (Baldwin *et al*, 1975) and electron microscopy (EM) (Thoma *et al*, 1979). The results obtained indicated that the nucleosomes in the fibre are arranged in a left-handed coil (Williams *et al*, 1986). More recent work has determined possible structural models for the 30 nm chromatin fibre and provided insights into the mechanisms by which the degree of chromatin compaction might be regulated by histone composition and PTMs (Robinson & Rhodes, 2006).

In scattering techniques, radially-symmetric diffraction of light, X-rays or neutrons from solutions of bio-molecular assemblies gives the observed total intensity, S(q), as a function of the scattering parameter $q=2\sin\theta/\lambda$, where θ is the scattering semi-angle and λ is the wavelength. Figure 2a (Dunn *et al*, 1986) shows a set of radial-distribution functions, D(r), (derived from transformation of S(q) plots) from X-ray scattering of 23-nucleosome particles, cleaved by micro-coccal nuclease (Lasters *et al*, 1985; Muyldermans *et al*, 1980) from the nuclear matrix of chicken-erythrocyte-cell nuclei. The data show how reproducible X-ray scattering can be, even comparing unfixed poly-nucleosomes with those that have been fixed using a short cross-linking agent and also with a long cross-linker. It is probable that the 23-mers straighten out to a 23-nucleosome short fibre on release from the nuclear matrix. The highest radius at which D(u) in Figure 2a reaches zero gives the largest distance between two points on the surface of the structure and the radius at which D(u) is a maximum is the most probable distance that can be drawn between volume elements in the structure. The data is in absolute units, so the values of D(u) are structurally meaningful. Only a limited range of structures for the 23mers fit the reproducible data (Figure 2b) and those are compatible with other low-resolution studies of the chromatin fibre, with a 33nm radius and approximately six nucleosomes per 10nm distance along the length of the fibre (Suau *et al*, 1979; Thoma *et al*, 1979; Ramakrishnan, 1997).

The high-resolution structure of the chromatin fibre, as it occurs in cell nuclei, is one of the most important problems remaining in bio-molecular structure determination. Probably the only possibility will be to crystallise straightened-out poly-nucleosome structures cleaved from the nuclear matrix and the most important result will be to determine how nucleosomes, and the DNA connecting them zig-zag fashion (Thoma *et al*, 1979), interlace: for example, is the chromatin fibre a single-helical arrangement of nucleosomes or does the linker DNA transit to and fro across the fibre? The reproducibility of the data of figure 2a suggests that crystals could be obtained from native chromatin.

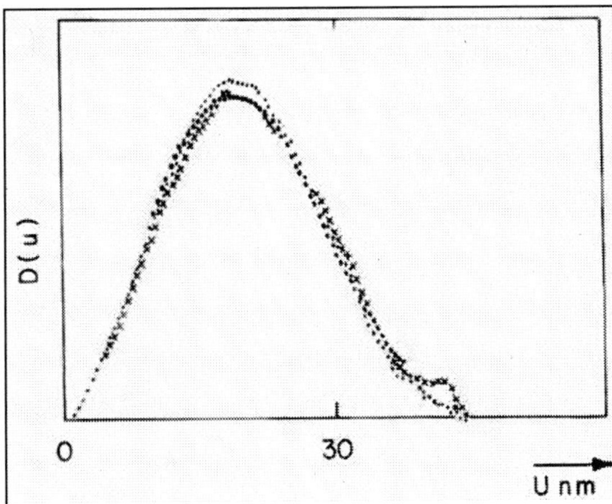

Figure 2a

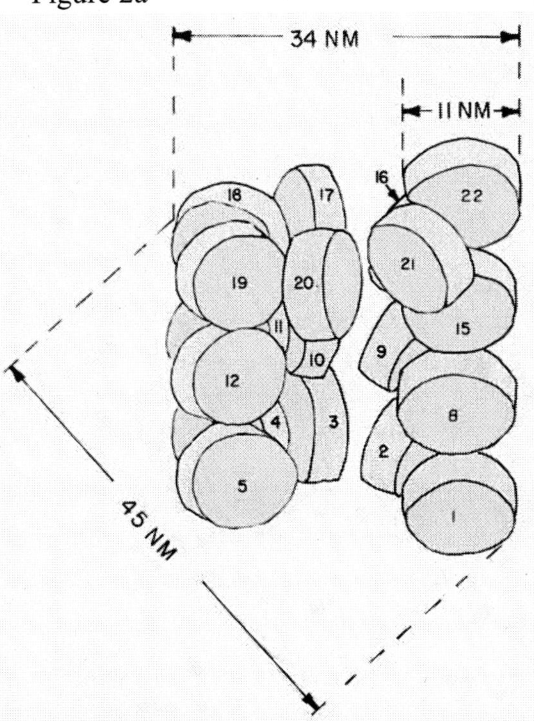

Figure 2b

Figure 2. (a) A pair-distribution function D(u) vs U for 23-nucleosome multimers. The samples are in a buffer containing 1.2 mM magnesium chloride. x: uncrosslinked 23-mers; •: crosslinked in glutaraldehyde; ●: crosslinked with EDC, a short-distance crosslinking agent. (b) A model for the arrangement of 23 nucleosomes, which fits the data, provided the central cavity is filled with scattering material. Nucleosomes are represented schematically by disks with the core histone octamers shaded yellow and each surrounded by a pale band representing the DNA. Nucleosomes 6, 7, 13 and 14 have been removed for ease of display. Figure 2 is adapted from Dunn *et al*, 1986.

Neutron diffraction studies of chromatin fibres at different humidities (Baldwin *et al*, 1975; Bradbury & Baldwin, 1986) showed that a nominal 10 nm equivalent-d-spacing diffraction maximum (which was semi-meridional) progressed in to 7nm as the fibres were dried and the 10 nm maximum was interpreted as due to the packing of nucleosomes along the length of the fibre. The chromatin fibre and its folding pathways as a function of mono-valent and di-valent cations have been studied by many groups using SAXS and SANS (Langmore & Schutt, 1980; Langmore & Paulson, 1983; Bordas *et al*, 1986).

Small-angle scattering from bio-molecular assemblies containing multiple components, such as lipid with protein or nucleic acids with protein, is a very powerful technique when the solvent scattering-length density is varied. The low-resolution structure of a bio-molecular assembly for diffraction is visualised as a distribution of "scattering-length density" for the radiation in question (optical polarisation, electron density for X-rays, nuclear potential for neutrons). However the real visualisation is in contrast (ρ) to the solvent scattering-length density and that can be varied for each $S(q,\rho)$ data set. In neutron scattering, the contrast is varied by changing the ratio of D_2O/H_2O in the solvent. At 40% D_2O in the solvent, in a DNA-Protein assembly, the average protein neutron scattering is contrasted out and so the $S(q,40)$ and $D(u,40)$ for low q applies to the DNA component. Similarly the DNA is contrasted out at 65% D_2O.

The theory of small-angle contrast variation by Stuhrmann (1974) has been applied using neutrons to several chromatin structures, including methods where proteins, deuterated at non-exchangable sites, were studied in contrast to other proteins in the assembly. This latter approach was used by Ramakrishnan to show that the linker histone lies on the inside of the chromatin fibre (Graziano *et al*, 1994).

The low-resolution structure of the nucleosome core particle came first from neutron-scattering contrast variation, including the use of the Stuhrmann fundamental-scatter-function analysis (Pardon *et al*, 1975; Hjelm *et al*, 1977; Suau *et al*, 1977), showing that the DNA wrapped around the outside of the histone octamer in the particle. Studies of chromatosomes indicated that the linker histone is not right on the outside of the core-particle component and is asymmetric (Lambert *et al*, 1991).

Recently, 23 nucleosome multinucleosomes have very successfully been reconstructed and analysed from histone octamers, linker histone and defined-sequence DNA (Robinson *et al*, 2006; Robinson & Rhodes, 2006). Neutron contrast variation could solve the problem of nucleosome interlacing in such chromatin fibres, or in native ca 23mer fibres, if low-resolution crystallography was possible. Use of deuterated linker DNA would facilitate analysis of the data.

Another biophysical technique employed in chromatin structure and function is circular dichroism (CD), either on a conventional laboratory source or using synchrotron radiation (SRCD; Clarke & Jones, 2004), which has the advantages of improved signal-to-noise across the energy range and access to data at wavelengths below 190 nm that are inaccessible with conventional xenon-lamp CD instruments. CD spectroscopy measures the differences in absorption between left- and right-hand polarised ultra violet (UV) radiation in optically-active substances (chiral molecules) as a function of wavelength. Secondary structure elements each have a distinctive UV absorption spectrum in CD: α-helices have a positive peak at 195 nm and negative troughs around 208 and 222 nm; β-sheet and strand structures are reflected in the amplitude of the 195 nm peak and a small minimum at 212 nm. The region accessible only by SRCD (160-190 nm) gives structural information on anti-parallel/parallel β-sheet, β-turns and aperiodic groups of amino-acid residues. The resultant CD absorption spectra, together with reference to structural databases such as *Dichroweb* (Whitmore & Wallace, 2004), provide information regarding the structural content of a protein, i.e. the proportion of each secondary structure element present. This is useful for characterising molecules that will not crystallise and for studying conformational changes in proteins during functional remodelling and generic protein folding and mis-folding, which have implications for the molecular mechanisms behind disease.

Conventional CD has been applied to study the N-terminal tail of histone H4 in its native and acetylated forms (Wang *et al*, 2000). These results indicated that the α-helical content of the H4 tail increased from approximately 12% to 24% with increasing acetylation. Secondary structure predictions show that there are regions of high α-helical probability on all of the core-histone tails, suggesting possible conformational changes upon binding of the histone octamer to the modifying enzymes and chromatin remodelling complexes (Nicholson *et al*, 2004). CD has also been applied to the study of the C-terminal globular-domain fragments of the linker histones H1 and H5, indicating that these domains preferably adopt a left-handed helical conformation (Makarov *et al*, 1992).

SRCD has recently been applied to studying intact histone octamers, H2A-H2B dimers and $(H3\text{-}H4)_2$ tetramers under various conditions of ionic strength, pH, buffer and oxidation (Sodngam, 2007). These studies established the principle of analysing intact histone structures with SRCD at wavelengths below 190 nm in low salt conditions. The SRCD of isolated histone tetramers $(H3\text{-}H4)_2$ showed a loss of α-helical content compared to the tetramer within the intact histone octamer, indicating possible conformational changes within the histones upon octamer formation.

A Bright Future

Developments in biophysical techniques and new large-scale third-generation synchrotron sources and planned future fourth-generation sources continue to push the boundaries of achievable science in biomolecular structure and dynamics. Continued use and development of mass spectrometry (MS) in chromatin research will lead to better insights into the synergy and cross-talk of histone modifications.

Ongoing developments in MS including increased sensitivity, cycling time, mass resolution and accuracy in conjunction with novel fragmentation approaches and further developments in separation technology are required to understand the dynamic interactions and functions of histone proteins in relation to their post translational modifications. The emergence of top-down MS/MS analysis of intact proteins enables the determination of modifications as they occur in combination and generates additional information regarding the PTM hierarchies. With improved instrumentation such approaches are likely to become compatible with chromatographic timescales, therefore enabling direct separation and top-down analysis in an automated high throughput manner. Further developments are required to understand the dynamic interactions and functions of histone proteins in relation to their post-translational modifications.

In X-ray crystallography the development of more intense third-generation synchrotron beamlines appears to have reached a limit both in terms of the realistic radiation dose biological crystals can absorb before they suffer critical damage, and the provision of such facilities and available access time to the user communities in the USA and Europe. The supply of crystals has not yet matched the availability of synchrotron beamlines. Automation of crystallisation screening and the many more microfocus synchrotron facilities available does mean that tiny crystals, which would have been discarded or ignored a few years ago, are now producing valuable data. However, the concurrent effect of the automation of X-ray data collection, phasing, model building and refinement, together with large, faster CCD detectors, greatly outweighs the improved yield of crystals. Macromolecular crystallography has now become a mature, routine biophysical technique that is more a means to an end than the complete end product in itself but crystallisation remains the bottleneck in the process.

Applications of the X-rays generated by free-electron lasers (FELs) are limited to X-ray spectroscopy methods for analysing time correlations of atoms within biological crystals and some applications in MAD phasing (Doniach, 2000). These sources are unlikely to compete with third-generation synchrotron facilities in macromolecular structure determination.

The field of single molecule research is relatively new in biology. As a consequence there are not many instruments available worldwide with which to gather proof-of-principle single molecule data and there have not so far been many applications of single-molecule methods to chromatin structure and function. ASMID (www.srs.dl.ac.uk/VUV/asmid.html) is a relatively new facility based at STFC Daresbury Laboratory (UK), providing laser sorces and other instrumentation for investigating the structure and dynamics of single molecules (Webb *et al*, 2008).

Fluorescence resonance energy transfer (FRET) provides information on the distance between fluorescent molecules, and so is a very powerful probe of protein-protein interactions, or protein conformation (intra-molecular FRET) if two dyes are placed on one protein. These measurements can be made either in living cells or on isolated proteins and protein complexes. Confocal microscopy can be used for fluorescence lifetime imaging (FLIM), which can again be used to monitor FRET in cells, or specific conditions within cells by using environmentally sensitive probes.

Other single-molecule techniques, such as atomic force microscopy (AFM) have been applied to studies of the chromatin fibre (Leuba *et al*, 2004) and to probe real-time chromatin assembly (Ladoux *et al*, 2000). These studies indicated a sequential process of chromatin assembly with at least three steps, including DNA wrapping as the final event.

More recent single-molecule studies using both AFM and fluorescence microscopy have shown that vitamin C has a protective effect against DNA double-strand breaks in reconstituted chromatin (Yoshikawa *et al*, 2006). It is anticipated that the application of single-molecule techniques, such as FRET and AFM, will become more widespread in chromatin research with the establishment of more facilities like ASMID worldwide.

SRCD is a useful tool for studying biological secondary structure and future developments are likely to focus on the development of more beamlines at third-generation synchrotron sources, the increased scope of structural databases used to analyse SRCD spectra and the application of stopped-flow devices for studying time-resolved macromolecular conformation changes *in vivo*. Future sources combining accelerator technologies with free-electron laser (FEL) and conventional laser sources are expected to extend the CD technique into the Raman and terahertz (THz) energy regimes to probe rapid timescales. These new techniques will be able to resolve fast dynamics on the picosecond (ps: 10^{-12} s) to nanosecond (ns: 10^{-9} s) timescale to study, for example, membrane channel dynamics. Currently such observations are based on molecular dynamics modelling between two end-state protein structures usually determined by X-ray crystallography. The molecular hinges or pores controlling membrane channels open and close at fast frequencies (ps - ns), which will only be observable by exploiting the intense THz (far infrared) radiation likely to be available on future light sources, such as the ALICE facility currently under development at STFC Daresbury Laboratory in the UK (www.4gls.ac.uk/ERLP.htm).

Other new scientific techniques that are likely to be exploited on future light sources include: pump-probe experiments utilising near-IR and THz radiation to study macromolecular dynamics including large molecular complexes and inter-complex energy transfer; non-linear two-dimensional IR spectroscopy for studying fast molecular dynamics; sub-cellular imaging with non-ionising IR and UV radiation, including faster timescales for single-molecule fluorescence imaging. For a full review of the possible applications in chromatin structure and function research on planned future light sources see Nicholson & Wood (2006).

With the development of these and other new facilities for biophysical science, the future for chromatin structure and function research promises to be bright and fruitful.

Acknowledgements

The chromatin research group is supported by the UK Science & Technology Facilities Council (STFC). JMN, CDR, CMW and JPB are founders of the company HistoneTech Ltd, which markets histones and other chromatin-related products through the Abcam catalogue. HistoneTech Ltd is a joint venture between CLIK at STFC Daresbury Laboratory and Liverpool John Moores University. The authors thank Harrison's Poultry of Antrobus, Cheshire (UK) for the provision chicken blood.

References

Ali, T., Coles, P., Stevens, T.J., Stott, K. & Thomas, J.O. Two homologous domains of similar structure but different stability in the yeast linker histone, Hho1p. *J. Mol. Biol. 338 (1)* (2004), 139-148.

Baldwin, J.P., Boseley, P.G., Bradbury E.M. & Ibel, K. The subunit structure of the eukaryotic chromosome. *Nature 253* (1975), 245-249.

Berman, H.M., Westbrook, J., Feng, Z., Gilliland, G., Bhat, T.N., Weissig, H., Shindyalov, I.N. & Bourne, P.E. The Protein Data Bank. *Nucleic Acids Research 28* (2000), 235-242.

Bordas, J., Perez-Grau, L., Koch, M.H., Vega, M.C. & Nave, C. The superstructure of chromatin and its condensation mechanism. II. Theoretical analysis of the X-ray scattering patterns and model calculations. *Eur. Biophys. J. 13(3)* (1986.), 175-185.

Boyne, M.T. II, Pesavento, J.J., Mizzen, C.A. & Kelleher, N.L. Precise characterization of human histones in the H2A gene family by top down mass spectrometry. *J. Proteome Res. 5 (2)* (2006), 248–253.

Bradbury, E.M. & Baldwin, J.P. Neutron scatter and diffraction techniques applied to nucleosome and chromatin structure. *Cell Biophys. 9(1-2)* (1986), 35-66.

Burlingame, A.L., Zhang, X., Chalkley, R.J. Mass spectrometric analysis of histone posttranslational modifications. *Methods 36 (4)* (2005), 383-94.

Chandonia, J.M. & Brenner, S.E. The impact of structural genomics: Expectations and outcomes. *Science 311 (5759)* (2006), 347-351.

Chantalat, L., Nicholson, J.M., Lambert, S. J., Reid, A. J., Donovan, M.J., Reynolds, C.D., Wood C.M. & Baldwin, J.P. (2003). Structure of the histone-core octamer in KCl/phosphate crystals at 2.15 Å resolution. *Acta Cryst. D59,* 1395-1407.

Clarke, D.T. & Jones, G.R. CD12: a new high-flux beamline for ultraviolet and vacuum-ultraviolet circular dichroism on the SRS, Daresbury. J*ournal of Synchrotron Radiation 11* (2004), 142-149.

Comisarov, M.B. & Marshall, A.G. Fourier transform ion cyclotron resonance spectroscopy. *hem. Phys. Lett. 25* (1974), 282–283.

Davey, C.A., Sargent, D.F., Luger, K., Maeder, A.W. & Richmond, T.J. (2002*).*_Solvent Mediated Interactions in the Structure of the Nucleosome Core Particle at 1.9 Å Resolution *J. Mol. Biol. 319*, 1097-1113.

Doniach, S. Fourth-generation X-ray sources: some possible applications to biology *J. Synch. Rad. 7 (3)* (2000), 116-120.

Dunn, S.P., Baldwin, J.P., Wyns, L., Muyldermans, S., Lasters, I., Poland, G.A., Staynov, D.Z., Rattle, H.W.E. & Wood, M.J. Neutron Scattering of Chromatin Multinucleosomes. *Physica 136B* (1986), 265-267.

Finn, R.D., Mistry, J., Schuster-Böckler, B., Griffiths-Jones, S., Hollich, V., Lassmann, T., Moxon, S., Marshall, M., Khanna, A., Durbin, R., Eddy, S.R., Sonnhammer E.L.L. & Bateman, A. Pfam: clans, web tools and services. *Nucleic Acids Research* Database Issue *34* (2006), D247-D251.

Graziano, V., Gerchman, S.E., Schneider, D.K., Ramakrishnan, V. Histone H1 is located in the interior of the chromatin 30-nm filament. *Nature 368 (6469)* (1994), 51-54.

Grossmann, J.G. Biological solution scattering: recent achievements and future challenges. *J. Appl. Cryst.* (2007). *40* (Supplement), s217-s222.

Hjelm, R.P., Kneale, G.G., Sauau, P., Baldwin, J.P., Bradbury, E.M. & Ibel, K. Small angle neutron scattering studies of chromatin subunits in solution. *Cell 10 (1)* (1977), 139-151.

Jacobs, S.A. & Khorasanizadeh, S. Structure of HP1 chromodomain bound to a lysine 9-methylated histone H3 tail. *Science 295* (2002), 2080-2083.

Ladoux, B., Quivy, J.P., Doyle, P., du Roure, O., Almouzni, G. & Viovy, J.L. Fast kinetics of chromatin assembly revealed by single-molecule videomicroscopy and scanning force microscopy. *Proc. Nat. Acad. Sci. USA 97 (26)* (2000), 14251-14256.

Lambert, S.J., Baldwin, J.P., Kilner, J., Ibel, K. & Wijns, L. Neutron scattering studies of chromatosomes. *Biochem. Biophys. Res. Comm. 179* (1991), 810-816.

Langmore, J.P. & Paulson, J.R. Low angle x-ray diffraction studies of chromatin structure in vivo and in isolated nuclei and metaphase chromosomes. *J. Cell Biol. 96(4)* (1983), 1120-1131.

Langmore, J.P. & Schutt, C. The higher order structure of chicken erythrocyte chromosomes in vivo. *Nature 288 (5791)* (1980), 620-622.

Lasters, I., Wyns, L., Muyldermans, S., Baldwin, J.P., Poland, G.A. & Nave, C. Scatter Analysis of Discrete-sized Chromatin Fragments Favours a Cylindrical Arrangement. *Eur. J. Biochem. 151 (2)* (1985), 283-289.

Leuba, S.H., Bennink, M.L. & Zlatanova, J. Single-molecule analysis of chromatin. *Method. Enzymol. 376 (B)* (2004), 73-105.

Luger, K., Mader, A.W., Richmond, R.K., Sargent, D.F. & Richmond, T.J. Crystal structure of the nucleosome core particle at 2.8 Å resolution. *Nature 389* (1997), 251-260.

Makarov, A.A., Lobachov, V.M., Adzhubei. I.A. & Esipova N.G. *FEBS Lett. 306 (1)* (1992), 63-65.

McLafferty, F.W., Horn, D.M., Breuker, K., Ge, Y., Lewis, M.A., Cerda, B., Zubarev, R.A. & Carpenter, B.K. Electron capture dissociation of gaseous multiply charged ions by Fourier-transform ion cyclotron resonance. *J. Am. Soc. Mass Spectrom. 12 (3)* (2001), 245–249.

Muyldermans, S., Lasters, I., Wyns, L. & Hamers, R. Upon the observation of superbeads in chromatin. *Nucleic Acids Res. 8(10)* (1980), 2165-72.

Namboodiri, V.M.H., Akey, I.V., Schmidt-Zachmann, M.S., Head, J.F. & Akey, C.W. The structure and function of Xenopus NO38-core, a histone chaperone in the nucleolus. *Structure 12 (12)* (2004), 2149-2160.

Nicholson, J.M. & Wood, C.M. (Editors). Chromatin Structure and Function. (2006) ISBN: 81-7895-234-3. Transworld Research Network.

Nicholson, J.M., Wood, C.M., Reynolds, C.D., Brown, A., Lambert, S.J., Chantalat, L. & Baldwin, J.P. Histone Structures: Targets for Modifications by Molecular Assemblies. *Ann. NY Acad. Sci. 1030* (2004), 642-653.

Nicholson, J.M., Wood, C.M., Reynolds, C.D., Lambert, S.J., Chantalat, L. & Baldwin, J.P. Structural molecular biology of the histone octamer. *Recent Research Developments in Molecular Biology 1* (2003), 165-187. ISBN 81-271-0032-3.

Olczak, A., Cianci, M., Hao, Q., Rizkallah, P. J., Raftery, J. & Helliwell, J.R. S-SWAT (softer single-wavelength anomalous technique): potential in high-throughput protein crystallography. *Acta Cryst. A59* (2003), 327-334.

Owen, D.J., Ornaghi, P., Yang, J.C., Lowe, N., Evans, P.R., Ballario, P., Neuhaus, D., Filetici, P. & Travers, A.A. The structural basis for the recognition of acetylated histone H4 by the bromodomain of histone acetyltransferase gcn5p. *EMBO Journal 19* (2000), 6141-6149.

Pardon, J.F., Worcester, D.L., Wooley, J.C., Tatchell, K., Van Holde K.E. & Richards, B.M. Low-angle neutron scattering from chromatin subunit particles. *Nuc. Acid. Res. 2(11)* (1975), 2163-2176.

Pesavento, J.J., Mizzen, C.A. & Kelleher, N.L. Quantitative analysis of modified proteins and their positional isomers by tandem mass spectrometry: human histone H4. *Anal. Chem. 78 (13)* (2006), 4271–4280.

Ramakrishnan, V. Histone H1 and chromatin higher-order structure. *Crit. Rev. Eukaryot. Gene Expr. 7(3)* (1997), 215-230.

Ramakrishnan, V., Finch, J.T., Graziano, V., Lee, P.L. & Sweet, R.M. Crystal-Structure of Globular Domain of Histone H5 and its Implications for Nucleosome Binding. *Nature 362* (1993), 219-223.

Ravelli, R.B.G. & Garman, E.F. Radiation damage in macromolecular cryocrystallography. *Curr. Op. Struct. Biol. 16 (5)* (2006), 624-629.

Ren, C., Zhang, L., Freitas, M.A., Ghoshal, K., Parthun, M.R. & Jacob ST. Peptide mass mapping of acetylated isoforms of histone H4 from mouse lymphosarcoma cells treated with histone deacetylase (HDACs) inhibitors. *J. Am. Soc. Mass Spectrom. 16 (10)* (2005), 1641–1653.

Reynolds, C.D., Nicholson, J.M. & Baldwin, J.P. Chromatin: cracking the chemical code. *European Pharmaceutical Review 6 (3)* (2008), in press.

Robinson, P.J., Fairall, L., Huynh, V.A & Rhodes, D. EM measurements define the dimensions of the "30-nm" chromatin fiber: evidence for a compact, interdigitated structure. *Proc. Natl. Acad. Sci. USA 103 (17)* (2006), 6506-6511.

Robinson, P.J.J. & Rhodes, D. Structure of the '30 nm' chromatin fibre: A key role for the Linker histone. *Current Opinion in Structural Biology 16 (3)* (2006), 336-343.

Shen, W.Q., Xu, C., Huang, W., Zhang, J.H., Carlson, J.E., Tu, X.M., Wu, J.H. & Shi, Y.Y. Solution structure of human Brg1 bromodomain and its specific binding to acetylated histone tails. *Biochemistry 46 (8)* (2007), 2100-2110.

Siuti, N. & Kelleher, N.L. Decoding protein modifications using top-down mass spectrometry. *Nature Methods 4 (10)* (2007), 817-21.

Siuti, N., Roth, M.J., Mizzen, C.A., Kelleher, N.L. & Pesavento, J.J. Gene-specific characterization of human histone H2B by electron capture dissociation. *J. Proteome Res. 5 (2)* (2006), 233–239.

Sodngam, S. (2007). Purification and structural studies of histone and histone-associated proteins. PhD thesis, Liverpool John Moores University (United Kingdom).

Somoza, J.R., Skene, R.J., Katz, B.A., Mol, C., Ho, J.D., Jennings, A.J., Luong, C., Arvai, A., Buggy, J.J., Chi, E., Tang, J., Sang, B.C., Verner, E., Wynands, R., Leahy, E.M., Dougan, D.R., Snell, G., Navre, M., Knuth, M.W., Swanson, R.V., McRee, D.E. & Tari L.W. Structural snapshots of human HDAC8 provide insights into the class I histone deacetylases. *Structure 12 (7)* (2004), 1325-1334.

Stuhrmann, H.B. Neutron small-angle scattering of biological macromolecules in Solution. *J. Appl. Cryst. 7* (1974), 173-178.

Su, X., Ren, C. & Freitas, M.A. Mass spectrometry-based strategies for characterization of histones and their post-translational modifications. *Expert Rev Proteomics. 4 (2)* (2007), 211-225.

Suau, P, Bradbury, EM, Baldwin, JP. Higher-order structures of chromatin in solution. *Eur. J. Biochem. 97(2)* (1979), 593-602.

Suau, P., Kneale, G.G., Braddock, G.W., Baldwin, J.P. & Bradbury, E.M. A low resolution model for the chromatin core particle by neutron scattering. *Nuc. Acid. Res. 4 (11)* (1977), 3769-3786.

Syka, J.E., Coon, J.J., Schroeder, M.J., Shabanowitz, J. & Hunt, D.F. Peptide and protein sequence analysis by electron transfer dissociation mass spectrometry. *Proc. Natl Acad. Sci. USA 101 (26)* (2004), 9528–9533.

Thomä, N.H., Czyzewski, B.K., Alexeev, A.A., Mazin, A.V., Kowalczykowski, S.C. & Pavletich, N.P. Structure of the SWI2/SNF2 chromatin-remodeling domain of eukaryotic Rad54. *Nat. Struc. Mol. Biol. 12* (2005), 350-356.

Thoma, F., Koller, T. & Klug, A. Involvement of H1 in the organization of the nucleosome and of salt-dependent superstructures of chromatin. *J. Cell Biol., 83* (1979), 403-427.

Thomas, C.E., Kelleher, N.L. & Mizzen, C.A. Mass spectrometric characterization of human histone H3: a bird's eye view. *J. Proteome Res. 5 (2)* (2006), 240–247.

Turner, B.M.. (2001). Chromatin and Gene Regulation: Molecular Mechanisms in Epigenetics. ISBN 0865427437, Blackwell Science.

van Ingen, H., Baltussen, M.A.H., Aelen, J. & Vuister, G.W. Role of structural and dynamical plasticity in Sin3: The free PAH2 domain is a folded module in mSin3B. *J. Mol. Biol. 358 (2)* (2006), 485-497.

Wang, X.Y., Moore, S.C., Laszckzak, M. & Ausio, J. Acetylation increases the alpha-helical content of the histone tails of the nucleosome. *J. Biol. Chem. 275 (45)* (2000), 35013-35020.

Webb, S.E.D., Roberts, S.K., Needham, S.R., Tynan, C.J., Rolfe, D.J., Winn, M.D., Clarke, D.T., Barraclough, R. & Martin-Fernandez, M. Single molecule imaging and FLIM show different structures for high and low-affinity EGFRs in A431 cells. *Biophys. J. 94* (2008), 803-819 .

Whitmore, L. & Wallace, B.A. DICHROWEB, an online server for protein secondary structure analyses from circular dichroism spectroscopic data. *Nucleic Acids Research 32* (2004), 668-672.

Williams, S.P., Athey, B.D., Muglia, L.J., Schappe, R.S., Gough, A.H., Langmore, J.P. Chromatin fibers are left-handed double helices with diameter and mass per unit length that depend on linker length, *Biophys J., 49(1)* (1986), 233-48.

Witze, E.S., Old, W.M., Resing, K.A. & Ahn, N.G. Mapping protein post-translational modifications with mass spectrometry. *Nature Methods 4 (10)* (2007), 798-806.

Wolffe, A.P. (1999). Chromatin Structure. ISBN 0127619151, Academic Press.

Wood, C.M., Nicholson, J.M., Reynolds, C.D., Lambert, S.J., Chantalat, L. & Baldwin, J.P. High-resolution (1.9 Å) structure of the native histone octamer. *Acta Cryst. F61* (2005), 541-545.

Yoshikawa, Y., Hizume, K., Oda, Y., Takeyasu, K., Araki, S. & Yoshikawa, K. Protective effect of vitamin C against double-strand breaks in reconstituted chromatin visualized by single-molecule observation. *Biophys. J. 90 (3)* (2006), 993-999.

Zhang, K., Williams, K.E., Huang, L., Yau, P., Siino, J.S., Bradbury, E.M., Jones, P.R., Minch, M.J. & Burlingame, A.L. Histone acetylation and deacetylation: identification of acetylation and methylation sites of HeLa histone H4 by mass spectrometry. *Mol. Cell. Proteomics 1 (7)* (2002), 500–508.

Zubarev, R.A., Horn, D.M., Fridriksson, E.K., Kelleher, N.L., Kruger, N.A., Lewis, M.A., Carpenter, B.K. & McLafferty, F.W. Electron capture dissociation for structural characterization of multiply charged protein cations. *Anal. Chem. 72 (3)* (2000), 563–573.

Zubarev, R.A., Kelleher, N.L. & McLafferty FW. Electron capture dissociation of multiply charged protein cations. A nonergodic process. *J. Am. Chem. Soc. 120 (13)* (1998), 3265–3266.

Index

A

B

C

D

E

F

G

H

I

J

K

L

M

N

O

P

Q

R

S

T

U

V

W

X

Y

Z

Β

B

B

C

D

E

F

G

H

I

M

N

O

P

Q

R

S

T

U

V